Statistics for Business

Statistics for Business explains the fundamentals of statistical analysis in a lucid, pragmatic way. A thorough knowledge of statistics is essential for decision making in all corners of business and management. By collecting, organizing, and analyzing statistical data you can express what you know, benchmark your current situation, and estimate future outcomes.

Based entirely on Microsoft Excel, this book covers a spectrum of statistical fundamentals from basic principles, to probability, sampling, hypothesis testing, forecasting, statistical process control, and six-sigma management. This second edition is packed with features to aid understanding and help ensure that every aspect of your knowledge of statistics is applicable to practice, including:

- Icebreakers introducing each chapter that relate statistics to the real world, drawn from management and hospitality situations
- Detailed worked examples in each chapter
- Over 140 case-exercises complete with *objective*, *situation*, *requirements*, and *answers*
- A complete glossary of key terminology and formulas, mathematical relationships, and Excel relationships and functions
- A brand new companion website containing slides, worked-out-solutions to the case-exercises, and a test bank.

With a clear and accessible style this book makes statistics easier to understand. It is ideal for business, management, tourism and hospitality students who want to learn how to apply statistics to the real world.

Derek L. Waller is an Adjunct Professor at the Institut Paul Bocuse and the CEFAM University, France. A chemical engineer by training and a chartered engineer in the UK, Derek has also published books on operations and supply chain management.

'*Statistics for Business* successfully navigates fundamental statistical methods through a variety of stimulating real-world examples, clearly demonstrating the applicability of many statistical techniques to the business world.'

Dr James Abdey, *Lecturer in Statistics, London School of Economics and Political Science, UK*

Statistics for Business
Second edition

Derek L. Waller

LONDON AND NEW YORK

Please visit the companion website at
www.routledge.com/cw/waller

First published 2017
by Routledge
2 Park Square, Milton Park, Abingdon, Oxon OX14 4RN

and by Routledge
711 Third Avenue, New York, NY 10017

Routledge is an imprint of the Taylor & Francis Group, an informa business

British Library Cataloguing in Publication Data
A catalogue record for this book is available from the British Library

Library of Congress Cataloging in Publication Data
Names: Waller, Derek L., author.
Title: Statistics for business/Derek L. Waller.
Description: Second edition. | Abingdon, Oxon; New York, NY: Routledge,
2017. | Includes bibliographical references and index.
Identifiers: LCCN 2016010192 | ISBN 9780415703758 (hardback) | ISBN
9780415703710 (pbk.) | ISBN 9781315890043 (ebook)
Subjects: LCSH: Commercial statistics. | Management–Statistical methods.
Classification: LCC HF1017.W35 2017 | DDC 519.5–dc23
LC record available at https://lccn.loc.gov/2016010192

ISBN: 978-0-415-70375-8 (hbk)
ISBN: 978-0-415-70371-0 (pbk)
ISBN: 978-1-315-89004-3 (ebk)

Typeset in Bembo
by Sunrise Setting Ltd, Brixham, UK

Printed and bound in the United States of America by
Edwards Brothers Malloy on sustainably sourced paper

This book is dedicated to the memory of five people who I will never forget for their guidance, encouragement, and patronage during my education and career.

John Cooper, the headmaster of Chater Primary School, Watford, UK in the 1950s, for his implicit confidence in a pupil starting out in the world. He trusted me to go into his home, collect and organize his personal documents.

Betty Harry (now Preston), the world's best chemistry teacher at Rickmansworth Grammar School, UK in the 1950s and 1960s, for her patience and tenacity. Of her own volition she came to my home and persuaded my parents that I should do A-level studies after O-levels, then two years later that I should go to university—a step that my father, a former Yorkshire coal miner, could never imagine and thought impossible for one of his children. And money was a problem. Betty is an absolute jewel!

Frank Manning, a director of Marchon Products in the Lake District, UK, for his direction and career encouragement to a teenager. It was he who suggested in the 1960s that I study chemical engineering at university since I had a strong background in mathematics, physics, and chemistry. He treated me as the son he never had.

Dr John Thornton, Professor of Chemical Engineering at Newcastle-upon Tyne University, UK, who after much reluctance on my part persuaded me to undertake a PhD at Newcastle rather than at Berkeley California, where I had been offered a place—in retrospect a wise decision. I received my PhD in May 1968; at that time the Berkeley campus was in turmoil from student protests because of the Vietnam War. I probably would never have graduated!

Hugh Baird, the former president of C. F. Braun and Co. California, USA, my employer, for his enthusiasm and support, particularly during the difficult period of the late 1980s in the severe downtown and lay-offs in the engineering, chemical, and petroleum industry.

… and of course to my family
  … my parents, Albert and Hilda Waller
   … my brother, Kenneth Waller
    … my wife, Christine
… my children, Delphine and Guillaume.

Contents

ICEBREAKER: International cost of living

Types of data • Quantitative data • Qualitative data • Mix qualitative/quantitative data
Central tendency • Arithmetic mean • Median • Mode • Midrange • Weighted averages • Questionnaires and weighted averages • Geometric mean • Politics and central tendency
Dispersion • Data spread • Range • Variance • Standard deviation • Calculation of variance and standard deviation • Deviation about the mean • Coefficient of variation • Quartiles • Percentiles

ICEBREAKER: How good is your performance?

Frequency distributions • Frequency distribution table • Absolute frequency histogram • Relative frequency histogram • Frequency polygons • Ogives • Stem–and–leaf display
Visuals of quartiles and percentiles • Box and whisker plot • The percentile histogram
Line graphs • Candid presentations • Geometric mean as a line graph
Visuals of categorical data • Pie chart • Categorical histogram • Bar chart • Contingency tables • Stacked histograms • Pareto diagram • Spider web diagram • Gap analysis • Pictograms

ICEBREAKER: The haven to perhaps make a fortune

• Learning curve • Natural logarithm • Waiting lines
Beta distribution • Activity time • Project management

6   Methods and theory of statistical sampling (appropriate testing can
    improve reliability)   127

ICEBREAKER: The sampling experiment was badly designed!

The sampling process • Bias in sampling • Randomness • Excel and random
numbers • Systematic sampling • Stratified sampling • Several stratas of interest
• Cluster sampling • Quota sampling • Consumer surveys • Primary and
secondary data
Sampling for the mean • Sample size • Central limit theorem • Variability and
sample size • Sample standard error • Sampling size and population characteristics
• Infinite population • Application case-exercise: *Safety valves* • Finite population
• Application case-exercise: *Work week*
Sampling for the proportion • Measuring the sample proportion • Sampling
distribution of the proportion • Application case-exercise: *Part-time workers*

7   Estimating population characteristics (giving confidence to your evaluations)   158

ICEBREAKER: Turkey and the margin of error

Estimating the mean value • Point estimates • Interval estimates • Confidence
level and reliability • Confidence interval for an infinite population • Application
case-exercise: *Paper* • Magnitude of sample size • Application case-exercise: *Jars of
coffee* • Confidence interval for a finite population • Application case-exercise: *Printing*
Student's *t* distribution • Concept • Degrees of freedom • Profile • Confidence intervals
• Excel and the Student's *t* distribution • Application case-exercise: *Kiwi fruit* • Sample size
• Re-look at the application case-exercise *Kiwi fruit* using the normal distribution
Estimating and auditing • Estimating the population amount • Application
case-exercise for large populations: *Tee-shirts* • Application case-exercise for a finite
population: *Paperback books*
Estimating the proportion • Large samples • Sample size for large samples
• Application case-exercise: *Circuit boards*
Margin of error and confidence levels• Margin of error • Confidence levels

8   Hypothesis testing from a single population (giving assurance to assumptions
    that are made)   185

ICEBREAKER: Radiation and exposure: you need to be objective

Concept of hypothesis testing • Significance level • Null and alternative
hypotheses
Presenting hypothesis testing for the mean • Two-tailed test • One-tailed,
right-hand test • One-tailed, left-hand test

Case-exercises

Figures

Chapter Figures

Case-exercises Figures

About this book

This second edition of the textbook *Statistics for Business* has been updated. Three new chapters have been added to make thirteen: *Decision and risk in business*, *Statistical process control* and *Six-sigma management* (the chapter on *Indexing* has been dropped). The entire book is based on Microsoft Excel as the basic working tool.

There are three principal sections in this new edition:

Subject material

Each of the 13 chapters presents a subject area: 1 *Characterizing and defining data*, 2 *Presenting and organizing data*, 3 *Essentials of probability and counting rules*, 4 *Discrete data* and *probability*, 5 *Continuous distributions and probability*, 6 *Methods and theory of statistical sampling*, 7 *Estimating population character-istics*, 8 *Hypothesis testing from a single population*, 9 *Hypothesis testing from different populations*, 10 *Forecasting from correlated data*, 11 *Business decisions and risk*, 12 *Statistical process control*, and 13 *Six-sigma management*. Each chapter begins with an icebreaker, the objective of which is to relate statistics to the real world. Each chapter contains application case-exercises, complete with worked-out solutions to aid comprehension. The appropriate Excel functions are always indicated.

Case-exercises

The 13 chapters are followed by over 140 pragmatic case-exercises based on my experience and the business world. These case-exercises are organized by chapter and referenced according to the statistical method of their business application. All of them detail an objective and describe a situation before asking the reader what is required and providing answers.

Appendices

Glossary

Like many business subjects, statistics contains definitions, jargon, terminology, and equations. These are compiled in Appendix I as a glossary.

Basic mathematics

You may feel a little rusty about the basic mathematics that you did in secondary school. Appendix II covers arithmetical terms, relationships, and conversion factors that provide the basis for statistical analysis and many other applications.

Microsoft Excel

Microsoft Excel contains all the statistical mathematical functions, including the normal and Student-t, binomial, Poisson, and chi-squared distributions. As you work through the chapters you will find references to the statistical functions employed. Appendix III is a guideline of the Excel relationships used, including a table, in English and French, of the most commonly employed functions.

(Note, in the text you may find that if you perform the application examples and test exercises on a calculator you may find slightly different answers than those presented in the textbook. This is because all the examples and exercises have been calculated using Excel, which carries up to 14 figures after the decimal point. A calculator will round numbers.)

Statistics

Statistics includes the collecting, organizing, and analysis of data for describing situations and for decision making, often in business. Usually, the data collected are quantitative, or numerical, but information can also be categorical or qualitative. However, qualitative data can be made quantitative by using a numerically scaled questionnaire, where subjective responses correspond to an established number scale.

Statistical analysis is fundamental in the management and business environment as logical decisions are based on quantitative data. Quite simply, if you cannot express what you know, your current situation, or the future outlook in the form of numbers, you really do not know much about it. And, if you do not know much about it, you cannot manage it. Without numbers, you are just another person with an opinion! This is where statistics plays a role and why it is important to study the subject. For example, by simply displaying statistical data in a visual form you can convince your manager, your client, or your co-worker. By using probability analysis you can test your company's strategy and, importantly, evaluate financial risk. Market surveys are useful to evaluate the success of new products and innovative processes. Operations managers in services and manufacturing use statistical process control for monitoring and controlling performance. Historical data are used to develop sales forecasts, budgets, capacity requirements, and personnel needs. In finance, managers analyze company stocks, performance, and the economic outlook for investment purposes. For many firms, six-sigma management, founded on statistics, is part of the company culture!

International

The business environment is global. This textbook recognizes this by using icebreakers, examples, exercises, and case-exercises from various countries where the $US, euro, £ sterling, and other currencies are used.

About the author

Derek Waller is a chemical engineer by training. He was born in England, where he gained BSc and PhD degrees at the University of Newcastle-upon-Tyne. He first worked in the oil industry in London and Turkey before moving to the United States. Here he earned an MBA (Finance), a BA (French), and a Business Certificate for Technical Personnel from the University of California, Los Angeles. For over 20 years he was involved in design and operations of projects in the oil, petrochemical, and renewable energy industries in the United States, Europe, Africa, and the Middle East, initially in engineering and then in marketing, sales, finance, and strategic planning. On many of these projects he used statistical and quantitative methods for system design, planning, and optimization. In the late 1980s he moved into the academic field, first as professor in business management in California and then for some 20 years at E.M. Lyon, a Business School in southeastern France. He taught applied statistics, quantitative methods, and operations management, and he worked on many industrial and service-related projects. He is currently an adjunct professor at the Institut Paul Bocuse, a hospitality management school, and the CEFAM University in partnership with US East Coast universities: Northeastern, Pace, Temple, Rider, and Siena College. Derek has also published books on operations and supply chain management. He is a Chartered Engineer in the UK and lives in Lyon, France.

To the student (and your professor)

Often students become afraid when they realize that they have to take a course in statistics as part of their college or university curriculum. I often hear remarks like: "I will never pass this course"; "I am no good at maths and so I am sure I will fail the exam."; "I don't need a course in statistics as I am going to be in marketing."; and "What good is statistics to me, I plan to take a job in human resources?"

All these remarks are unwarranted. The knowledge of statistics is vital in all areas of business. The subject is made easier (and more fun) by using Microsoft Excel with its cell relationships, mathematical functions, and graphical applications. The book explains the fundamentals clearly in a readable, step-by-step approach, beginning with basic concepts and then moving into more advanced areas. It avoids working through tedious mathematical computations, sometimes found in other statistical texts, that I find confuses students. The case-exercises reinforce understanding.

You should not have any qualms about studying statistics; it really is not a difficult subject to grasp. Please, if you need any further information, have comments on the book, or questions to ask, do not hesitate to get in touch with me at the following electronic address:

derek.waller@wanadoo.fr

Derek L Waller

1 Characterizing and defining statistical data

ICEBREAKER: International cost of living

How do you compare the international cost of living? One way is to measure how much it costs to eat out. However, here the question comes up, what type of meal? The UK-based magazine *The Economist* has devised a simple innovative approach by comparing the price of a McDonald's Big Mac hamburger in various countries worldwide. It uses this as a basis since most of us at some point have eaten a McDonald's hamburger, wherever that may have been. This is a study and analysis that *The Economist* has been performing since 1986. Statistical data, in local currencies, for January 2012 is given in Table 1.0 for 44 different countries.

Using the $US currency exchange rate at this period and considering, for example, the BRIC countries, we note that the cost of living in Brazil is 35 percent higher than in the USA. However, it is 39 percent, 61 percent and 42 percent lower in Russia, India, and China respectively. If you are a globetrotter and a hamburger is your staple diet, you will learn that the worldwide average price of the hamburger is $US 3.59; half of the hamburgers cost more than $US 3.36; the price range of the hamburgers is $US 5.17; the most expensive hamburger is in Norway at $US 6.79; and the cheapest "hamburger" is in India at $US 1.62, though in India a hamburger is made from chicken and not beef!

It is these characteristics and properties of statistical data that are covered in this chapter.

Figure 1.0 The ubiquitous hamburger.

Source: author.

Table 1.0 Big Mac index: *The Economist*, January 14, 2012

Country	Local currency units	Price, local currency	Country	Local currency units	Price, local currency
USA	$US	4.20	Malaysia	Ringgit	7.35
Argentina	Peso	20.00	Mexico	Peso	37.00
Australia	$Aus	4.80	New Zealand	$New Zealand	5.10
Brazil	Real	10.25	Norway	Nor.Kroner	41.00
Britain	£	2.49	Pakistan	Rupee	260
Canada	$Can	4.73	Peru	Sol	10.00
Chile	Peso	2,050	Philippines	Peso	118
China (a)	Yuan	15.40	Poland	Zloty	9.10
Colombia	Peso	8,400	Russia	Rouble	81.00
Costa Rica	Colones	2,050	Saudi Arabia	Riyal	10.00
Czech Republic	Koruna	70.22	Singapore	$Singapore	4.85
Denmark	Dan. Kroner	31.50	South Africa	Rand	19.95
Egypt	Pound	15.50	South Korea	Won	3,700
Eurozone (b)	Euro	3.49	Sri Lanka	Rupee	290
Hong Kong	$Hong Kong	16.50	Sweden	Swe. Kroner	41.00
Hungary	Forint	645	Switzerland	Swiss Franc	6.50
India (c)	Rupee	84.00	Taiwan	$NT	75.00
Indonesia	Rupia	22,534	Thailand	Baht	78.00
Israel	Shekel	15.90	Turkey	Lira	6.60
Japan	Yen	320	UAE	Dirhams	12.00
Latvia	Lats	1.65	Ukraine	Hryvnia	17.00
Lithuania	Litas	7.80	Uruguay	Peso	90.00

(a) Average of five cities.
(b) Weighted average price in Eurozone.
(c) Maharaja Mac (made with chicken instead of beef).

Chapter subjects

✓ **Types of data** • Quantitative data • Qualitative data • Mix qualitative/quantitative data

✓ **Central tendency** • Arithmetic mean • Median • Mode • Midrange • Weighted averages • Questionnaires and weighted averages • Geometric mean • Politics and central tendency

✓ **Dispersion** • Data spread • Range • Variance • Standard deviation • Calculation of variance and standard deviation • Deviation about the mean Coefficient of variation • Quartiles • Percentiles

---

In God we trust; everyone else bring data

Mike Bloomberg, Mayor of New York, 2001–13[1]

*To characterize and describe people we use terms like tall, slim, small, blond, blue-eyed, intelligent, and the like. In this way we develop an image of the person in question. Similarly in business it is important to characterize and describe statistical data so that we can understand situations using perhaps average, range, median, standard deviation, mode, etc. Statistical data are a collection of information from which conclusions or logical decisions can be drawn. From the data **benchmark** values can be determined, which then provide a reference point such as the world average price of a hamburger as given in the icebreaker.*

Types of data

A collection of data is a **dataset** where the concept of "set" arises from referring to a set of objects: a set of stamps, a set of coins, or a set of drawings. A dataset is also called a **data array**. Data may be quantitative, qualitative, or a combination qualitative/quantitative. A dataset may describe a **population** where all data values are included: the population of China, the hotel staff population, or the elephant population in Africa. If the dataset is not a population it is a **sample** of a population: the back office staff would be a sample of the hotel population, a blood sample, or a water sample from the Rhine river. Population values are often considered **parameters** whereas sample values are taken as **variables**.

Quantitative data

Quantitative data may be **integer** values, or whole numbers, which is information that comes from the counting process: 1, 2, 3, etc. The data may be fractions. The **common fraction**, which is a number less than unity, meaning the numerator (top part of the fraction) is less than the denominator (bottom part of the fraction), as for example $^3/_4$. The data may be as an **improper fraction** where the value is greater than unity meaning that the numerator is greater than denominator as for example 14/4 that simplified becomes 3½. The data may be in **decimal** form, which is a fraction where the denominator is any power of 10. For example: 7/10 = 0.70; 9/100 = 0.09; 7,051/1,000 = 7.051. The data may be presented as a percentage, that is, a value per 100: 20 per 100 is 20 percent, 35 per 100 is 35 percent, etc. All numerical data have **units** such as for volume, weight, length, etc. **Metric units:** liters, kilograms, or meters are used in Europe. **Imperial units:** gallons, pounds, or feet are used in the USA and to a certain extent other English-speaking countries

A dataset might contain just **univariate data** meaning there is just one type of variable unconnected with others as shown in Table 1.1, which is a dataset of whole numbers. This is **raw** information as the data is in no particular order. The same information is in Table 1.2 but now the data is **ordered** by rows from highest to lowest.

Table 1.1

14	5	10	41
74	2	90	64
12	17	108	85
23	94	145	11

Table 1.2

2	5	10	11
12	14	17	23
41	64	74	85
90	94	108	145

The data might be raw univariate data of decimals, as in Table 1.3 or sorted data as in Table 1.4.

Table 1.3

14.72	5.46	10.24	41.56
74.23	2.78	90.58	64.58
12.56	17.69	108.75	85.72
23.98	94.56	145.69	11.29

Table 1.4

2.78	5.46	10.24	11.29
12.56	14.72	17.69	23.98
41.56	64.58	74.23	85.72
90.58	94.56	108.75	145.69

[*In Excel to organize data either from lowest to highest use the command:* **SORT**. *To do this all the data must be grouped in a single column or a single line*]

In datasets there is always a lowest value—the **minimum** and a highest value—the **maximum**. This is easily seen from sorted data: in Table 1.2 where the minimum value is 2 and the maximum 145; in Table 1.4 the values are 2.78 and 145.69 respectively.

[*In Excel to find lowest and highest values use functions:* **MIN** *and* **MAX**. *Sorting is not necessary.*]

A dataset can be **bivariate** with two associated variables, (x, y) as shown in Table 1.5. In this dataset it means that the value of the variable x is related to its corresponding variable y as, for example, "x" represents a particular month and "y" represents sales revenues. This data can be plotted on a two-dimensional graph (*see Chapter 2*).

Table 1.5

| x | 7 | 5 | 3 | 2 | 8 | 9 | 10 | 4 | 1 | 15 | 12 |
| y | 14.2 | 15.32 | 16.78 | 8.95 | 14.58 | 20.78 | 19.56 | 12.56 | 9.78 | 8.45 | 2.56 |

Data values have units. They may be financial units such as Dollars, Pounds, Yen, Euros or some other currency terms. If they are measurement units they might be in the metric or imperial system (*See Appendix III*). With data, units should be clearly expressed and known to persons using the data. Not a statement that should be considered lightly: In 1999 NASA, of the USA, lost a $125 million Mars orbiter because one engineering team used metric units while another used imperial units for a key spacecraft operation. For that reason, information failed to transfer between the Mars Climate Orbiter spacecraft team at Lockheed Martin in Colorado and the mission navigation team in California. Edward Weiler, NASA's Associate Administrator for Space Science said in a statement that, "People sometimes make errors but the problem here was not the error, but the failure of NASA's systems engineering, and the checks and balances in the processes to detect the error. That is why we lost the spacecraft." That being said, the people involved in the operation were high-salaried, highly qualified, and competent technical personnel![2] (See: *icebreaker, Chapter 13*)

Qualitative data

Data might be just qualitative with appropriate **labels** according to a desired description as in Table 1.6.

Table 1.6

Color	red	blue	yellow	green	orange
Food type	beef	poultry	pastries	vegetables	fish
Country	USA	France	UK	Germany	Turkey
Opinion	very good	good	satisfactory	poor	very poor
Name	Peter	Paul	Mary	Susan	Helen

Mix qualitative/quantitative data

Qualitative data might be associated with reference values that then give a stronger identity to the data. Table 1.7 repeats Table 1.6 but this time numerical information is included.

Table 1.7

Color	Red	Blue	Yellow	Green	Orange
Price/m^2 (Clothing fabric)	4.25	4.02	3.75	2.98	4.00
Food type	Beef	Poultry	Pasta	Potatoes	Fish
Price, €/kg	18.92	12.75	2.75	3.75	15.75
Country	USA	France	UK	Turkey	China
Population, millions, 2012^3	314	66	63	80	1,343
Opinion	Unacceptable	Poor	Satisfactory	Good	Excellent
Score (For questionnaires)	1	2	3	4	5
Name	Peter	Paul	Mary	Susan	Helen
Age (Firm's personnel data)	26	41	27	52	33

Central tendency

Central tendency refers to clustering of datasets around a pivotal value. This is useful since once a pivotal value has been determined datasets can be compared, or benchmarked, according to this pivotal value to see if performance levels are being met; if adjustments need to be made; or what conclusions can be drawn. Common central tendency measurements are arithmetic mean, median, mode, midrange, weighted average, and the geometric mean.

Arithmetic mean

The arithmetic mean, also known as the **mean** or **average value**, and written as $\bar{x}$, is the most common measure of central tendency. It is calculated by summing all the values in the dataset $\sum x$ and dividing by the number of values, N.

$$\bar{x} = \frac{\sum x}{N}$$

1(i)

Table 1.8 gives the weekly revenues in Euros at a restaurant.

Table 1.8

| 40,250 | 50,390 | 35,725 | 20,625 | 27,190 | 21,460 | 28,290 |

The total of these value is €223,930 and thus the weeks average revenues are €31,990 (€223,930/7). If this pattern is representative of the restaurant's operation we may say that average daily revenues are about €32,000. This value can be used as a benchmark for performance. When average weekly revenues are above €32,000 the restaurant is doing well; below €32,000 the performance is not so good. Note, that the arithmetic mean is influenced by extreme values or **outliers**. If in Table 1.8 the last value was €158,350 as in Table 1.9, the mean would then be €50,570 (€353,990/7) or greater than all of the above individual values, except the last. The mean value is often used as the **target** value of an operation or experiment.

Table 1.9

| 40,250 | 50,390 | 35,725 | 20,625 | 27,190 | 21,460 | 158,350 |

[*In Excel to find the mean value use the function*: **AVERAGE**]

Median

The median is a measure of central tendency that divides **sorted** data into two equal parts. Table 1.8 sorted from highest to lowest gives the dataset in Table 1.10.

Table 1.10

| 20,625 | 21,460 | 27,190 | 28,290 | 35,725 | 40,250 | 50,390 |

The median value is the **middle** value of this **ordered** dataset, or €28,290. Thus half of the week's revenues are below €28,290 and half are above. In Table 1.10 there are seven values, or an odd number, so the median value is easy to spot. In Table 1.11 there are six values, or an even number. In this case the median value is the average of the third and fourth values or €27,740, i.e. (€27,190 + €28,290)/2.

Table 1.11

| 20,625 | 21,460 | 27,190 | 28,290 | 35,725 | 40,250 |

The rule for calculation of the median in a dataset of n values is as follows: if a dataset has an odd number of values as in Table 1.10, then:

$$\text{Median} = \frac{(n+1)}{2} \qquad \text{1(ii)}$$

In Table 1.10 the median is (7 + 1)/2 or the fourth value.
If a dataset has an even number of values, then:

$$\text{Median is average of } \frac{n}{2} \text{ and } \frac{(n+2)}{2} \qquad \text{1(iii)}$$

For Table 1.11 there are 6 values so the median is the average of 6/2 and 8/2 or the average of the third and fourth values.

Where is the median used? In 2012 the median age of the male population in Japan was 44.1 years out of a total population of 127 million. This means that half the male population is above 44.1 years and half below. This implies that in the future many people will reach retirement age with a consequent drain on social services. In contrast, the median age of the male population in Nigeria is 17.5 years out of a total population of 170 million; half the male population is below 17.5 years and half above. Very soon an explosion of young people will be looking for employment. Will there be enough available?[4] Between August and October 2012 the median house price in Beverly Hills, California was $2,050,000: half the houses were priced above $2,050,000 and half below:[5] This underscores what is already known, that to live in Beverly Hills you better have pots of money! In driving we talk about the median of a road. This is the white line that divides the road into two equal parts with the same number of lanes on each side.

[*In Excel to find the median value use the function*: **MEDIAN**]

Mode

The mode is another measure of central tendency, and is that value that occurs most frequently in a dataset. The mode may be of interest because a value that occurs most frequently is worthy of further attention. As an example, Table 1.12 gives the number of customer complaints that are recorded in a hotel in a 16-day period.

Table 1.12

Sun	Mon	Tue	Wed	Thu	Fri	Sat	Sun	Mon	Tue	Wed	Thu	Fri	Sat	Sun	Mon
7	2	1	4	2	5	3	7	6	1	2	3	1	1	7	2

The mode is 7 since it occurs three times and it also occurs on a Sunday! What is special about Sunday? Perhaps it is insufficient personnel, poorly trained employees, or subcontracted personnel. The reason should be investigated. In a quality control study that I was performing for a firm it was noted that in the three factory work shifts: night shift, 22:00–06:00; morning shift, 06:00–14:00; and afternoon/evening shift, 14:00–22:00, there were more product quality problems during the night shift. On investigation the cause was related to lack of proper management supervision, alcohol, and also drugs!

Unlike the mean and median the mode can be used for purely qualitative as well as for quantitative data. For example Table 1.13 shows the results of a restaurant questionnaire where clients were asked to give their preferred main dish.

Table 1.13

Salmon	Duck	Cod	Beef	Chicken	Pigeon	Tripe	Salmon
Beef	Veal	Salmon	Turkey	Cod	Salmon	Lamb	Turkey
Chicken	Salmon	Mussels	Veal	Salmon	Haddock	Cod	Lamb

The modal value is "Salmon" since this response occurred six times. Thus the restaurant should pay attention to always keeping a sufficient inventory of salmon for its customers. In another survey, of color preferences, Table 1.14 gave the responses.

The modal value is "Blue" since it occurs eight times. This information would be useful for a firm preparing paint stock, curtain fabrics, carpeting, automobile chassis etc. Datasets might be bimodal

Table 1.14

Red	White	Green	Mauve	Mauve	Gray	Blue	Brown
Green	Yellow	Blue	Blue	Blue	Yellow	Indigo	Yellow
Blue	White	Orange	Black	Blue	Blue	Purple	Blue

where two values occur more often; trimodal, three values occurring more often; and higher orders of modality.

> [*In Excel to find the modal value use the function*: **MODE**. *To determine how many times the modal value occurs, use the function* **COUNTIF**]

Midrange

The midrange, also a measure of central tendency, is the average of the smallest and largest observation in a dataset or halfway between the lowest and the highest value. In Table 1.15 the midrange is (20,625 + 50,390)/2 or 35,507.50.

Table 1.15

40,250	50,390	35,725	20,625	27,190	21,460	28,290

The midrange is not a reliable measure of central tendency as it only considers two numbers and thus can be distorted by extreme values. For example, Table 1.16 is the same as Table 1.15 except for the last value.

Table 1.16

40,250	50,390	35,725	20,625	27,190	21,460	158,350

The midrange is now (20,625 + 158,350)/2 or 89,487.50. A big difference from 35,507.50 of Table 1.15!

The midrange is sometimes used as an "average value". For example a weather report may note an average daily temperature of $20°$C when the average of the maximum temperature of $28°$C and a minimum of $12°$C has been used—or the midrange. The average daily price of a certain stock may be quoted as $52.95 when the average of the minimum price of $50.20 and a maximum of $55.70 has been used—the midrange. Or the average daily balance of a customer's checking account may be recorded as $4,525.84 when the average of the minimum balance of $4,025.36 and a maximum balance of $5,026.32 has been used in the calculation—the midrange. When the differences are not extreme this is not an unreasonable way to use the midrange.

> [*In Excel to determine the midrange add the maximum and minimum values and divide the result by two.*]

Weighted averages

The weighted average is also a measure of central tendency though it takes into account the importance, or **weighting**, of values in the dataset where pairs of data have a relationship. Consider a kitchen restaurant where there are three staff members: chef, assistant chef; and an apprentice. Their salaries and time for preparing meals are in Table 1.17. Here the pairing is between salary and time.

Table 1.17

	Chef	Assistant chef	Apprentice
Salary, £/hour	37.50	24.50	12.20
Times spent per 10 meals, hours	1.50	2.00	1.75

Using weighted averages the labor cost per meal is:

37.50*1.50 + 24.50*2.00 + 12.20*1.75 = £126.60 per 10 meals or £12.66 per meal.

If we used simply the average hourly wage, the hourly labor cost would be:

(37.50 + 24.50 + 12.20)/3 = £24.73/hour.

It takes 5.25 hours to prepare 10 meals (1.50 + 2.00 + 1.75). Thus labor cost per 10 meals using the average hourly labor cost is:

£24.73*5.25 = 129.83 or £12.98/meal.

It is wrong to use this value. We must consider the time, or the importance each member of the kitchen staff spends on meal preparation.

Questionnaires and weighted averages

Often in business, particularly when services are involved, **opinions** of customers are solicited about quality. As an example, in a hotel opinions may be asked about the restaurant meal; breakfast service; hotel room; spa; and other services. For this a **questionnaire** is used to determine a customer's subjective (qualitative) opinion. These opinions are then given a quantitative score as shown in Table 1.18 (The same as the fourth row of Table 1.7).

Table 1.18

Opinion	Unacceptable	Poor	Satisfactory	Good	Excellent
Score	1	2	3	4	5

The results from the questionnaire are then analyzed using **weighted averages**. In this case we look for a central value that considers the importance of each opinion. An illustration is shown in Table 1.19 that is the opinion of 15 clients on the hotel breakfast service.

The weighted average of the response is calculated by:

$$\text{Weighted average} = \sum \frac{\text{Number of responses}^*\text{score}}{\text{Total responses}}$$

From Table 1.19: $\text{Weighted average} = \dfrac{2^*1 + 1^*2 + 1^*3 + 5^*4 + 6^*5}{15} = 3.80.$

This translates into saying the breakfast service is considered between acceptable and good, and closer to being good.

[*in Excel for a matrix such as this the function,* **SUMPRODUCT**]

Table 1.19

Category Score	Unacceptable 1	Poor 2	Acceptable 3	Good 4	Excellent 5	Total Responses
Client 1					✓	1
Client 2				✓		1
Client 3		✓				1
Client 4				✓		1
Client 5	✓					1
Client 6			✓			1
Client 7					✓	1
Client 8				✓		1
Client 9					✓	1
Client 10	✓					1
Client 11					✓	1
Client 12					✓	1
Client 13				✓		1
Client 14					✓	1
Client 15				✓		1
Total	2	1	1	5	6	15

Geometric mean

The geometric mean is a measure of central tendency for data, such as revenues, costs, interest rates, prices, productivity etc., that are **changing over time**. Consider Table 1.20 that gives food sales over six years.

Table 1.20

Year	1	2	3	4	5	6
Sales, €	876,000	912,140	1,023,500	999,850	1,023,650	1,115,500

The information from this table has been expanded to calculate the percentage annual change and the growth factor each year as shown in Table 1.21. For example, the **annual percentage** change for Year 2 is (912,140 – 876,000)/876,000 or 0.0413 or 4.13%. The corresponding **growth factor** for Year 2 is 1.0413 (1 + 0.0413). The growth factor is the amount by which the value of the previous period has to be multiplied in order to give the current value. If the percentage change is negative, then the growth factor is less than 1 as for example in Year 4.

Table 1.21

Year	1	2	3	4	5	6
Sales, €	876,000	912,140	1,023,500	999,850	1,023,650	1,115,500
% change		4.13%	12.21%	–2.31%	2.38%	8.97%
Growth rate		1.0413	1.1221	0.9769	1.0238	1.0897

The compounded average growth rate, or geometric mean for a dataset with "n" time periods, is:

$$\sqrt[n]{(\text{product of growth rates})}$$

In this case the geometric mean growth rate for the data in Table 1.21 is:

$$\sqrt[5]{1.0413*1.1221*0.9769*1.0238*1.0897} = 1.0495$$

The geometric mean growth is 4.95% (1.0495 − 1). This means that food sales over a five years from Year 1 to Year 6 have grown by a mean geometric amount of 4.95%. This is the central tendency for the given time frame. We can check this value:

- sales in Year 1 are €876,000
- going from Year 1 to Year 6 there are five years changes of sales, (6 − 1)
- geometric growth rate for this period is 1.0495
- sales in Year 6 are $876,000*1.0495^5 = 1,115,500$. The same as in the last column of Table 1.21.

(See *Chapter 2* for a graphical representation of the geometric mean.)

Note that it would be wrong to use the simple average of the growth rates, 5.08% (4.13 + 12.21 − 2.31 + 2.38 + 8.97)/5 as this assumes a linear change. The compound growth rate, the term used in finance, is the same as the geometric mean.

[*In Excel to find the geometric mean use the function*: **GEOMEAN** *using growth rates for input data*]

Politics and central tendency

An interesting qualitative analogy of central tendency is the attitude of many political parties and their leaders. In most democracies politicians expound, often vehemently, their policies. The labor, or left wing, parties demand more social spending, higher taxes on the wealthy, more concern for the environment, more rights for the unionized labor force, and the like. On the other hand, the conservative, or right wing, parties demand less taxation, adherence to capitalistic policies, less union control, etc. However when one of these parties gets elected they have a "central tendency" or move to the middle of their doctrine. Examples are David Cameron of the UK who is conservative and was elected prime minister in May 2015; he is moving to the left of traditional conservative policies and more to the center.[6] Alexis Tsipras, the prime minister of Greece, an ardent leftist when first elected prime minister, fought violently against austerity measures. Now, as re-elected prime minister in September 2015, he has moved to the center and accepted "more conservative" principals to government.[7] Justin Trudeau, a liberal who was elected prime minister of Canada in October 2015, has a more leftist approach than his conservative predecessor Steven Harper. However, observers indicate that he will have to move closer to the center when he really starts to govern.[8] Then there is the classic illustration of Francois Mitterrand who was elected President of France in 1981 on a strong left-wing platform.[9] However, not long into his mandate he was obliged to take a 180 degree turn and adopted many middle-of-the road or centrist policies.

Dispersion

A second benchmark data characteristic is **dispersion**. This is how data is separated, spread out, or varies from other values in the dataset. Very often the dispersion of data is more important than the mean.

Data spread

Assume that you are hiking in Yosemite, California, and arrive at a fast-flowing river. The only way to reach the other side is to wade across with your backpack. Your guidebook indicates that the average depth at this point of the river is 70cm. At this measure you feel that crossing is feasible. Then your colleague who knows this area says "hang on, the minimum depth is 5cm and the maximum is 1m 35cm" [equivalent to an average of 70cm (135+5)/2]. Now you are not so sure about crossing! Alternatively you know that average body temperature is 37.5°C. This might be achieved by sticking your feet in a bucket of ice at 0°C and your head in an oven at 75°C (an average of 37.5°C). However, you would be uncomfortable! In business knowing dispersion or **variation** is important, as a wide dispersion in a dataset means that the dataset is less reliable for analytical purposes. Further, when clients are involved, situations that show wide variation are unacceptable: a meal that is good this week but not so good the following week; service quality that was impeccable yesterday but poor today; food delivery that was on time last week but was late this week. The range, variance, standard deviation, quartiles, and percentiles are benchmark measurements that indicate the spread of data.

Range

The range is a measure of dispersion and is the difference between the maximum, and the minimum value in a dataset. Table 1.22 (the same as Table 1.15) gives restaurant revenues.

Table 1.22

40,250	50,390	35,725	20,625	27,190	21,460	28,290

The range is €29,765 (50,390 – 20,625). Like the midrange, the range only considers the two extreme dataset values. Again it is distorted by extreme values. Consider now Table 1.23 that is the same as Table 1.22 except for the last value.

Table 1.23

40,250	50,390	35,725	20,625	27,190	21,460	158,350

The range is €137,725 (158,350 – 20,625). A big difference from €29,765! As such the range is *not an honest measure* of dispersion as it only considers the two extreme values. More dependable measures of dispersion are the **variance** and its square root, **standard deviation**.

[*In Excel, find the maximum and minimum and subtract the two to determine* **RANGE**]

Variance

The variance considers all the values in the dataset, not just the lowest and the highest, and thus the impact of extreme values is mitigated. There is a variance for both a population and a sample. The **population variance**, σ_x^2, is the sum of the squared difference between each observation, x, and the mean or average value, μ, divided by the number of values, N:

$$\sigma_x^2 = \frac{\sum (x - \mu_x)^2}{N}$$

1(iv)

- For each value of x, the mean value μ_x is subtracted from this number. This indicates how far this x-value is from the mean, or the **range** this value is from the **mean**.
- By squaring each of the differences obtained, the negative signs are removed.
- Dividing by N gives an average value of the population variance.

The **sample variance**, s^2, is analogous to the population variance and is given by:

$$s^2 = \frac{\sum (x - \bar{x})^2}{(n - 1)} \qquad \text{1(v)}$$

In the sample variance, $\bar{x}$, or x-bar, the average of the values of x, replaces μ_x, the population variance, and $(n-1)$ replaces N, the population size. A reason for the different denominator is that a principal use of statistical analysis is to take a sample from the population and make estimates of the population parameters based only on sample measurements. (Measuring population parameters may be long, costly, and perhaps nearly impossible.) By convention, using the symbol "n" means a sample of size n taken from the population of size N. Using $(n - 1)$ in the denominator reflects the fact that we have used $\bar{x}$ in the formula and so we have lost one degree of freedom in our calculation. For example, consider you have a sum of $1,000 to distribute to your six restaurant co-workers based on certain criteria. To the first five you have the freedom to give any amount, say $200, $150, $75, $210, and $260. To the sixth co-worker you have no degree of freedom of the amount to give: it has to be the amount remaining from the original $1,000, which in this case is $105. When we perform sampling experiments to estimate the population parameter with $(n - 1)$ in the denominator of the sample variance formula we have an unbiased estimate of the true population variance. Note that if the sample size, n, is large, then using n or $(n - 1)$ will give results that are very close.

[*In Excel to find the population variance use function* **VARP**. *To find the sample variance use* **VAR**]

Standard deviation

The variance is not always a convenient measure of dispersion to use as the units are squared—meters^2, liters^2, kilograms^2, etc. In this case we use the standard deviation, which is the square root of the variance and thus has the same units as the data used in the measurement—meters, liters, kilograms, etc. The standard deviation is the most often used measure of dispersion in analytical work and is the typical distance from a point to the center of the dataset. The **population standard deviation, σ_x** is:

$$\sigma_x = \sqrt{\sigma_x^2} = \sqrt{\frac{\sum (x - \mu_x)^2}{N}} \qquad \text{1(vi)}$$

and the **sample standard deviation**, s, is:

$$s = \sqrt{s^2} = \sqrt{\frac{\sum (x - \bar{x})^2}{(n - 1)}} \qquad \text{1(vii)}$$

The smaller the standard deviation then the less dispersed, or spread out, are the data values.

[*In Excel to find the population standard deviation use function* **STDEVP** *or* **STDEV.P**. *To find the sample variance use function* **STDEV**]

Calculation of variance and standard deviation

The following shows the calculation procedure and the relationship between variance and the standard deviation. Assume that we are interested in the age characteristics of nine employees, the population, in a hotel restaurant. Thus $N = 9$. The age of these employees in years, x, is given in the first line of Table 1.24.

Table 1.24

Age, x (years)	19	29	33	18	31	44	50	22	33
$x - \mu$	−12.00	−2.00	2.00	−13.00	0.00	13.00	19.00	−9.00	2.00
$(x - \mu)^2$	144.00	4.00	4.00	169.00	0.00	169.00	361.00	81.00	4.00

- The average of these ages is:

$$\frac{19 + 29 + 33 + 18 + 31 + 44 + 50 + 22 + 33}{9} = 31.00.$$

- From each of the individual ages (x) we subtract the average age of 31.00 or $(x - \mu)$. This gives the second line of the Table 1.24.
- The third line of the table is $(x - \mu)^2$, and this removes all negative signs.
- The sum of $(x - \mu)^2$ is written $\sum(x - \mu)^2 = 936.00$.
- Population variance, Equation 1(iv) is: $\sigma_x^2 = \frac{\sum(x - \mu_x)^2}{N} = \frac{936.00}{9} = 104.00$ year^2.
- Standard deviation, σ, is the square root of the population variance or $\sqrt{(104.00)} = 10.1980$ years.

Thus we say that the average age of the population of nine restaurant employees is 31.00 years with a standard deviation of 10.1980 years.

- If the restaurant employees are considered a sample, n, of the population of the all the hotel employees then sample variance from Equation 1(v): $s^2 = \frac{\sum(x - \bar{x})^2}{(n-1)}$.
- Here $(n - 1) = (9 - 1) = 8$.
- The value of the numerator remains unchanged since the term, $\bar{x}$, is still the mean value, μ. Thus the sample variance = 936.00/8 = 117.00 year^2.
- Sample standard deviation is the square root of the sample variance or $\sqrt{117.00} = 10.8167$ years.

Thus in a sample experiment we say that in a sample of nine hotel employees the average age is 31.00 years and the sample standard deviation is 10.8167 years.

Note that for any given dataset sample variance and sample standard deviation are always **less** than the corresponding population variance and population standard deviation since the denominator $(n - 1)$ is smaller than the population denominator N.

Deviation about the mean

The deviation about the mean of all observations, x, about the mean value $\bar{x}$, is zero or mathematically,

$$\sum(x - \bar{x}) = 0 \hspace{4cm} \text{1(viii)}$$

In the second line of Table 1.24:

$$\sum(x - \mu) = (-12.00) + (-2.00) + 2.00 + (-13.00) + 0.00 + 13.00$$
$$+ 19.00 + (-9.00) + 2.00 = 0.00$$

This is perhaps a logical conclusion since the mean value is calculated from all the dataset values.

Coefficient of variation

The standard deviation as a measure of dispersion on its own is not easy to interpret. In general terms a small value for the standard deviation indicates that the dispersion of the data is low and conversely the dispersion is large for a high value of the standard deviation. However, the magnitude of these values depends on what you are analyzing. Further, how small is small, and what about the units? If you say that the standard deviation of the total travel time, including waiting, to fly from London to Vladivostok is 2 hours, the number 2 is small. However, if you convert that into minutes the value is 120, and, if you use seconds, a high 7,200. But in any event, the standard deviation has not changed! A way to overcome the difficulty in interpreting the standard deviation is to include the value of the mean of the dataset and use the **coefficient of variation.** The coefficient of variation is a relative measure of the standard deviation of a distribution, σ, to its mean, μ. The coefficient of variation can be either expressed as a proportion or a percentage of the mean. It is defined by:

$$\text{coefficient of variation} = \frac{\sigma}{\mu}$$

Thus in using the flying time example if the average flying time is 5 hours then the coefficient of variation is 2/5 or 40%. If instead of hours we use minutes then the ratio is 120/300 or unchanged at 40%. The coefficient of variation is most useful to compare two sets of data. Consider a restaurant that has two kitchen teams. The lunchtime team prepares on average 25 meals/hour with a standard deviation of 4 meals/hour. The dinnertime team prepare on average 39 meals/hour with a standard deviation of 8 meals/hour. The coefficient of variance for the lunchtime team is 16.00% (4/25); for the dinnertime team it is 20.51% (8/39). On this basis the variation of the lunchtime team is less and the team is more consistent in its meal preparation.

Quartiles

Quartiles are values that divide ordered data into four equal quarters each containing 25% of the dataset. There are **five boundary limits** for quartiles. Q_0 is the lower boundary limit and is the minimum value in the dataset; Q_1 is the first quartile boundary limit; Q_2 is the second quartile boundary limit and also the median value; Q_3 is the third quartile boundary limit. The upper boundary limit is Q_4 and is also the maximum value in the dataset. With these five divisions, quartiles measure **dispersion** of the dataset. They are useful as they give a logical grouping of a dataset. Table 1.25 gives a hotel's weekly operating costs in $US for the last year.
   Here:

$$Q_0 = 81,340.00; \ Q_1 160,621.75; \ Q_2 = 188,423.50; \ Q_3 = 233,119.25; \text{ and}$$
$$Q_4 = 332,212.00.$$

Table 1.25

217,076	86,157	293,373	171,135
149,886	274,856	167,175	142,382
187,173	215,564	248,146	228,577
238,840	217,177	122,211	157,775
187,173	110,336	188,977	332,212
203,411	159,262	298,256	161,075
185,377	210,573	81,340	237,524
106,155	187,124	224,276	303,466
180,676	253,076	225,880	155,251
231,651	120,415	188,426	241,171
182,677	132,424	249,651	164,249
210,682	251,251	188,421	210,536
161,372	127,076	177,226	275,936

Other characteristics of quartiles are the **inter-quartile range**, or **mid-spread.** This is the difference between the third and the first quartile in a dataset or $(Q_3 - Q_1)$. It measures the **range** of the middle 50% of the data. One half of the inter-quartile range, $(Q_3 - Q_1)/2$ is the **quartile deviation** and this measures the average range of one half of the data. The smaller the quartile deviation, the greater is the concentration of the middle half of the observations in the dataset. The **mid-hinge** is $(Q_3 + Q_1)/2$ and is a measure of central tendency. It is analogous to the midrange of a dataset.

Here, inter-quartile range = $\$72,497.50$, quartile deviation = $\$36,248.75$, and mid-hinge = $\$196,870.50$.

Although, like the range, these additional quartile properties only use two values in their calculation, distortion from extreme values is somewhat limited as the quartile values are taken from an ordered dataset that can be of a considerable size.

[*In Excel to find the quartile values use function:* **QUARTILE**]

Percentiles

Percentiles divide data into 100 equal parts and thus give a more precise positioning and grouping of a dataset than quartiles. The percentiles for hotel operating costs of the previous paragraph are given in Table 1.26. With percentiles the zeroth percentile is the minimum value in the dataset and the lower boundary limit of the quartiles; the twenty-fifth percentile is the same value of the first quartile; the fiftieth percentile is the same value of the second quartile and the median value; the seventy-fifth percentile is the same value of the third quartile. The hundredth percentile is the maximum value and the upper boundary limit of the quartiles.

[*In Excel to find the percentile values use the function:* **PERCENTILE**]

Percentiles are used in the medical field—for example pediatricians measure the height and weight of small children and indicate how the child compares with others in the same age range using a percentile measurement. If the pediatrician says that for your child's height he is in the tenth percentile, this means that only 10% of all children in the same age range have a height less than your child; 90% have a height greater than that of your child. This information can be used as an indicator of the child's growth pattern. Another use of percentiles is in education for exam grading to determine in what percentile, a student is positioned.

Table 1.26

Percentile	Value	Percentile	Value	Percentile	Value	Percentile	Value
0%	81,340.00						
1%	83,796.67	26%	161,152.22	51%	188,431.51	76%	236,114.48
2%	86,556.96	27%	161,303.69	52%	188,712.52	77%	237,879.32
3%	96,755.94	28%	162,177.56	53%	189,410.02	78%	238,550.48
4%	106,322.24	29%	163,644.83	54%	196,771.36	79%	239,515.99
5%	108,454.55	30%	165,126.80	55%	203,767.25	80%	240,704.80
6%	110,940.74	31%	166,619.06	56%	207,401.00	81%	243,333.25
7%	116,081.03	32%	168,442.20	57%	210,538.59	82%	246,890.50
8%	120,558.68	33%	170,461.80	58%	210,557.46	83%	248,642.65
9%	121,474.64	34%	173,205.94	59%	210,582.81	84%	249,410.20
10%	122,697.50	35%	176,312.35	60%	210,638.40	85%	250,211.00
11%	125,178.65	36%	178,468.00	61%	211,219.02	86%	251,027.00
12%	127,717.76	37%	180,227.50	62%	213,708.84	87%	251,926.25
13%	130,445.24	38%	181,436.38	63%	215,760.56	88%	252,857.00
14%	133,818.12	39%	182,456.89	64%	216,531.68	89%	261,570.20
15%	138,896.70	40%	183,757.00	65%	217,091.15	90%	272,678.00
16%	143,582.64	41%	185,134.00	66%	217,142.66	91%	275,298.80
17%	147,409.68	42%	186,110.74	67%	218,383.83	92%	275,849.60
18%	150,851.70	43%	187,001.71	68%	222,004.32	93%	283,433.91
19%	153,587.85	44%	187,145.56	69%	224,580.76	94%	292,326.78
20%	155,755.80	45%	187,170.55	70%	225,398.80	95%	295,570.35
21%	157,043.04	46%	187,173.00	71%	226,446.37	96%	298,060.68
22%	158,102.14	47%	187,173.00	72%	227,821.84	97%	300,704.70
23%	158,860.51	48%	187,772.04	73%	229,284.02	98%	303,361.80
24%	159,697.12	49%	188,408.52	74%	230,851.76	99%	317,551.54
25%	160,621.75	50%	188,423.50	75%	233,119.25	100%	332,212.00

Moving on

This chapter has detailed the significance of numbers highlighting the most common ways of characterizing, defining, and measuring data. The next chapter describes ways that data can be presented visually in order to get the message across to your manager, co-workers, or customers.

Notes

1 "Don't screw it up" *The Economist*, November 9, 2013, p. 15.
2 *NASA's metric confusion caused Mars orbiter loss*, CNN report, September 1999.
3 www.factfish.com/statistic/inhabitants [accessed November 2012].
4 US Central Intelligence Agency: www.cia.gov/library/publications/the-world-factbook/fields/2177.html [accessed November 2012].
5 www.trulia.com/real_estate/90210-Beverly_Hills/market-trends/ [accessed November 2012].
6 "The prime minister plays the part of a triumphant Tory. But he faces problems inside his party, especially over Europe," *The Economist*, October 10, 2015, p. 35.
7 "Greece's elections: The prime minister pivots away from leftism, and his party follows," *The Economist*, September 26, 2015, p. 25.
8 "Justin Trudeau has proved he can campaign, but can he govern?", *The Economist*, October 24, 2015, p. 46.
9 "François Mitterrand est président Chronique du 20e siècle", *La Rousse* 1990, ISBN 2-03-503-269-5, p. 1201.

2　Presenting and organizing data

ICEBREAKER: How good is your performance?

In the travel and hospitality industries management needs to ascertain that fixed assets are used to the fullest. A key performance measure is to measure capacity utilization, or load factor, for airplanes and trains; room occupancy for hotels; or place settings filled in a restaurant. These measures are a ratio of units actually sold to the number of units available. For planes or trains this is the number of seats filled per journey to the number available; in a hotel the number of rooms sold each night to the number in the hotel; and in a restaurant the number of clients eating to the number of settings available. The higher the ratio then the better is the firm's productivity and the higher the revenues. This is a key part of the

Figure 2.0　Hotel in China.

Source: Michael Wintenberger, France (with permission).

Table 2.0

200	181	110	175	211	293	319	360	297	291	291	308	358	353	343
347	345	185	172	185	216	172	166	155	97	116	134	216	237	192
147	138	155	241	209	179	164	228	198	222	239	282	319	265	317
226	284	302	304	310	317	328	213	261	269	347	321	289	315	190
235	261	310	313	308	261	123	244	235	278	246	256	252	172	276
310	433	392	295	386	203	289	310	386	332	308	459	211	287	302
287	222	252	263	377	259	412	377	351	509	511	325	429	506	511
511	511	509	466	502	457	509	500	351	416	388	416	422	308	293
254	336	181	216	315	315	310	323	317	263	401	440	481	502	506
442	343	427	397	366	375	319	347	250	321	334	250	244	304	224
198	440	375	427	427	388	308	295	308	310	390	369	433	259	409
472	364	435	444	386	293	250	263	394	433	274	261	237	384	407
379	397	360	425	319	254	276	366	371	377	377	341	364	267	369
407	360	332	302	362	425	425	420	315	263	306	364	386	392	345
325	353	287	390	444	463	416	455	429	302	250	334	315	463	405
338	272	351	427	306	336	427	401	196	336	332	416	498	487	489
379	300	295	427	377	293	323	263	431	431	453	483	438	394	284
259	356	317	345	463	470	295	310	429	496	468	360	295	315	358
252	216	261	241	190	284	194	325	304	341	405	438	427	433	341
358	366	453	431	364	502	491	489	478	498	381	278	325	377	366
463	466	476	401	435	405	470	446	476	366	323	489	476	474	278
313	429	425	422	440	394	459	343	407	332	392	390	306	321	414
356	185	317	388	496	435	313	353	207	280	325	349	377	308	334
218	256	336	364	412	414	293	194	222	269	323	248	308	377	216
261	261	166	235	294										

revenue benchmark measure but another of course is the price paid for the service—the plane ticket, the hotel room, or the meal consumed. The travel and hospitality industries track these performance measures daily. As an illustration, the data in Table 2.0 is the room occupancy for a 500-room, five-star hotel in a major city in China (Figure 2.0), for one-year. There are 365 values reading across from January 1 to December 31—the hotel is open all year. The data was collected by Anthony Metral an M.Sc. student from the Institute Paul Bocuse, a Hotel and Restaurant Management School in France, who was performing an internship in 2011 on revenue management for the hotel.[1]

Anthony was very aware that his top managers were busy people and did not have the patience, time, or even perhaps the intellectual level to clearly understand these raw numbers as given in Table 2.0. What they needed was a quick, clear, neat, and precise visual presentation of the hotel's performance. Anthony pondered over the most appropriate presentation: an absolute frequency histogram; a relative frequency histogram; a polygon; an ogive; a stem-and-leaf display; a box and whisker plot; a line graph ... ? The purpose of this *Chapter 2* is to demonstrate some ways to visually present data so that the correct message gets transmitted.

Chapter subjects

✓ **Frequency distributions** • Frequency distribution table • Absolute frequency histogram • Relative frequency histogram • Frequency polygons • Ogives • Stem-and-leaf display
✓ **Visuals of quartiles and percentiles** • Box and whisker plot • The percentile histogram
✓ **Line graphs** • Candid presentations • Geometric mean as a line graph
✓ **Visuals of categorical data** • Pie chart • Categorical histogram • Bar chart • Contingency tables • Stacked histograms • Pareto diagram • Spider web diagram • Gap analysis • Pictograms

---

Using Microsoft Excel's graphic capabilities, data can be transposed into visual displays that make interpretation and subsequent decision making easier. Most of us, even with the best education, cannot quickly and meaningfully interpret data such as presented in the Icebreaker. However, presented as a graph, histogram, or bar chart infor-mation can more easily be understood. All media publications: The Economist; The New York Times, Wall Street Journal; Time; Fortune *and others either in a print version or on the Web use visual displays. They are relatively easy to understand.*

Frequency distributions

Frequency distributions are groupings of data values into class limits to see if a pattern exists. The first step is to develop the group in a table and then from this plot the various visual aids.

Frequency distribution table

A frequency distribution table groups dataset values into unique **categories** according to how often, i.e. the **frequency**, that grouped data appear in a given category. The purpose is to see if this grouped data exhibit a pattern or a trend from which perhaps important business decisions can be made. When we have univariate numerical data, the groups are organized by **class ranges** of data where each **class range**, or **width**, is of the same dimension. A class range or class width is given by:

$$\text{Class range or width} = \frac{\text{Range of the complete dataset}}{\text{Number of groups selected}} \qquad \text{2(i)}$$

The range is the difference between the maximum and minimum value in the dataset. The number of groups selected depends on the pattern and type of data. More groups separates data values; fewer bundles data. Consider as an example Table 2.1, weekly hotel operating costs in $US. (It is the same as Table 1.25 in *Chapter 1*.)

Here the minimum value is $81,340 and the maximum value is $332,212 or a range of $250,872. However, if we select rational values that encompass the entire dataset, such as $80,000 as the minimum value, and $340,000 as the maximum, then the complete dataset values of Table 2.1 lie within these values. The chosen minimum and maximum are the closest, to the nearest $10,000, of the real minimum and maximum and are appropriate for eventually constructing graphs. If we select 13 groups then from Equation 2(i):

$$\text{Class range or class width} = \frac{340,000 - 80,000}{13} = \frac{260,000}{13} = 20,000$$

Table 2.1

217,076	86,157	293,373	171,135
149,886	274,856	167,175	142,382
187,173	215,564	248,146	228,577
238,840	217,177	122,211	157,775
187,173	110,336	188,977	332,212
203,411	159,262	298,256	161,075
185,377	210,573	81,340	237,524
106,155	187,124	224,276	303,466
180,676	253,076	225,880	155,251
231,651	120,415	188,426	241,171
182,677	132,424	249,651	164,249
210,682	251,251	188,421	210,536
161,372	127,076	177,226	275,936

Table 2.2 then shows the corresponding frequency distribution table of the data in Table 2.1. Class 1 has a range of >$80,000 to ≤$100,000; Class 2 a range of >$100,000 to ≤$120,000; etc.

Column 3, "Number of values" gives the number of data values according to the class range; Column 4, "Proportion of values" gives the percentage of data out of a total of 52 for the class range. This is a **closed-ended** frequency table since all values of Table 2.1 are included in the frequency table. The frequency distribution table is the basis for construction of the absolute and relative frequency histograms and polygons, and ogives.

Table 2.2

Column 1 Class number	Column 2 Boundary limits	Column 3 Number of values	Column 4 Proportion of values
0	80,000	0	0.00%
1	100,000	2	3.85%
2	120,000	2	3.85%
3	140,000	4	7.69%
4	160,000	5	9.62%
5	180,000	6	11.54%
6	200,000	9	17.31%
7	220,000	7	13.46%
8	240,000	6	11.54%
9	260,000	5	9.62%
10	280,000	2	3.85%
11	300,000	2	3.85%
12	320,000	1	1.92%
13	340,000	1	1.92%
	Total	52	100.00%

[*In Excel the function* **FREQUENCY** *is used. Using the information in Table 2.1 and Table 2.2 as an example, select a virgin column adjacent to, and of the same dimension, as Column 2 of Table 2.2. Call up the function* **FREQUENCY**. *It asks for the data file. This would be Table 2.1. It asks for the class limits. This is the Column 2 of Table 2.2. Execute the function and this will give the data of Column 3 of Table 2.2.*].

Absolute frequency histogram

The information from Table 2.2 is shown as an absolute closed-ended frequency histogram in Figure 2.1. This display shows the pattern of the data, in particular that the highest number of revenues lie in the range >$180,000 to ≤$200,000. This display is called an **absolute frequency** distribution since it shows the actual number of data values, out of a total of 52, that appear in each class range.

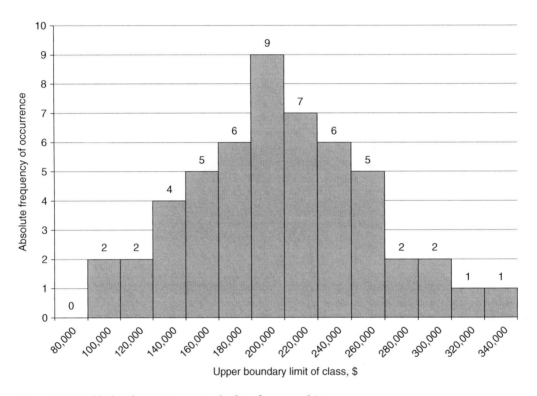

Figure 2.1 Weekly hotel operating costs: absolute frequency histogram.

Relative frequency histogram

The information from Table 2.2 is shown as a relative closed-ended frequency histogram in Figure 2.2.

This display shows the pattern of the percentage of data out of the total sample of 52 weeks. Again the highest percentages of revenues lie in the range >$180,000 to ≤$200,000. This display is called a **relative frequency** distribution since it shows the percentage of data values **relative** to the total number sampled. This type of display is important in probabilities since if we can say that the sample data is representative of future conditions we might say that the probability is about 17% (17.31%) that revenues will lie in the range $180,000 to $200,000. Note that the format of the absolute frequency distribution and the relative frequency distribution are the same. The only difference is the scale of the y-axis.

Frequency polygons

The absolute frequency histogram, or the relative frequency histogram, can be converted into a line graph or **frequency polygon.** Figure 2.3 gives the absolute frequency polygon for the hotel weekly

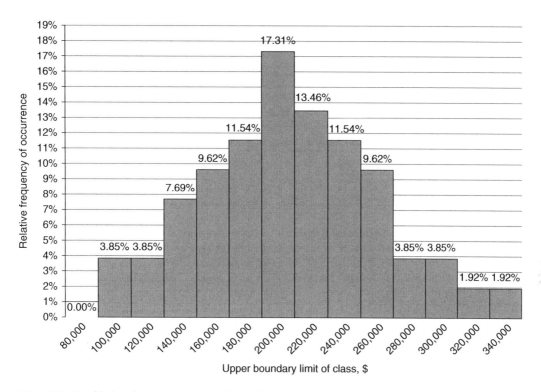

Figure 2.2 Weekly hotel operating costs: relative frequency histogram.

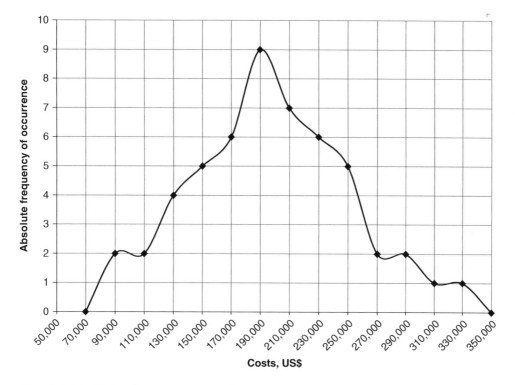

Figure 2.3 Weekly hotel operating costs: polygon of absolute numbers.

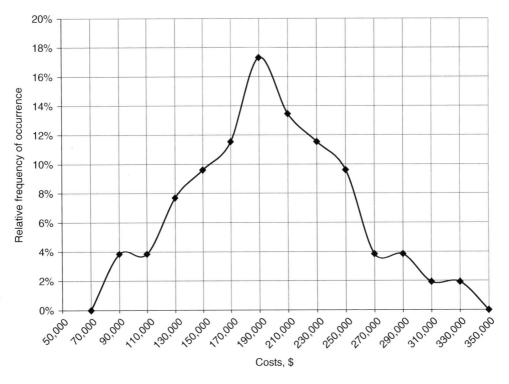

Figure 2.4 Weekly hotel operating costs: polygon of relative occurrence.

operating costs and Figure 2.4 gives the relative frequency polygon. Again the formats of the two polygons are the same. The difference is the scale of the *y*-axis. The frequency polygon shows the continuation of frequency with *x* values whereas the histograms show the data as rectangular blocks.

The frequency polygon is developed by determining the midpoint of the class range in the respective histogram. The **midpoint** of a class range is:

$$\frac{(\text{maximum value } + \text{ minimum value})}{2}$$

For example, the midpoint of the class range 180,000 to 200,000 is:

$$\frac{(180,000 + 200,000)}{2} = \frac{380,000}{2} = 190,000.$$

Ogives

An **ogive** is an adaptation of a frequency distribution where data values are progressively totalled, or cumulated, such that the resulting table indicates how many, or the proportion of, observations lie above or below certain limits. There is a **less than ogive** that indicates the amount of data below certain limits and a **greater than ogive** that gives data above certain values. Table 2.3 gives the weekly hotel operating cost data developed from Table 2.2 in the "less than" and "greater than" format. As Table 2.3 indicates there are ogives for absolute and relative values.

Table 2.3

Limits	Number of values less than limit	Number of values greater than limit	Proportion of values less than limit	Proportion of values greater than limit
80,000	0	52	0.00%	100.00%
100,000	2	50	3.85%	96.15%
120,000	4	48	7.69%	92.31%
140,000	8	44	15.38%	84.62%
160,000	13	39	25.00%	75.00%
180,000	19	33	36.54%	63.46%
200,000	28	24	53.85%	46.15%
220,000	35	17	67.31%	32.69%
240,000	41	11	78.85%	21.15%
260,000	46	6	88.46%	11.54%
280,000	48	4	92.31%	7.69%
300,000	50	2	96.15%	3.85%
320,000	51	1	98.08%	1.92%
340,000	52	0	100.00%	0.00%

The graphs for the absolute ogives are given in Figure 2.5. From Figure 2.5 there are 39 of the total 52 weeks that have revenues greater than $160,000 and 48 weeks when revenues are less than $280,000.

The relative ogives for the same data is in Figure 2.6. From this Figure 2.6, 15.38 percent of the revenue are less than $140,000 and 3.85 percent of revenues are greater than $300,000.

The less than ogive has a positive slope, where the *y*-values increase as the *x* values increase. The greater than ogive has a negative slope, where the *y*-values decrease as the *x* values increase.

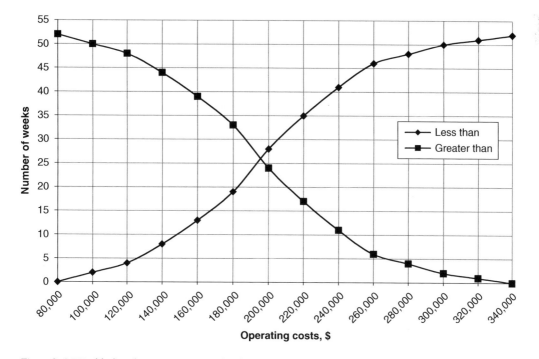

Figure 2.5 Weekly hotel operating costs: absolute ogives.

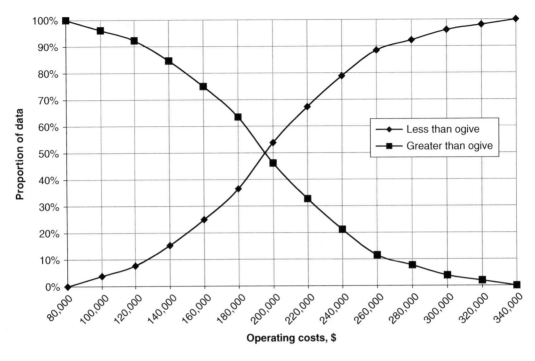

Figure 2.6 Weekly hotel operating costs: relative ogives.

Stem-and-leaf display

Another way of presenting univariate data according to the frequency of occurrence is a **stem-and-leaf display.** This display separates **sorted data** into two parts: leading digits, or **stems**, and trailing digits, or **leaves**. The choice of deciding the leading digits and the stems is up to the analyst. Consider Table 2.4 that is the **sorted data**, from lowest to highest by line, of the hotel weekly operating costs already presented in Table 2.1. I have chosen sequentially each group of ten thousand, starting from 80,000, as my leading digits or stems and the residual amount in the group as the leaves. The stem-and leaf display is given in Figure 2.7, showing all the 52 leaves (there are 52 weeks in the data). The first stem is 80 and this corresponds to 80,000. The first leaf is 1,340, corresponding to the data value of

Table 2.4

81,340	86,157	106,155	110,336
120,415	122,211	127,076	132,424
142,382	149,886	155,251	157,775
159,262	161,075	161,372	164,249
167,175	171,135	177,226	180,676
182,677	185,377	187,124	187,173
187,173	188,421	188,426	188,977
203,411	210,536	210,573	210,682
215,564	217,076	217,177	224,276
225,880	228,577	231,651	237,524
238,840	241,171	248,146	249,651
251,251	253,076	274,856	275,936
293,373	298,256	303,466	332,212

80	1,340	6,157
90		
100	6,155	
110	336	
120	415	2,211	7,076
130	2,424	
140	2,382	9,886
150	5,251	7,775	9,262
160	1,075	1,372	4,249	7,175
170	1,135	7,226
180	676	2,677	5,377	7,124	7,173	7,173	8,421	8,426	8,977
190		
200	3,411	
210	536	573	682	5,564	7,076	7,177
220	4,276	5,880	8,577
230	1,651	7,524	8,840
240	1,171	8,146	9,651
250	1,251	3,076
260		
270	4,856	5,936
280		
290	3,373	8,256
300	3,466	
310		
320		
330	2,212	
340		

Figure 2.7 Weekly hotel operating costs: stem-and-leaf diagram.

81,340 (80,000 + 1,340). The second leaf for this same stem is 6,157, corresponding to the data value of 86,157 (80,000 + 6,157). There are no leaves for a stem of 90 (90,000) and the next stem is 100 corresponding to 100,000. For this stem there is one leaf: 6,155 corresponding to the data value 106,155 (100,000 + 6,155). The dominant stem is 180 (180,000) where there are nine leaves: 676 (180,676); 2,677 (182,677); ... through to 8,977 (188,977).

An advantage of the stem–and-leaf display over the frequency distribution for univariate data is that all data values are shown. With the frequency distribution the data values are grouped in the given class ranges so actual values are hidden. A disadvantage with the stem–and leaf display is that for large datasets the stem–and-leaf display is unwieldy to develop.

Visuals of quartiles and percentiles

Chapter 1 developed the quartiles and percentiles as a measure of dispersion. These measures can also be shown visually.

Box and whisker plot

Quartile data can be presented in a box and whisker plot (after the face of cat if you use your imagination—though my cats have round faces!), or simply a **box plot** that clearly identifies the quartile and the boundary limits. The box and whisker plot for the hotel operating costs of Table 2.1

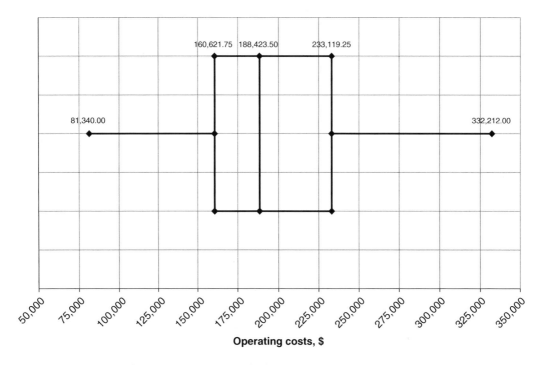

Figure 2.8 Weekly hotel operating costs: box and whisker plot.

is shown in Figure 2.8. This identifies the minimum value Q_0 (81,340.00), the three quartiles Q_1 (160,621.75), Q_2 (188,423.50), Q_3 (233,119.25), and the maximum value of the dataset Q_4. (332,212.00). There is 50% of the dataset in the area of the box; 25% between the whisker Q_0 and Q_1; and 25% between the whisker Q_3 and Q_4. This particular dataset, whose mean value is 197,509.94, is **right-skewed**. The test for this is as follows:

- A box and whisker plot is **symmetrical** if the mean and median values are equal; the distance from Q_0, to the median, Q_2, and the distance from Q_2 to Q_4, are the same; the distance from Q_0, to Q_1 equals the distance from Q_3 to Q_4; and the distance from Q_1 to Q_2 equals the distance from the Q_2 to Q_3.

- A box and whisker plot is **right-skewed** if the mean value is greater than the median; the distance from Q_2 to Q_4 is greater than the distance from Q_0 to Q_2; the distance from Q_3 to Q_4 is greater than the distance from Q_0 to Q_1; A right-skewed box and whisker plot means that data values to the right of the median are more dispersed than those to the left of the median.

- A box and whisker plot is **left-skewed** if the mean value is less than the median; the distance from Q_2 to Q_4 is less than the distance from Q_0 to Q_2; the distance from Q_3 to Q_4 is less than the distance from Q_0 to Q_1; This means that the data values to the left of the median are more dispersed than those to the right.

The percentile histogram

Percentile data can be shown as a histogram illustrating the progression from the lowest to the highest percentile. The percentile histogram for the hotel weekly operating costs of Table 2.1 is shown in Figure 2.9.

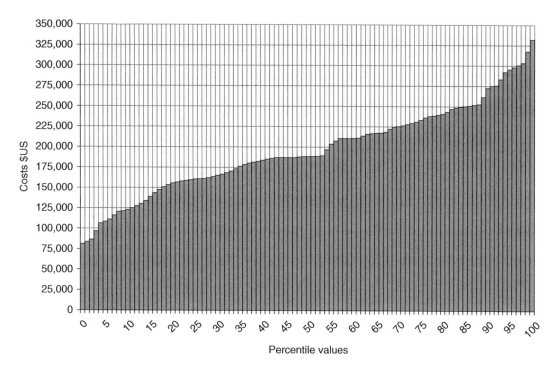

Figure 2.9 Weekly hotel operating costs: percentiles.

Line graphs

The line graph is perhaps the most common presentation of bivariate data where a dependent variable y on the vertical axis is plotted against the independent variable x on the horizontal axis.

Candid presentations

Consider the line graphs of Figure 2.10 for Hotel A and Figure 2.11 for Hotel B that show the sales revenues for the years 2002 through 2013.

At first glance it appears that Hotel B has a more rapid growth of revenues since the upward slope is steeper. However, the data is identical. The difference is that the scales on the y-axes are not the same! Attention must be paid to the scales on the axes as the appearance of the graph can change and decision making may be distorted. (The media sometimes plays this game by changing the scale of the axis in an effort to present an alarming situation.)

Geometric mean as a line graph

In *Chapter 1*, using sales revenue data in Table 1.21, we showed the calculation of the geometric mean, the average growth rate over a given time period. The geometric mean, or the average change from Year 1 to Year 6, was 4.95 percent. Figure 2.12 is a line graph illustrating this concept. The yearly actually percentage changes fluctuates and the geometric value is linear having smoothed out the actual annual changes.

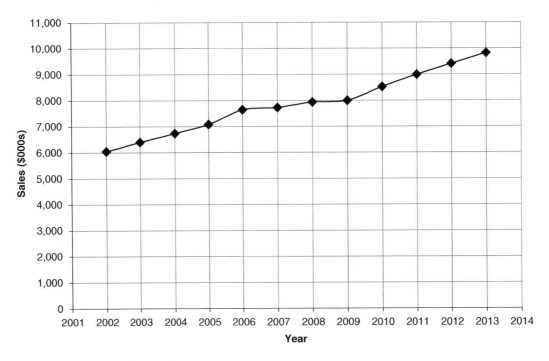

Figure 2.10 Hotel A: sales growth over 11 years.

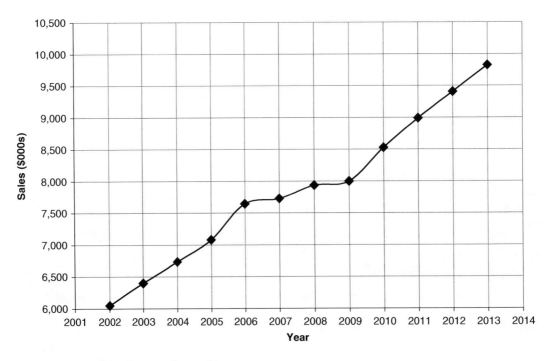

Figure 2.11 Hotel B: sales growth over 11 years.

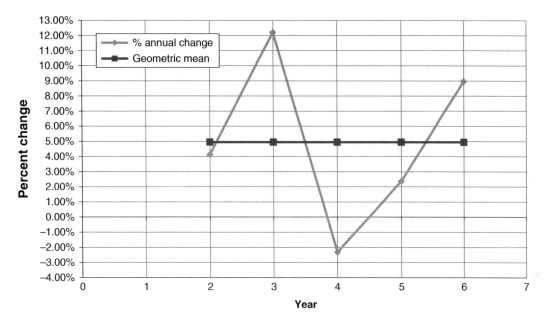

Figure 2.12 Annual percentage change and corresponding geometric mean.

Visuals of categorical data

Categorical data (i.e. data that contains labels), can conveniently be presented as a pie chart, (*camembert*—France; *tortilla*—*Mexico*), histogram, or bar chart.

Pie chart

Numerical data with labels or categories can be presented as a pie chart. This is a circle representing the numerical data, divided into segments like portions of a pie. Each segment of the pie has a label and the size of the portion is a percentage of the total values in the dataset. The pie is considered the "whole," meaning it represents all the dataset or 100%. The usefulness of the pie chart is that we can see clearly the importance of each category. As an illustration Table 2.5 gives the revenues in Euros for a restaurant chain in several countries of Europe.

Table 2.5

Country	2012	2013
Austria	522,065	622,065
Belgium	1,266,054	1,566,054
Finland	741,639	941,639
France	2,470,257	2,770,257
Germany	2,876,431	2,276,431
Italy	2,086,829	2,186,829
Netherlands	1,091,779	1,391,779
Portugal	1,161,479	1,861,479
Sweden	3,884,566	3,584,566

The pie chart in Figure 2.13 illustrates this information. Here the UK is the major contributor with 21.59% of the revenues and Austria contributes the least with 2.54%.

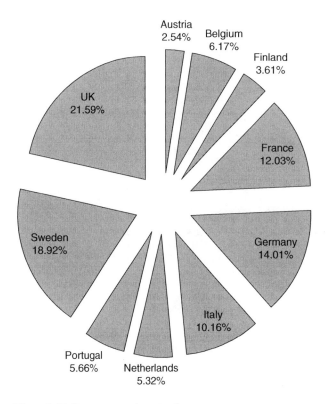

Figure 2.13 Restaurant sales: pie chart.

Categorical histogram

An alternative to a pie chart is a vertical histogram where the vertical bars on the *y*-axis show either absolute values or the percentage of data, and the *x*-axis the categories. Figure 2.14 gives a vertical histogram of the restaurant sales data from Table 2.5 both for 2010 and 2011. This is a **side-by-side histogram** and its utility is that you can compare the values from one year to another. Note, in these vertical histograms, that the bars are separated, as one category does not directly flow to another—they are discrete. For example, Austria and Belgium are discrete categories. This is not the case for histograms showing univariate data where there is no space between each class range, since the ranges are sequential; they follow on from each other.

Bar chart

An alternative to the vertical histogram is a **horizontal bar chart** as shown in Figure 2.15. This gives the same information as in Figure 2.14 except in a horizontal format. Which one to use is a matter of personal preference. Publications such as *The Economist*, *The Wall Street Journal*, and *The New York Times* use both without any apparent logic!

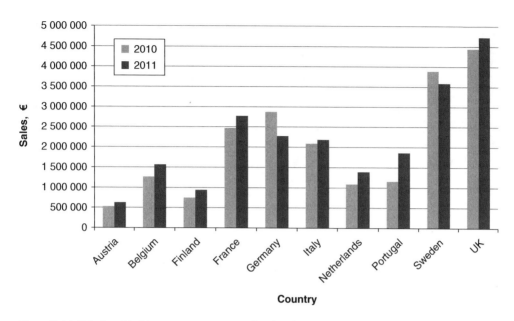

Figure 2.14 Side-by-side histogram: restaurant sales data for two consecutive years.

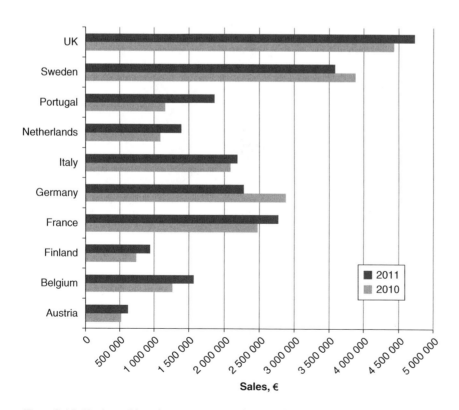

Figure 2.15 Horizontal bar chart: restaurant sales data for two consecutive years.

Contingency tables

Contingency, or cross-classification tables, organize related data such that connections can be made between variables. As an illustration, Table 2.6 gives the weight of different ingredients in a 30 cm diameter "standard pizza" made in five different countries. From this table we might say that 100 g of ham in a pizza is **contingent** of being made in Australia. Alternatively we could say that 121 g of tomatoes in a standard pizza is contingent of being sold in France.

Contingent means that values are dependent or conditioned on another factor. They are useful in comparing information that has been obtained from sampling experiments and from which decisions can be made. They are also useful in determining probability (*see Chapter 3*).

Table 2.6

Ingredients: g/pizza	Australia	New Zealand	France	Netherlands	Belgium
Flour	400	350	250	300	275
Cheese	50	50	77	74	73
Ham	100	117	275	111	113
Mushrooms	20	11	11	19	29
Tomatoes	60	44	121	43	87
Oil	18	26	61	28	22
Total weight	648	598	794	575	600

Stacked histograms

Stacked histograms are developed from cross-classification or contingency tables and visually indicate relationships. Figure 2.16 gives a stacked histogram of the data in Table 2.6 of the proportion by

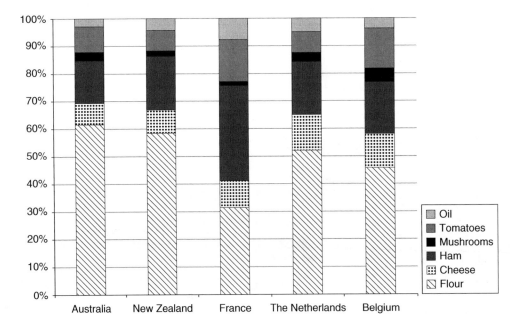

Figure 2.16 Stacked histogram: proportion by weight of ingredients in a pizza by country.

weight in a "standard pizza" according to the dominant criteria; "country". The legend gives the product in the pizza.

An alternative is Figure 2.17 that gives a stacked histogram of the proportion by weight in a "standard pizza" according to the dominant criteria "product". The legend gives the country for this data.

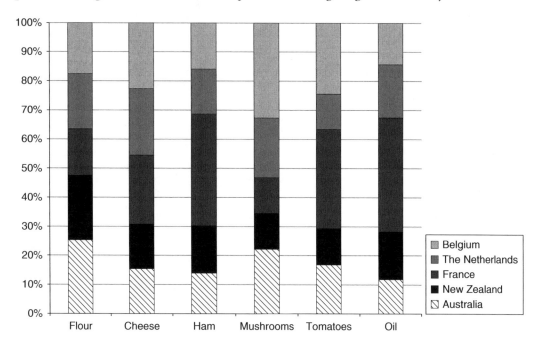

Figure 2.17 Stacked histogram: proportion by weight of ingredients in a pizza.

Pareto diagram

A Pareto diagram is a combined line graph with a categorical histogram often used to display information obtained from an operational audit. As an illustration Table 2.7 is a frequency check sheet of customer complaints in a restaurant taken over a period of three months. For example, during this three-month audit period, there were 24 customer complaints indicating that the waiting time to be seated was too long.

Table 2.7

Area of dissatisfaction	Frequency of occurrence
Cramped seating	21
Food overcooked	7
Food sometimes cold	8
Meal portions small	31
Menu choice limited	19
Prices high	11
Salad not fresh	14
Service time long	5
Table ware not clean	2
Waiting to be seated long	24
Total	142

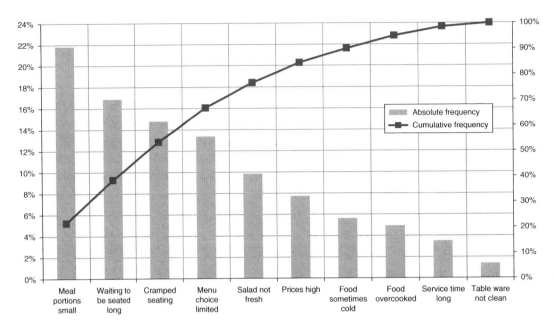

Figure 2.18 Pareto diagram of customer complaints in a restaurant.

The Pareto diagram, named after the Italian economist Vilfredo Pareto (1848–1923), known for coining the 80/20 rule used in business, is in Figure 2.18. The *x*-axis gives the area of dissatisfaction; the left-hand *y*-axis gives the individual percentage frequency of each area to the total number of complaints; and the right-hand *y*-axis gives the cumulative occurrence of all the complaints. The presentation shows the histogram bars in descending order from those that occur most often on the left to those occurring least on the right. The reason for organizing the data in this way is that those that occur most often should be those that are dealt with and corrected first.

Spider web diagram

A spider web diagram, also known as a radar diagram in Excel, is a Pareto type of analysis that shows qualitative information, correlated with a numerical score. Figure 2.19 illustrates where quality attributes for a production operation are presented.

Here there are eight criteria for measuring appropriate operating conditions: quality of material; availability of material; practical design; modern technology; just-in-time operation; experienced operators; documentation availability; and clear instructions. For this particular operation the criteria have been analyzed and given a numerical score from 0 to 5, with 5 being the best. The resulting graph looks like a spider web. The larger the area of the web, the better operating performance.

Gap analysis

A gap analysis is an alternative way of presenting spider web information by indicating the "gap" between the maximum possible attainable score and the measured value. Figure 2.20 shows the gap analysis (a tool in the ISO-9001:2000 quality standards) for the spider web diagram of the previous paragraph. The gap indicates what is "missing" for perfect performance.

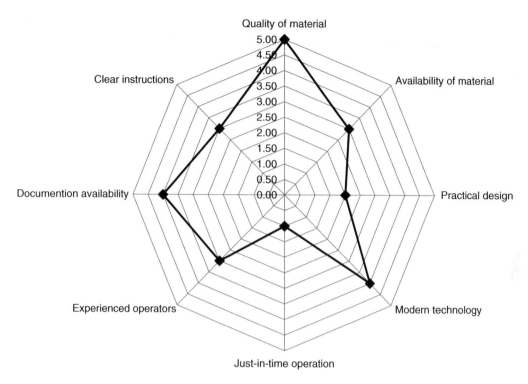

Figure 2.19 Spider web diagram.

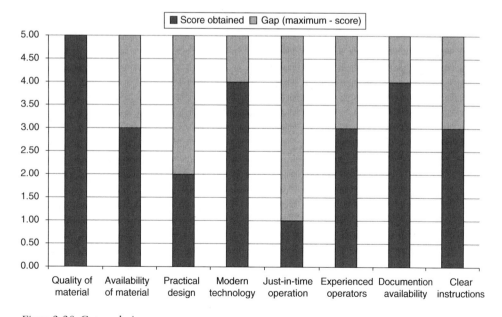

Figure 2.20 Gap analysis.

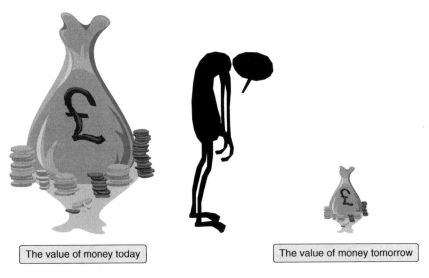

The value of money today   The value of money tomorrow

Figure 2.21 A pictogram to illustrate inflation.

Pictograms

A **pictogram** or **pictograph** is a picture, **icon**, or sketch that represents quantitative data but in a categorical, qualitative or comparative manner, such as shown in Figure 2.21.

Pictograms are not accurate in how they show data as they do not show factual information. They are often used by the media to exaggerate a situation. In Figure 2.21 has our money been reduced by 50%, 100%, or 200%? We cannot say. Pictograms are not covered further in this textbook.

Moving on

This chapter has shown ways to present data in a clear, precise manner. The next chapter takes some of the concepts developed here and also in Chapter 1 to introduce probability and counting rules, where data analysis can be extended to measure risk, chance, and outcomes.

Note

1 "Hotel performance at a five-star hotel in China," Anthony Metral, 2011 M.Sc. Thesis, Institute Paul Bocuse, Ecully, France.

3 Essentials of probability and counting rules

ICEBREAKER: The haven to perhaps make a fortune

For many, gambling casinos are exciting establishments. The one-arm bandits are colorful machines with flashing lights that require no intelligence to operate. When there is a "win" coins drop noisily into an aluminium receiving tray and blinking lights indicate to the world the amount that has been won. The gaming rooms for poker, or blackjack, and the roulette wheel have an air of mystery about them. The dealers and servers are beautiful people; smartly dressed; say very little and give an aura of superiority. Throughout the casinos there are no clocks or windows so you don't see the time passing. Drinks are cheap, or maybe free, so having "a few" encourages you to take risk. The carpet patterns are busy so that you look at where the action is rather than looking at the floor. When you want to go to the toilet you have to pass by rows of slot machines and perhaps on the way you try your luck!

Figure 3.0 Slot machines in the Trump Taj Mahal, Atlantic City, New Jersey.

(Permission granted under terms of the GNU Free Documentation License.)

Poker in all its forms, such as Draw and Stud Poker and their variations, Texas Hold 'Em, and Omaha, is one of the most popular card games in gambling casinos. In their heyday the Mississippi riverboats became notorious for poker games where passengers were fleeced by professional gamblers. Poker was invented in the USA but took its main elements from Bouillotte, a nineteenth century French game. In Bouillotte four players were each dealt a three-card hand from a 20-card pack (Ace, King, Queen, 9 and 8 in four suits). The thirteenth card was turned face up so that all of the players could use it to improve their hands. The players would bet on their hands in a way similar to modern poker: ranking was four of a kind (best), three of a kind, or two of a kind. The highest hand won the pot, and four or three of a kind with or without the up-card won bonus chips from the other players. One of poker's best-known legends is that of "Wild Bill" Hickok, who made the mistake of playing poker in a gambling saloon in Deadwood in 1876. For once he didn't take a seat with his back to the wall. A gambler called Jackie "Crooked Nose" McCall, who had been hired by other crooked gamblers to kill Hickok because they feared he might be made marshal of Deadwood and clean up the game, shot "Wild Bill" in the back of the head with a Colt 45. Bill Hickok died clutching his poker hand, which was found to be two Aces, and two 8s (some say all black), with the odd card either the Queen or Jack of Diamonds. From that day, pairs of Aces and 8s have been known as "Dead Man's Hand." By 1850 the 52-card pack had superseded the 20 card game. It allowed players to improve their hands by changing some card if they wanted, and the "draw" became a standard part of the game—hence Draw Poker.

Gambling used to be a by-word for racketeering. Now it has cleaned up its act and is more profitable than ever. Today the gambling industry is run by respectable corporations instead of by the "Mob" and it is confident of winning public acceptance. In 2004 in the USA, some 54.1 million people, or more than one-quarter of all American adults, visited a casino, on average six times each. Poker is a particular growth area and some 18 percent of Americans played poker in 2004, which was a 50 percent increase over 2003. Together, the USA's 445 commercial casinos, excluding those owned by Indian tribes, had revenues in 2004 of nearly $29 billion. Further, it paid state gaming taxes of $4.74 billion or almost 10 percent more than in 2003. A survey of 201 elected officials and civic leaders, not including any from gambling dependent Nevada and New Jersey, found that 79 percent believed casinos had a positive impact on their communities. Europe is no different. The company Partouche owns and operates very successful casinos in Belgium, France, Switzerland, Spain, Morocco, and Tunisia. And, let's not forget the famed casino in Monte Carlo. Just about all casinos are associated with hotels and restaurants and many others include resort settings, spas, and cruise ships. Las Vegas immediately springs to mind. In 2014 their card tables and slot machines generated some $9.6 billion in revenues. Over one-third of Las Vegas's workers are employed directly serving visitors. This makes the whole combination, gambling casinos, hotels, resorts, spas, and cruise ships a significant part of the service industry!

Gambling, whether it be one-arm bandits, poker, the roulette wheel, horse racing or betting on a soccer match, is where statistics plays an important role. You have a probability of winning; but also a greater probability of losing your shirt! As an illustration, according to the Global Betting and Gaming Consultancy, Italians lost €17.2 billion in 2014, almost three times more than in 2001. Italians overtook Spaniards as southern Europe's most ardent punters in 2005. As of 2015 Italy prints one-fifth of the world's scratch cards and hosts a third of its video lottery terminals.[1-5]

Chapter subjects

✓   **Probability** • Basics • Technology, probability, and risk • Subjective probability • Relative frequency probability • Classical probability • Addition rules in classical probability • Venn diagram • Application case-exercise: *Sourcing agents* • Joint probability • Gambling, odds, and probability

✓   **Conditional probabilities under statistical dependence** • Dependency relationship • Bayes' theorem • Application case-exercise: *Arriving to work* • Probability using contingency tables • Application case-exercise: *Market survey*

✓   **System reliability and probability** • Series arrangement • Parallel arrangement

✓   **Counting rules** • Single type of event • Different types of events • Arrangement of different objects • Permutation rule • Combination rule • Probability and odds with permutations and combinations.

---

*Probability implies there is a **chance** something happens or does not happen. It is **binomial** if there are only two possible outcomes. If you toss a coin you obtain heads or tails; if you gamble you win or lose; if you have a baby it will be a boy or girl; if you take a card from a pack it will red or black. In this binomial situation it does not mean there is a 50/50 chance of being right or wrong or a 50/50 chance of winning. If you toss a coin, you have a 50% chance of obtaining heads or 50% chance of throwing tails. If you buy one ticket in a fund raising raffle, then you will either win or lose. If there are 2,000 tickets that have been sold you have only a 1/2000 or 0.05% chance of winning and a 1999/2000 or a 99.95% chance of losing! Counting rules, the last part of this chapter, forms the backbone to many phases of probability.*

Probability

Basics

A principal objective of statistics, known as **inferential statistics**, is to make, or infer, decisions concerning a **population** by measuring **sample** data taken from this population. For example, we are interested to know how people will vote in a presidential election. We sample the opinion of 7,500 of the electorate and we use this result to estimate the opinion of the population of 60 million. Since we extend sample results beyond measured data there is no certainty of being correct, only a **probability**. The corollary to this is that there is a chance of being incorrect. Probability is the likelihood something happens or does not happen. In statistics it is denoted by the capital letter **P** and is measured on an inclusive numerical scale from zero to one or from 0% to 100%. If you are a citizen of the United States, but were not born in that country, the probability of becoming president is 0% (under current law). The probability that you will die is 100%—though hopefully a long way in the future. Between the two extremes of 0 and 1 something might occur or might not occur. The meteorological office may announce that there is a 30% chance of rain today, which also means that there is a 70% chance that it will not rain. In contrast to probability is **deterministic**, where the outcome is certain on the assumption that the input data is reliable. For example, if restaurant revenues are £10,000 and food and labor costs are £7,000 then the gross income is £3,000 (£10,000 − £7,000).

An extension of probability is **risk** when we might put an economic value on outcomes. If you open a new restaurant, your business plan might indicate that there is a 20% probability of not breaking even in the first year with the risk of losing €40,000. If you are a skier, and you take a black run in bad weather, there is a high probability of having an accident and breaking a leg, an event that happened to the author in California. (The economic consequence was that he was off work for a period with no income!) Statistically the probability of an automobile driver between 18 and 25 having an accident is greater than for people in higher age groups. Thus, to insurance companies, young people present a high risk and so their premiums are high. If you drink and drive the probability of you having an accident is high. In this case you risk having an accident, or risk killing yourself, or worse somebody else. In the USA, based on traffic deaths per 100,000, **statistically** the risk of a fatal accident is higher in Montana than in most other states.[6] Thus in these cases the "value" on the outcome is more than monetary.

In probability we have an **event**, which is the result from an activity or experiment. If you obtain heads on the toss of a coin, "obtaining heads" would be an event. If you draw the King of Hearts from a pack of cards, then "drawing the King of Hearts" would be an event. If a client becomes ill after eating a meal, "becoming ill" is an event. If you obtain an A on your management course, then "obtaining an A grade" would be an event. If on a particular day your hotel is full then "hotel is full" would be an event. Having an accident is an event.

Technology, probability, and risk

Our economic environment improves in part because of technological inventions either as new products, improved manufacturing processes, or enhancements in performing service functions. In the industrial revolution of the eighteenth century artisan weavers were swept aside by the Spinning Jenny for mechanical weaving; the steam engine, perfected by James Watt, revolutionized transportation; preserving jars for food was perfected in France. Since the 1980s the digital revolution has replaced typists, ticket agents, and bank tellers in services; drilling, machining, and soldering in manufacturing. All these innovations have cost people their jobs, though indications are that many of those displaced workers have been able to move into new activities. However, in 2014 there is uncertainty as to whether, as a result of computer-based changes, it might lead to a permanent cadre of unemployed people. A study by Frey and Osborne (see data shown in Figure 3.1 based on 2013 estimates), illustrates the probability of certain sectors' jobs being lost in the next two decades.[7]

Many of these jobs that involve repetitive work (machinists and typists) or where the activity can be performed by the internet (retail sales and real estate agents) have a greater than 50% probability that they will disappear. Jobs that are more complex or unique (chemical engineers and editors) have a much lower probability of being lost. And those that require one-on-one personal attention (athletic trainers and dentists) essentially will remain. From this then the risk that you might be out of work in the next 20 years is high when the probability of your job disappearing is high.

Subjective probability

Subjective probability is qualitative, sometimes emotional, and simply based on the belief or the "gut" feeling of the person making the judgment. You ask Michael, a single 22-year-old student, what the probability is of him getting married next year. His response is 0%. You ask his friend, John, what he thinks is the probability of Michael getting married next year and his response is 50%; his friend Helen thinks it is closer to 100%. These are subjective responses as the people really do not know. There is no basis for their answers. The particular situation has never occurred before (Michael by his own admission has never been married). Subjective probability may be a function of a person's experience.

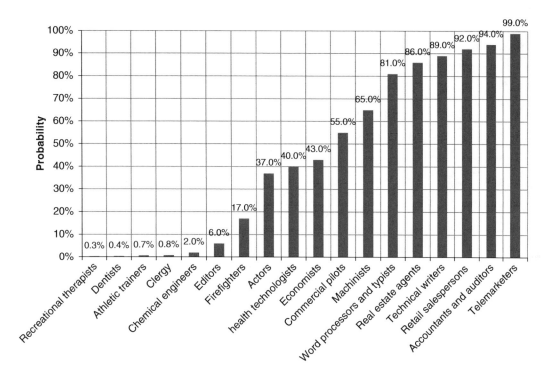

Figure 3.1 Probability in 2013 that in the next 20 years technology will lead to job losses.

The hotel general manager is 90% certain that his hotel will be full on the night of the soccer game. The hotel's operations manager says he is only 50% certain that his hotel will be full on that night. Both are basing their arguments on subjective probability. A manager who knows his employees well may be able to give a subjective probability of his department succeeding in a particular project. This probability might differ from that of an outsider assessing the probability of success. Usually the subjective probability of success announced by those who are **risk takers**—optimistic or gung-ho individuals, is higher than those who are **risk averse**.

Relative frequency probability

Probability based on data collected from situations that have occurred previously is relative frequency probability. This was introduced in *Chapter 2* by the histogram shown in Figure 3.2 (identical to Figure 2.2) where past data of hotel weekly operating costs can be used to assess probability outcomes on the basis that past data can be considered representative of the future. From this figure we could say that there is about a 17% (17.31%) probability that operating costs will be between $180,000 and $200,000 or the probability of operating costs being more than $320,000 is about 2% (1.92%).

In determining relative frequency probabilities, reliability is higher if conditions from which the data has been collected are stable and a large amount of data has been measured. Relative frequency probability is also called **empirical probability** as it is based on previous experimental work. The data collected is sometimes also referred to as **historical data** as it is information that after it is assembled is history!

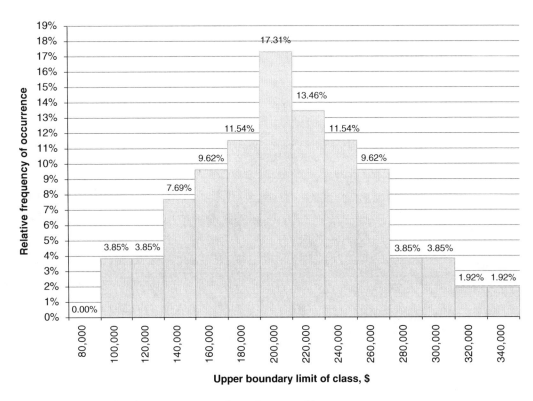

Figure 3.2 Weekly hotel operating costs: relative frequency histogram.

Classical probability

A probability measure that is the basis for gambling or betting, and thus useful if you have a casino in your hotel complex, is **classical probability.** Classical probability, also known as **simple probability** or **marginal probability** is defined as:

$$\text{Classical probability} = \frac{\text{Number of outcomes where the event occurs}}{\text{Total number of possible outcomes}} \qquad 3(i)$$

For this expression to be valid, the probability of the outcomes, as defined by the numerator (upper part of the ratio), must be equally likely. In the tossing of a coin the total number of possible outcomes is two, heads or tails. Thus, even before a coin is tossed we know that the probability of obtaining heads is:

$$P(\text{Heads}) = \frac{1}{2} = 50.00\%$$

The probability of obtaining tails is the same.

In the throwing of a single dice there are six possible outcomes, 1, 2, 3, 4, 5, or 6. Thus, even before the dice is tossed we know that the probability of obtaining the number 2 for example is:

$$P(\text{Number } 2) = \frac{1}{6} = 16.67\%$$

The probability of obtaining any of the other numbers is the same.

Now consider a full pack of 52 playing cards. The format of each of the 52 cards is given in Table 3.1.

Table 3.1

Suit															Total cards
Hearts	Ace	1	2	3	4	5	6	7	8	9	10	Jack	Queen	King	13
Clubs	Ace	1	2	3	4	5	6	7	8	9	10	Jack	Queen	King	13
Spades	Ace	1	2	3	4	5	6	7	8	9	10	Jack	Queen	King	13
Diamonds	Ace	1	2	3	4	5	6	7	8	9	10	Jack	Queen	King	13
Total	**4**	**4**	**4**	**4**	**4**	**4**	**4**	**4**	**4**	**4**	**4**	**4**	**4**	**4**	**52**

The total number of possible outcomes is 52, the number of cards in the pack. We know in advance that the probability of drawing an Ace of Spades, or in fact any one single card, is 1/52 or 1.92%. These three examples of classical probability are also referred to as **a priori probability** since we know the probability of an event in advance without the need to perform any experiments or trials.

Addition rules in classical probability

In probability situations we may have a **mutually exclusive** event, which means that there is no connection between one event and another. They exhibit **statistical independence**. Obtaining heads on a coin toss is mutually exclusive from obtaining tails: you have either heads or tails, but not both. If you obtain heads on one toss of a coin this event has no impact of the following event when the coin is tossed again. In many chance situations each time you make an experiment the outcome resets itself back to zero. If your first born child is a girl the gender of this first child has no bearing on the gender of the second; if you won a game of tennis today this win has no bearing on the outcome of your tennis game tomorrow; the fact that your restaurant was full on Monday has no impact on the number of customers you have on Tuesday. When two events are mutually exclusive then the probability of A or B occurring is expressed by the following **addition rule for mutually exclusive events**:

$$P(A \text{ or } B) = P(A) + P(B)$$ 3(ii)

In the drawing of one card from a pack, the probability of drawing the Ace of Spades, A_S **or** the Queen of Hearts, Q_H, is calculated by Equation 3(ii) as 3.85 percent:

$$P(A_S \text{ or } Q_S) = \frac{1}{52} + \frac{1}{52} = \frac{1}{26} = 3.85\%$$

This is because we have defined two specific cards and have said **either/or** is acceptable. There is a 1/52 chance of obtaining Ace of Spades and a 1/52 chance of obtaining the Queen of Hearts. Since either one is acceptable then we double the probability. If in the experiment neither of the two cards is drawn, and we replace the drawn card and repeat the experiment, the probabilities are the same. This action is called **sampling with replacement**.

Assume now that we redo the experiment again and **do not replace** the first card that was withdrawn. This first card was neither the Ace of Spades nor the Queen of Hearts. Then the new probability of drawing the Ace of Spades, A_S **or** the Queen of Hearts, Q_H, in the second draw is:

$$P(A_S \text{ or } Q_S) = \frac{1}{51} + \frac{1}{51} = \frac{2}{51} = 3.92\%$$

That is, a slightly higher probability than with a full pack of cards. This action is **sampling without replacement**.

If two events are **non-mutually exclusive**, this means that it is possible for both events to occur. Consider the probability of drawing either an ace or a spade from a deck of cards. The event ace and spade can occur together since it is possible to draw the Ace of Spades. Thus an ace and a spade are not mutually exclusive events. In this case, Equation 3(ii) for mutually exclusive events must be adjusted to avoid double counting, or to reduce the probability of drawing an ace, or a spade, by the chance we could draw both of them together, that is, the Ace of Spades. The modified **addition rule for non-mutually exclusive events** is:

$$P(A \text{ or } B) = P(A) + P(B) - P(AB) \qquad\qquad 3(iii)$$

For example, what is the probability of drawing an ace or a spade?

- There are four aces in the card pack of 52, thus probability of drawing an ace is: $\frac{4}{52}$
- There are 13 spades in the card pack of 52, thus probability of drawing a spade is: $\frac{13}{52}$
- One of these cards will be the Ace of Spades where the probability of obtaining this is: $\frac{1}{52}$. This has to be deducted to avoid double counting.

Thus applying Equation 3(iii) the probability of drawing an ace or a spade is:

$$P(\text{Ace or Spade}) = \frac{4}{52} + \frac{13}{52} - \frac{1}{52} = \frac{16}{52} = 30.77\%$$

Venn diagram

A Venn diagram, named after John Venn, an English mathematician (1834–1923), is a way to visually show mutually exclusive and non-mutually exclusive events. A surface area, such as a circle or rectangle, represents an entire sample space, and a particular outcome of an event is represented by a part of this surface. If two events, A and B, are mutually exclusive, their areas will not overlap, as shown in the upper part of Figure 3.3 that is a visual representation for a pack of cards using a rectangle for the surface. Here the number of boxes is 52, which is entire sample space, or 100.00%.

Consider two cards: the Ace of Spades that occupies a box and the Queen of Spades that occupies another box. The sum of occupied areas is two boxes or 2/52 = 3.85%. This is the same result obtained using Equation 3(ii). If two events are not mutually exclusive their areas would overlap as shown in the lower scheme of Figure 3.3. Here again the number of boxes is 52, or the entire sample space. Here we consider the suit spades and all the aces. Each of the cards, 13 spades and 4 aces would normally occupy one box each or a total of 17 boxes. However, one card is common to both, the Ace of Spades, and so the sum of occupied areas is 17 – 1 boxes or 16/52 = 30.77%. This is the same result as derived from Equation 3(iii).

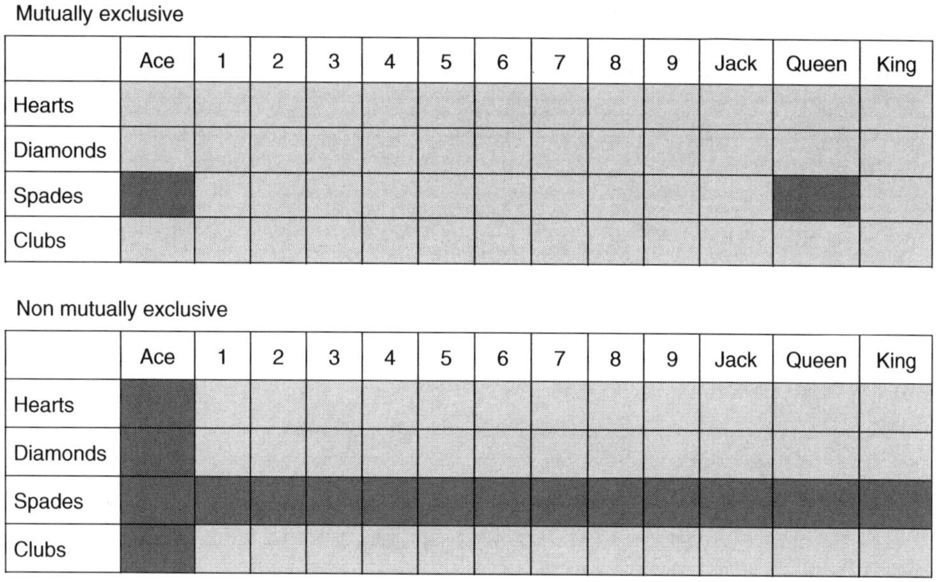

Mutually exclusive

	Ace	1	2	3	4	5	6	7	8	9	Jack	Queen	King
Hearts													
Diamonds													
Spades	■											■	
Clubs													

Non mutually exclusive

	Ace	1	2	3	4	5	6	7	8	9	Jack	Queen	King
Hearts	■												
Diamonds	■												
Spades	■	■	■	■	■	■	■	■	■	■	■	■	■
Clubs	■												

Figure 3.3 Venn diagram.

Basic probability with the use of the Venn diagram is illustrated by the following application case-exercise: *Sourcing agents*.

Application case-exercise: Sourcing agents

Situation

A large international retailer has sourcing agents worldwide to search out suppliers of products according to the best quality/price ratio for goods that it sells in its stores in the USA. The retailer has a total of 131 sourcing agents internationally. Of these 51 have a specialization in textiles, 32 have a specialization in footwear, and these numbers include 17 with a joint specialization in both textiles and footwear. The remainder are general sourcing agents with no particular specialization. All sourcing agents are in a general database with a common email address. When a purchasing manager from retail stores needs information on its sourced products they send an email to the general database address. Any one of the 131 sourcing agents is able to respond to the email.

Required

1   Illustrate the category of the specialization of the sourcing agents on a Venn diagram.
2   From the Venn diagram, what is the probability that at any time an email is sent it will be received by a sourcing agent specializing in textiles?
3   From the Venn diagram, what is the probability that at any time an email is sent it will be received by a sourcing agent specializing in both textiles and footwear?
4   From the Venn diagram, what is the probability that at any time an email is sent it will be received by a sourcing agent with no speciality?

5   Given that the email is received by a sourcing agent specializing in textiles then from the Venn diagram, what is the probability that the agent also has a speciality in footwear?
6   Given that the email is received by a sourcing agent specializing in footwear then from the Venn diagram, what is the probability that the agent also has a speciality in textiles?

Worked out solutions

1   The Venn diagram is given in Figure 3.4. This is developed from the given information. There are a total of 131 sourcing agents; 51 specialize in textiles; 32 in footwear; 17 in both textiles and footwear. The remainder have no specialization.

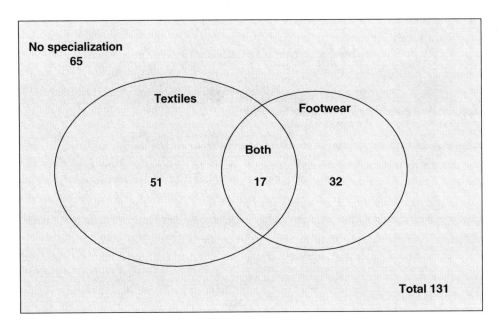

Figure 3.4 Venn diagram for sourcing agents.

- Specializing in just textiles is 34 (51 − 17)
- Specializing in just footwear is 15 (32 − 17)
- No specialization is 65 (131 − 51 − 32 + 17)
- Check: total agents is 131 (34 +15 +17 +65)

2   Probability that at any time an email will be received by a sourcing agent with a specialization in textiles is 38.93% (51/131).
3   Probability that at any time an email will be received by a sourcing agent with a specialization in both textiles and footwear is 12.98% (17/131).
4   Probability that at any time an email will be received by a sourcing agent no specialization is 49.62% (65/131).
5   Given that the email is received by a sourcing agent specializing in textiles, the probability that the agent also has a specialization in footwear is 33.33% (17/51).
6   Given that the email is received by a sourcing agent specializing in footwear, the probability that the agent also has a specialization in textiles is 53.13% (17/32).

Joint probability

The probability of two or more independent events occurring **together** or in **succession** is called joint probability and this is calculated by the product of the individual marginal probabilities:

$$P(AB) = P(A)*P(B) \tag{3(iv)}$$

Here P(AB) is the joint probability of events A and B occurring together or in succession. P(A) is the marginal probability of A occurring, and P(B) is the marginal probability of B occurring. The joint probability is always lower than the marginal probability since we are determining the probability of more than one event occurring together in our experiment; we are multiplying together two fractions and the result will be less than either of the original fractions. Consider a gambling game where we are using one pack of cards. The classical or marginal probability of drawing the Ace of Spades from a pack is 1/52 or 1.92 percent. The probability of drawing the Ace of Spades both times, **with replacement**, in two successive draws is:

$$\frac{1}{52}*\frac{1}{52} = \frac{1}{2,704} = 0.037\%$$

Here the value of 0.037% for drawing the Ace of Spades twice in two draws is much lower than the marginal productivity of 1.92% of drawing the Ace of Spades once in a single drawing.

In another gambling game, two dice are thrown together, and their combined score calculated, as illustrated in Table 3.2. The most frequently occurring number is 7 as given by throws 6, 11, 16, 21, 26, and 31. From joint probability the chance of throwing a six and a one together (throw 6), is:

$$\frac{1}{6}*\frac{1}{6} = \frac{1}{36} = 2.78\%$$

The chance of throwing a five and a two together (throw 11) is:

$$\frac{1}{6}*\frac{1}{6} = \frac{1}{36} = 2.78\%$$

The chance of throwing a four and a three together (throw 16) is:

$$\frac{1}{6}*\frac{1}{6} = \frac{1}{36} = 2.78\%$$

Similarly, the joint probability for throwing a three and four together (throw 21), a two and three, (throw 26), and a one two and a six together (throw 31) is always 2.78%. Thus, the probability that all six can occur is, from the addition rule:

$$2.78\% + 2.78\% + 2.78\% + 2.78\% + 2.78\% + 2.78\% = 16.67\%$$

This is the same as the result obtained using the criteria of classical or marginal probability of Equation 3(i):

$$\text{Classical probability} = \frac{\text{Number of outcomes where the event occurs}}{\text{Total number of possible outcomes}} \tag{3(i)}$$

Table 3.2

Throw number	Dice No.1	Dice No.2	Total score	Throw number	Dice No.1	Dice No.2	Total score
1	1	1	2	19	1	4	5
2	2	1	3	20	2	4	6
3	3	1	4	21	3	4	7
4	4	1	5	22	4	4	8
5	5	1	6	23	5	4	9
6	6	1	7	24	6	4	10
7	1	2	3	25	1	5	6
8	2	2	4	26	2	5	7
9	3	2	5	27	3	5	8
10	4	2	6	28	4	5	9
11	5	2	7	29	5	5	10
12	6	2	8	30	6	5	11
13	1	3	4	31	1	6	7
14	2	3	5	32	2	6	8
15	3	3	6	33	3	6	9
16	4	3	7	34	4	6	10
17	5	3	8	35	5	6	11
18	6	3	9	36	6	6	12

Here, the number of possible outcomes where the number seven occurs is six. The total number of possible outcomes is 36 by the joint probability of 6*6. Thus, the probability of obtaining a seven on the throw of two dice is as before:

$$\frac{6}{36} = 16.67\%$$

Similarly to obtain the number five with two dice the combinations that must come up together are (from Table 3.2) throws 4, 9, 14, and 19. The probability that all four can occur is then from the addition rule:

$$2.78\% + 2.78\% + 2.78\% + 2.78\% = 11.12\% \,(11.11\% \text{ by rounding at end of the calculation})$$

And again from marginal probabilities this is $\frac{4}{36} = 11.11\%$

Again this is a priori probability since in the throwing of two dice we know in advance that the probability of obtaining a five is 4/36 or 11.11%.

In gambling with slot machines or **one–arm bandits**, often the winning situation is obtaining three identical objects on the pull of a lever as for example in Figure 3.5, where three apples are illustrated.

Figure 3.5 Three identical images in a one-arm-bandit game.

The probability of winning is calculated using joint probability and given by:

$$P(A_1.A_2.A_3) \; = \; P(A_1)^*P(A_2)^*P(A_3) \hspace{4cm} 3(v)$$

If there are six different objects on each wheel and each wheel has the same six objects, then the marginal probability of obtaining one object is $1/6 = 16.67\%$. Then the joint probability of obtaining all three objects together is:

$$0.1667^*0.1667^*0.1667 \; = \; 0.0046 \; = \; 0.4630\%$$

With eight objects on each wheel, marginal probability for each wheel is $1/8 = 0.1250$ and joint probability is:

$$0.1250^*0.1250^*0.1250 \; = \; 0.0020 \; = \; 0.1953\%$$

With 20 objects on each wheel, marginal probability for each wheel is $1/20 = 0.0500$ and joint probability is:

$$0.0500^*0.0500^*0.0500 \; = \; 0.0001 \; = \; 0.1250\%$$

These are very low probabilities. This illustrates that the odds in the long run are against the gambler being a winner.

Gambling, odds, and probability

Up to this point in the chapter you might argue that much of the previous analysis is related to gambling and then you might say, "But business is not just gambling." That is true, but don't put gambling aside; as alluded to in the Icebreaker of this chapter the service sector of hotels and restaurants are associated with gambling casinos; Las Vegas, Atlantic City, Monte Carlo, Hong Kong, Macau, and the like. In addition, online gambling occurs worldwide. This activity represents a non–negligible part of the world economy. In gambling or betting we refer to the **odds of winning**. Though we can also consider the **odds of losing**. Although the odds are related to probability they are a way of looking at **risk**. The probability is the number of favorable outcomes divided by the total number of possible outcomes. The odds of winning are the ratio of the chances of winning to the chances of losing. Earlier we illustrated that the probability of obtaining the number 7 in tossing of two dice is 6 out of 36 throws. The probability of not obtaining the number 7 is 30 out of 36 throws $(36 - 6)$. The odds of "winning," or obtaining the number 7, are:

Odds of obtaining the number 7 are: $\dfrac{6/36}{30/36} = \dfrac{6}{30} = \dfrac{1}{5}$. That is odds of a 1 in 5 chance, or low.

Odds of not obtaining the number 7 are: $\dfrac{30/36}{6/36} = \dfrac{30}{6} = \dfrac{5}{1}$. That is odds of a 5 in 1 chance, or high.

Suppose you are planning a meeting and you choose a date at random. The probability of this being an "open day" is 5/7. Probability of this date being a weekend is 2/7 (there are five work days, and two weekend days).

Odds of date selected being a work day are $\dfrac{5/7}{2/7} = \dfrac{5}{2}$ or odds of 5 to 2, or relatively high.

Odds of date selected being a weekend are $\dfrac{2/7}{5/7} = \dfrac{2}{5}$ or odds of 2 to 5, or relatively low.

Although odds depend on probability, it is odds that matter when you are placing a bet or taking a risk!

Conditional probabilities under statistical dependence

Statistical dependence implies that the probability of a certain event is dependent also on the occurrence of another event.

Dependency relationship

Consider a blind wine tasting event where there are ten glasses of wine as shown in Figure 3.6.

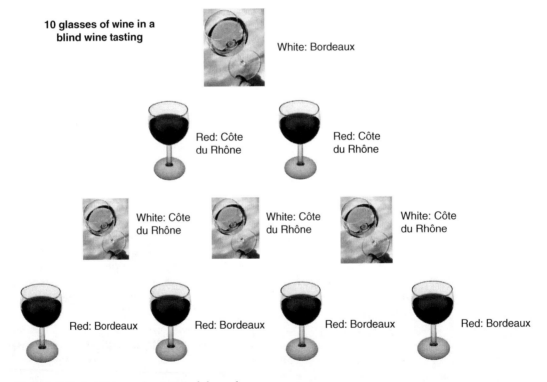

Figure 3.6 Probabilities under statistical dependence.

There are four different wine choices: one glass is a white Bordeaux; two glasses are a red Côte du Rhône; three glasses are a white Côte de Rhône; and four are a red Bordeaux. As there are ten glasses then there are ten possible events. The probability of selecting any one glass at random is 10%. There is statistical dependence since a red wine can be a Bordeaux or a Côte du Rhône; a white wine can also be a Bordeaux or a Côte du Rhône. These possible outcomes are shown in Table 3.3 according to the choice of each selection.

This information can be simplified into a two by two **cross-classification** or **contingency** table as in Table 3.4. (Contingency tables were introduced in *Chapter 2* in the development of stacked histograms).

Table 3.3

Selection	Probability of selection	Wine type
1	10.00%	White: Bordeaux
2	10.00%	Red: Côte du Rhône
3	10.00%	Red: Côte du Rhône
4	10.00%	White: Côte du Rhône
5	10.00%	White: Côte du Rhône
6	10.00%	White: Côte du Rhône
7	10.00%	Red: Bordeaux
8	10.00%	Red: Bordeaux
9	10.00%	Red: Bordeaux
10	10.00%	Red: Bordeaux

Table 3.4

Region	Red	White	Total
Côte du Rhône	2	3	5
Bordeaux	4	1	5
Total	6	4	10

The contingency table shows that there are two glasses of red from the Côte du Rhône region; four glasses of red from the Bordeaux region; three glasses of white from the Côte du Rhône region; and one glass of white from the Bordeaux region. Assume that in the blind wine tasting a glass is selected at **random,** meaning that each glass has an equal chance of being chosen, then the probability outcomes are given in Table 3.5.

Table 3.5

Probability outcome	Probability ratio	Percent
Red	6/10	60.00%
White	4/10	40.00%
Côte du Rhône	5/10	50.00%
Bordeaux	5/10	50.00%
Red and Bordeaux	4/10	40.00%
White and Bordeaux	1/10	10.00%
Red and Côte du Rhône	2/10	20.00%
White and Côte du Rhône	3/10	30.00%

Now assume the wine taster selects a red wine: what is the probability of this being Bordeaux red? There six red wines and four Bordeaux red wines, so the probability is 4/6 or 66.67%. Or assume that the wine taster selects a Côte du Rhône wine: what is the probability that it is white? There are five Côte du Rhône wines, three of which are white, and thus the probability is 3/5 or 60.00%. All the outcomes given a first selection, red, white, Bordeaux, or Côte du Rhône then the probability outcomes are in Table 3.6.

This conditional probability under statistical dependence is written as:

$$P(B \,|\, A) \;=\; \frac{P(BA)}{P(A)} \hspace{6cm} 3(vi)$$

This is means that the probability of B occurring given that A has occurred is equal to the joint probability of B and A happening together, or in succession, divided by the marginal probability of A.

Table 3.6

First selection	Probability outcome	Probability ratio	Percent
Red	Bordeaux	4/6	66.67%
White	Bordeaux	1/4	25.00%
Red	Côte du Rhône	2/6	33.33%
White	Côte du Rhône	3/4	75.00%
Bordeaux	Red	4/5	80.00%
Côte du Rhône	Red	2/5	40.00%
Bordeaux	White	1/5	20.00%
Côte du Rhône	White	3/5	60.00%

Using the relationship from Equation 3(vi) and referring to Tables 3.5 and 3.6:

$$P(\text{Bordeaux, given Red}) = \frac{P(\text{Bordeaux and Red})}{P(\text{Red})} = \frac{4/10}{6/10} = \frac{4}{6} = 66.67\%$$

That is the same value given in Table 3.6. All the outcomes using this relationship are shown in Table 3.7.

Table 3.7

Probability	Probability equivalent	Probability ratios	Simple ratio	Percent
P(Bordeaux, given Red)	[P(Bordeaux and Red)]/[P(Red)]	(4/10)/(6/10)	4/6	66.67%
P(Bordeaux, given White)	[P(Bordeaux and White)]/[P(White)]	(1/10)/(4/10)	1/4	25.00%
P(Côte du Rhône, given Red)	[P(Côte du Rhône and Red)]/[P(Red)]	(2/10)/(6/10)	2/5	33.33%
P(Côte du Rhône, given White)	[P(Côte du Rhône and White)]/[P(White)]	(3/10)/(4/10)	3/4	75.00%
P(Red given Bordeaux)	[P(Red and Bordeaux)]/[P(Bordeaux)]	(4/10)/(5/10)	4/5	80.00%
P(White given Bordeaux)	[P(White and Bordeaux)]/[P(Bordeaux)]	(1/10)/(5/10)	1/5	20.00%
P(Red given Côte du Rhône)	[P(Red and Côte du Rhône)]/[P(Côte du Rhône)]	(2/10)/(5/10)	2/5	40.00%
P(White, given Côte du Rhône)	[P(White and Côte du Rhône)]/[P(Côte du Rhône)]	(3/10)/(5/10)	3/5	60.00%

Further, from the addition rule:

- P(Côte du Rhône, given Red) + P(Bordeaux, given Red) = 33.33% + 66.67% = 100.00%
- P(Côte du Rhône, given White) + P(Bordeaux, given White) = 75.00% + 25.00% = 100.00%
- P(Red given Côte du Rhône) + P(White, given Côte du Rhône) = 40.00% + 60.00% = 100.00%
- P(Red given Bordeaux) + P(White, given Bordeaux) = 80.00% + 20.00% = 100.00%

Bayes' theorem

The relationship of Equation 3(vi) for conditional probability under statistical dependence is called Bayes' Theorem after the englishman Reverend Thomas Bayes (1702–1761). It is also referred to as

Bayesian decision making. It illustrates that if you have additional information, or based on the fact that *something has occurred*, certain probabilities may be revised to give *posterior* probabilities (*post* meaning afterwards). For example, in this wine tasting situation, from the ten wine glasses, the probability of selecting a Bordeaux red wine is 40%. However, if a red wine has been selected then the probability of it being a Bordeaux red increases to 66.67%. Assume that a wedding party of 80 people have booked dinner at your restaurant and you need to decide on what to serve for the main meal. Closer to the date you learn that 70 of the guests would like a fish course. This new information simplifies the planning of your meal preparation and limits the number of choices you need to prepare.

If Bayes' rule is used correctly it may be unnecessary to collect vast amounts of data over time in order to make the best decisions based on probabilities. Or, another way of looking at Bayes' posterior rule is applying it to the often-used phrase of Hamlet, "He who hesitates is lost." This phrase implies that we should make a decision based on the information we have at hand; buy stock in Company A; purchase the house you visited; or take the job you were offered in Algeria[8]. However, if we wait until new information comes along Bayes' Theorem comes into play: Company A's financial accounts turn out to be inflated; the house you thought about buying turns out is in the path of the construction of a new highway; or new elections in Algeria make the political situation there unstable with a security risk. In these cases, procrastination may be the best approach and, "He who hesitates comes out ahead."

The following application case-exercise, *Arriving to work*, illustrates basic probability and Bayes' theorem.

Application case-exercise: Arriving to work

Situation

David works for Marriott hotel in Washington DC in front desk/reception. When the weather is nice he walks to work and sometimes he bicycles. In bad weather he takes the metro or he car pools with friends. Based on past habits there is a 35% probability that David walks, 40% he uses his bike, 10% he car pools, and 15% of the time he takes the metro. If David walks, there is a 5% probability of being late to the office; if he cycles there is a 10% chance of being late; a 75% chance of being late if he car pools because of traffic; and a 20% chance of being late if he takes the metro.

Required

1   On any given day, what is the probability of David being late to work?
2   On any given day, what is the probability of David being on time to work?
3   Given that David is late one day, what is the probability that he car pooled?
4   Given that David is on time for work one day, what is the probability that he walked?
5   Given that David takes the metro one day, what is the probability that he will arrive on time?
6   Given that David walks to work one day, what is the probability that he will arrive on time?
7   Show the data in a cross-classification or contingency table using 100 days as the sample data basis.

Worked-out solutions

The given probability data is in Table 3.8.
1   By the addition rule:
    Probability of being late = 35%*5% + 40%*10% +10%*75% +15%*20% = 16.25%.
2   Probability of not being late is 100% − 16.25% = 83.75%.

Table 3.8

Transportation mode	Walks	Cycles	Car pool	Metro
Probability of this transport mode	35.00%	40.00%	10.00%	15.00%
Probability of being late with this mode	5.00%	10.00%	75.00%	20.00%

3   Using Bayes' theorem:

$$P(B \mid A) = \frac{P(BA)}{P(A)}$$

This means that the probability of B occurring given that A has already happened is the joint probability of B and A occurring together, or in succession, divided by the probability of A occurring:

$$P(\text{Car pooled} \mid \text{Given late}) = \frac{P(\text{Car pooled}^*\text{Late})}{P(\text{Late})} = \frac{(10.00\%^*75.00\%)}{16.25\%} = 46.15\%$$

4   Using Bayes' theorem: $P(\text{Walked} \mid \text{Given on time}) = \frac{P(\text{Walked}^*\text{On time})}{P(\text{On time})} = \frac{(35.00\%^*(1-5\%))}{83.75\%} =$ 39.70%

5   Using Bayes' theorem: $P(\text{On time} \mid \text{Metro}) = \frac{P(\text{On time}^*\text{metro})}{P(\text{Metro})} = \frac{(1-20\%^*(15\%))}{15.00\%} = 80.00\%$

6   Using Bayes' theorem: $P(\text{On time} \mid \text{Walks}) = \frac{P(\text{On time}^*\text{walks})}{P(\text{Walks})} = \frac{(1-5\%^*(35\%))}{35.00\%} = 95.00\%$

7   Table 3.9 gives the cross-classifications identified by a grid.

Table 3.9

	A	B	C	D	E	F
1		Walks	Cycles	Car pool	Metro	Total
2	Late	1.75	4.00	7.50	3.00	16.25
3	On time	33.25	36.00	2.50	12.00	83.75
4	Total	35.00	40.00	10.00	15.00	100.00

- Cell F4 is the sample number of days of 100.
- Days late, Cell F2 is 100*16.25% = 16.25.
- Days on time, Cell F3 is 100*83.75% = 83.75 or 100 − 16.25.
- Using the probability for each transportation mode from the given data: Cell B4 is 35%*100 = 35; Cell C4 is 40%*100 = 40; Cell D4 is 10%*100 = 10; Cell E4 is 15%*100 = 15.
- Cell B2 is 35*5.00% = 1.75; Cell B3 is 35.00 − 1.75 = 33.25: Cell C2 is 40*10.00% = 4.00; Cell C-3 is 40.00 − 4.00 = 36.00; Cell D2 is 10*75.00% = 7.50; Cell D3 is 10.00 − 7.50 = 2.50: Cell E2 is 15*20.00% = 3.00; Cell E3 is 15.00 − 3.00 = 12.00.

Probability using contingency tables

In *Chapter 2* we introduced cross-classification or contingency tables, which illustrate the relationship between variables that might be obtained from a sampling experiment. On the basis that the data obtained can represent future conditions, contingency tables can be used for probability analysis, as demonstrated by the following application case-exercise, *Market survey*.

Application case-exercise: **Market survey**

Situation

A business publication in Europe performs a survey of some of its readers and classifies the survey responses according to the person's country of origin and their type of work. This information according to the number or respondents is given in the following contingency Table 3.10.

Table 3.10

Country	Consultancy	Engineering	Banking	Marketing	Architecture
Denmark	852	232	541	452	385
France	254	365	842	865	974
Spain	865	751	695	358	845
Italy	458	759	654	587	698
Germany	598	768	258	698	568

Required

1   What is the probability that a survey response taken at random comes from a reader in Italy?
2   What is the probability that a survey response taken at random comes from a reader in Italy and who is working in engineering?
3   What is the probability that a survey response taken at random comes from a reader who works in consultancy?
4   What is the probability that a survey response taken at random comes from a reader who works in consultancy and is from Germany?
5   What is the probability that a survey response taken at random from those who work in banking comes from a reader who lives in France?
6   What is the probability that a survey response taken at random from those who live in France is working in banking?
7   What is the probability that a survey response taken at random from those who live in France is working in engineering or architecture?
8   What is the probability that a reader is from France or is working in banking?

Worked out solutions

1   The first step is to calculate the totals for the rows and the columns. This is in Table 3.11.

Table 3.11

Country	Consultancy	Engineering	Banking	Marketing	Architecture	Total
Denmark	852	232	541	452	385	2,462
France	254	365	842	865	974	3,300
Spain	865	751	695	358	845	3,514
Italy	458	759	654	587	698	3,156
Germany	598	768	258	698	568	2,890
Total	3,027	2,875	2,990	2,960	3,470	15,322

Total in the market survey is 15,322; total from Italy is 3,156. Probability that a reader comes from Italy is 3,156/15,322 = 20.60%.

2   Number working in engineering and from Italy is 759. Thus probability in the sample survey that a reader is from Italy *and* is in engineering is 759/15,322 = 4.95%.
Alternatively using Baye's theorem: P(from Italy *and* in engineering) = (Probability of being in engineering, given living in Italy) *(marginal probability of living in Italy).
Probability of being in engineering, given living in Italy = 759/3,156 = 24.05% or 0.2405 as a fraction. Marginal probability of living in Italy = 3,156/15,322 = 20.60% or 0.2060 as a fraction. P(from Italy *and* in Engineering) = 0.2405*0.2060 = 0.0495 or 4.95%.

3   Number who work in consultancy is 3,027. Thus probability that a reader works in consultancy is 3,027/15,322 = 19.76%.

4   Number who works in consultancy and are from Germany is 598. Thus probability that a reader work in consultancy *and* is from Germany is 598/15,322 = 3.90%.
Alternatively using Baye's theorem: P(in consultancy *and* from Germany) = (Probability of being from Germany, given in consultancy) *(marginal probability of in consultancy).
Probability being from Germany, given in consultancy = 598/3,027 = 19.76% or 0.1976 as a fraction. Marginal probability of being in consultancy = 3,027/15,322 = 19.76% or 0.1976 as a fraction. P(from Germany *and* in consultancy) = 0.1976*0.1976 = 0.0390 or 3.90%.

5   Total number who work in banking is 2,990 and of those there are 842 from France. Thus given that a reader works in banking *and* is from France then the probability is 842/2,990 = 28.16%.
Alternatively using Baye's theorem: given in banking, P(from France) = P(from France and in banking)/P(in banking)
Probability (from France and in banking) = 842/15,322 = 5.50% or 0.0550 as a fraction. Marginal productivity of (in banking) = 2,990/15,322 = 19.51% or 0.1951 as a fraction. Given in banking, P(from France) = 0.0550/0.1951 = 28.16%

6   Total number who come from France is 3,300 and of those 842 are in banking. Thus given that a reader is from France *and* is in banking then the probability is 842/3,300 = 25.52%.
Alternatively using Baye's theorem: given from France, P(in banking) = P(from France and in banking)/P(from France).
Probability from France and working in banking = 842/15,322 = 5.50% or 0.0550 as a fraction. Marginal probability from France = 3,300/15,322 = 21.54% or 0.2154 as a fraction. Given from France, Probability in banking = 0.0550/0.2154 = 25.52%

7   Total number who come from France is 3,300 and of those 365 are in engineering and 974 are in architecture. Thus given that a reader is from France and is in engineering *or* architecture then the probability is (365 + 974)/3,300 = 40.58%.

8   Total number who come from France is 3,300. Total number of readers who are working in banking is 2,990. Thus the probability in the sample survey that a reader is from France *or* is working in banking is (3,300 + 2,990 − 842)/15,322 = 35.56%. (We must deduct 842 as if not this amount is included twice in the calculation, once in the row and then again in the column.)

System reliability and probability

A system includes all the interacting components, activities, or processes necessary to arrive at an end result or the production of a product. Probability can be used to evaluate the reliability of a system. In the system the reliability is the confidence given to a product, process, service, work team, or individual, such that it can operate under prescribed conditions without failure, or stopping, in order to produce the required output. In the supply chain of a firm, for example, reliability might be applied to whether the trucks delivering raw materials arrive on time; whether the suppliers produce quality components; whether the operators turn up for work; or whether packing machines operate without breaking down. The more components or activities in a product or a process, the more complex is the system and thus the greater is the risk of failure, or **unreliability**.

Series arrangements

A product or a process might be organized in a series arrangement as illustrated in the upper flow scheme in Figure 3.7.

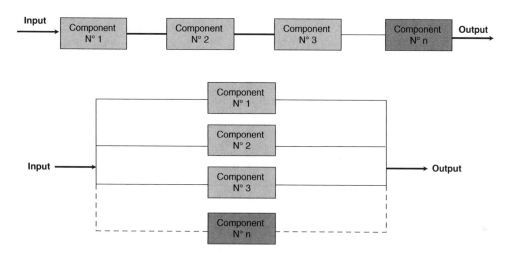

Figure 3.7 Reliability: series and parallel systems.

This is a general structure with *n* products, or *n* processes, where *n* can take on any integer value. For a series arrangement to operate from input to output, activities must pass sequentially through Component 1, Component 2, and Component 3 and eventually through Component n. For example, when an electric heater is on, the input electric current comes from the main power supply (Component 1), through a cable (Component 2), to a resistor (Component 3), from which the output heat is generated. The reliability of a series system, R_s, is the joint probability of the number of interacting components, *n*, according to the relationship.

$$R_S = R_1 {}^* R_2 {}^* R_3 {}^* R_4 {}^* ..... Rn \qquad\qquad 3(vii)$$

Here R_1, R_2, $R_{3....}$ R_n represents the reliability of the individual components expressed as a fraction or percentage. The relationship in Equation 3(vii) assumes that each component is independent of the others and that the reliability of one does not depend on the reliability of the others. In the electric heater example, the main power supply, the electric cable, and the resistor are all independent. However, the complete electric heating system does depend on all the components functioning, or in the system they are interdependent. If one component fails, then the system fails. For the electric heater, if the power supply fails, or the cable is cut, or the resistor is broken then the heater will not function. The reliability of the series system, or the value of R_S, will always be less than 100%. Consider the system between the input and output in the series scheme of Figure 3.7 and that there are four components whose reliability as a fraction are R_1 0.99; R_2 0.98; R_3 0.97, and R_4 a reliability of 0.96. The system reliability is then:

$$R_S = R_1 * R_2 * R_3 * R_4 = 0.99*0.98*0.97*0.96 = 0.9035 \text{ or } 90.35\%$$

That is a value less than the reliability of each of the individual components. By corollary, the probability of system failure is 9.65% (100.00% − 90.35%). If components have the same reliability then the system reliability is given by the following relationship, where *n* is the number of components:

$$R_S = R^n \hspace{6cm} 3\text{(viii)}$$

For example, if there are four components each with a reliability of 99%, then the system reliability is 96.06 $(0.99)^4$. If there were ten components then the system reliability drops to 90.44%. Table 3.12 illustrates this rapid decline in system reliability according to the number of components. Here each component is considered to have a reliability of 99%.

Table 3.12

Number of components	1	5	10	25	50	75	100	200
System reliability, R_S	99.00%	95.10%	90.44%	77.78%	60.50%	47.06%	36.60%	13.40%

For a more complete picture, Figure 3.8 gives a family of curves showing the system reliability for various values of the individual component reliability from 90 to a hypothetical 100%.

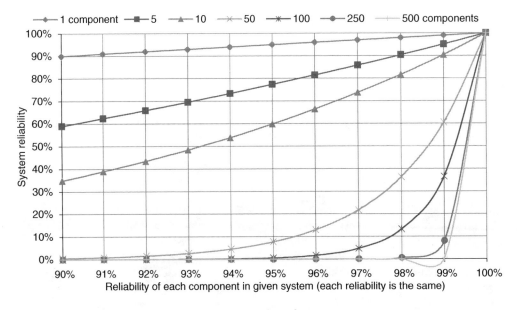

Figure 3.8 Series system reliability according to number of components.

These curves illustrate the rapid decline in the system reliability as the number of components increases. For example when there are:

- 10 components in the system, each with a reliability of 94%, system reliability is $0.94^{10} = 53.86\%$
- 50 components in the system, each with a reliability of 94%, system reliability is $0.94^{50} = 4.53\%$
- 100 components in the system, each with a reliability of 94%, system reliability is $0.94^{100} = 0.21\%$

The implication is that in manufactured products the more components there are, the less reliable is the product, or the higher the probability the product could malfunction—computer printer, automobile, electric circuits, etc. In services such as hotel and food services the more *components* there are in an activity then the higher is the probability of overall customer satisfaction being low. For example, a client that stays in a hotel complex for one week will use system components many times over—the room, the restaurants, the swimming pool, the spa, the bar, front desk services, etc. Thus if the reliability of these services is less than 100% the overall reliability or customer satisfaction at the end of the one week stay could be low. Alternatively if a client comes to a hotel and stays for one night and just has breakfast then the number of "components" used is much lower and the system reliability would be higher.

Parallel arrangement

The parallel or backup arrangement is illustrated in the lower diagram of Figure 3.7. Here, for a system to operate, there is a choice to pass through Component 1, Component 2, Component 3, or eventually Component n. Consider two components in a parallel system, where R_1 is the reliability of the main component and R_2 is the reliability of the **backup** or auxiliary component. The logic is that the backup Component R_2 would only be used if the main Component R_1 fails. In this case the system reliability R_S is expressed as follows:

$$R_s = \text{Reliability of } R_1 + \text{Reliability of } R_2 (\text{if used}) * \text{Probability of needing } R_2$$

The probability of needing R_2 is when R_1 fails or $(1 - R_1)$. Thus,

$$R_S = R_1 + R_2(1 - R_1) \qquad \qquad 3(\text{ix})$$

Reorganizing Equation 3(ix):

$$R_S = R_1 + R_2 - R_2 * R_1$$
$$R_S = 1 + R_1 + R_2 - R_2 * R_1 - 1$$
$$R_S = 1 - (1 - R_1 - R_2 + R_2 * R_1)$$
$$R_S = 1 - (1 - R_1)(1 - R_2) \qquad \qquad 3(\text{x})$$

If there are *n* components in a parallel arrangement then the system reliability becomes

$$R_S = 1 - (1 - R_1)(1 - R_2)(1 - R_3)(1 - R_4)..........(1 - R_n) \qquad \qquad 3(\text{xi})$$

where R_1, R_2 ... ... R_n represents the reliability of the individual components. This equation implies that the more the number of backup units, then the greater is the system reliability. However, this increase in reliability comes at an increased cost since we are adding backups that may rarely or never be used. When there are n backup components with an **equal reliability**, then the system reliability is given by the relationship

$$R_S = 1 - (1 - R)^n \qquad\qquad\qquad 3(xii)$$

Consider the four components of the series system are now in a parallel arrangement where the principal Component R_1 has a reliability of 99%, R_2 the first backup component has a reliability of 98%, and R_3 the second backup component having a reliability of 97% and the third backup Component R_4 has a reliability of 96% The system reliability is then from Equation 3(xi):

$$R_S = 1 - (1 - R_1)(1 - R_2)(1 - R_3)(1 - R_4)$$

$$R_S = 1 - (1 - 0.99)(1 - 0.98)(1 - 0.97)(1 - 0.96)$$

$$R_S = 1 - 0.01 * 0.02 * 0.03 * 0.04$$

$$R_S = 1 - 0.00000024 = 0.9999997600 \text{ or } 99.999976\%$$

That is, a system reliability greater than with a single component.
Even if there is only the first backup unit, R_2 then the system reliability is:

$$R_S = 1 - (1 - R_1)(1 - R_2)$$

$$R_S = 1 - (1 - 0.99)(1 - 0.98)$$

$$R_S = 1 - 0.01 * 0.02$$

$$R_S = 1 - 0.0002 = 0.9998 \text{ or } 99.98\%$$

Again this reliability is greater than the reliability of the individual components. Since the components are assembled in parallel this makes them **backup** units. The more the number of backup units, then the greater is the system reliability, as illustrated in Table 3.13 that gives the system reliability according the number of backup units for a given component reliability. The assumption is that the backup unit has the same reliability of the principle unit.

Table 3.13

Component reliability	No backup	1 backup	2 backups	3 backups	4 backups
90%	90.0000%	99.0000%	99.9000%	99.9900%	99.9990%
91%	91.0000%	99.1900%	99.9271%	99.9934%	99.9994%
92%	92.0000%	99.3600%	99.9488%	99.9959%	99.9997%
93%	93.0000%	99.5100%	99.9657%	99.9976%	99.9998%
94%	94.0000%	99.6400%	99.9784%	99.9987%	99.9999%
95%	95.0000%	99.7500%	99.9875%	99.9994%	100.0000%
96%	96.0000%	99.8400%	99.9936%	99.9997%	100.0000%
97%	97.0000%	99.9100%	99.9973%	99.9999%	100.0000%
98%	98.0000%	99.9600%	99.9992%	100.0000%	100.0000%
99%	99.0000%	99.9900%	99.9999%	100.0000%	100.0000%

Hospitals have backup energy systems in case of failure of the principal power supply. Most banks and other financial institutions have backup computer systems containing client data should one system fail. Supply distribution platforms such as for Walmart, Carrefour, or Tesco have backup computer information systems in case their main computer malfunctions. Without such systems firms would be unable to organize delivery of products to their retail stores. Aeroplanes have backup units in their design such that in the event of a failure of one component or subsystem there is recourse to a backup. A Boeing 747 can fly on one engine, although at a much reduced efficiency. To a certain extent the human body has a backup system as it can function with only one lung, though again at a reduced efficiency. In August 2004 my wife and I were in a motor home in St. Petersburg, Florida when we were informed that Hurricane Charlie was about to touch down. We were told of four possible escape routes to get out of the path of the hurricane. The emergency services had designated several backup exit routes—thankfully! In hotel and food service operations there may be backup systems for the hotel reservation system, backup cooking equipment such as refrigerators and ovens. There might also be personnel that are on call, "*backup*" in medical and fire services. When backup systems are in place this implies redundancy since the backup units are not normally operational.

Figure 3.9 gives a graphical representation for the reliability of a system using a parallel arrangement when the reliability of the individual component may be quite low. Even at a component reliability of just 50% five such components (four backups) connected in parallel give a system reliability of over 95%.

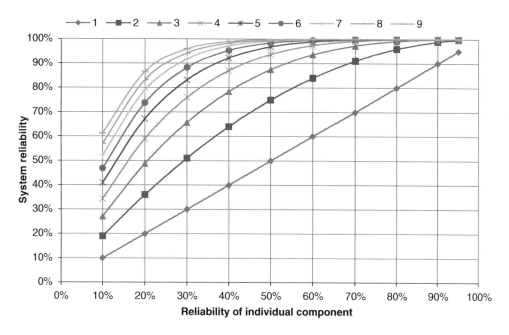

Figure 3.9 Reliability of parallel arrangement according to number of components from 1 to 9.

This is determined by Equation 3(xii):

$$R_S = 1 - (1 - R)^n = 1 - (1 - 0.50)^5 \qquad \text{3(xii)}$$

$$= 1 - 0.5^5 = 96.88\%$$

Counting rules

Counting rules describe possible outcomes or results of experiments. They give precise answers to many basic design situations and are similar to probability rules as they have a defined mathematical framework. In a sense they are **a priori** since required information is known before an experiment is performed. The difference with this counting theory is that probabilities are not involved. However, the counting rules can serve as the basis for probability. Outcomes from counting theory are **deterministic**.

Single type of event

With a single type of event the number of events is k, and the number of trials or experiments is n, then

$$\text{Total possible outcomes} = k^n \qquad\qquad 3(\text{xiii})$$

Suppose that a coin is tossed four times, then the number of trials, n, is four and the number of events, k, is two since heads or tails are the only two possible events. The events, obtaining heads or tails, are mutually exclusive since only heads or tails can be obtained in one throw of a coin. The **collectively exhaustive** outcomes are 2^4, or 16, as shown in Table 3.14 (H = Heads, T = Tails).

Table 3.14

Outcome	1	2	3	4	5	6	7	8	9	10	11	12	13	14	15	16
1st toss	H	T	T	T	T	H	H	H	H	H	T	T	H	T	T	T
2nd toss	H	H	T	T	T	T	H	H	T	T	H	H	H	H	H	H
3rd toss	H	H	H	T	T	T	T	H	H	T	H	T	T	H	T	T
4th toss	H	H	H	H	T	T	T	T	H	H	T	H	T	H	T	T

Different types of events

Another rule involves **different types of events.** Here there are k_1 possible events on the first trial or experiment, k_2 possible events on the second trial, k_3 possible events on the third trial, and k_n possible events on the nth trial. In this case the total possible outcomes of different events is calculated by the following relationship:

$$\text{Total possible outcomes} = k_1 {}^*k_2 {}^*k_3 \ldots\ldots\ldots k_n \qquad\qquad 3(\text{xiv})$$

As an illustration, my first car when I was a student in Newcastle on Tyne, England was an old Austin A40 with the license plate number of 212TPV. Thus, at that the time the pattern of the licenses issued by the Driving and Vehicle Center was 123ABC or three numbers followed by three letters. For numbers, there are ten possible outcomes; the numbers from 0 to 9. For letters there are 26 possible outcomes; from A to Z. Thus the total possible different combinations, or the number of feasible license plates, is 17,566,000 ($10*10*10*26*26*26$) assuming that zero is possible in the first place. If zero is not permitted in the first place, then the outcome is 15,818,000 ($9*10*10*26*26*26$). This is not a large number compared to the population and explains why formats have to be changed, or more characters have to be added, when the car-owning population increases. The same logic applies to telephone numbers.

Arrangement of different objects

Another rule concerns the **arrangement of different objects**. If there are *n* different objects then the number of ways they can be arranged is n factorial:

Number of ways to arrange *n* objects is $n! = n(n-1)(n-2)(n-3).....1$ 3(xv)

The last term in equation is $(n-n)!$ or $0!$ (In the factorial relationship, $0! = 1$). The number of ways that the four colors, red (R), yellow (Y), green (G) and blue (B) can be arranged is 24 ($4! = 4*3*2*1$) as shown in Table 3.15.

Table 3.15

R	R	R	R	R	R	B	B	B	B	B	B	G	G	G	G	G	G	Y	Y	Y	Y	Y	Y
B	G	Y	G	B	Y	R	G	Y	G	R	Y	R	B	Y	B	R	Y	B	G	R	G	B	R
G	Y	B	B	Y	G	G	Y	R	R	Y	G	B	Y	R	R	Y	B	G	R	B	B	R	G
Y	B	G	Y	G	B	Y	R	G	Y	G	R	Y	R	B	Y	B	R	R	B	G	R	G	B

[*In Excel use the Function* **FACT**]

Permutation rule

The permutation rule is a combination of data arranged in a **particular order**. The number of ways, or permutations, of arranging *x* objects, selected in order, from a total of *n* objects is:

$$^nP_x = \frac{n!}{(n-x)!}$$ 3(xvi)

Suppose John, Carol, Eric, and Helen, are restaurant personnel; then the number of ways one person could be the chef and the other the manager is 12 using Equation 3(xvi):

$$^4P_2 = \frac{4!}{(4-2)!} = 12$$

The possible arrangements are in Table 3.16.

Table 3.16

Arrangement	1	2	3	4	5	6	7	8	9	10	11	12
Chef	John	John	John	Carol	Carol	Carol	Eric	Eric	Eric	Helen	Helen	Helen
Manager	Carol	Eric	Helen	John	Eric	Helen	John	Carol	Helen	John	Carol	Eric

In permutations the same two people can be together, providing they have different functions.
[*In Excel use the Function* **PERMUT**]

Combination rule

The combination rule is a selection of distinct items **regardless of order**. The number of ways, or combinations, of arranging *x* objects, regardless of order, from *n* objects is given by:

$$^nC_x = \frac{n!}{x!(n-x)!}$$ 3(xvii)

Again, suppose John, Carol, Eric, and Helen are restaurant personnel; then the number of ways one person could be the chef and the other the manager without the same two people working together is 6 using Equation 3(xvii). The arrangement is shown in Table 3.17.

$$^4C_2 = \frac{4!}{2!(4-2)!} = 6$$

Table 3.17

Arrangement	1	2	3	4	5	6
Chef	John	John	John	Carol	Carol	Eric
Manager	Carol	Eric	Helen	Eric	Helen	Carol

Note that permutations differ from combinations by the value of $x!$ in the denominator. For a given set of items the number of permutations will always be more than the number of combinations because with permutations the order of the data is important, whereas it is unimportant for combinations.

[*In Excel use the Function* **COMBIN**]

Probability and odds with permutations and combinations

Suppose that you are gambling on a horse race at Ascot. In one particular event there are ten horses in the race. You place a bet on three horses, as in Table 3.18, indicating how they will be positioned.

Table 3.18

Position	First place	Second place	Third place
Horse name	Maple	Astrid	Isis

This is **permutations**. The sample size, n, is 10, and the variable, x, is 3. Thus from the relation for permutations, Equation 3(xvi) the number of different ways that three horses from the ten in the race can arrive in a first, second, and third position is:

$$^{10}P_3 = \frac{10!}{(10-3)!} = \frac{10!}{7!} = \frac{10*9*8*7*6*5*4*3*2*1}{7*6*5*4*3*2*1} = \frac{10*9*8}{1} = 720$$

Thus probability of winning is 1/720 or 0.1389%, or odds of winning are 1 to 719: $(1/720)/(720-1)/(720)$ or very low.

If your bet is to have these three horses in any of the first three places then this is **combinations**. And, using the factorial rule the number of different ways these horses can be positioned in the first three places is 6. (3! or $3*2*1 = 6$). These possibilities are shown in to Table 3.19.

Thus from the relation for combinations, Equation 3(xvii), the number of different ways that three horses from the ten in the race can arrive in the first three in any order, is:

$$^{10}C_3 = \frac{10!}{3!(10-3)!} = \frac{10*9*8*7*6*5*4*3*2*1}{3*2*1*(7*6*5*4*3*2*1)} = \frac{10*9*8}{3*2*1} = \frac{720}{6} = 120$$

Table 3.19

Position	First place	Second place	Third place
Horse name	Maple	Astrid	Isis
Horse name	Maple	Isis	Astrid
Horse name	Astrid	Isis	Maple
Horse name	Astrid	Maple	Isis
Horse name	Isis	Astrid	Maple
Horse name	Isis	Maple	Astrid

Thus the probability of winning is $1/120$ or 0.8333%, or odds of 1 to 119: $(1/120)/(120-1)/(120)$. Again this is low but a better chance than with permutations. This explains why the financial returns in betting in permutations are more advantageous than with combinations.

Moving on

This chapter has covered the essentials of probability relationships to estimate outcomes, and also counting rules that precisely indicate results. The next chapter covers probability in more detail, applying it to discrete data or whole number values.

Notes

1 "Las Vegas: Viva again," *The Economist*, July 18, 2015, p. 33.
2 "Italian punters: Gambling is booming, helping the government, but feeding addiction," *The Economist*, October 3, 2015, p. 30.
3 "The gambling industry," *The Economist*, September 24, 2005.
4 http://ww.partouche.fr [accessed September 27, 2005].
5 Arnold, Peter, "*How to play poker*," Bounty Books, London 2005.
6 "Traffic accidents: Road to kill," *The Economist*, July 4, 2015, p. 37.
7 C. Frey and M. Osborne, "The future of employment: How susceptible are jobs to computerization?" (Data published in *The Economist* January 18, 2014, p. 20.)
8 Based on a real situation for the author in the 1980s.

4 Discrete data and probability

ICEBREAKER: The shopping mall

There are shopping malls worldwide. These vast emporiums of retail stores together with restaurants are in every city and town. Depending on where you live you may find a Hugo Boss, C&A, Zara, Macy's, Tiffany's, Walmart, Louis Vuitton . . . the list is almost infinite. For many people a shopping mall is a weekend retreat. They were a staple of the US culture in the 1980s and 1990s and ever since there are few countries in the world that do not have a shopping mall. Shopping malls are very labor intensive. Each store and restaurant has its own sales and operating personnel and, since most malls are open from morning to late evening, there will be a rotating

Figure 4.0 The shopping mall.

Source: Author.

staff. Planning is not easy. A store may be practically empty on some days, and at other times have more customers than the sales staff can handle. Labor represents a significant cost.

How often do you go to the shopping mall: every day, once a week, or perhaps just once a month? When do you go? Perhaps after work, after dinner, in the morning when you think you can beat the crowds, or on the weekends? Why do you go? It might be that you have nothing else better to do; it's a gray, dreary day and it is always bright and cheerful in the mall, you need a new pair of shoes, you need a new coat, you fancy buying a couple of CDs, you are going to meet some friends, you want to see a film in the evening so you go to the mall a little early and just have a look around; you can shop; have a meal with your friend; go to the cinema. Now shopping habits are changing; more people are buying online and this is drastically reducing the mall foot traffic. It is reported that 15 percent of US shopping malls are expected to fail or be converted to non-retail in the next decade.[1]

Thus there are a multitude of variables that come into play as to when, and for what reason, people go to the mall, and to which store they will visit. These variables represent a complex random pattern of potential customers. How does the mall manage this randomness? How does the retailer plan staff levels and expected sales? There are potential customers in the mall; do they come to your store; yes or no? This is binomial. Even if they come into your store, do they buy? Again this is binomial. If they buy then how much do they spend? You need to estimate your expected values of revenues. Perhaps in the shopping mall there is a supermarket; Walmart (USA); Carrefour (France); Tesco (Britain). It is Saturday, and the supermarket is full. How do you manage the waiting time or the queue at the cashier desk? This planning is based on a Poisson distribution.

This chapter covers the application of some of these concepts—discrete random variables, the binomial and Poisson distributions.

Chapter subjects

✓ **Discrete data** • Random variables • Tossing of two dice • Application case-exercise: *Selling of wine* • Expected values and the law of averages

✓ **Bernoulli process** • Binomial distribution • Mathematical relationships of the binomial function • Application case-exercise: *Having children* • Deviations from binomial validity

✓ **Poisson distribution** • Mathematical relationships for the Poisson distribution • Application case-exercise: *Bar-café*

Discrete data

*Discrete data is statistical information composed of **integer values**, or **whole numbers** that originate from the counting process. We could say that 9 machines are shut down; 29 bottles of wine have been sold; 8 units are defective; 5 hotel rooms are vacant; or 3 students are absent. It makes no sense to say that 9½ machines are shut down, 29¾ bottles of wine have been sold; 8½ units are defective; 5½ hotel rooms are empty; or 3¼ students are absent. Discrete data is disjointed. There is a clear separation of information rather like the staccato effect of notes in music. Suppose in an airport you wish to transfer from Arrivals to Departure on separate levels. You have a possibility of walking up the stairs using discrete steps; one, two three, etc; or if you have luggage you can use the ramp where there is continuous progression from one level to another.*

Random variables

If the values of discrete data occur in no special order, and there is no explanation of their configuration or distribution, then they are considered **discrete random variables**. This means that, within a given range, every value has an equal chance of occurring. In gambling the value obtained by throwing a single dice is random; the drawing of a card from a full pack is random. Besides gambling there are other situations that occur randomly and it is useful to understand the pattern of randomness in order to make appropriate decisions. As illustrated in the icebreaker, the number of people arriving at a shopping mall in any particular day is random. If we understood the pattern it would help to better plan staff needs. The number of cars on a particular stretch of road on any given day is random. Knowing the pattern would help to decide on where to install stop signs, roundabouts, or traffic signals, for example. The number of people seeking medical attention at a hospital emergency center is random. Again, understanding the loading pattern helps in scheduling medical staff and equipment. It is true that in some cases of randomness factors like the weather, day of the week, or hour of the day influence the magnitude of the data. However, often even if we know these influences, the data is still random.

Like any dataset, random variables can be described by a mean, variance, and standard deviation. The mean is the weighted average of all the possible outcomes of the random variable, given by the expression:

$$\text{Mean value, } \mu = \Sigma(x*P(x)) = E(x) \qquad \qquad 4(\text{i})$$

- x is any value of the discrete random variable
- $P(x)$ is the probability, or the chance, of obtaining the value x
- $E(x)$ is the **expected value**, or also the **mean value**, μ of the dataset.

$$\text{Variance, } \sigma^2 = \Sigma(x - \mu)^2 * P(x) \qquad \qquad 4(\text{ii})$$

The variance expressed in 4(ii) is analogous to variance of a population given in *Chapter 1* except that instead of dividing by the number of data values, which gives a straight average, here we multiply by *P(x)* to give a **weighted average** according to probabilities.

$$\text{Standard deviation, } \sigma, \text{ is the square root of the variance} = \sqrt{(\Sigma(x - \mu)^2 * P(x))} \qquad 4(\text{iii})$$

The use of these relationships can be illustrated by the throwing of two dice.

Tossing of two dice

In gambling games two dice are often used, with the sum of the two values thrown being the score. Since each dice has six possible outcomes, with two dice there are 36 possible outcomes (6*6). Table 4.1 gives the 36 possible combinations and the total value that can be obtained on the throw of these two dice. It indicates that of the 36 possible combinations there are just 11 different outcomes; the values 2 through 12. The number of ways that these 11 outcomes can be achieved is summarized in Table 4.2, Column 2 and the probability *P(x)* of obtaining these outcomes is in Column 3. For example, the total times the number 5 can occur is four. There are 36 possible combinations so the probability of obtaining a total of 5 is 4/36 = 11.11%. This is the relative frequency as it is relative to the total possible outcomes. The weighted average of this probability is *x*P(x)* and for obtaining a score of 5 this is 5*11.11% or 0.5556 as shown in Column 4 of this Table 4.2.

- From Equation 4(i) the mean value of throwing two dice is 7.00, the last line of Column 4
- From Equation 4(ii) the variance is 5.8333, the last line of Column 7
- From Equation 4(iii) the standard deviation is the square root of the variance or $\sqrt{5.8333} = 2.4152$.

An alternative way of calculating the expected or mean value is from Column 1. The total value of all the possible throws is $\Sigma x = 77$. There are 11 possible combinations and so the mean value is 77/11 = 7.00.

Table 4.1

Throw No.	1	2	3	4	5	6	7	8	9	10	11	12
Value Dice A	1	2	3	4	5	6	1	2	3	4	5	6
Value Dice B	1	1	1	1	1	1	2	2	2	2	2	2
Total	2	3	4	5	6	7	3	4	5	6	7	8

Throw No.	13	14	15	16	17	18	19	20	21	22	23	24
Value Dice A	1	2	3	4	5	6	1	2	3	4	5	6
Value Dice B	3	3	3	3	3	3	4	4	4	4	4	4
Total	4	5	6	7	8	9	5	6	7	8	9	10

Throw No.	25	26	27	28	29	30	31	32	33	34	35	36
Value Dice A	1	2	3	4	5	6	1	2	3	4	5	6
Value Dice B	5	5	5	5	5	5	6	6	6	6	6	6
Total	6	7	8	9	10	11	7	8	9	10	11	12

Table 4.2

Column 1 Total value of throw x	Column 2 No. times value of x occurs n	Column 3 Frequency or probability of this throw $P(x) = n/\Sigma n$	Column 4 Weighted value of x $[x*P(x)]$	Column 5 $(x - \mu)$	Column 6 $(x - \mu)^2$	Column 7 $P(x)*(x - \mu)^2$
2	1	2.78%	0.0556	−5.00	25.00	0.6944
3	2	5.56%	0.1667	−4.00	16.00	0.8889
4	3	8.33%	0.3333	−3.00	9.00	0.7500
5	4	11.11%	0.5556	−2.00	4.00	0.4444
6	5	13.89%	0.8333	−1.00	1.00	0.1389
7	6	16.67%	1.1667	0.00	0.00	0.0000
8	5	13.89%	1.1111	1.00	1.00	0.1389
9	4	11.11%	1.0000	2.00	4.00	0.4444
10	3	8.33%	0.8333	3.00	9.00	0.7500
11	2	5.56%	0.6111	4.00	16.00	0.8889
12	1	2.78%	0.3333	5.00	25.00	0.6944
$\Sigma x = 77$	$\Sigma n = 36$	$\Sigma P(x) = 100.00\%$	$\Sigma x*P(x) = 7.00$			$\Sigma P(x)*(x - \mu)^2$ $= 5.8333$

The histogram of the relative frequency of obtaining a given value of x, and thus as a corollary, the probability, is given in Figure 4.1. Note this is a symmetrical distribution.

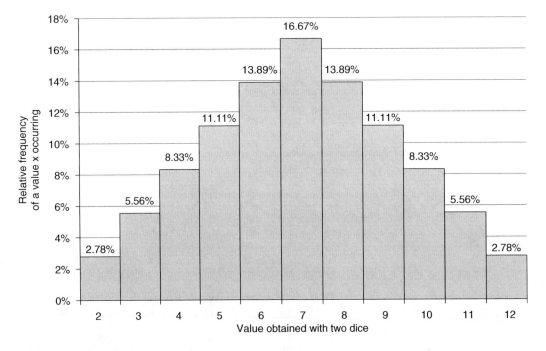

Figure 4.1 Throwing of two dice: relative frequency (probability) of obtaining a certain value.

The following application case-exercise: *Selling of wine*, illustrates the application of random variables and expected values.

Application case-exercise: **Selling of wine**

Situation

Matthew, a wholesale wine distributor, sells wine by cases of six bottles. He prices the wine such that each case generates €8.00 in profit. The sale of his wine is considered random. Matthew has records to indicate the cases of wine sold each day for the last consecutive 240 days and these are in the range of 10–20 cases as indicated by Table 4.3.

Table 4.3

Cases sold per day x	No. days this value x is sold, n	Cases sold per day x	No. days this value x is sold, n
10	12	16	25
11	19	17	15
12	30	18	14
13	35	19	10
14	40	20	8
15	32	Total days	240

Required

1   Estimate the average number of cases of wine sold per day. Determine the variance and the standard deviation of the number of cases of wine sold per day.
2   Show the relative frequency probability of the sales of wine.
3   Estimate the average profit per day generated by the sale of wine and the average annual profit based on 240 days.

Worked out solutions

1   Table 4.3 is expanded to give Table 4.4 using Equations 4(i) to 4(iii).

Table 4.4

Column 1 Cases sold per day, x	Column 2 No. of days this value x sold, n	Column 3 Probability this number of cases x sold $P(x) = n/\Sigma n$	Column 4 Weighted value of x x*P(x)	Column 5 $(x - \mu_x^2)$	Column 6 $(x - \mu_x^2)*P(x)$
10	12	5.00%	0.50	18.8139	0.9407
11	19	7.92%	0.87	11.1389	0.8818
12	30	12.50%	1.50	5.4639	0.6830
13	35	14.58%	1.90	1.7889	0.2609
14	40	16.67%	2.33	0.1139	0.0190
15	32	13.33%	2.00	0.4389	0.0585
16	25	10.42%	1.67	2.7639	0.2879
17	15	6.25%	1.06	7.0889	0.4431
18	14	5.83%	1.05	13.4139	0.7825
19	10	4.17%	0.79	21.7389	0.9058
20	8	3.33%	0.67	32.0639	1.0688
Total	240	100.00%	14.34		6.3319

Column 1 gives the value x, or the number of cases sold per day; Column 2 gives the number of days out of the annual 240 that this quantity of x is sold; For example in 35 days out of 240 days, 13 cases of wine are sold. Column 3 is the relative frequency of a value of x being sold and is given by the number of days this quantity of x is sold divided by the total number of days in the analysis, or 240. For example, if we consider 13 cases then this amount is sold in 35 days of the total 240 days recorded to give a relative frequency of 14.58 percent (35/240). By corollary this is also the probability if we are using use this data for future estimates. Column 4 gives the weighted value of x with the total $\Sigma x * P(x)$ equal to 14.34 as shown. This is the expected or mean value of the number of cases of wine sold per day. The histogram of the relative frequency of the sale of wine is in Figure 4.2.

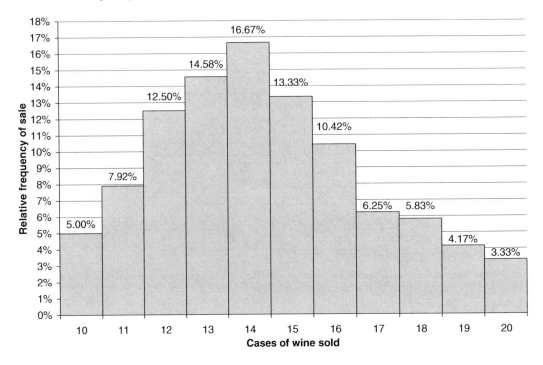

Figure 4.2 Selling of wine: relative frequency distribution.

2   The calculations for the variance are given in Columns 5 and 6 of Table 4.4 using Equation 4(ii). The variance is 6.3319 as shown in the last line of Column 6. The standard deviation, which is the square root of the variance, is 2.5163 ($\sqrt{6.3319}$) using Equation 4(iii).

3   Thus for Matthew for estimating profits:

- Mean number of cases sold per day is 14.34 where each case generates €8.00 in profits
- Average profit per day is €114.72 (14.34*8.00) [114.70 using Excel]
- Estimated annual profit on the basis that 240 days represents annual activity is €27,532.80 (240*114.72) [27,528.00 using Excel]

Expected values and the law of averages

When we talk about the mean, or expected value, in probability situations, this is not the value that will occur next, tomorrow, or even in the following week. It is the value that is expected to be obtained in the long run. In the short term we really do not know what will happen. When you play the slot machines, or one-armed bandits, in a casino you may win a few games. In fact, quite a lot of the money put into slot machines does flow out as jackpots but about six percent rests with the house.[2] Thus, if you continue playing, in the long run you will lose because the gambling casinos have set their machines so that the casino will be the long-term winner. If, not they would go out of business! Gambling is a fool's business. With probability, it is the **law of averages** that governs. This law says that the average value obtained in the long term will be close to the expected value, which is the weighted outcome based on each of the probability of occurrence.

The long-term result corresponding to the law of averages can be explained by Figure 4.3. This illustrates the tossing of a coin 1,000 times (if you have the time) where there is a 50% probability of obtaining heads and a 50% probability of obtaining tails. The *y*-axis of the graph is the cumulative frequency of obtaining heads and the *x*-axis is the number of times the coin is tossed. In the early throws, as the coin is tossed the cumulative number of heads obtained may be more than the cumulative number of tails, as illustrated. For example after 50 throws 76% or 38 throws are heads whereas 24% or 12 are tails. However, as the tossing of the coin is continued the law of averages comes into play and the cumulative number of heads obtained approaches the cumulative number of tails obtained. After 1,000 throws there will be approximately 50% heads or 500 heads and 50% or 500 tails. This illustration supports Rule 1 of the counting process given in *Chapter 2*.

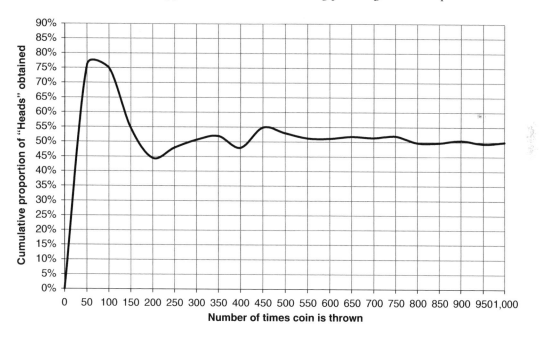

Figure 4.3 Tossing a coin.

You can perhaps apply the law of averages on a non-quantitative basis to behavior in society. We are educated to be honest, respectful, and ethical. This is the norm, or the average, of society's behavior. There are a few people who might cheat, steal, be corrupt, or be violent. In the short term these people may get away with it. However, often in the long run the law of averages catches up with them. They get caught, lose face, are punished or may be removed from society! Some examples: in the Great Train Robbery, Bruce Reynolds, with careful planning using a 15-member gang, attacked a Royal Mail Train in August 1963 in Buckinghamshire, England. They got away with £2.6 million and expected to be comfortable for the rest of their lives. However, most of the gang were arrested and sentenced to 30 years in prison. The bulk of the money was never recovered.[3] Bernard Madoff founded and ran Madoff Investment Securities from 1960 until he was arrested in 2008. He was convicted of operating a mammoth Ponzi scheme that defrauded thousands of investors out of billions of dollars. Other family members were involved. In June 2009 Madoff was sentenced to 150 years in prison. One son committed suicide.[4] A more recent example in 2015 illustrating non-respect for society's norms is the German automobile company, Volkswagen. Apparently for many years they used hidden software to deceive American regulators measuring emissions from their diesel-engine vehicles. The damage to VW, the world's biggest carmaker, has been cataclysmic. The company's shares have since collapsed by one-third and they face $billions in fines.[5]

Bernoulli process

A Bernoulli process, named after the Swiss Jacque (or Jacob) Bernoulli (1654–1705) means that there are only two possible outcomes from each trial of an experiment. That is binomial. The tossing of a coin is binomial; only heads or tails is possible; a light bulb either illuminates or it does not; you either like a product or you do not. The baby is born; it is either a boy or a girl; you either pass a university exam, or you fail. If you know that a situation exhibits a binomial characteristic then a distribution can be developed from which probabilities may be estimated and appropriate decisions made.

Binomial distribution

A binomial distribution is a table or a graph that shows the outcomes of a binomial-type experiment. For the distribution to be valid it is considered that each observation is selected from an infinite population (or one of a very large size) without replacement or if the population is finite (or small) then selection is with replacement. Since there are only two possible outcomes we say that the probability of obtaining the desired outcome or "*success*" is p, and the probability of obtaining the opposite outcome "*failure*", is q. Failure means the opposite of what you are testing, expecting, or hoping for but it does not mean failure in the classic sense. As the situation is binomial $p + q = 1$. Table 4.5 gives some ideas of a binomial relationship. Remember "failure" here is just the opposite of "success." Note, as shown, that the standard denoting whether electrical appliances are on or off is that on = 1; off = 0.

Table 4.5

Criteria	Gambling	Exam	Coin Toss	Electric Heater	Door	Reunion	Election Vote	No. of theater seat	Child birth
Success, p	Win	Pass	Heads	On = 1	Open	Present	Yes	Odd	Girl
Failure, q	Lose	Fail	Tails	Off = 0	Closed	Absent	No	Even	Boy

Other criteria for the binomial distribution are that the probability, p, of obtaining an outcome must be fixed over time and that the outcome of any result must be independent of a previous result. In all of the illustrations in Table 4.5 a success in the first attempt—gambling, exam, coin toss, electric heater, etc. has no bearing on the outcome on the second or subsequent attempts. In gambling situations, for example, in the tossing of a coin, the probability of obtaining heads or tails remains always 50% and obtaining a head on one toss has no effect on what face is obtained on subsequent tosses. In the throwing of a dice, an odd or even number can be thrown, again with a probability outcome of 50%. For each result one throw has no bearing on another throw. In the drawing of a card from a full pack, the probability of obtaining a black card (spade or clubs) or obtaining a red card (heart or diamond) is again 50%. If a card is replaced after the drawing, and the pack shuffled, the results of subsequent drawings are not influenced by previous drawings. In these three illustrations we have the following relationship:

$$\text{Probability, } p = (1 - p) = q = 0.5 \text{ or } 50\% \qquad \qquad 4(\text{iv})$$

Mathematical relationships of the binomial function

It was Jacques Bernoulli who from experiments postulated the following expression to describe a binomial distribution that is also referred to as a **Bernoulli process:**

$$\text{Probability of } x \text{ successes, in } n \text{ trials} = \frac{n!}{x!(n-x)!}*\left[p^{x}*q^{(n-x)}\right] \qquad \qquad 4(\text{v})$$

- p is the characteristic probability, or the probability of *success*
- $q = (1 - p)$ or the probability of *failure*
- $x =$ is a random variable of the number of successes
- $n =$ number of trials undertaken, or the sample size

The binomial random variable, x, can have any integer value ranging from 0 to n, the number of trials. If p is 50%, then q is 50% and the resulting binomial distribution is symmetrical regardless of the sample size, n. This is the case in the coin toss experiment as illustrated in Figure 4.1; obtaining an even or odd number on throwing a dice; or selecting from a pack a black or red card. When p is not 50% the distribution is skewed.

In the binomial expression of 4(v) the expression:

$$p^{x}*q^{(n-x)} \qquad \qquad 4(\text{vi})$$

is the probability of obtaining exactly x successes out of n observations in a particular sequence. The relationship

$$\frac{n!}{x!(n-x)!} \qquad \qquad 4(\text{vii})$$

is how many combinations of the x successes, out of n observations, are possible. This expression is discussed in the counting process of *Chapter 3*.

The **expected value** of the binomial distribution, $E(x)$, or the mean value, μ, is the product of the number of trials and the **characteristic probability**:

$$\mu = E(x) = n*p \qquad \qquad 4(\text{viii})$$

The **variance of the binomial distribution** is the product of the number of trials, the characteristic probability of *success*, and the characteristic probability of *failure*:

$$\sigma^{2} = n*p*q \qquad \qquad 4(\text{ix})$$

The **standard deviation of the binomial distribution** is the square root of the variance:

$$\sigma = \sqrt{\sigma^2} = \sqrt{n*p*q} \qquad\qquad 4(x)$$

Assume a coin is tossed 40 times:

- Mean or expected value is 20 (40*50%), from Equation 4(viii)
- Variance is 10.00 (40*50%*50%) from Equation 4(ix)
- Standard deviation is 3.16 ($\sqrt{10}$) from Equation 4(x).

The binomial distribution is demonstrated by the following application case-exercise, *Having children*. (A situation familiar, or will be, to most of us!)

Application case-exercise: **Having children**

Situation

George and Susan live together and wish to have seven children. In the genetic makeup of both George and Susan the chance of having a boy or girl is equally possible. In their family history there is no incidence of twins or other multiple births.

Required

1   Develop a binomial distribution table and histogram of the gender possibilities of the children of George and Susana on the basis that the probability of having a boy or girl is 50%.
2   Develop a binomial distribution table and histogram of the gender possibilities of the children of George and Susana on the basis that the probability of having a boy is 52%. (Statistics show that this is the case.)

Worked out solutions

1   Sample size, or number of trials of having a baby, *n*, is 7.

- x is the random variable. It means that the number of boys (or the number of girls) can take on the values, 0, 1, 2, 3, 4, 5, 6, and 7. For example there can be zero boys, and seven girls.
- The probability of having a boy (or girl) on any trial is 50%.
- As an illustration, probability $P(x)$ of George and Susan having two boys and five girls in the seven children is calculated from 4(xii) where $x = 2$; $n = 7$; $p = 50\%$ or 0.5.

$$P(x = 2) = \frac{7!}{2!(7-2)!}*0.5^2*0.5^{(7-2)}$$

$$= \frac{5,040}{2*120}*0.2500*0.031250$$

$$= 21*0.007813 = 0.164063 \text{ or } 16.41\%$$

Table 4.6

Column 1 Random variable, x	Column 2 P(x) exactly	Column 3 P(x) cumulated
0	0.78%	0.78%
1	5.47%	6.25%
2	16.41%	22.66%
3	27.34%	50.00%
4	27.34%	77.34%
5	16.41%	93.75%
6	5.47%	99.22%
7	0.78%	100.00%
Total	100.00%	

The probabilities for all conditions of having 0, 1, 2 ... to 7 boys (or girls) is in Column 2 of Table 4.6. The Column 3 gives the probability values cumulated. For example, for a value of x of 2 the cumulative probability is 22.66% (0.78 + 5.47 + 16.41).

[*In Excel to generate this information use function:* **BINOMDIST** and insert the values of x and probability] The histogram showing these probabilities is in Figure 4.4.

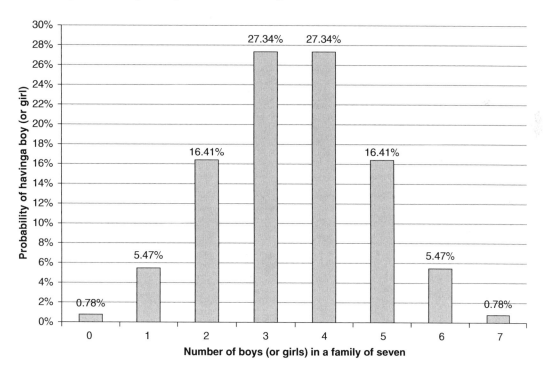

Figure 4.4 Probability of the gender of children when $P(x) = 50\%$ for having a boy (or girl).

Table 4.6 is interpreted as follows:

- Probability of having exactly two boys is 16.41%
- Probability of having more than two boys (3, 4, 5, 6, or 7 boys) is 77.34% (27.34 + 27.34 + 16.41 + 5.47 + 0.78) from Column 2: alternatively (100.00 − 22.66) from Column 3)

- Probability of having at least two boys (2, 3, 4, 5, 6 or 7 boys) is 93.75% (16.41 + 27.34 + 27.34 + 16.41 + 5.47 + 0.78) from Column 2: alternatively or (100.00 − 6.25) from Column 3
- Probability of having fewer than two boys (0 or 1 boy) is 6.25% (0.78 + 5.47), or 6.25% directly from Column 3
- Mean or expected value is $n * p = 7 * 50\% = 3.50$ boys (though not a feasible value).
- Standard deviation is $\sqrt{n * p * q} = \sqrt{7 * 0.50 * 0.50} = 1.3229$.

2   Historical data shows that the probability of having a boy is about 52% rather than 50% and thus the binomial relationships are modified as follows:

- Sample size, or number of trials of having a baby, n, is 7
- x, the random variable, means the number of boys can take on the values, 0, 1, 2, 3, 4, 5, 6, and 7. For example there can be zero boys, and seven girls
- Probability of having a boy on any try is 52%. Correspondingly the probability of having a girl is 48%

The probability $P(x)$ of George and Susan having two boys and five girls in the seven children for this modified situation is calculated from 4(xii) where $x = 2$; $n = 7$; $p = 52\%$ or 0.52.

$$P(x = 2) = \frac{7!}{2!(7 - 2)!} * 0.52^2 * 0.48^{(7 - 2)}$$

$$= \frac{5,040}{2*120} * 0.27040 * 0.02548$$

$$= 21 * 0.00689 = 0.14468 \text{ or } 14.47\%$$

The results for all conditions are in Table 4.7.

Table 4.7

Column 1 *Random variable, x*	*Column 2* P(x) *exactly*	*Column 3* P(x) *cumulated*
0	0.59%	0.59%
1	4.45%	5.04%
2	14.47%	19.51%
3	26.12%	45.63%
4	28.30%	73.93%
5	18.40%	92.33%
6	6.64%	98.97%
7	1.03%	100.00%
Total	100.00%	

- The probability of having exactly two boys is 14.47%
- The probability of having more than two boys is 80.49%
- The probability of having at least two boys is 94.96%
- The probability of having fewer than two boys is 5.04%
- Mean or expected value is $n * p = 7 * 52\% = 3.64$ boys (though not a feasible value)
- Standard deviation is $\sqrt{n * p * q} = \sqrt{7 * 0.52 * 0.48} = 1.3218$

The histogram showing these probabilities is in Figure 4.5.

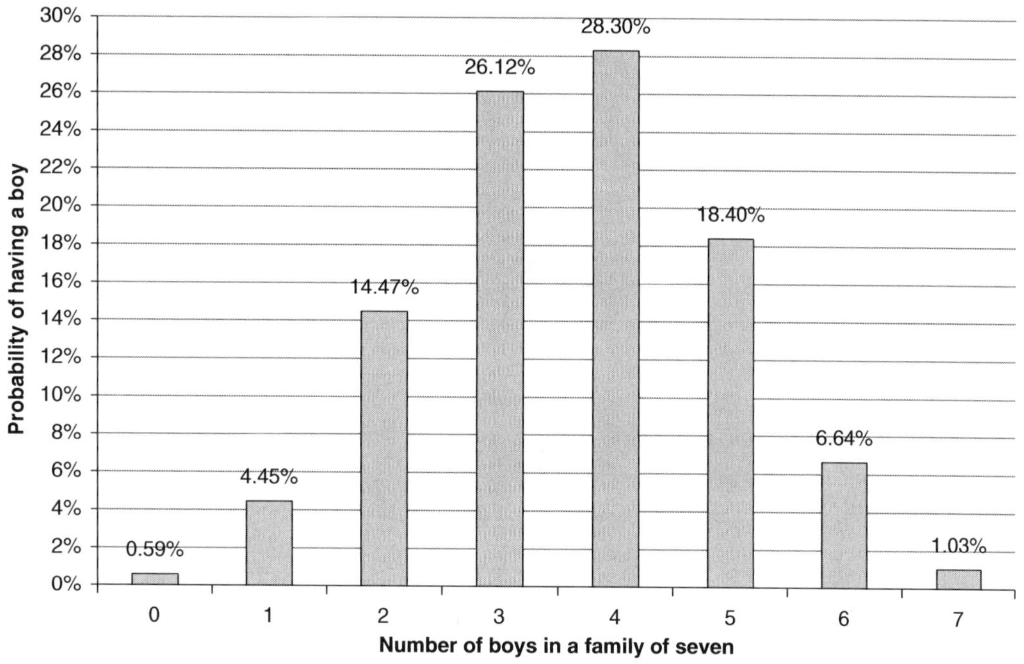

Figure 4.5 Probability of having a boy when [*P(x)* – boy] = 52%.

In summary, the probability of having 4, 5, 6 or 7 boys in a family of 7, when the probability of having a boy is 52%, is higher than when the probability is 50%. And the mean value is higher, 3.64 as opposed to 3.50.

Deviations from binomial validity

Many situations may appear to be binomial, meaning that the probability outcome is fixed over time, and the result of one outcome has no bearing on another. However, in practice these two conditions might be *violated*. Consider, for example, a manager interviewing, in succession, twenty candidates for one position in his firm. One of the candidates has to be chosen. Each candidate represents discrete and unique information where their experience and ability are independent of each other. Thus, the interview process is binomial—either a particular candidate is selected or is not. As the manager continues the interviewing process he makes a subliminal comparison of competing candidates, so if one candidate is rated positively this results perhaps in a less positive rating of another candidate. Thus, the evaluation is not entirely independent. Further, as the day goes on, if no candidate has been selected, the interviewer gets tired and may be inclined to offer the post to one of the last few remaining candidates out of sheer desperation!

In another situation, consider driving your car to work each morning. When you get into the car, either it starts or it does not. This is binomial and your expectation is that your car will start every time. The fact that your car started on Tuesday morning should have no effect on whether it starts on Wednesday and should not have been influenced by the fact that it started on Monday morning. However, over time, mechanical, electrical, and even electronic components wear. Thus, one day you turn the ignition in your car and it does not start.

Poisson distribution

The Poisson distribution, named after the Frenchman, Denis Poisson (1781–1840), is another discrete probability distribution used to describe events that usually occur during a given time interval. Illustrations might be the number of cars/hour arriving at a tollbooth; the number of patients/day arriving at the emergency center of a hospital; the number of airplanes in a given period waiting in a holding pattern to land at a major airport; or the number of customers waiting in line at the cash checkout, as highlighted in the Icebreaker.

Mathematical relationships for the Poisson distribution

The equation describing the Poisson probability of occurrence, *P(x)* is:

$$P(x) = \frac{\lambda^x e^{-\lambda}}{x!}$$ 4(xi)

- λ (lambda, the Greek letter l) is the mean number of occurrences
- *e* is the base of the natural logarithm, or 2.71828
- *x* is the Poisson random variable
- *P(x)* is the probability of exactly *x* occurrences

The standard deviation of the Poisson distribution is the square root of the mean number of occurrences, or

$$\sigma = \sqrt{(\lambda)}$$ 4(xii)

In applying the Poisson distribution the assumptions are that the mean value can be estimated from past data. Further, if we divide the time period into seconds then the following apply:

- The probability of exactly one occurrence per second is a small number and is constant for every one-second interval.
- The probability of two or more occurrences within a one-second interval is small and can be considered zero.
- The number of occurrences in a given one-second interval is independent of the time at which that one-second interval occurs during the overall prescribe time period.
- The number of occurrences in any one-second interval is independent on the number of occurrences in any other one-second interval.

The following application case-exercise, *Bar-café* illustrates the Poisson distribution.

Application case-exercise: **Bar-café**

Situation

The owner of a bar-café in a village in the Rhone region of France knows that on a Sunday an average eight people per hour come in for service. Sunday is not a busy day and only the owner herself attends to the customers. Some periods are slow but at other times she is busy and then client service is slow. The owner of the coffee shop proposes to employ an assistant if she has confidence that there is a greater than 10% probability that in any one-hour period there might be at least 12 customers. This relationship is considered to approximate a Poisson distribution.

Required

1 Develop the Poisson distribution to illustrate whether the owner should employ an assistant for Sunday.

Worked out solution

1 Here the mean value, λ, is 8 customers. Customer level of interest, x, is 12.
   The Poisson distribution table to describe this situation is in Table 4.8. Column 2 gives the probability $P(x)$ for the exact value x; Column 3 gives the cumulative values of $P(x)$. The corresponding histogram is in Figure 4.6. Note that in the Poisson distribution the sample size n is not defined. In theory this can be infinite but in practice there is a sample size when the cumulative probabilities reach 100%.
   [*In Excel to generate this information use function:* **POISSON**]

Table 4.8

Column 1 Number of customers x	Column 2 Probability of exactly x P(x)	Column 3 Cumulative probability of P(x)	Column 1 Number of customers x	Column 2 Probability of exactly x P(x)	Column 3 Cumulative probability of P(x)
0	0.03%	0.03%	12	4.81%	93.62%
1	0.27%	0.30%	13	2.96%	96.58%
2	1.07%	1.38%	14	1.69%	98.27%
3	2.86%	4.24%	15	0.90%	99.18%
4	5.73%	9.96%	16	0.45%	99.63%
5	9.16%	19.12%	17	0.21%	99.84%
6	12.21%	31.34%	18	0.09%	99.93%
7	13.96%	45.30%	19	0.04%	99.97%
8	13.96%	59.25%	20	0.02%	99.99%
9	12.41%	71.66%	21	0.01%	100.00%
10	9.93%	81.59%	22	0.00%	100.00%
11	7.22%	88.81%	Total	100.00%	

This distribution is interpreted as follows:

- Probability of exactly 12 customers entering in a given hour = 4.81%
- Probability of more than 12 customers entering in a given hour = $(100 - 93.62) = 6.38\%$
- Probability of no more than 12 customers entering in a given hour = 93.62%
- Probability of at least 12 customers entering in a given hour = $(100 - 88.81) = 11.19\%$
- Probability of less than 12 customers entering in a given hour = 88.81%

Since the probability of at least 12 customers in a given hour is 11.19%, or greater than the 10% ceiling, the owner should decide to hire an assistant for Sunday.
   Equation 4(xviii) can be used to check the calculation for exactly 12 customers, where x is 12 and λ is 8:

$$P(12) = \frac{\lambda^{12}e^{-8}}{12!} = \frac{68,719,476*0.000335463}{479,001,600} = 4.81\%$$

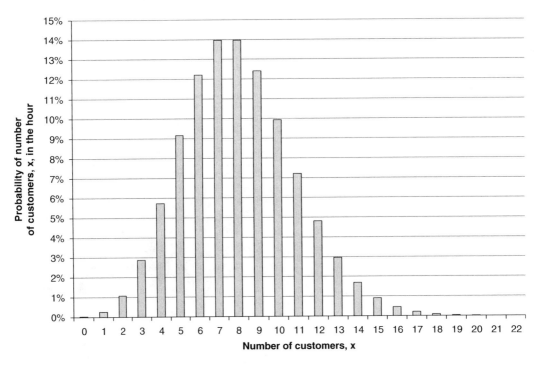

Figure 4.6 Bar-café: Poisson distribution.

Moving on

This chapter has presented discrete data and its associated probability. The next chapter looks at continuous data and probabilities using the normal distribution.

Notes

1 *Financial Times*, March 16, 2015.
2 Henriques, Diana B., "On bases, problem gamblers battle the odds," *International Herald Tribune*, October 20, 2005, p. 5.
3 *Wikipedia*, the free encyclopedia [accessed October 19, 2015].
4 *Wikipedia*, the free encyclopedia [accessed October 19, 2015].
5 "A mucky business: systematic fraud by the world's biggest carmaker threatens to engulf the entire industry and possibility reshape it," *The Economist*, September 26, 2015, p. 19.

5 Continuous distributions and probability

ICEBREAKER: Your can of beer, bar of chocolate, or length of fabric

We shop at the local store, supermarket, or hypermarket every day. We buy manufactured processed products by volume—salad dressing, beer; wine; milk, etc. On the can of beer the label indicates "*Volume 33 cl.*" So you have a volume of exactly 33 cl. in the can, right? You are almost certainly wrong as this implies a volume of 33.000000 cl. Or, we purchase by weight—chocolate, butter, coffee, tea, etc. When you buy a bar of chocolate written on the label is: "*net weight 100 grams.*" Again, it is highly unlikely that you have 100.000000 grams of chocolate. In your hypermarket you also buy three meters of curtain material. It is highly unlikely that you have exactly 3,000 mm.

All these types of products are developed using some type of machine operation. The target or machine setting is to obtain a prescribed value: 33 cl., 100 grams, 3 meters. However, it is practically impossible to always obtain exactly this target quantity. Some values will be higher,

Figure 5.0 Your can of beer, bar of chocolate, or length of fabric.

Source: Author's own photograph.

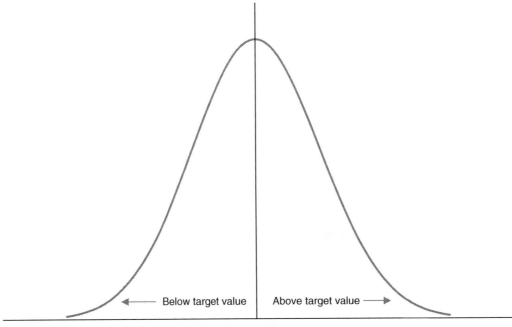

Target value

Figure 5.0.1 Normal distribution.

and some will be lower, just because of the process variation in the filling of the cans of beer, the molding operation for the chocolate, or the cutting process for the fabric. These variations are a consequence of a continuous distribution or, more precisely, the normal distribution, as illustrated in Figure 5.0.1.

The volume of the beer, the weight of the chocolate, or the length of the fabric should not be consistently high since over time this would be costly to the producing firm. Conversely, the volume, weight, or length cannot be always too low as the firm will not be respecting the marketing information given on the label and clearly this would be unethical.

This is the purpose of this chapter—to examine several continuous distributions with an emphasis on the normal distribution.

Chapter subjects

✓ **The normal distribution** • Description • Number of standard deviations, z • Boundary limits • Configurations of the normal distribution • Activities that follow a normal distribution • Excel and the normal distribution • Application case-exercise: *Lightbulbs*

✓ **Asymmetrical or non-normal distributions** • Skewed data • Frequency distributions • Box and whisker plots.

✓ **Verifying that a dataset approximates a normal distribution** • Visual displays • Characteristics of the dataset • Ogive comparisons • Constructing a normal probability plot

✓ **Approximation for the binomial distribution** • Conditions for a normal–binomial approximation • Application case-exercise: *Ceramic plates* • Sample size needed to use normal-binomial approximation

✓ **Exponential distributions** • Exponential progression • Bacteria growth • Financial investment and exponential growth • Magazine circulation and exponential decline • Learning curve • Natural logarithm • Waiting lines

✓ **Beta distribution** • Activity time • Project management

---

*Continuous distributions are developed from random variables that take fractional or decimal values such that numbers run into each other. As illustrated in the Icebreaker, the nominal volume of beer in a can is indicated on the label as 33 cl. However, the actual volume may be 32.8785, 32.9856, or 33.0528 cl., that is a number close to, but not exactly, 33 cl. Similarly, the actual weight of the bar of chocolate may be 99.7285, 100.1285, or 100.2459 grams, or the length of the fabric may be 2,998; 3,002; or 3,105 mm. For all these values of volume, weight, and length there is no distinct cut-off point between data values and they can overlap into other class ranges. This is different from discrete data where the numbers 1, 2, 3, 4, ... etc. might represent the number of employees, the number of manufactured products, or the number of hotel clients. Here there is a distinct separation of data values. There are many families of continuous distribution and those considered in this chapter are the **normal distribution** and a variant of this, the **asymmetrical distribution**; **exponential distribution**; and the **beta distribution**.*

The normal distribution

The normal distribution is widely used in statistical analysis to describe a continuous random variable. It was conceived by the German Karl Friedrich Gauss (1777–1855) and as such is also known as the Gaussian distribution.

Description

The shape of the normal distribution is illustrated in Figure 5.1. Here the x-axis is the value of the random variable, and the y-axis the frequency of occurrence of this random variable. On the left side of the curve, the frequency of occurrence of the random variable is small, it increases to a maximum, and then declines on the right side. The curve has the following characteristics:

- It is bell, mound, or humped shaped and symmetrical around this hump. This implies that the left side is a mirror image of the right side.
- The central point or the hump of the distribution, as measured at the x-axis, has equal values of the mean, median, mode, and midrange.

- The range is close to six times the standard deviation.
- The inter-quartile range is equal to 1.33 standard deviations.
- The values of the two parameters, mean, μ, and standard deviation, σ, uniquely characterize a given normal distribution.
- The left and right extremities, or the two tails of the normal distribution, may extend far from the central point. This implies that the associated random variable, x, can have a range $-\infty > x < +\infty$.

If the frequency of occurrence can represent future outcomes, then the normal distribution can be used as a measure of probability and this makes it a useful aid in decision making. The following is the equation for the normal distribution, and from which the continuous curve of Figure 5.1 is developed:

$$y = f(x) = \left(\frac{1}{\sqrt{2\pi\sigma_x}}\right) e^{-\left(\frac{1}{2}\right)\left[\frac{(x-\mu_x)}{\sigma_x}\right]^2}$$

5(i)

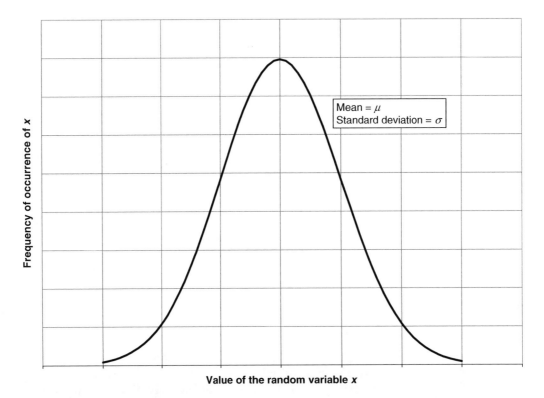

Figure 5.1 Shape of the normal distribution.

- $f(x)$ is the probability density function
- π is the constant pi equal to 3.14159
- σ_x is the standard deviation
- e is the base of the natural logarithm equal to 2.71828

- x is the value of the random variable
- μ_x is the mean value of the distribution

The probability density function, $f(x)$, describes the relative likelihood for a random variable, x, to take on a given value. The probability for the random variable to fall within a particular range is given by the integral of this variable's density over this range. The integral over the entire range is unity. This means that no matter what the values of the mean or the standard deviation, the area under the curve is always equal to 1.00 or that the area under the curve represents 100% of the data used to construct the distribution. And, since the curve is symmetrical, 50% of the area is to the left of the mean; 50% is to the right.

The normal distribution is a polygon, whose development is discussed in *Chapter 2*, is a line graph superimposed on the histogram of the same data array as shown in Figure 5.2.

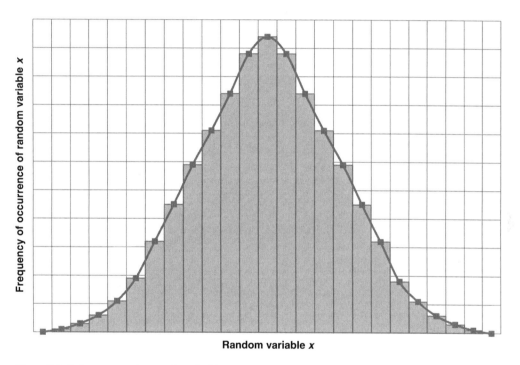

Figure 5.2 Polygon superimposed on its histogram.

Number of standard deviations, z

The units of the random variable x in the continuous distribution might be volume—centiliters of beer; weight—grams of chocolate; length—meters of fabric; time—hours for the duration of a lightbulb; etc. A normal distribution representing these measurement units will have different means and standard deviations but all these datasets can be transformed into a standard normal distribution using the following normal distribution **transformation equation:**

$$z = \pm \frac{x - \mu_x}{\sigma_x}$$

5(ii)

- x is the value of the random variable in given units
- μ_x is the mean of the distribution of the random variables in the given units
- σ_x is the standard deviation of the distribution in the given units
- z is the **number of standard deviations** from x to the mean of this distribution.

Since the numerator and the denominator (top and bottom parts of the equation respectively) have the same units, these cancel out such the value of z is unitless. It is simply a number. Further, since the value of x can be more, or less, than the mean value, then z can be either plus or minus, as indicated by Equation 5(ii).

Consider that in beer production, for a certain container format, the mean value of beer in a can is 33 cl. and from past data we know that the standard deviation of the bottling process is 0.50 cl. (The filling operation has been carried out umpteen times.) Assume that a single can of beer is taken at random from the bottling line and its volume is 33.75 cl. In this case, using Equation 5(ii):

$$z = \frac{x - \mu_x}{\sigma_x} = \frac{33.75 - 33.00}{0.50} = \frac{0.75}{0.50} = 1.50$$

Alternatively, the mean value of a certain size chocolate bar is 100 g. and from past data we know that the standard deviation of a production lot of these chocolate bars is 0.40 g. Assume one slab of chocolate is taken at random from the production line and its weight is 100.60 g. In this case, again using Equation 5(ii):

$$z = \frac{x - \mu_x}{\sigma_x} = \frac{100.60 - 100.00}{0.40} = \frac{0.60}{0.40} = 1.50$$

Again, assume that the mean value of the life of a certain model automobile tire is 35,000 km and from past data we know that the standard deviation of the life of a tire is 1,500 km. Then suppose that one tire is taken at random from the production line and tested on rolling machine. The tire lasts 37,250 km. Then, again using Equation 5(ii):

$$z = \frac{x - \mu_x}{\sigma_x} = \frac{37,250 - 35,000}{1,500} = \frac{2,250}{1,500} = 1.50$$

Thus in each of the three examples the number of standard deviations, z, is the same whereas the units and values of mean and standard deviation from which z is derived are different. It is with this transformation relationship that the normal distribution can be used to determine probability.

Equation 5(ii) can be reorganized to make x, the random variable, the subject of the equation:

$$z = \pm \frac{x - \mu_x}{\sigma_x}$$

$$z * \sigma_x = \pm (x - \mu_x)$$

$$x = \mu_x \pm z\sigma_x \qquad\qquad 5(\text{iii})$$

This relationship given by Equation 5(iii) is useful if the number of standard deviations is known or required and it is of interest to calculate the value of the random variable.

Boundary limits

Using the transformation relationship, the relationship between, z, x, μ_x, and σ_x can be shown on the normal distribution. Consider again a slab of chocolate of net weight 100 grams (let's say Cadbury's milk chocolate, which is what I grew up on and made me fat!). Since Cadbury's (sadly now Kraft Foods) have produced millions of 100 gram chocolate bars they know that the standard deviation of the weight of the chocolate bars as a result of the molding and cooling of the bars is 0.4 grams. Let's say that an operator randomly takes from the production line one slab of chocolate and it weighs 101.20 grams. Using Equation 5(ii) the number of standard deviations, z, of this weight is:

$$z = \frac{x - \mu_x}{\sigma_x} = \frac{101.20 - 100.00}{0.40} = \frac{1.20}{0.40} = 3.00$$

Say that now another slab of chocolate is randomly taken from the production line and it weighs 99.20 grams. Again using Equation 5(ii) the number of standard deviations, z, of this weight is:

$$z = \frac{x - \mu_x}{\sigma_x} = \frac{99.20 - 100.00}{0.40} = -\frac{0.80}{0.40} = -2.00$$

Assume that we do this experiment seven times and just by chance x, the weight of these randomly taken chocolate bars, is 98.80, 99.20, 99.60, 100.00, 100.40, 100.80, and 101.20 grams. Then, using Equation 5(ii), Table 5.1 gives the corresponding number of standard deviations, according to x. The values of μ_x and σ_x are also indicated but they do not change for this experiment.

Table 5.1

μ_x grams	100.00	100.00	100.00	100.00	100.00	100.00	100.00
σ_x, grams	0.40	0.40	0.40	0.40	0.40	0.40	0.40
x, grams	98.80	99.20	99.60	100.00	100.40	100.80	101.20
z	−3.00	−2.00	−1.00	0	+1.00	+2.00	+3.00

With these values of z, the number of standard deviations, the normal distribution has the boundary limits as illustrated in Figure 5.3. These boundary limits are used as the basis for statistical analysis:

- 68.27%, or about 68 percent, of all given data falls within ± 1 standard deviations from the mean. This means that the boundary limits of this 68.27% of the data are $\mu \pm 1\sigma$.
- 95.45%, or about 95%, of all given data falls within ± 2 standard deviations from the mean. This means that the boundary limits of this 95.45% of the data are $\mu \pm 2\sigma$.
- 99.73%, or about 100%, of all given data falls within ± 3 standard deviations from the mean. This means that the boundary limits of this almost 100% of the data are $\mu \pm 3\sigma$.

Although 99.73% of data lies between ± 3 standard deviations from the mean we might say that this is close to 100% and that it is satisfactory for an operation. However, there is still 0.27% of data (100 − 99.73) that is outside the measurements. What if this small amount applies to the operation of an aircraft engine, a surgical procedure, or testing the structural strength of a building? The resulting outcome could be catastrophic. This is the basis behind six-sigma management presented in *Chapter 13*.

The value of z does not have to be a whole number but can take on any numerical value, for example, −0.45, 0.78, or 2.35, which gives areas under the curve from the left-hand tail to the z value of 32.64%, 78.23%, and 99.06% respectively. When z is negative it means that the area under the curve from the left is less than 50% and when z is positive it means that the area from the left of the curve is greater than 50%. These area values can then also be interpreted as **probabilities** and this is how the normal distribution is used for measuring chance estimates. Thus for any data of any continuous units, such as weight, volume, speed, length, decibels, etc., all intervals containing the same number of standard deviations, z, from the mean will contain the same proportion of the total area under the curve for any normal probability distribution. There is a **standard normal distribution** that has a mean value, μ, of zero and in this case the profile is the same as in Figure 5.3 except now the x-axis would now be the values shown by z.

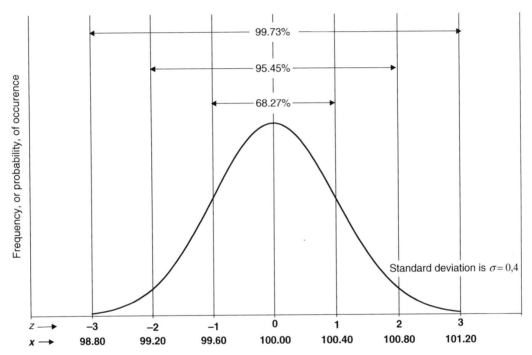

Figure 5.3 Areas under a standard normal distribution.

Configurations of the normal distribution

The mean measures the central tendency of data, and the standard deviation measures its spread or dispersion. Datasets forming a normal distribution can exhibit many configurations, for example the same mean but different standard deviations, as shown in Figure 5.4. Here there are three distributions with the same mean, μ, but with standard deviations of σ_1, σ_2, and σ_3. The distribution that has the smallest standard deviation of σ_1 is narrower and the data congregates around the mean. The one with the highest standard deviation of σ_3 is flatter and the deviation around the mean is greater.

Figure 5.5 shows three normal distributions with different means μ_1, μ_2, and μ_3 but the same standard deviation σ_1.

Different means and also different standard deviations are as illustrated in Figure 5.6. Here the flatter curve has a mean of μ_1 and a large standard deviation of σ_3. The middle curve has a mean of μ_2 and a standard deviation of σ_2. The sharper curve has a mean of μ_3 and a smaller standard deviation of σ_1.

In conclusion, the shape of the normal distribution is determined by its standard deviation and the mean establishes its position on the x-axis. As such, there is an infinite combination of curves

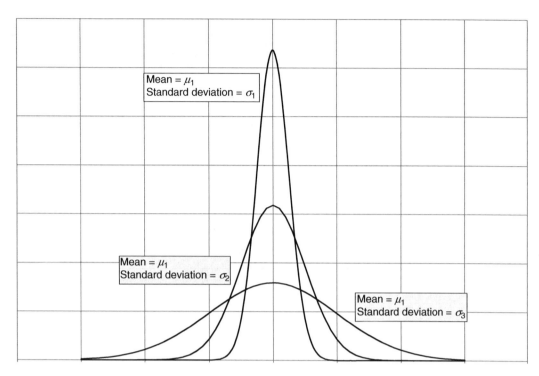

Figure 5.4 Normal distributions: equal means but different standard deviations.

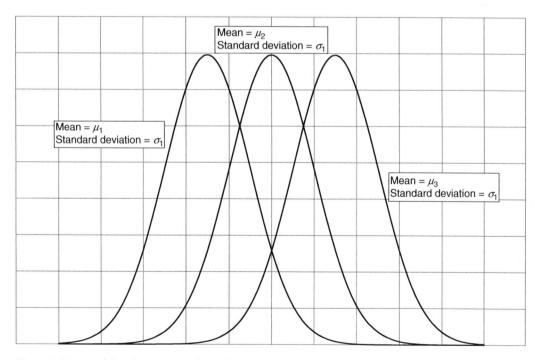

Figure 5.5 Normal distributions: equal standard deviations but different means.

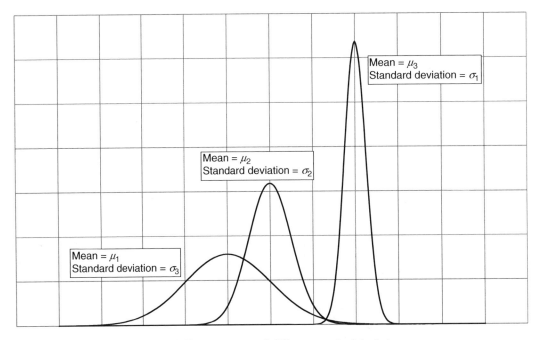

Figure 5.6 Normal distributions: different means and different standard deviations.

according to their respective means and standard deviations. However, a given set of data is uniquely defined by its mean and standard deviation. Since continuous normal distributions may have the same mean, but different standard deviations, the different standard deviations alter the sharpness, or hump, of the peak of the curve, as illustrated by the three normal distributions given in Figure 5.4. This difference in shape is the **kurtosis,** or the characteristic of the peak of a frequency distribution curve. The curve that has a small standard deviation, σ_1, is **leptokurtic** after the Greek word *lepto* meaning slender where the peak is sharp. The curve that has a standard deviation, σ_3, is **platykurtic** after the Greek word *platy* meaning broad, or flat which is the form of this curve. The intermediate curve of standard deviation, σ_2, is called **mesokurtic** (from the Greek meso meaning intermediate), where the sharpness of the peak of the curve is in between the other two. Understanding the shape of a normal distribution is relevant as a leptokurtic form is more reliable for analytical purposes, as the data shows less deviation from the mean, which is a target in six-sigma management, discussed in *Chapter 13*.

Activities that follow a normal distribution

The outputs of operations using machines or robots follow a normal distribution and it is the machine setting that sets the mean or target value. However, machines are a complex mix of computer commands driving mechanical equipment so that the output of each activity has a variation, albeit very small. Further material used in the operation may show a variation regarding temperature, viscosity, or composition, although again this variation would be very slight. For example, in a beverage-filling operation for beer or soft drinks the setting may be for a volume of 33 cl. This is the average volume of the cans in a very large production run but some cans have volumes below this amount and some above. This situation is the same for a whole host of food-related products, such as jelly, yogurt, margarine, toothpaste, wine, ketchup, etc. You will never have the exact net weight of the product but it will be within a range that is considered acceptable for the production operation. Machines undergo periodic maintenance in order that machine settings are respected and it may be close to the time that the maintenance is scheduled that off-specification outputs are obtained. Robots and/or machines used in other manufacturing activities, such as soldering, stamping, or painting in auto-mobile assembly, normally perform according to the control setting but there will always be slight exceptions.

The life of the use of many standard products follows pretty much a normal distribution: auto-mobiles, the tires of a truck or an automobile, batteries, washing machines, refrigerators, lightbulbs, etc. There is an average life, usually measured in hours, and some may last longer, and some shorter. For automobiles or trucks the random variable is the distance in kilometers or miles before the engine needs a complete overhaul. There is an average distance but some may last longer, some less. This may be a function of the way the vehicle is driven and under what sort of road conditions.

Activities where the proportion of labor or people is high will not be as regular as machine-based operations since people show more variability in their behavior. However, if human activity is repeated a sufficient number of times the action will approximate normality, as for example the time to make a cheeseburger, the time to take a shower, the time to test a blood sample, etc. The com-muting time from your home to place of work, providing they remain the same, follows a normal distribution over the long term. In the short term there will be variations due to traffic density, road conditions, and weather but over many commutes there will be a mean value and a standard deviation from all of the commuting times measured.

Excel and the normal distribution

Some books on statistics and quantitative methods publish standard tables for determining z. These give the area of the curve either to the right or the left side of the mean and from these tables probabilities can be estimated. Instead of tables, this book uses the Microsoft Excel function for the normal distribution that has a database of z values. The logic of the z values in Excel is that the area of the curve, and by corollary the probability, starts at 0% on the left of the curve increasing to reach 100% at the right of the curve. The four normal distribution functions in Excel used for the case-exercises in this textbook are:

- **NORMDIST** (Excel 2007) or **NORM.DIST** (Excel 2010 and later) determines the area under the curve from left to right, or probability $P(x)$, given the value of the random variable x, the mean value, μ, of the dataset, and the standard deviation, σ. In **NORM.DIST** there is a parameter "cumulative". Here you enter "1".
- **NORMINV** (Excel 2007) or **NORM.INV** (Excel 2010 and later) computes the value of the random variable, x, given the area under the curve from left to right or the probability, $P(x)$, the mean value, μ, and the standard deviation σ.
- **NORMSDIST** (Excel 2007) or **NORM.S.DIST** (Excel 2010 and later) gives the value of the area from left to right or probability, $P(x)$ given z.
- **NORMSINV** (Excel 2007) or **NORM.S.INV** (Excel 2010 and later) calculates the value of z given the area from left to right or probability, $P(x)$.

It is not necessary to learn by heart which function to use because, as for all Excel functions, when they are selected, they indicate what values to insert to obtain the result. Thus, knowing the information that you have available tells you what normal function to use. The kurtosis value can be determined in Excel by using **KURT**.

The use of the normal distribution is illustrated by the following application case-exercise, *Lightbulbs*.

Application case-exercise: **Lightbulbs**

Situation

A manufacturer of lightbulbs in Germany has past data concerning the life of a particular 36 watt neon lightbulb that shows, on average, it lasts 2,500 hours before failure. This is the mean value μ. The standard deviation of illumination time, σ, is 725 hours. Illumination time of lightbulbs follows a normal distribution.

Required

1   What is the probability that a lightbulb selected at random will last no more than 3,250 hours?
2   What is the probability that a lightbulb selected at random will last at least 3,250 hours?
3   What is the probability that a randomly selected lightbulb will last no more than 2,000 hours?
4   What is the probability that a lightbulb selected at random will last between 2,000 and 3,250 hours?
5   What are the symmetrical lower and upper limits for which 75% of the lightbulbs will last?
6   If the company has 50,000 of these lightbulbs in stock, how many bulbs would be expected to fail at 3,250 hours or less?
7   If the company has 50,000 of these lightbulbs in stock, how many bulbs would be expected to fail between 2,000 and 3,250 hours?

Worked out solutions

1   Using Equation 5(ii), where the random variable, x, is 3,250:

$$z = \frac{3,250 - 2,500}{725} = \frac{750}{725} = 1.0345$$

Using **NORMSDIST** (Excel 2007) or **NORM.S.DIST** (Excel 2010 and later) and entering "1" for the cumulative parameter, the area under the curve from left to right for $z = 1.0345$, is 84.95%. Thus the probability of one lightbulb taken randomly from the production line has a 84.95% probability of lasting not more than 3,250 hours. The concept is shown on the normal distribution of Figure 5.7.

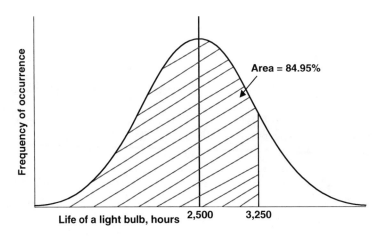

Figure 5.7 Probability that the life of a lightbulb lasts no more than 3,250 hours.

2   Here the interest is the area of the curve to the right where x is at least 3,250 hours. This area is 15.05% (100% − 84.95%). Thus we can say that there is a 15.05% probability that a single lightbulb taken from the production line will last at least 3,250 hours. This is shown on the normal distribution in Figure 5.8.

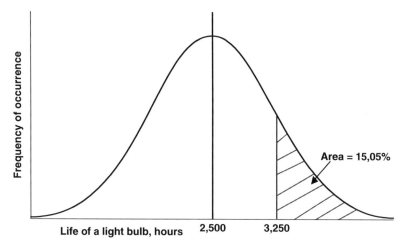

Figure 5.8 Probability that the life of a lightbulb lasts at least 3,250 hours.

3   Using Equation 5(ii), where the random variable, x, is now 2,000 hours:

$$z = \frac{2,000 - 2,500}{725} = -\frac{500}{725} = -0.6897$$

The fact that z has a negative value implies that the random variable lies to the left of the mean; which it does since 2,000 hours is less than 2,500 hours. From **NORMSDIST** (Excel 2007) or **NORM.S.DIST** (Excel 2010 and later) the area of the curve for $z = -0.6897$ is 24.52%. Thus there is a 24.52% probability that a single lightbulb taken randomly will last no more than 2,000 hours. The situation is shown on the normal distribution curve in Figure 5.9.

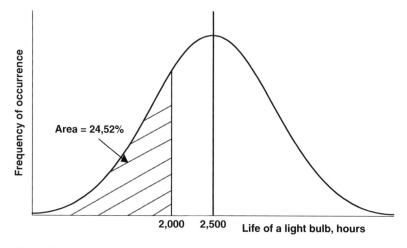

Figure 5.9 Probability that the life of a lightbulb lasts no more than 2,000 hours.

4   Here the interest is the area of the curve between 2,000 and 3,250 hours where 2,000 hours is to the left of the mean and 3,250 is greater than the mean. We can determine this probability by several methods.

Method 1
- Area of the curve 2,000 hours and below is 24.52% from answer to Question 3.
- Area of the curve 3,250 hours and above is 15.05% from answer to Question 2.
- Thus, area between 2,000 and 3,250 hours is 60.43% (100.00 – 24.52 – 15.05).

Method 2
Since the normal distribution is symmetrical, the area of the curve to the left of the mean is 50.00% and also the area of the curve to the right of the mean is 50.00%. Thus,

- Area of the curve between 2,000 and 2,500 hours is 25.48% (50.00 – 24.52) from Question 3.
- Area of the curve between 2,500 and 3,250 hours is 34.95% (50.00 – 15.05) from Question 2.
- Thus, area of the curve between 2,000 and 3,250 hours is 60.43% (25.48 + 34.95).

Method 3
- Area of the curve at 3,250 hours and below is 84.95%
- Area of the curve at 2,000 hours and below is 24.52%
- Thus, area of the curve between 2,000 and 3,250 hours is 60.43% (84.95 – 24.52).

This situation is shown on the normal distribution curve in Figure 5.10.

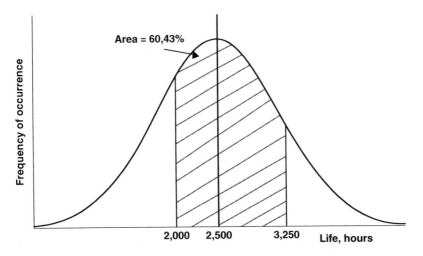

Figure 5.10 Probability that the life of a lightbulb lasts between 2,000 and 3,250 hours.

5   Here the interest is the 75% middle area of the curve. The area of the curve outside this is 25% (100.00 – 75.00). Since the normal distribution is symmetrical, the area on the left side of the limit, or the left tail, is 25/2 or 12.50% and similarly, the area on the right of the limit, or the right tail, is also 12.50%.
Using **NORMSINV** (Excel 2007) or **NORM.S.INV** (Excel 2010 and later) the value of z at 12.5% has a numerical value of 1.1503. Again, since the curve is symmetrical the value of z on the left side is –1.1503 and on the right side it is +1.1503.

From Equation 5(iii) where z at the upper limit is 1.1503, $\mu_x = 2{,}500$ and σ_x is 725,

x (upper limit) $= 2{,}500 + 1.1503{*}725 = 3{,}334$ hours

At the lower limit z is -1.1503, so

x (lower limit) $= 2{,}500 - 1.1503{*}725 = 1{,}666$ hours

These values are also shown on the normal distribution curve in Figure 5.11.

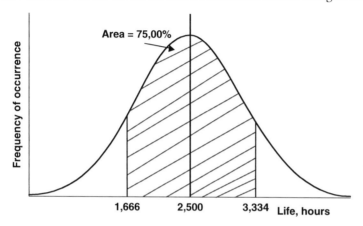

Figure 5.11 Symmetrical limits between which 75% of the lightbulbs last.

6   In this case we multiply the population N, 50,000, by the area under the curve determined by the answer in Question 1:

$50{,}000{*}84.95\% = 42{,}477.24$ or 42,477 lightbulbs rounded to the nearest whole number

7   Again, we multiply the population N, 50,000, by the area under the curve determined by the answer to Question 4:

$50{,}000{*}60.43\% = 30{,}216.96$ or 30,217 lightbulbs rounded to the nearest whole number

In all these calculations the values have been determined by first finding the value of z. A quicker route in Excel is to use the **NORMDIST** (Excel 2007) or **NORM.DIST** (Excel 2010 and later) where the mean, standard deviation, and the value of x are entered and this gives the area or probability directly. It is a matter of preference which functions to use. If z is first calculated it is easy to position the problem on the normal distribution curve and this can be helpful.

Asymmetrical or non-normal distributions

The normal distribution means that the left side can be superimposed on the right-hand side. That is, they are mirror images. However, a distribution may be skewed, or asymmetrical, and in this case the normal relationships do not apply.

Skewed data

In a continuous dataset when the mean and median are significantly different then the probability distribution is not normal but is asymmetrical or **skewed**. A distribution is skewed because values in the frequency plot are concentrated at either the low end or left side of the *x*-axis, or at the high end or right side of the *x*-axis. When the mean value of the dataset is greater than the median then the distribution of the data is positively or **right-skewed** and the curve tails off to the right as there are more values in the dataset lower than the mean. This is because the mean is the most affected by extreme values and is pulled over to the right. Here the distribution of the data has its mode, the hump, or the highest frequency of occurrence, at the left end of the *x*-axis where there is a higher proportion of relatively low values and a lower proportion of high values. The median is the middle value and lies between the mode and the mean. If the mean value is lower than the median, then the data is negatively or **left-skewed** such that the curve tails off to the left. This is because it is the mean that is the most affected by extreme values and is pulled back to the left. Here the distribution of the data has its mode, the hump, or the highest frequency of occurrence, at the right end of the *x*-axis where there is a higher proportion of large values and lower proportion of relatively small values. The median is again the middle value and lies between the mode and the mean. This logic is demonstrated as follows.

Frequency distributions

Suppose three Firms A, B, and C each have 1,000 employees and the frequency distributions of the annual salaries in \$US for each firm are as illustrated in Figure 5.12 together with their statistical data in Table 5.2. The minimum and maximum salaries are the same for all three firms. Thus the range and midrange are the same. For Firm A the mean (\$79,466), median (\$79,533), and mode (\$75,840) are relatively close and the frequency distribution curve for this firm in Figure 5.12 shows that the data pretty much follows a normal distribution. For Firm B the mean (\$64,820) is greater than the

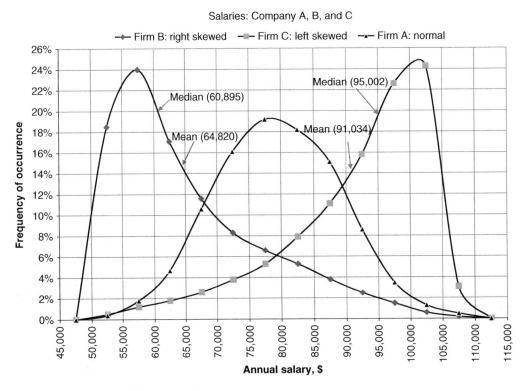

Figure 5.12 Symmetry of continuous data.

Table 5.2

Firm	A	B	C
Minimum, $	50,180	50,180	50,180
Maximum, $	108,760	108,760	108,760
Mean, $	79,466.44	64,820.26	91,034.03
Median, $	79,532.50	60,895.00	95,002.50
Mode, $	75,840	57,140	100,625
Number of mode values	4	8	5
Midrange, $	79,470	79,470	79,470
Range, $	58,580	58,580	58,580
Standard deviation, $	9,731.78	12,510.65	10,826.17
Range/standard deviation	6.02	4.68	5.41
σ/μ	12.25%	19.30%	11.90%
Distribution shape	Normal	Right-skewed	Left-skewed
No. values greater than mean	501	418	575
No. values less than mean	499	482	425
No. values less than median	500	500	500

median ($60,895), and the median lies between the mean and the mode ($57,140). As Figure 5.12 illustrates, the distribution is right-skewed as the curve tails off to the right; there are more salaries lower than the mean. For Firm C the mean ($91,034) is lower than the median ($95,003), and again the median lies between the mean and the mode ($100,625). As Figure 5.12 illustrates, the distribution is left-skewed; there are more salaries greater than the mean salary.

What these distributions in Figure 5.12 illustrate is that for Firm A, exhibiting a normal distribution, the level of salaries is pretty evenly distributed among the 1,000 employees. For Firm B, which has a right-skewed distribution, there are more salaries lower than the mean salary. For Firm C, which has a left-skewed distribution, the situation is the opposite; there are more salaries greater than the mean salary. Thus if you are an employee in one of these firms, positioning your salary with the rest of the employees depends on the shape of the distribution, even though the minimum and maximum salaries are the same. Another illustration of this asymmetrical concept can be applied to the hospitality industry. If, in a restaurant attached to a large hotel, the distribution of the value of wine in the cellar is left-skewed, that is the mean is less than the median, then it implies that there are more expensive bottles of wine. However, there are fewer values to the right of the hump, or mode. Alternatively, say that a hotel with several brands (Accor for example) has a distribution of all its hotel room prices that is right-skewed, that is the mean is greater than the median; this implies that there are more low-priced hotel rooms in its total portfolio of hotels. The mean is less than the median and there are more values to the left of the hump, or mode.

Box and whisker plots

In *Chapter 2* the development of a box and whisker plot was presented as a visual aid to illustrate the shape of data. As an illustration the salary data of companies A, B, and C from the previous section are presented as box and whisker plots in Figure 5.13, together with their corresponding mean and median values. Just by observation the normal distribution box and whisker plot is very symmetrical. The right-skewed data has a long whisker to the right and the right box is longer than the left box, demonstrating that the range of the 50% data above the median is high; conversely the range of the 50% data below the median is low. In contrast the left-skewed data has a longer whisker to the left and the left box is longer than the right box, demonstrating that the range of the 50% data below the

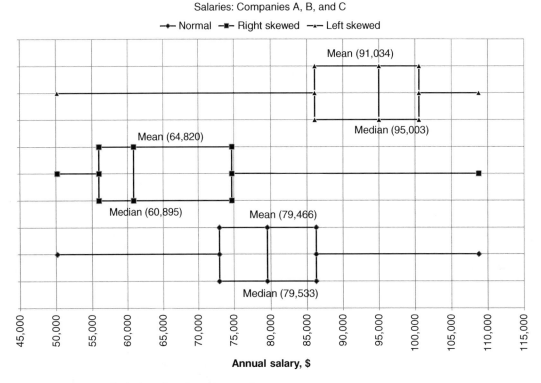

Figure 5.13 Box and whisker plot of continuous data.

median is high; conversely the range of the 50% data above the median is low. These observations are quantified by the statistics in Table 5.3. For the normal distribution, Firm A, the length of the whiskers, $(Q_1 - Q_0)$ and $(Q_4 - Q_3)$ are very close in value and similarly the length of the two boxes

Table 5.3

Firm	A	B	C
Q_0	50,180.00	50,180.00	50,180.00
Q_1	72,938.75	55,957.50	86,085.00
Q_2	79,532.50	60,895.00	95,002.50
Q_3	86,268.75	74,726.25	100,555.00
Q_4	108,760.00	108,760.00	108,760.00
$Q_2 - Q_0$	29,352.50	10,715.00	44,822.50
$Q_4 - Q_2$	29,227.50	47,865.00	13,757.50
$Q_1 - Q_0$	22,758.75	5,777.50	35,905.00
$Q_4 - Q_3$	22,491.25	34,033.75	8,205.00
$Q_2 - Q_1$	6,593.75	4,937.50	8,917.50
$Q_3 - Q_2$	6,736.25	13,831.25	5,552.50
Inter-quartile range, $Q_3 - Q_1$	13,330.00	18,768.75	14,470.00
Quartile deviation, $(Q_3 - Q_1)/2$	6,665.00	9,384.38	7,235.00
Mid-hinge $(Q_3 + Q_1)/2$	79,603.75	65,341.88	93,320.00
Inter-quartile range/Standard deviation	1.37	1.50	1.34
Distribution shape	Normal	Right-skewed	Left-skewed

$(Q_2 - Q_1)$ and $(Q_3 - Q_2)$ are also very close in value. Further, the ratio of the inter-quartile range to the standard deviation is 1.37 (13,330/9,731.78), close to the 1.33 characteristic of the normal distribution. For the right-skewed distribution, Firm B, the length of the right whisker $(Q_4 - Q_3)$ is greater than the left whisker $(Q_1 - Q_0)$ and the length of the right box $(Q_3 - Q_2)$ is greater than the length of the left box $(Q_2 - Q_1)$. It is the opposite for the left-skewed distribution, Firm C: the length of the left whisker $(Q_1 - Q_0)$ is greater than the right whisker $(Q_4 - Q_3)$ and the length of the left box $(Q_2 - Q_1)$ is greater than the length of the right box $(Q_3 - Q_2)$.

Verifying that a dataset approximates a normal distribution

With continuous data it is necessary to establish that the data approximates the profile of a normal distribution such that rules applying to normality can be used with confidence. If this is not the case the data is skewed and if normal probability standards are applied results will be erroneous. The following gives some approaches for verification.

Visual displays

An obvious way to establish if a dataset approximates a normal distribution is to simply plot the data and examine its profile for symmetry. The type of displays to construct, all discussed in *Chapter 2*, might be:

- A stem-and-leaf display
- A frequency polygon or histogram
- A box and whisker plot.

For example, consider the data in Table 5.4 that are revenues in $US for the last 200 days for a hypermarket.

Table 5.4

35,378	170,569	104,985	134,859	120,958	107,865	127,895	106,825	130,564	108,654
109,785	184,957	96,598	121,985	63,258	164,295	97,568	165,298	113,985	124,965
108,695	91,864	120,598	47,865	162,985	83,964	103,985	61,298	104,987	184,562
89,597	160,259	55,492	152,698	92,875	56,879	151,895	88,479	165,698	89,486
85,479	64,578	103,985	81,980	137,859	126,987	102,987	116,985	45,189	131,958
73,598	161,895	132,689	120,654	67,895	87,653	58,975	103,958	124,598	168,592
95,896	52,754	114,985	62,598	145,985	99,654	76,589	113,590	80,459	111,489
109,856	101,894	80,157	78,598	86,785	97,562	136,984	89,856	96,215	163,985
83,695	75,894	98,759	133,958	74,895	37,856	90,689	64,189	107,865	123,958
105,987	93,832	58,975	102,986	102,987	144,985	101,498	101,298	103,958	71,589
59,326	121,459	82,198	60,128	86,597	91,786	56,897	112,854	54,128	152,654
99,999	78,562	110,489	86,957	99,486	132,569	134,987	76,589	135,698	118,654
90,598	156,982	87,694	117,895	85,632	104,598	77,654	105,987	78,456	149,562
68,976	50,128	106,598	63,598	123,564	47,895	100,295	60,128	141,298	84,598
100,296	77,498	77,856	134,890	79,432	100,659	95,489	122,958	111,897	129,564
71,458	88,796	110,259	72,598	140,598	125,489	69,584	89,651	70,598	93,876
112,987	123,895	65,847	128,695	66,897	82,459	133,984	98,459	153,298	87,265
72,312	81,456	124,856	101,487	73,569	138,695	74,583	136,958	115,897	142,985
119,654	96,592	66,598	81,490	139,584	82,456	150,298	106,859	68,945	122,654
70,489	94,587	85,975	138,597	97,498	143,985	92,489	146,289	84,592	69,874

The stem-and-leaf display for the data of Table 5.4 is given in Figure 5.14. The stems, or principal digits, are the ten thousands, $30, $40, $50 thousand, etc. The leaves, or trailing digits, are the $thousands, for example for the stems of $40 thousand the leaves are: 5,189; 7,865; 7,895.

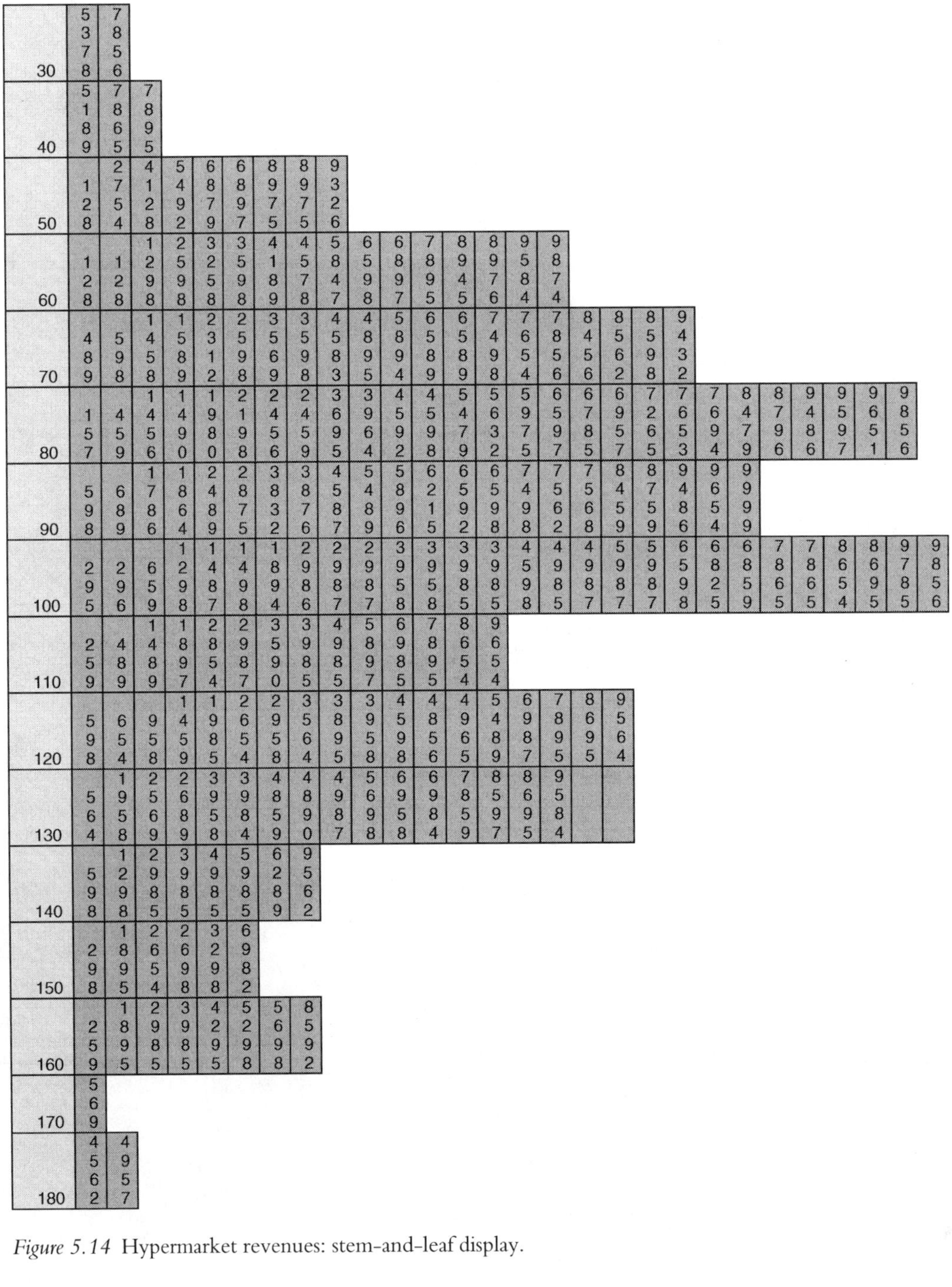

```
      5 | 7
      3 | 8
      7 | 5
 30   8 | 6

      5 | 7 7
      1 | 8 8
      8 | 6 9
 40   9 | 5 5

        | 2 4 5 6 6 8 8 9
      1 | 7 1 4 8 8 9 9 3
      2 | 5 2 9 7 9 7 7 2
 50   8 | 4 8 2 9 7 5 5 6

        | 1 2 3 3 4 4 5 6 6 7 8 8 9 9
      1 | 1 2 5 2 5 1 5 8 5 8 8 9 9 5 8
      2 | 2 9 9 5 9 8 7 4 9 9 9 4 7 8 7
 60   8 | 8 8 8 8 8 9 8 7 8 7 5 5 6 4 4

        | 1 1 2 2 3 3 4 4 5 6 6 7 7 7 8 8 8 9
      4 | 5 4 5 3 5 5 5 5 8 5 5 4 6 8 4 5 5 4
      8 | 9 5 8 1 9 6 9 8 9 8 9 5 5 5 6 9 3
 70   9 | 8 8 9 2 8 9 8 3 5 4 9 9 8 4 6 6 2 8 2

        | 1 1 1 2 2 2 3 3 4 4 5 5 5 6 6 6 7 7 7 8 8 9 9 9 9
      1 | 4 4 4 9 1 4 4 6 9 5 5 4 6 9 5 7 9 2 6 6 4 7 4 5 6 8
      5 | 5 5 9 8 9 5 5 9 6 9 9 7 3 7 9 8 5 6 5 9 7 9 8 9 5 5
 80   7 | 9 6 0 0 8 6 9 5 4 2 8 9 2 5 7 5 7 5 3 4 9 6 6 7 1 6

        | 1 1 2 2 3 3 4 5 5 6 6 6 7 7 7 8 8 9 9 9
      5 | 6 7 8 4 8 8 5 4 8 2 5 5 4 5 4 7 4 6 9
      9 | 8 8 6 8 7 3 7 8 8 9 1 9 9 9 6 6 5 5 8 5 9
 90   8 | 9 6 4 9 5 2 6 7 9 6 5 2 8 8 2 8 9 9 6 4 9

        | 1 1 1 1 2 2 2 3 3 3 3 4 4 4 5 5 6 6 6 7 7 8 8 9 9
      2 | 2 6 2 4 8 9 9 9 9 9 9 9 9 9 9 5 8 8 8 6 7 7 8 8 9
      9 | 9 5 9 8 9 8 8 5 5 8 8 9 8 8 8 8 9 2 5 6 6 5 9 8 5
100   5 | 6 9 8 7 8 4 6 7 7 8 8 5 5 8 5 7 7 7 8 5 9 5 5 4 5 6

        | 1 1 2 2 3 3 4 5 6 7 8 9
      2 | 4 8 8 9 5 9 9 8 9 8 6 6
      5 | 8 8 9 5 8 9 8 8 9 8 9 5 5
110   9 | 9 9 7 4 7 0 5 5 7 5 5 4 4

        | 1 1 2 2 3 3 3 4 4 4 5 6 7 8 9
      5 | 6 9 4 9 6 9 5 8 9 4 9 6 7 8 9
      9 | 5 5 5 8 5 5 6 9 5 9 5 6 8 8 9 9 6
120   8 | 4 8 9 5 4 8 4 5 8 8 6 5 9 7 5 5 4

        | 1 2 2 3 3 4 4 4 5 6 6 7 8 8 9
      5 | 9 5 6 9 9 8 8 8 9 6 9 9 8 5 6 5
      6 | 5 6 8 5 8 5 9 8 9 5 8 5 9 9 8
130   4 | 8 9 9 8 4 9 0 7 8 8 4 9 7 5 4

        | 1 2 3 4 5 6 9
      5 | 2 9 9 9 9 2 5
      9 | 9 8 8 8 8 8 6
140   8 | 8 5 5 5 5 9 2

        | 1 2 2 3 6
      2 | 8 6 6 2 9
      9 | 9 5 9 9 8
150   8 | 5 4 8 8 2

        | 1 2 3 4 5 5 8
      2 | 8 9 9 2 2 6 5
      5 | 9 8 8 9 9 9 9
160   9 | 5 5 5 5 8 8 2

      5
      6
170   9

      4 4
      5 9
      6 5
180   2 7
```

Figure 5.14 Hypermarket revenues: stem-and-leaf display.

With a dataset of 200 pieces of data, the development of a stem-and-leaf display is somewhat cumbersome. This type of visual display is more appropriate for smaller datasets. The Figure 5.15 gives the polygon for the same data of Table 5.4 and Figure 5.16 gives the corresponding box and whisker plot.

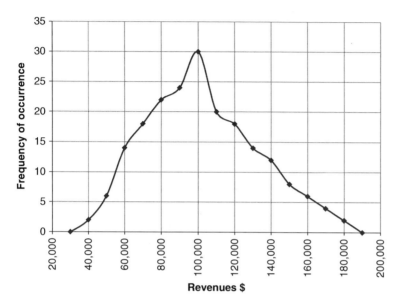

Figure 5.15 Hypermarket revenues: polygon.

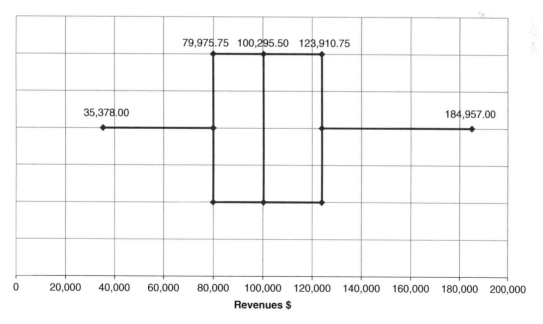

Figure 5.16 Hypermarket revenues: box and whisker plot.

In summary these three visuals indicate that there is reasonable symmetry around a center value though it does trail off to the right.

Characteristics of the dataset

Another verification of the normal assumption is to determine the characteristics of the dataset to see if they correspond to normal distribution criteria. If they do then the following relationships should be close:

- The mean is equal to the median value.
- The inter-quartile range is equal to 1.33 times the standard deviation.
- The range of the data is equal to six times the standard deviation.
- There is 68.27% of the area of the distribution between ± 1 standard deviations of the mean.
- There is 95.45% of the area of the distribution between ± 2 standard deviations of the mean.
- There is 99.73% of the area of the distribution between ± 3 standard deviations of the mean.

To test the characteristics of the revenue data of Table 5.4 are given in Table 5.5.

Table 5.5

Property	Value
Amount of data	200
Mean	102,666.67
Median	100,295.50
Maximum	184,957.00
Minimum	35,378.00
Range	149,579.00
σ (population)	30,888.20
Q_3	123,910.75
Q_1	79,975.75
$Q_3 - Q_1$ (inter-quartile range)	43,935.00
6σ	185,329.17
1.33σ	41,081.30
Area between ± 1σ	64.50%
Area between ± 2σ	96.00%
Area between ± 3σ	100.00%

For the last three rows, the percentage areas between given sigma limits in the distribution are calculated first, using Equation 5(iii) to find the x limits for given values of z using the data mean and standard deviation:

$$x = \mu_x \pm z\sigma_x$$

- Mean value $\mu = 102,667$
- Standard deviation is 30,888
- Values of x when $z = \pm 1$ are $102,667 \pm 1*30,888$ or 71,778 and 133;555
- Values of x when $z = \pm 2$ are $102,667 \pm 2*30,888$ or 40,890 and 164,443
- Values of x when $z = \pm 3$ are $102,667 \pm 3*30,888$ or 10,002 and 195,331

The dataset is then sorted to determine the amount of data, and subsequently the percentage quantity, lying between these values.

- Between values x of 71,778 and 133;555 there are 129 values, or 64.50% of the total data (129/200).
- Between values x of 40,890 and 164,443 there are 192 values, or 96.00% of the total data (192/200).
- Between values x of 10,002 and 195,331 there are 200 values, or 100.00% of the total data (200/200).

Thus in comparing the revenue dataset with the normal distribution criteria:

- The mean value is greater than the median by an amount of 2.36%.
- The standard deviation is 30,888.20 and the inter-quartile range is 43,935.75, or the ratio of the standard deviation to the inter-quartile range is 1.422. This is 6.95% more than the standard of 1.33.
- The range of the data is 149,579. The ratio of the range to the standard deviation is 4.84 or 19.29% less than the standard of 6.00.
- The area of the distribution between ± 1 standard deviations of the mean is 64.50% or 5.52% less than the standard of 68.27%.
- The area of the distribution between ± 2 standard deviations of the mean is 96.00% or 0.58% more than the standard of 95.45%.
- The area of the distribution between ± 3 standard deviations of the mean is 100.00% or 0.27% more than the standard of 99.73%.

From the properties, or characteristics, of the sales data, the normal assumption for the hypermarket revenues data seems reasonable.

Ogive comparisons

We can demonstrate the normal distribution relationship further by using the ogives of the revenue that are presented in Figure 5.17. (Ogives are discussed in *Chapter 2*.)

- From the greater than ogive, 80.00% of the sales revenues are at least $75,000.

Assuming a normal distribution, then at least 80% of the sales revenue lies in the area of the curve, as illustrated in Figure 5.18. The value of z at the point x from the normal Excel functions **NORMSINV** (Excel 2007) or **NORM.S.INV** (Excel 2010 and later) is –0.8416. Using this, the mean, and standard deviation values for the sales data and using Equation 5(iii) gives:

$$x = 102,667 + (-0.8416) * 30,888 = \$76,671$$

The value of $75,000 from the ogive is only 2.18% less than this value of $76,671 determined from the normal distribution relationship.

- Alternatively, from the less than ogive, 90.00% of the revenues are no more than $145,000.

Similarly, assuming a normal distribution then 90% of the sales revenue appears in the area of the curve, as illustrated in Figure 5.19. The value of z at the point x with the Excel normal distribution

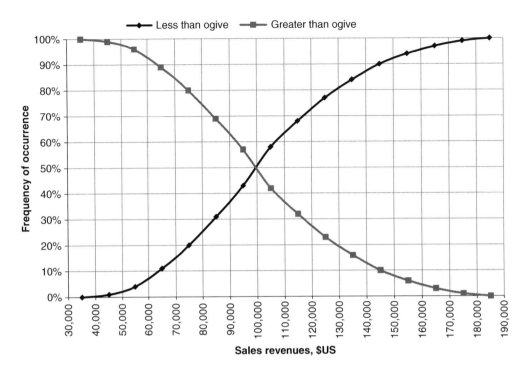

Figure 5.17 Hypermarket revenues: ogives.

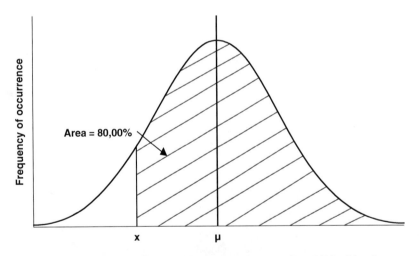

Figure 5.18 Area of the normal distribution containing at least 80% of the data.

function is +1.2816. Using this and the mean, and standard deviation values for the sales data using Equation 5(iii) we have:

$$x = 102,667 + 1.2816 * 30,888 = \$142,251$$

The value of $145,000 from the ogive is only 1.93% more than this value of $142,251 determined from the normal distribution relationship.

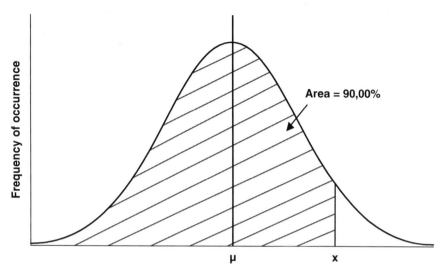

Figure 5.19 Area of the normal distribution giving upper limit of 90% of the data.

Thus comparing the ogives of the data with the normal distribution calculations there is a pretty good association.

Constructing a normal probability plot

Another way to establish symmetry of data is to construct a normal probability plot. The procedure is as follows, using the 1,000 pieces of salary data in the previous section, *"Asymmetrical or non-normal distributions"* for the three firms A, B, and C that have salary data that is normal, right-skewed, and left-skewed:

- Organize the data into an ordered data array from lowest to highest.
- For each data point determine the area under the curve given by each boundary point, on the assumption that the data follows a normal distribution. For example, here there are 1,000 salary values or 1,000 data points. The median value, using the relationship of *Chapter 1*, when there is even set of data, is $(n + 1)/2$ or here $(1,000 + 1))/2 = 500.50$. Thus on the standard normal distribution at 500.50 the z value is 0.00 and the area to the left of curve is 50.00% of the total data. That means at data point 500.50 the area of the curve to the left is given by: $500.50/(1,000 + 1) = 50.00\%$.
- By the same logic, for data point 1 the area to the left is $1/(1,000 + 1) = 0.10\%$; for data point 2 the area to the left is $2/(1,000 + 1) = 0.20\%$; for data point 3 the area to the left is $3/(1,000 + 1) = 0.30\%$; etc.
- Determine the number of standard deviations, z, corresponding to these areas using the Excel Function **NORMSDIST** (Excel 2007) or **NORM.S.DIST** (Excel 2010 and later). For data points 1, 2, and 3 the values of z are respectively -3.0905; -2.8785; and -2.7481 as shown in Table 5.6. The area, the z values for the beginning, middle, and ending salary amounts are also given in Table 5.6.
- The next step is to plot the data values, in this case the salaries on the y-axis against the z values on the x-axis. This is given in Figure 5.20.

Table 5.6

Data point (salary amount)	Area below	No. standard deviations z	Salary: symmetrical data	Salary: Right-skewed data	Salary: Left-skewed data
1	0.10%	−3.0905	50,180	50,180	50,180
2	0.20%	−2.8785	50,940	50,180	50,370
3	0.30%	−2.7481	53,915	50,180	50,395
4	0.40%	−2.6524	54,885	50,180	50,420
5	0.50%	−2.5762	55,110	50,180	50,445
6	0.60%	−2.5125	55,230	50,180	55,495
7	0.70%	−2.4576	55,800	50,180	57,500
8	0.80%	−2.4093	56,270	50,185	57,500
9	0.90%	−2.3660	57,075	50,190	57,750
10	1.00%	−2.3267	57,140	50,195	58,000
. . . . . .	. . . . . .	. . . . . .	. . . . . .	. . . . . .	. . . . . .
496	49.55%	−0.0113	79,385	60,850	94,650
497	49.65%	−0.0088	79,405	60,860	94,745
498	49.75%	−0.0063	79,415	60,870	94,950
499	49.85%	−0.0038	79,435	60,880	95,000
500	49.95%	−0.0013	79,530	60,890	95,000
501	50.05%	0.0013	79,535	60,900	95,005
502	50.15%	0.0038	79,555	60,910	95,005
503	50.25%	0.0063	79,560	60,920	95,020
504	50.35%	0.0088	79,590	60,930	95,035
505	50.45%	0.0113	79,600	60,940	95,050
506	50.55%	0.0138	79,600	60,950	95,060
. . . . . .	. . . . . .	. . . . . .	. . . . . .	. . . . . .	. . . . . .
991	99.00%	2.3267	102,960	99,230	107,050
992	99.10%	2.3660	103,490	99,340	107,100
993	99.20%	2.4093	103,705	100,400	107,150
994	99.30%	2.4576	103,725	102,960	107,200
995	99.40%	2.5125	104,440	103,490	107,250
996	99.50%	2.5762	105,180	103,705	107,300
997	99.60%	2.6524	106,480	103,725	107,350
998	99.70%	2.7481	106,965	104,440	107,400
999	99.80%	2.8785	108,145	106,480	108,145
1000	99.90%	3.0905	108,760	108,760	108,760

- Observe the profile of the graph. If the graph is essentially a straight line with a positive slope then the data follows a normal distribution. If the graph is non-linear of a concave format then the data is right-skewed. If the graph has a convex format then the data is left-skewed.
- This is exactly the case for this salary data. Where the distribution of the salaries is symmetrical the plot is essentially a straight line with a positive slope; the right-skewed data is non-linear with a concave format; the left-skewed data is non-linear with a convex format.

Approximation for the binomial distribution

Chapter 4 covered the binomial distribution. Under certain conditions the binomial distribution can be approximated by the continuous normal distribution that means for sampling experiments of discrete data we can employ the normal distribution. This is useful in statistical process control, SPC, covered in *Chapter 12*.

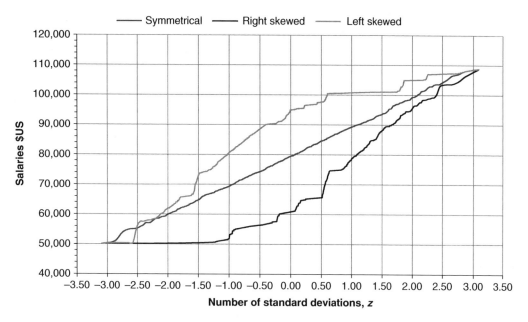

Figure 5.20 Salaries: normal probability plot.

Conditions for a normal-binomial approximation

The conditions for using the normal distribution to approximate the binomial distribution are that the product of the sample size, *n*, and the probability of *success, p,* is greater, or equal to five and at the same time the product of the sample size and the probability of *failure* is also greater or equal than five. That is:

$$n * p \geq 5 \qquad\qquad 5\text{(iv)}$$

$$n * (1 - p) \geq 5 \qquad\qquad 5\text{(v)}$$

From *Chapter 4*, Equation 4(viii), the mean or expected value of the binomial distribution is:

$$\mu_x = E(x) = n * p$$

And from Equation 4(x) the standard deviation of the binomial distribution is:

$$\sigma = \sqrt{\sigma^2} = \sqrt{n * p * (1 - p)} = \sqrt{(n * p * q)} \qquad\qquad 4$$

Substituting the mean and standard deviation in the transposition relationship gives the following **normal-binomial approximation,**

$$z = \frac{x - \mu_x}{\sigma_x} = \frac{x - n.p}{\sqrt{n.p.q}} = \frac{x - n.p}{\sqrt{n.p.(1 - p)}} \qquad\qquad 5\text{(vi)}$$

The following application case-exercise, *Ceramic plates,* illustrates this application.

Application case-exercise: **Ceramic plates**

Objective

The purpose of this case-exercise is to demonstrate how the binomial distribution can be approximated by the normal distribution and also the application of the continuity correction factor.

Situation

A firm has a continuous production operation to mold, glaze, and fire ceramic plates. It knows from historical data that in the operation 3% of the plates are defective and have to be sold at a marked-down price. This is a binomial situation: either the plates are defective, or they are not. The quality control manager takes a random sample of 500 of these plates and makes an inspection.

Required

1   Can the normal distribution be used to approximate the binomial distribution?
2   Using the binomial distribution, what is the probability that exactly 20 of the plates are defective?
3   Using the normal-binomial approximation, what is the probability that exactly 20 of the plates are defective?
4   Using a continuity correction factor, what is the probability that exactly 20 of the plates are defective?

Worked out solutions

1   Sample size *n* is 500; probability *p* is 3.00%.

   - $n * p = 500 * 0.03 = 15$ or a value > 5. The average number of plates defective is 15.
   - $n * (1 - p) = 500*(1 - 0.03) = 500*0.97 = 485$; again > 5. The average number of "good" or not defective plates is 485.

   Thus both conditions are satisfied and so we can correctly use the normal distribution as an approximation of the binomial distribution.
2   Using, in Excel, **BINOMDIST** (Excel 2007) or **BINOM.DIST** (Excel 2010 and later) Table 5.7 (an extract of the binomial distribution) gives the probability of exactly 20 plates being defective as 4.16%.

Table 5.7

x	Exactly P(x)	Cum P(x)
20	4.16%	92.02%

3   Mean value for data that follows a binomial distribution is:

$$\mu_x = n * p = 500 * 0.003 = 15$$

Standard deviation for data that follows a binomial distribution is:

$$\sigma = \sqrt{n * p * q} = \sqrt{(500 * 0.003 * 0.997)} = 3.8144$$

Using, in Excel, **NORMDIST** (Excel 2007) or **NORM.DIST** (Excel 2010 and later)

   - Variable value *x* is 20
   - Mean value is 15

- Standard deviation is 3.8144
- Cumulative value is 0 (we want the exact value)

This gives the probability of exactly 20 plates being defective as 4.43%; This is a value 6.49% higher than the 4.16% obtained in Question 2. Pragmatically, if we round the numbers, it is 4% in both cases.

4   The normal distribution is continuous and shown by a line graph whereas the binomial distribution is discrete and illustrated by a histogram. Another way to make the normal-binomial approximation is to apply a **continuity correction factor** so that we encompass the range of the discrete values, recognizing that we are superimposing a histogram onto a continuous curve. In this example, if we apply a correction factor of 0.5 to the defective value 20, the random variable x, then on the lower side is $x_1 = 19.5$ ($20 - 0.5$) and $x_2 = 20.5$ ($20 + 0.5$) on the upper side. The concept is illustrated in Figure 5.21.

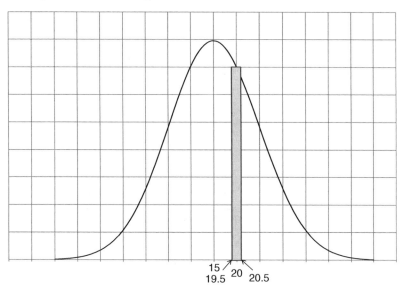

Figure 5.21 Continuity correction factor.

Using Equation 5(vi) for these two values of x gives:

$$z_1 = \frac{x_1 - n.p}{\sqrt{n.p.(1-p)}} = \frac{19.5 - 500 * 0.03}{\sqrt{500 * 0.03(1 - 0.03)}} = \frac{19.5 - 15}{\sqrt{14.55}} = \frac{4.5}{3.8144} = 1.1797$$

$$z_2 = \frac{x_2 - n.p}{\sqrt{n.p.(1-p)}} = \frac{20.5 - 500 * 0.03}{\sqrt{500 * 0.03(1 - 0.03)}} = \frac{20.5 - 15}{\sqrt{14.55}} = \frac{5.5}{3.8144} = 1.4419$$

Using, in Excel, **NORMSDIST** (Excel 2007) or **NORM.S.DIST** (Excel 2010 and later):

- When $z = 1.1797$, the area under the curve from the left to $x = 19.5$ is 88.09%
- When $z = 1.4419$, the area under the curve from the left to $x = 20.5$ is 92.53%

The difference between these two areas is 4.44% (92.53 − 88.09). This value is close to the answers of Question 2 and Question 3.

In summary, Table 5.8 gives the probability that out of the sample of 500 that 20 of the plates are defective according to the three methods.

Table 5.8

Method	Probability 20 plates are defective
Binomial distribution	4.16%
Normal-binomial approximation	4.43%
Continuity correction factor	4.44%

Rounded, the values are 4%; that is pretty much the same.

Sample size needed to use the normal-binomial approximation

Whether the conditions that Equations 5(iv) and 5(v) are met depends on the values of n and p. When p is large then for a given value of n the product $n*p$ is large; conversely $n(1-p)$ is small. If for example p is 99%, then the minimum sample size in order to apply the normal distribution assumption is 500:

- $p = 99\%$ and thus, $n * p = 500 * 99\% = 495$
- $(1-p) = 1\%$ and thus, $n(1-p) = 500 * 1\% = 5$

The minimum sample size such that the normal-binomial approximation can be applied is 10 and in this case the probability must be 0.50 or 50%:

$$n * p = 10 * 0.50 = 5.00; \qquad n*(1-p) = 5 * (1 - 0.5) = 5.00$$

The curve in Figure 5.22 gives the minimum sample size, n, to validate the normal-binomial approximation for values of p from 5% to 95% in order to satisfy both Equations 5(iv) and 5(v).

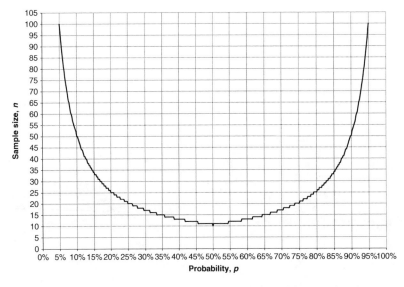

Figure 5.22 Sample size and probability to validate normal-binomial approximation.

The curve is not smooth since *n*, the minimum sample size, can only be a whole number. For example:

- When probability is 11%, sample size must be 46: $n*p = 5.06$ (46*0.11) and $n(1 - p) = 40.94$ (46*0.89).
- When probability is 35%, sample size must be 15: $n*p = 5.25$ (15*0.35) and $n(1 - p) = 9.75$ (15*0.65).
- When probability is 55%, sample size must be 12: $n*p = 6.60$ (12*0.55) and $n(1 - p) = 5.40$ (12*0.45).
- When probability is 70%, sample size must be 17: $n*p = 11.90$ (17*0.70) and $n(1 - p) = 5.10$ (17*0.30).

Exponential distributions

An exponential distribution is another continuous relationship that demonstrates rapid growth or decline and that can be represented by a mathematical relationship. Compared to the normal distribution, the exponential distribution is one-directional, either increasing or decreasing, whereas the normal distribution increases up to a maximum frequency and then declines.

Exponential progression

The spread of untreated diseases such as AIDS, cholera, smallpox, tuberculosis, Ebola, and measles display a rapid exponential-type growth. Bacterial spread on a wound, a gardener's compost pile, or in a person's stomach in the digestion process is exponential. The growth of a forest, seaweed, or rabbit population is an exponential-type, as is the population growth in the Indian subcontinent. In theory these growths can extend to infinity if unchecked. However, growth is slowed or stopped by some intervention: vaccines in the case of disease; myxomatosis decimated the rabbit population; and lack of sunlight prevents forest growth. An exponential change can also be declining. In the nuclear industry the reduction of the activity of elements such as uranium, radium, or radon is exponential. Here the term *half life period* is used, which is the time required for the activity to be diminished to one-half of its initial value. The half life of uranium I is 4.56×10^9 years, and that explains the problem of its disposal from nuclear reactors. In business there is the *learning curve* that is an exponential curve that represents the reduction in unit time to complete an activity when it is repeated many times. This curve is an indicator of productivity and a selected value at the extreme end of the curve is used as the labor standard for that task. The flow rate of beer from the spout at the bottom of the keg is exponential. The flow rate depends on the pressure at the entrance of the spout. This pressure is a function of the height of beer in the barrel. When the barrel is full and the spout is first opened the flow is very rapid. As the volume and thus the height of liquid in the barrel decreases, the flow rate declines. Exponential changes are different from linear progression, whose movement is at a constant rate.

Bacteria growth

Bacteria are a group of minute organisms that spread rapidly by simple division. The *Bacillus subtilis* can divide into two every 20 minutes, which means that, if left, unchecked, in eight hours there will be over 16 million bacteria. This progression for the first three hours is given in Table 5.9 and shown graphically in Figure 5.23.

Table 5.9

Time, minute	0	20	40	60	80	100	120	140	160	180
Time, hours	0.00	0.33	0.67	1.00	1.33	1.67	2.00	2.33	2.67	3.00
No. of bacteria	1	2	4	8	16	32	64	128	256	512

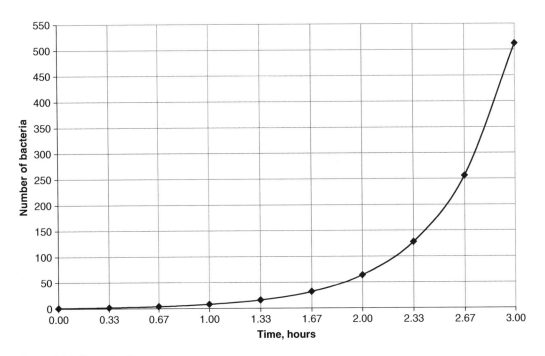

Figure 5.23 Growth of bacteria.

This sequence is an exponential distribution, sometimes referred to as the *function of natural growth*, with a mathematical relationship where the dependent variable, y, changes according to the power, or **exponent**, of a given independent value, in this case the time t. The progression can be described by the relationship:

$$y_t = y_0(1 + r)^t \qquad\qquad 5(\text{vii})$$

where:

- y_t is the number of bacteria at a time t
- y_0 the number of bacteria at time zero
- t is an integer value, 0, 1, 2, 3, etc
- r is a factor representing growth rate

In order to calculate the growth rate:

- At time zero there is one bacterium
- After time of 1 hour (60 min) there are eight bacteria thus:

$$8 = 1(1 + r)^1$$
$$8 = 1 + r$$
$$r = (8 - 1) = 7$$

Thus the equation describing the growth from 5(vii) is:

$$y_t = 8^t \qquad\qquad\qquad 5(\text{viii})$$

- After 1 hour, the number of bacteria is $y^1 = 8^1 = 8$.
- After 2 hours, the number of bacteria is $y^2 = 8^2 = 64$.
- After 3 hours, the number of bacteria is $y^3 = 8^3 = 512$.
- After 4 hours, the number of bacteria is $y^4 = 8^4 = 4{,}096$.
- After 8 hours, the number of bacteria is $y^8 = 8^8 = 16{,}777{,}216$.

The general formula of the exponential equation when quantity y is a function of time is:

$$y = a * b^{t/r} \qquad\qquad\qquad 5(\text{vix})$$

- y is the value required at time t
- a is the value of y at time zero
- b is the growth factor
- r is the time constant or the time for y to increase by one factor of b

For example, for bacteria:

- a is 1 (there is one bacterium at the beginning)
- b is 2 (the bacteria double every 20 minutes)
- r is 20 minutes (the time to grow from 1 to 2 bacteria; 2 to 4; 4 to 8 etc.

After 160 minutes the number of bacteria y is:

$$y = 1 * 2^{160/20} = 2^8 \text{ or } 256 \text{ as shown in Table 5.9}$$

After 180 minutes the number of bacteria y is:

$$y = 1 * 2^{180/20} = 2^9 \text{ or } 512 \text{ as shown in Table 5.9 and the same from the relationship } y^3 = 8^3 = 512$$

Note that Equations 5(viii) and 5(vix) are different in form. In 5(viii) the time t is an integer value, 1, 2, 3, etc. and in this case hours; in 5(vix) t is the same units as r, in this case minutes and t must be an exact multiple of r as otherwise it does not make sense. You cannot have fractions of a bacterium!

Financial investment and exponential growth

The investment of a sum of money in say a certificate of deposit at a constant interest rate compounds and grows exponentially according to Equation 5(vii). For example, assume that you have a sum of $2,000 to invest at a constant interest rate of 6.50% annual yield. The monetary value at the end of each year for the first eight years is according to Table 5.10.

Table 5.10

End of year	1	2	3	4	5	6	7	8
Amount, $	2,130.00	2,268.45	2,415.90	2,572.93	2,740.17	2,918.28	3,107.97	3,309.99

At the end of 12 years the investment would be:

$$y_t = y_0(1 + r)^t$$
$$y_{12} = 2000.00 * (1 + 0.065)^{12} \hspace{3cm} 5(x)$$
$$= \$4,258.19 \text{ or over twice the original investment.}$$

Alternatively this can be written:

$$y = a * b^t \hspace{3cm} 5(xi)$$

- y is the value required at any time t
- a is the value of y or y_0 at time zero. Here equal to $2,000
- b is the growth rate or $(1 + r)$. Here equal to 1.065

The value of the investment over 20 years is illustrated in Figure 5.24. This is the same logic as presented in *Chapter 1* covering the *Geometric Mean*.

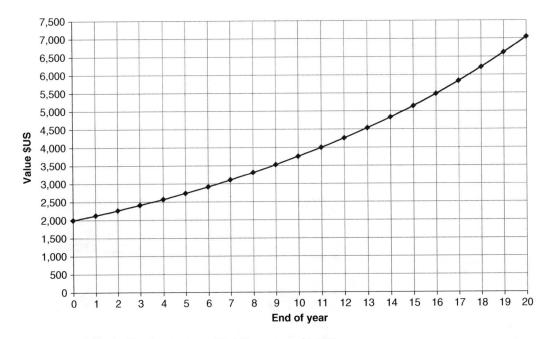

Figure 5.24 Growth of an investment of $2,000 at a rate of 6.50%.

Magazine circulation and exponential decline

The exponential decline progression is similar to Equation 5(x) except that r is now negative, or:

$$y_t = y_0[1 + (-r)]^t \qquad\qquad\qquad 5(\text{xii})$$

which is the same as:

$$y_t = y_0(1 - r)^t \qquad\qquad\qquad 5(\text{xiii})$$

- y_t is the value required at any time t
- y_0 is the value of y at time zero
- r is the decline factor

Alternatively this can be written as Equation 5(xi):

$$y = a * b^t$$

- y is the value required at any time t
- a is the value of y or y_0 at time zero
- b is the "growth rate" or $(1 - r)$. It can take a value greater than zero but less than 1 $(0 < b < 1)$ and r is the decline factor

For example, assume that a magazine publisher notices that on average readership declines 15% per year as more and more people use the Internet. (*Newsweek* ceased publication in December 2012 and went digital for this very reason: fewer and fewer people are reading newsprint). In one particular year the readership was 260,000. Thus:

- y_0 is the value of y at time zero and equal to a and the value is 260,000
- r is the decline factor, or 15%
- b is the "growth rate" or $(1 - r)$, or 0.85
- t is the time in years.

The progression for this decline is shown in Figure 5.25 for the first fifteen years. Table 5.11 gives the numerical information for the first five years.

Note that the value of y, the magazine circulation, never reaches zero but falls to a low value. This decline would translate into lower profits and eventually a financial loss, causing the owners to cease publication of the print version.

Learning curve

An exponential progression may not be a function of time but could depend upon another variable. For example, in a manufacturing or service operation there is the concept of a learning curve that is the speed (rate) at which an operator learns to perform a certain task. The curve indicates the unit time it takes to double output from 1 to 2 to 4 to 8, etc., rather like the bacteria progression. Here the time per unit to perform the task is a function of the number of units produced and the relationship describing this is:

$$T_n = T_1 * \left(n^b\right) \qquad\qquad\qquad 5(\text{xiv})$$

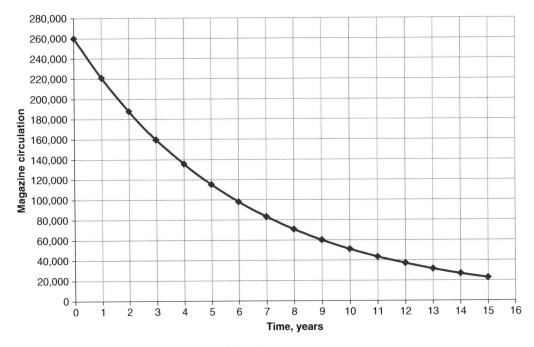

Figure 5.25 Magazine circulation: exponential decline.

Table 5.11

Year	0	1	2	3	4	5
Circulation	260,000	221,000	187,850	159,673	135,722	115,363

- T_n = labor hours/unit for the processing of n units
- T_1 = labor hours to process the first unit
- n = the number of units processed
- b = is a constant representing the slope of the curve

An 80% learning curve means that the number of hours/unit to double the production quantity is equal to 80% of the previous quantity. Assume that to produce a product for the first time it takes 50 hours then, at an 80% learning rate, the time to produce the second unit is 40 hours (50*80%); the time to produce the fourth unit is 32 hours (40*80%); the time to produce the eighth unit is 25.6 hours (32*80%), etc. The progression is shown in Table 5.12 and illustrated in Figure 5.26.

The curve decreases rapidly at first, and then evens out declining very slowly as new units are produced. It is this almost horizontal level that is used to fix the labor standard for the particular item of work. The relationship describing this 80% learning rate is developed from Equation 5(xii):

- Learning rate is 80%
- Number of units produced, n, is 2
- Time to produce the first unit T_1 is 50.00 hours
- Time to double production or to make two units is 50.00*80% = 40.00 hours—this is the value T_2

Table 5.12

Units produced	1	2	4	8	16	32	64	128	256
Hours per unit	50.00	40.00	32.00	25.60	20.48	16.38	13.11	10.49	8.39

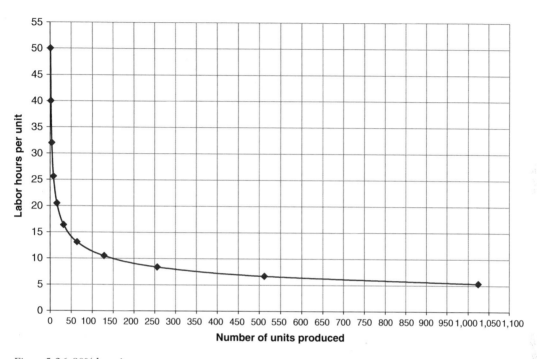

Figure 5.26 80% learning curve.

From Equation 5(xii):

$$40.00 = 50.00 * 2^b$$

$$40/50 = 2^b$$

$$0.80 = 2^b$$

Taking the logarithm of both sides of the equation:

$$\ln(0.8) = b * \ln(2)$$

$$b = \ln(0.80) * \ln(2) = -0.2231/0.6931$$

$$= -0.3219. \text{ This is the slope of the exponential function.}$$

Thus the exponential relationship for the 80% learning curve is:

$$T_n = 50.00 * n^{-0.3219}$$

5(xv)

For example to produce 256 units:

Hours/unit = $50.00*256^{-0.3219} = 50*0.1678 = 8.39$ hours/unit as given in Table 5.8.

Natural logarithm

A general relationship for the exponential distribution, and that found in Excel functions, uses e, the base of the natural logarithm and a random variable x. The value of e is 2.718281, and x can take on any positive or negative value. (We have seen the use of the base of the natural logarithm in the development of the Poisson distribution in *Chapter 4*.) The mathematical expression for the exponential distribution when x is positive is:

$$y = e^x \text{ or } y = 2.7183^x$$ 5(xvi)

This exponential distribution appears as in Figure 5.27.

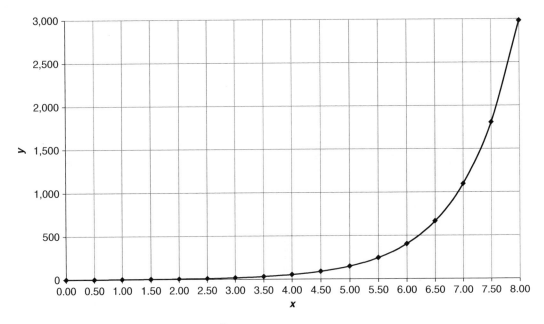

Figure 5.27 Exponential progression: $y = e^x$.

In theory the curve can extend to infinity. It has the same form as for bacterial growth. This exponential growth illustrates the sales of automobiles in the 1950s/1960s; the sale of mobile telephones in the 1990s; and the explosion of information technology (Apple, Google, Linkedin, Facebook, Twitter, etc.) The function implies that there is really no limit but the sales of any product reaches a saturation limit at some point purely because of demographics.

When x is negative the expression is:

$$y = e^{-x} \text{ or } y = 2.7183^{-x}$$

The continuous function appears as in Figure 5.28. This is similar to the curves for magazine circulation; the learning curve; average production costs. The decline is rapid at first and for large values of x the value of y approaches zero.

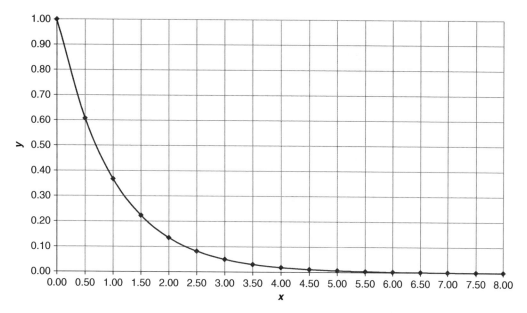

Figure 5.28 Exponential progression: $y = e^{-x}$.

Another mathematical relationship is:

$$y = 1 - e^{-x} \text{ or } y = 1 - 2.7183e^{-x}$$

Here the value of y reaches a maximum value according to the profile in Figure 5.29. This type of curve is similar to the Pareto analysis discussed in *Chapter 2* where the maximum of all occurrences is 100%. It is also the profile used in quality control. For example, in increasing marketing activity to improve hotel occupancy there is an upper limit of 100%.

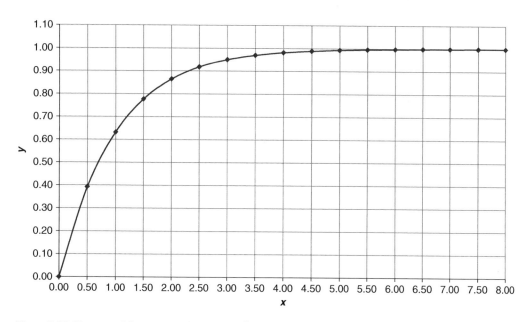

Figure 5.29 Exponential progression: $y = 1 - e^{-x}$.

Waiting lines

The exponential distribution is used to model waiting lines or queuing situations such as arrivals at supermarkets, gasoline stations, road tunnels, motor way service areas, etc. The probability density function, or equation, to describe this exponential distribution is:

$$y = f(x) = \frac{1}{\mu} * e^{-\frac{x}{\mu}}$$ 5(xvii)

- f(x) is the frequency, or probability of occurrence
- x is the value of the discrete random variable that can take on any value but must be ≥ 0
- μ is the mean or average value in time periods/unit. For example a customer arrives at a fast food restaurant on average every 2 minutes, or 2 minutes/unit
- e is a constant, the base of the natural logarithm with a value of 2.7183

Whereas the normal distribution needs two parameters to describe its shape, μ and σ, this waiting line exponential distribution is a one-parameter distribution. Only μ is required. Note that x is not a parameter but a variable in the equation. In the exponential distribution the standard deviation, σ, is also equal to the mean value μ. As the mean is in time periods per unit, the reciprocal of $1/\mu$ is units per time period. The arrival rate of a client every 2 minutes for the fast food restaurant is equivalent to an average of 30 customers per hour (60/2 since there are 60 minutes per hour). The average units per time period is called l (the Greek letter l) such that, $1/\mu = l$. The exponential density function thus can be rewritten as:

$$y = f(x) = \lambda.e^{-\lambda x}$$ 5(xviii)

and

$$\sigma = \mu = 1/\lambda$$

This relationship describing waiting lines has an analogy with the Poisson distribution, presented in *Chapter 4* that examines probability.

Beta distribution

A beta distribution is another continuous distribution used to describe the inherent variability in samples such as the activity time groups take to perform a specific activity.

Activity time

Activity time is the duration of an operation and varies according to circumstances. It is measured by considering sample groups, as for example:

- The time students take for a three-hour exam can be as low as five minutes (student has not studied for the subject, looks at the copy and walks out). Another student takes the full allotted three hours.
- The time people spend in a shopping mall. An individual arrives, buys his shirt, and leaves. Time taken is fifteen minutes (me). Another arrives in the morning, shops, has lunch, and shops again. Time taken is five hours.

- The time spent in a cinema watching a movie. An individual gets bored and leaves after twenty minutes (me). Another sees the whole film that lasts two hours and fifteen minutes.
- The time people sleep. Some people say they can live on just two hours; others say they need ten hours.
- The time taken to lay an oil pipeline in Alaska: three months if the weather is good; eleven months if there is a severe winter.

The distribution of these activity times is contained within finite limits; a lower and upper value. And the time distributions can be symmetrical or skewed to the right or the left according to the nature of the activity. Compare this to the normal distribution that is symmetrical and in theory as infinite limits.

Project management

A common use of the beta distribution, as illustrated in Figure 5.30, is in project management network diagrams using the PERT method (program evaluation and review technique) developed in the 1950s.

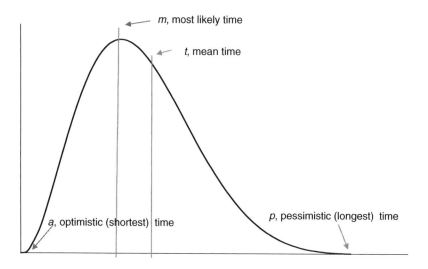

Figure 5.30 Beta distribution.

In this distribution there are three documented time estimates:

- p = pessimistic time: the longest time, when things go badly.
- a = optimistic time: the shortest time, when things go well.
- m = most likely time: takes into consideration the downside and upside of activities.

Then mean time, t, for an activity is given by:

$$t = \frac{a + 4m + b}{6}$$

5(xvix)

The importance is weighted to the most likely time or 2/3 compared to 1/3 for the pessimistic and optimistic times. The denominator numerical value of 6 makes similarities with the normal distribution standard deviation of $\pm$ 3 sigma or a range of 6-sigma.

$$\sigma^2 = \left[\frac{(b-a)}{6}\right]^2 = \frac{(b-a)^2}{36} \qquad\qquad 5(\text{xx})$$

The standard deviation for the activity is the square root of the variance:

$$\sigma = \sqrt{\sigma^2} = \left[\frac{(b-a)}{6}\right] \qquad\qquad 5(\text{xxi})$$

Unlike a normal distribution this beta distribution has the properties that all of the data is entirely contained within a finite interval. There will always be a finite optimistic time and a finite pessimistic time. The beta distribution has no predetermined shape, such as the bell shape of the normal curve, and takes the shape or is skewed, according to the activity time estimates.

The magnitude of the variance reflects the degree of uncertainty associated with the time of an activity. An activity with a variance of 20 would have more uncertainty with respect to its actual duration than one with a variation of 5. The standard deviation of each activity's time is estimated as one-sixth of the difference between the pessimistic and optimistic times. Again, this is analogous to all of the area under a normal distribution, which lies within plus or minus three standard deviations of the mean, or a range of six standard deviations. In project management networks, the standard deviation, or variance, can be computed for each path by summing the individual variances, or standard deviations, for each activity path. This enables project managers or schedulers to make probabilistic estimates of the project completion times as for example: the probability a project will be completed with 15 months of start is 92%; the probability that project will take longer than 18 months is 3%. These statements are based on the assumption that the path duration is a random variable that is normally distributed around the expected path time. This is true for large samples, and can be considered approximately true for smaller samples.

Moving on

This chapter has presented continuous distributions: normal, exponential, and beta distributions with an emphasis on probability occurrences in the normal distribution. The next chapter looks at sampling from both discrete and continuous distributions.

6   Methods and theory of statistical sampling

ICEBREAKER: The sampling experiment was badly designed!

Sampling is the process of taking a portion from a population with the objective of estimating the characteristics of that population. Through the media we often come across the sampling process when opinion polls are made of who, or what political party, will stand as the representative in upcoming elections. In 2015/2016 opinion polls in the USA were made as to who might be the nominee for the Republican party: Jeb Bush, Chris Christie, Donald Trump, or one of many others. Polls as to whether Hillary Clinton would be nominated as the Democratic representative. If so, would she be elected president? In Great Britain there were parliamentary elections in May 2015. Right up to the eve of the election on May 7, media polls indicated that the election results would be so close between Labour and Conservative parties that there would be a hung parliament similar to 2010. The polls were wrong. The conservatives, led by David Cameron, won an outright majority!

A well-designed sample survey can and should give fairly accurate predictions of the requirements, desires, or needs of a population. However, the accuracy of the survey lies in

Figure 6.0 November 3, 1948.

Source: Wikipedia

the phrase "*well-designed.*" A classic illustration of sampling gone wrong was when the media, in this case the *Chicago Tribune*, went so far as to publish the erroneous results. This occurred in 1948 during the US presidential election campaign. The two candidates were Harry Truman, the Democratic incumbent, and Governor Dewey of New York, the Republican candidate. The *Chicago Tribune*, who made the poll, was so sure of their sample and the election outcome that the headlines in their morning daily paper of November 3, 1948 as illustrated in Figure 6.0, announced "*Dewey defeats Truman.*" In fact Harry Truman won by a narrow but decisive victory of 49.5% of the popular vote to Dewey's 45%; and with an electoral margin of 303 to 189.

The *Chicago Tribune* withdrew the indicated edition of the morning paper but it was too late. The damage was done. The newspaper had egg on its face; they lost credibility. Something had gone terribly wrong with the design of their sample experiment![1,2]

The purpose of this chapter is to demonstrate appropriate techniques of statistical sampling and the theory behind sampling.

Chapter subjects

✓ **The sampling process** • Bias in sampling • Randomness • Excel and random numbers • Systematic sampling • Stratified sampling • Several stratas of interest • Cluster sampling • Quota sampling • Consumer surveys • Primary and secondary data

✓ **Sampling for the mean** • Sample size • Central limit theorem • Variability and sample size • Sample standard error • Sampling size and population characteristics • Infinite population • Application case-exercise: *Safety valves* • Finite population • Application case-exercise: *Work week*

✓ **Sampling for the proportion** • Measuring the sample proportion • Sampling distribution of the proportion • Application case-exercise: *Part-time workers*

---

*In business, and even in our personal life, we often make decisions based on limited data. What we do is take a **sample** from a **population** and then make an inference about the population characteristics, based entirely on the analysis of this sample. When you order a bottle of wine in a restaurant, the waiter pours a small quantity in your glass to taste. Based on that small quantity of wine you accept or reject the bottle of wine as drinkable. The waiter would hardly let you drink the whole bottle before you decide it is no good! The US Dow Jones Industrial Average consists of just 30 stocks, but this sample average is used as a measure of economic changes. In reality there are hundreds of stocks in the US market where $millions change hands daily. As the Icebreaker indicates, in political elections, samples of people's voting intentions are made and, based on the proportion that prefer a particular candidate, the expected outcome of the nation's election are presented beforehand. In manufacturing, lots of materials, assemblies, or finished products are sampled at random to see if pieces conform to appropriate specifications. If they do, the assumption is that the entire population, the production line or the lot from where these samples are taken, meet the desired specifications and so all the units can be put onto the market. How many months do we date our future spouse before we decide to spend the rest of our lives together!*

The sampling process

The purpose of the sampling process is to make reliable estimates about a population using either a numerical value, say the mean, or a proportion or percentage. It is usually impossible, too expensive, too time consuming, or just downright impossible to sample the whole population; thus when a sampling experiment is developed it should as closely as possible parallel the population conditions. As the Icebreaker indicates, the sampling experiment to determine voter intentions was obviously badly designed. This section looks at considerations when undertaking sampling.

Bias in sampling

When you sample to make estimates of a population you must avoid **bias** in the sampling experiment. Bias is favoritism, purposely or unknowingly present in the sample data that then gives lopsided, misleading, or unrepresentative results. For example, you wish to obtain the voting intentions of the people in the UK and you sample people who live in the West End of London. This would be biased as the West End is pretty affluent and the voters sampled are more likely to vote Tory (Conservative). To measure the average intelligence quotient (IQ) of all the 18-year old students in a country you take a sample of students from a private school. This would be biased because private school students often come from high-income families and their education level is higher (though not always). To measure the average income of residents of Los Angeles, California you take a sample of people who live in Santa Monica. This would be biased as people who live in Santa Monica are wealthy.

Randomness

A **random sample** is one where each item in the population has an equal chance of being selected. Assume a farmer wishes to determine the average weight of his 200 pigs. He samples the first 12 who come when he calls. They are probably the fittest—thus thinner than the rest! Or, a hotel manager wishes to determine the quality of the maid service in his 90-room hotel. The manager samples the first 15. If the maid works in order, then the first 15 probably were more thoroughly cleaned than the rest—the maid was less tired! These sampling experiments are not random and probably are not representative of the population.

In order to perform random sampling, you need a framework for your sampling experiment. For example, as an auditor you might wish to analyze 10% of the financial accounts of the firm to see if they conform to acceptable accounting practices. A business might want to sample 15% of its clients to obtain the level of customer satisfaction. A hotel might want to sample 12% of the condition of its hotel rooms to obtain a quality level of its operation. Or the farmer wishes to analyze 6% of his pigs in order to estimate the weight of the population.

Excel and random numbers

In Excel there are two functions for generating random numbers, **RAND** that generates a random number between 0 and 1, and **RANDBETWEEN** that generates a random number between the lowest and highest number that you specify. You first create a random number in a cell and copy this to other cells. Each time you press the function key F9 (with a PC) the random number will change.

Suppose that as an auditor you have 630 accounts in your population and you wish to examine 10% of these accounts, or 63. You number the accounts from 1 to 630. You then generate 63 random numbers between 1 and 630 and you examine those accounts whose numbers correspond to the numbers generated by the random number function. For example, the matrix in Table 6.1 is 63 random numbers within the range 1 to 630. Thus you would examine those accounts corresponding to those numbers.

Table 6.1

537	322	142	366	507	274	310	359	95
625	79	30	426	541	175	473	440	429
599	287	6	178	256	575	86	400	116
280	7	315	48	201	379	180	76	175
151	574	554	188	572	231	391	8	253
398	282	604	96	409	231	318	517	108
615	27	576	374	364	619	137	44	610

The same procedure would apply to the farmer and his pigs. Each pig would have identification, either a tag, tattoo, or embedded chip giving a numerical indications from 1 to 200. The farmer would generate a list of twelve random numbers between 1 and 200 as indicated in Table 6.2, and weigh those 12 pigs, or 6% of his population, that correspond to those numbers.

Table 6.2

23	115	149	55	94	81
91	133	116	74	131	73

Systematic sampling

When a population is relatively homogeneous and you have a listing of the items of interest such as invoices; a fleet of company cars; physical units such as products coming off a production line; inventory going into storage; a stretch of road; a row of houses; restaurants; or clients then **systematic sampling** may be appropriate. You first decide at what frequency you need to take a sample. For example if you want a 4% sample you analyze every twenty-fifth unit (4% of 100 is 25). If you want a 5% sample you analyze every twentieth unit (5% of 100 is 20). If you want a 0.5% sample you analyze every 200th unit (0.5% of 100 is 200), etc. Care must be taken in using systematic sampling that no bias occurs where the interval you choose corresponds to a pattern in the operation. For example, you use systematic sampling to examine the filling operation of soft drink machine. You sample every twenty-fifth can of drinks. It so happens that there are 25 filling nozzles on the machine. In this case you will always be sampling a can that has been filled from the same nozzle. The US population census, undertaken every 10 years, is a form of systematic sample where although every household receives a survey datasheet to complete, every tenth household receives a more detailed survey form to complete.

Stratified sampling

The technique of **stratified sampling** is useful when the population can be divided into relatively homogeneous groups, or strata, and random sampling is made only on the strata of interest. For example the strata may be students; people of a certain age range; male or female; married or single households; socio-economic levels; affiliated with the Labour or Conservative parties; etc. Stratified sampling is used because it more accurately reflects the characteristics of the target population. Single people of a certain socio-economic class are more likely to buy a sports car; people in the 20–25 age range have a different preference of music and different needs of smart phones than say those in the 50–55 age range. Stratified sampling is used when there is a small variation within each group, but a wide variation between each group. For example, teenagers in the age range 13–19 and their parents in the age range 40–50 differ very much in their tastes and ideas! Television advertising uses the concept of strata for advertising. Advertising in the mornings from around 07:00 to 09:00 covers breakfast foods and toys as this is when children are often watching TV. In the early evening from about 06:00 to 21:00 advertising is directed toward adults who have returned from work. Then advertising after midnight is directed toward another population!

Several stratas of interest

In a given population you may have several well-defined strata and perhaps you wish to take a representative sample from this population. Consider for example Table 6.3 that gives the number of company employees by function. Each function is a stratum since it defines a specific activity and is relatively homogeneous. Suppose we wish to obtain the employees' preference of changing from the current 8-hour per day, 5-day work week to a proposed 10-hour per day, 4-day work week. In order to limit the cost and the time of the sampling experiment we decide to only survey 60 of the employees. There are a total of 1,200 employees in the firm and so 60 represents 5% of the total workforce (60/1,200). Thus we would take a random sample of 5% of the employees from each of

Table 6.3

Department	Total employees	Sample Size at 5%
Accounting	60	3
Administration	160	8
Design	200	10
Information systems	140	7
Operations	300	15
Research & development	80	4
Sales	260	13
TOTAL	1,200	60

the departments or strata such that the sampling experiment parallels the population. The number that we would survey is given in the last column row of Table 6.3.

Cluster sampling

In cluster sampling the population is divided into groups, or clusters, and each cluster is then sampled at random. For example assume Birmingham, UK is targeted for preference of a certain consumer product. The city is divided into clusters using a city map and an appropriate number of clusters are selected for analysis. Cluster sampling is used when there is considerable variation in each group or cluster, but groups are essentially similar. Cluster sampling, if properly designed, can provide more accurate results than simple random sampling from the population.

Quota sampling

In market research, or market surveys, interviewers carrying out the experiment may use quota sampling, which means where they have a specific target quantity to review. In this type of sampling the population is often stratified according to some criteria so that the interviewer's quota is based within these strata. For example, the interviewer may be interested to obtain information regarding a ladies' fashion magazine. The interviewer conducts her survey in a busy shopping area such as London's Oxford Street. Using quota sampling, in her survey she would only interview females who are perhaps less than 40, and who are elegantly dressed. This *stratification* should give a reasonable probability that the selected candidates have some interest, and thus an opinion, regarding the fashion magazine in question. If you are in an area where surveys are being carried out, it could be that you do not fit the strata desired by the interviewer. For example you are male and the interviewer is targeting females, you appear to be over 50 and the interviewer is targeting the age group under 40, you are white and the interviewer is targeting other ethnic groups, etc.

Consumer surveys

If your sampling experiment involves opinions concerning a product, a concept, or a situation, then you might use a **consumer survey**, where responses are solicited from individuals who are targeted according to a well-defined sampling plan. The sampling plan would use one, or a combination of the methods above—simple random sampling, systematic, stratified, cluster, or quota sampling. The survey information is prepared on **questionnaires**, which might be sent through the mail, completed by telephone, sent by electronic mail, or requested in person. In the latter case this may either be by going door-to-door (rare these days with security concerns) or by soliciting the information in areas frequented by potential consumers, such as shopping malls or busy pedestrian areas. The collected survey data, or sample, is then analyzed and used to forecast or make estimates for the population from which the survey data was taken. Surveys are often used to obtain ideas about a new product, because required data is unavailable from other sources.

When you develop a consumer survey remember that it is perhaps you who has to analyze it afterwards. Thus you should structure it so that this task is straightforward with responses that are easy to organize. Avoid open-ended questions. For example, rather than asking the question, "How old are you?" give the respondent age categories, as for example in Table 6.4. Here these categories are all-encompassing.

Table 6.4

Under 25	25 to ≤ 34	34 to ≤ 44	44 to ≤ 55	55 to ≤ 65	> 65

Alternatively, if you want to know the job of the respondent then rather than asking, "What is your job?" rephrase the question to, "Which of the following best describes your professional activity?" giving possible responses according to Table 6.5. This is not all-encompassing but there is a category "Other" for activities that may have been overlooked.

Table 6.5

Construction
Consulting
Design
Education
Energy
Financial services
Government
Health care
Hospitality
Insurance
Legal
Logistics
Manufacturing
Media communications
Research
Retail
Telecommunications
Tourism
Other (Please describe)

Soliciting information from consumers is not easy, as "everyone is too busy". Postal responses have a very low response rate and their use has declined. Those people that do respond may not be representative in the sample. Telephone surveys give a higher return because voice contact has been obtained. However, the sample obtained may also not be representative as those contacted may be the unemployed, retirees, elderly people, or non-employed individuals who are more likely to be at home when the telephone call is made. The other segment of the population, usually larger, is not available because they are working (though if you have access to portable phone numbers this may not apply). Electronic mail is often the method of conducting surveys. They give a reasonable response, as it is very quick to send the survey back. However, the questionnaire only reaches those who have electronic mail, and then those who care to respond. Person-to-person contact gives a high response for consumer surveys since if you are stopped in the street it is not easy to refuse and a relatively large proportion of people will accept being questioned. Consumer surveys can be expensive. There is the cost of designing the questionnaire such that it is able to solicit the correct response; there is the operating side of collecting the data; and then the subsequent analysis. Often businesses use outside consulting firms that specialize in developing consumer surveys.

Primary and secondary data

In sampling, if we are responsible for carrying out the analysis, or at least responsible for designing the consumer surveys, then the data is considered **primary data**. If the sample experiment is well designed then this primary data can provide useful information. The disadvantage with primary data is the time, and the associated cost, of designing the survey and the subsequent analysis. In some instances it may be possible to use **secondary data** in analytical work. Secondary data is information that has been developed by someone else, but is used in your analytical work. Secondary data might be demographic information, economic trends, or consumer patterns, which is often available through the Internet and prepared by government agencies. The advantage with secondary data, provided that it is in the public domain, is that it costs less or at best is free. The disadvantage is that the secondary data may not contain all the information you require, the format may not be ideal, and/or it may be not be up-to-date. Thus, there is a trade-off between using less costly, but perhaps less accurate, secondary data, and more expensive, but more reliable, primary data.

Sampling for the mean

The usual purpose of taking and analyzing a sample is to make an estimate of the population parameter. This is **inferential statistics**. As the sample size is smaller than the population there is no guarantee of the value of population parameter that we are trying to measure, but from the sample analysis we draw conclusions. If we really wanted to guarantee our conclusion we would have to analyze the whole population. However, this is often impractical, too costly, takes too long, or is clearly impossible. An alternative to inferential statistics is **descriptive statistics** that involves the collection and analysis of the dataset in order to characterize just the sampled dataset.

Sample size

A question that arises in sampling is what the size of the sample should be in order to make a reliable conclusion on the population. Clearly the larger the sample size then the greater is the probability of being close to estimating the correct population parameter. Alternatively, the smaller the sample the greater is the risk of making an inappropriate estimate. To demonstrate the impact of the sample size, consider an experiment where there is a population of seven steel rods, as shown in Figure 6.1.

The total length of these seven rods is 35 cm (2 +3 +4 +5 +6 +6 +9). This translates into a mean value of the length of the rods of 5.00 cm (35/7). A summary of the numbered rods, their individual lengths, total length, and average length is in Table 6.6.

If we take samples of these rods from the population, with replacement, then from the counting relations in *Chapter 3*, Equation 3(xvi), the number of possible combinations of rods that can be taken, the same rod not appearing twice in the sample, is given by:

$$\text{combinations} = \frac{n!}{x!(n-x)!}$$

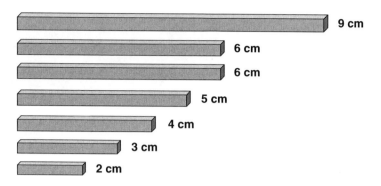

Figure 6.1 Seven steel rods and their length in centimetres.

Table 6.6

Rod no.	1	2	3	4	5	6	7	Total length	Mean length
Rod length, cm	2.00	3.00	4.00	5.00	6.00	6.00	9.00	35.00	5.00

Here, n, is the size of the population, or in this case seven, and x, is the size of the sample. For example, if we select a sample of size of three, the number of possible different combinations, from Equation 3(xvi) is:

$$\text{combinations} = \frac{7!}{3!(7-3)!} = \frac{7!}{3! * 4!} = \frac{7 * 6 * 5 * 4 * 3 * 2 * 1}{3 * 2 * 1 * 4 * 3 * 2 * 1} = 35$$

If we increase the sample sizes from one to seven rods, then from Equation 3(xvi) the total possible number of different samples is as shown in Table 6.7. Thus we sample from the population first with a sample size of one, two, three etc. right through to seven. Each time we select a sample we determine the sample mean value of the length of rods selected. For example, if the sample size is three and rods of length 2, 4, and 6 cm are selected, then the mean length, $\bar{x}$, of the sample is:

$$\frac{2+4+6}{3} = 4.00 \text{ cm}$$

Table 6.7

Sample size	1	2	3	4	5	6	7
No. of possible different samples	7	21	35	35	21	7	1

The possible combinations of rod sizes for the seven different samples are shown in Table 6.8. (Note that there are two rods of length 6 cm.)

For a particular sample size, the sum of all the sample means is then divided by the number of samples withdrawn to give the mean value of the samples or, $\bar{\bar{x}}$. For example, for a sample of size three, the sum of the sample means is 175 and this number, divided by the sample number of 35 gives 5.00. These values are given on the bottom row of Table 6.8. What we conclude is that the sample means are always equal to 5.00 cm, or exactly the same as the population mean.

Table 6.8

Sample mean cm	Sample size 1	Sample size 2	Sample size 3	Sample size 4	Sample size 5	Sample size 6	Sample size 7
2.00	1	0	0	0	0	0	0
2.25	0	0	0	0	0	0	0
2.50	0	1	0	0	0	0	0
2.75	0	0	0	0	0	0	0
3.00	1	1	1	0	0	0	0
3.25	0	0	0	0	0	0	0
3.50	0	2	1	1	0	0	0
3.75	0	0	3	2	0	0	0
4.00	1	3	3	2	2	0	0
4.25	0	0	0	3	1	0	0
4.50	0	3	4	4	1	1	0
4.75	0	0	4	3	2	0	0
5.00	1	2	4	4	5	3	1
5.25	0	0	0	4	3	1	0
5.50	0	3	3	4	3	2	0
5.75	0	0	4	3	2	0	0
6.00	2	2	3	3	2	0	0
6.25	0	0	0	1	0	0	0
6.50	0	1	2	1	0	0	0
6.75	0	0	2	0	0	0	0
7.00	0	1	1	0	0	0	0
7.25	0	0	0	0	0	0	0
7.50	0	2	0	0	0	0	0
7.75	0	0	0	0	0	0	0
8.00	0	0	0	0	0	0	0
8.25	0	0	0	0	0	0	0
8.50	0	0	0	0	0	0	0
8.75	0	0	0	0	0	0	0
9.00	1	0	0	0	0	0	0
N° of samples	7	21	35	35	21	7	1

Next, for each sample size, a frequency distribution of the mean length is determined. The data from which the frequency distributions are developed is given in Table 6.9. The right-hand column gives the sample mean and the other columns give the number of occurrences within a class limit according to the sample size. For example, for a sample size of four there are four sample means greater than 4.25 cm but less than or equal to 4.50 cm. This data is now plotted as frequency histograms as shown in Figure 6.2 through to Figure 6.8, where each of the seven histograms have the same scale on the x-axis.

From Figure 6.2 through to Figure 6.8 we can see that as the sample size increases from one to seven, the dispersion about the mean value of 5.00 cm becomes smaller or alternatively more sample means lie closer to the population mean. For the sample size of seven, or the whole population, the dispersion is zero. The mean of the sample means, $\bar{\bar{x}}$, is always equal to the population mean of 5.00 or they have the same central tendency. This experiment demonstrates the concept of the **central limit theorem.**

Central limit theorem

The foundation of sampling is based on the central limit theorem, which is the criterion by which information about a population parameter can be inferred from a sample. The central limit theorem

Table 6.9

Sample No.	Sample size 1		Sample size 2		Sample size 3		Sample size 4		Sample size 5		Sample size 6		Sample size 7	
	Rod length	Mean length	Rod length	Mean length	Rod length	Mean length	Rod length	Mean length	Rod length	Mean length	Rod length	Mean length	Rod length	Mean length
1	2	2.00	2 3	2.50	2 3 4	3.00	2 3 4 5	3.50	2 3 4 5 6	4.00	2 4 5 6 6 9	5.33	2 3 4 5 6 6 9	5.00
2	3	3.00	2 4	3.00	2 3 5	3.33	2 3 4 6	3.75	2 3 4 5 6	4.00	2 3 5 6 6 9	5.17		
3	4	4.00	2 5	3.50	2 3 6	3.67	2 3 4 6	3.75	2 3 4 5 9	4.60	2 3 4 6 6 9	5.00		
4	5	5.00	2 6	4.00	2 3 6	3.67	2 3 4 9	4.50	2 3 4 6 6	4.20	2 3 4 5 6 9	4.83		
5	6	6.00	2 6	4.00	2 3 9	4.67	2 3 5 6	4.00	2 3 4 6 9	4.80	2 3 4 5 6 9	4.83		
6	6	6.00	2 9	5.50	2 4 5	3.67	2 3 5 6	4.00	2 3 4 6 9	4.80	2 3 4 5 6 6	4.33		
7	9	9.00	3 4	3.50	2 4 6	4.00	2 3 5 9	4.75	2 3 5 6 6	4.40	3 4 5 6 6 9	5.50		
8			3 5	4.00	2 4 6	4.00	2 3 6 6	4.25	2 3 5 6 9	5.00				
9			3 6	4.50	2 4 9	5.00	2 3 6 9	5.00	2 3 5 6 9	5.00				
10			3 6	4.50	2 5 6	4.33	2 3 6 9	5.00	2 3 6 6 9	5.20				
11			3 9	6.00	2 5 6	4.33	2 4 5 6	4.25	2 4 5 6 6	4.60				
12			4 5	4.50	2 5 9	5.33	2 4 5 6	4.25	2 4 5 6 9	5.20				
13			4 6	5.00	2 6 6	4.67	2 4 5 9	5.00	2 4 5 6 9	5.20				
14			4 6	5.00	2 6 9	5.67	2 4 6 6	4.50	2 4 6 6 9	5.40				
15			4 9	6.50	2 6 9	5.67	2 4 6 9	5.25	2 5 6 6 9	5.60				
16			5 6	5.50	3 4 5	4.00	2 4 6 9	5.25	3 4 5 6 6	4.80				
17			5 6	5.50	3 4 6	4.33	2 5 6 6	4.75	3 4 5 6 9	5.40				
18			5 9	7.00	3 4 6	4.33	2 5 6 9	5.50	3 4 5 6 9	5.40				
19			6 6	6.00	3 4 9	5.33	2 5 6 9	5.50	3 4 6 6 9	5.60				
20			6 9	7.50	3 5 6	4.67	2 6 6 9	5.75	3 5 6 6 9	5.80				

(Continued)

Table 6.9 (Continued)

Sample No.	Sample size 1		Sample size 2		Sample size 3		Sample size 4		Sample size 5		Sample size 6		Sample size 7	
	Rod length	Mean length	Rod length	Mean length	Rod length	Mean length	Rod length	Mean length	Rod length	Mean length	Rod length	Mean length	Rod length	Mean length
21			6 9	7.50	3 5 6	4.67	3 4 5 6	4.50	4 5 6 6 9	6.00				
22					3 5 9	5.67	3 4 5 6	4.50						
23					3 6 6	5.00	3 4 5 9	5.25						
24					3 6 9	6.00	3 4 6 6	4.75						
25					3 6 9	6.00	3 4 6 9	5.50						
26					4 5 6	5.00	3 4 6 9	5.50						
27					4 5 6	5.00	3 5 6 6	5.00						
28					4 5 9	6.00	3 5 6 9	5.75						
29					4 6 6	5.33	3 5 6 9	5.75						
30					4 6 9	6.33	3 6 6 9	6.00						
31					4 6 9	6.33	4 5 6 6	5.25						
32					5 6 6	5.67	4 5 6 9	6.00						
33					5 6 9	6.67	4 5 6 9	6.00						
34					5 6 9	6.67	4 6 6 9	6.25						
35					6 6 9	7.00	5 6 6 9	6.50						
Total means		35.00		105.00		175.00		175.00		105.00		35.00		35.00
Avg. of mean		5.00		5.00		5.00		5.00		5.00		5.00		5.00

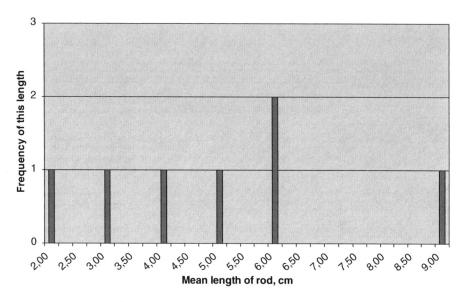

Figure 6.2 Samples of size 1 taken from a population of size 7.

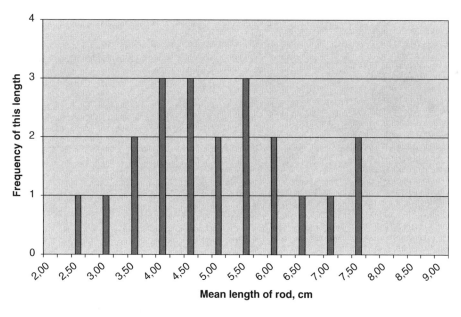

Figure 6.3 Samples of size 2 taken from a population of size 7.

states that in sampling, as the size of the sample increases, there becomes a point when the **distribution of the sample means,** $\bar{x}$, can be approximated by the normal distribution. This is so even though the distribution of the population itself may not necessarily be normal. The distribution of the sample means, also called **sampling distribution of the means,** is a probability distribution of all the possible means of samples taken from a population.

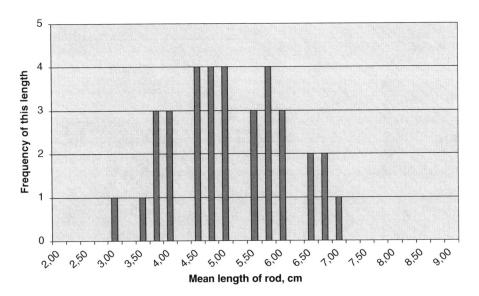

Figure 6.4 Samples of size 3 taken from a population of size 7.

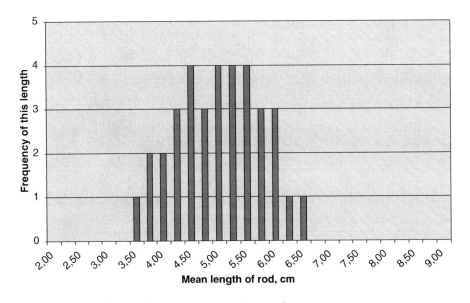

Figure 6.5 Samples of size 4 taken from a population of size 7.

This concept of sampling and sampling means is illustrated by Table 6.10 for the production of chocolate. Here the production line is producing 500,000 chocolate bars. This is the population value, N. The molding machine for the chocolate is set at 100 g. This is thus the nominal weight of the chocolate bar, is the population mean, μ, and is the weight that appears on the label. Now, for quality control purposes an inspector takes ten random samples from the production line in order to verify that the weight of the chocolate is the specified 100 g. Each sample contains fifteen chocolate bars. Each bar in the sample is weighed and these individual weights, and the mean weight of each sample, are recorded. For example, if we consider Sample 1, the weight of the first bar is 100.16 g, the weight of the

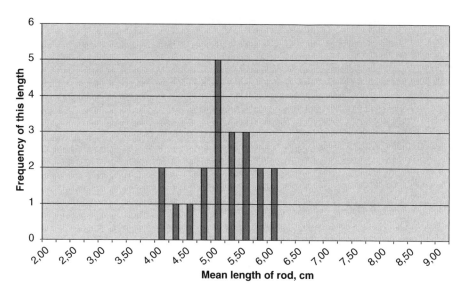

Figure 6.6 Samples of size 5 taken from a population of size 7.

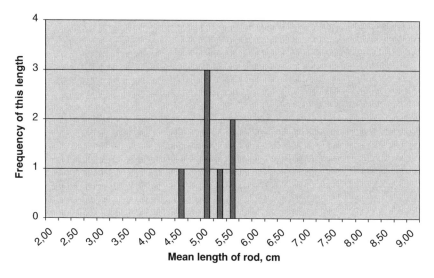

Figure 6.7 Samples of size 6 taken from a population of size 7.

second bar is 99.48 g and the weight of the fifteenth bar is 98.56 g. The mean weight of this first sample, $\bar{x}_1$ is 99.88 g. The mean weight of the tenth sample, $\bar{x}_{10}$, is 100.02 g. The mean value of the means of all the ten samples, $\bar{\bar{x}}$ (x-double bar) is 99.85 g. The values of $\bar{x}$ (x-bar) of all the possible samples plotted as a frequency distribution would give a sampling distribution of the means. (Here though, only ten values are insufficient to show a correct distribution.)

Variability and sample size

Consider a large organization such as a government unit that has over 100,000 employees. This is a large enough number so that for statistical purposes it can be considered infinite. Assume that the

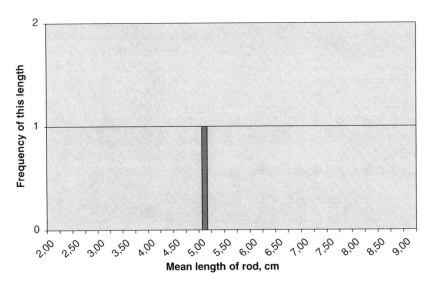

Figure 6.8 Samples of size 7 taken from a population of size 7.

Table 6.10

Chocolate bar no.	Sample 1	Sample 2	Sample 3	Sample 4	Sample 5	Sample 6	Sample 7	Sample 8	Sample 9	Sample 10
1	100.16	100.52	101.20	101.15	98.48	98.31	101.85	101.34	98.56	99.27
2	99.48	98.30	101.23	101.30	98.75	99.18	99.74	101.38	101.31	101.50
3	100.66	99.28	98.39	101.61	99.84	100.47	99.72	101.09	101.61	101.62
4	98.93	98.01	98.06	99.07	98.38	98.30	98.76	98.89	101.26	100.84
5	98.25	98.42	98.94	99.71	99.42	99.09	100.00	98.08	98.03	98.94
6	98.06	99.19	100.53	99.78	99.23	98.23	101.42	101.50	99.74	98.94
7	100.39	100.15	98.81	98.12	100.98	100.64	98.10	100.44	99.66	99.65
8	101.16	99.60	99.79	101.58	100.82	98.71	100.49	101.70	98.80	98.82
9	100.03	98.89	99.07	98.03	101.51	101.23	100.54	100.84	99.04	99.96
10	101.27	101.94	98.39	100.77	100.17	100.99	101.66	98.40	100.61	100.95
11	99.18	98.34	99.61	98.60	101.56	99.24	101.68	99.22	99.20	99.86
12	101.77	100.80	99.66	98.84	100.55	98.13	99.13	99.34	100.52	98.11
13	99.07	98.79	101.18	100.46	101.59	98.27	98.81	101.23	98.80	100.85
14	101.17	101.02	99.57	100.30	101.87	98.16	101.73	99.98	99.26	99.17
15	98.56	98.93	101.27	98.55	99.04	101.35	99.89	98.24	98.87	101.84
x–bar	99.88	99.48	99.71	99.86	100.15	99.35	100.23	100.11	99.68	100.02
x – double bar	99.85									

distribution of the employee salaries is considered normal with an average salary of $40,000. Sampling of individual salaries is made using random computer selection.

- Assume a random sample of just one salary value is selected that happens to be $90,000. This value is a long way from the mean value of $40,000.
- Assume now that random samples of two salaries are taken which happen to be $60,000 and $90,000. The average of these is $75,000 [(60,000 + 90,000)/2]. This is still far from $40,000 but closer than in the case of a single sample.

- Now random samples of five salaries $60,000, $90,000, $45,000, $15,000, and $20,000 come up. The mean value of these is $46,000 or closer to the population average of $40,000. What is happening is that low values are balanced out by high values.

Alternatively consider Figure 6.9 for random sampling from a pack of cards.

There are 52 cards. Each suit of cards has a total score of 91. For the pack of cards total score is 364. Dividing this by the number of cards gives 7.00 (364/52). As shown in Figure 6.9, if we select two cards at random whose values are 12 and 13, the average is 12.50, a long way from 7.00. As we go to four, six, eight cards etc. we approach the population mean of 7.00. Thus in conclusion, by taking larger samples there is a higher probability of making an estimate close to the population parameter. Alternatively, increasing the sample size reduces the spread or variability of the average value of the samples taken.

Suit	Ace	Two	Three	Four	Five	Six	Seven	Eight	Nine	Ten	Jack	Queen	King
Clubs	1	2	3	4	5	6	7	8	9	10	11	12	13
Diamonds	1	2	3	4	5	6	7	8	9	10	11	12	13
Hearts	1	2	3	4	5	6	7	8	9	10	11	12	13
Spades	1	2	3	4	5	6	7	8	9	10	11	12	13

Total score of population pack of cards is 364	Average score/card is 7.00 (364/52)

With replacement select at random cards with the following score													Average score/card
N° cards chosen													
Two cards		12	13										12,50
Four		13	12	3	12								10,00
Six		11	2	3	10	8	6						6,67
Eight		3	9	9	8	6	5	11	8				7,38
Ten		9	7	6	1	5	1	3	10	9	9		6,00
Twelve		7	13	9	12	13	3	7	7	1	5	11	5	8,00
Fifty two													7,00

Low scores balance out high values. Eventually average of population would be reached.

Figure 6.9 Random sampling from a pack of cards.

Sample standard error

The mean of a sample taken from a population is $\bar{x}$. The mean of all possible sample means withdrawn from the population is $\bar{\bar{x}}$. From the central limit theorem, the mean of the entire sample means taken from the population can be considered equal to the population mean, μ_x. (This is the situation as shown with the rod experiment.)

$$\bar{\bar{x}} = \mu_x \qquad\qquad 6(i)$$

And, because of this relationship in Equation 6(i), the arithmetic mean of the sample is said to be an **unbiased estimator** of the population mean. By the central limit theorem, the **standard deviation**

of the sampling distribution, $\sigma_{\bar{x}}$, is related to the population standard deviation, σ_x, and the sample size, n, by the following relationship:

$$\sigma_{\bar{x}} = \frac{\sigma_x}{\sqrt{n}} \qquad\qquad 6(\text{ii})$$

This implies that as the size of the sample increases, the standard deviation of the sampling distribution decreases. The standard deviation of the sampling distribution is more usually referred to as the **standard error of the sample means,** or more simply the **standard error** as it represents the error in our sampling experiment. For example, going back to our illustration of the salaries of the government employees, if we took a series of samples from the employees and measured the $\bar{x}$ value of salaries each time, we will almost certainly have different values each time simply because the chances are that our salary numbers in our sample will be different. That is, the difference between each sample among the several samples and the population causes variability in our analysis. This variability, as measured by the standard error of Equation 6(ii), is due to the chance or **sampling error** in our analysis between the samples we took and the population. The standard error indicates the magnitude of the chance error that has been made and also the accuracy when using a sample statistic to estimate the population parameter. A distribution of sample means that has less variability, or is less spread out, as evidenced by a small value of the standard error, is a better estimator of the population parameter than a distribution of sample means that is widely dispersed with a larger standard error. Note that the standard error is not an error in calculation, or the way the experiment has been carried out but the "error" between the sample value and the population value.

As a comparison to the standard error, we have the standard deviation of a population. This is not an error but a deviation that is to be expected since by their very nature populations have variation. We do not live in a perfect world! There are variations in the ages of people; variations in the volumes of liquid in cans of soft drinks; variations in the weights of a nominal chocolate bar; variations in the per capita incomes of individuals, etc. These comparisons are illustrated in Figure 6.10, which shows the

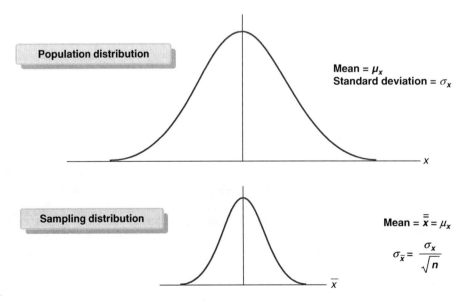

Figure 6.10 Population distribution and the sampling distribution.

shape of a normal distribution with its standard deviation and the corresponding profile of the sample distribution of the means with its standard error.

Sample size and population characteristics

Like many of the statistical experiments, sampling is based on the application of the normal distribution. But what is the sample size from which we can develop a sampling distribution of the means that is normal in shape? This depends on the characteristics of the population. From past statistical experiments the following has been demonstrated.

- For most population distributions, regardless of their shape, the sampling distribution of the means of samples taken at random from the population will be approximately normally distributed if samples of at least *size 30 units* are withdrawn.
- If the population distribution is itself symmetrical, the sampling distribution of the means of samples taken at random from the population will be approximately normal if samples of at least *size 15 units* are withdrawn.
- If the population is normally distributed, the sampling distribution of the means of samples taken at random from the population will be normally distributed *regardless of the sample size* withdrawn.

The practicality of these relationships with the central limit theorem is that by sampling, either from non-normal populations or normal populations, inferences can be made about the population parameters without having information about the shape of the population distribution other than the information obtained from the sample.

Infinite population

An infinite population is a collection of data so large that taking a sample from this population and discarding it after an analysis has been made has no statistical impact on the remaining population. If you take a sample of 50 cl. from the River Thames and test for nitrates then throw that sample away will not impact the concentration that remains in the river.

In Equation 5(ii) of *Chapter 5* we indicated that the relationship between the mean, μ_x, the standard deviation, σ_x, and the random variable x in a normal distribution is:

$$z = \frac{x - \mu_x}{\sigma_x}$$

From Figure 6.10 for the sampling distribution:

- the random variable, x of a normal distribution is replaced by the sample mean, $\bar{x}$
- the mean value μ_x of the population is replaced by the sample mean, $\bar{\bar{x}}$
- the standard deviation of the normal distribution, σ_x, is replaced by the standard deviation of the sample distribution, or the sample error, $\sigma_{\bar{x}}$.

The standard equation for the sampling distribution of the means now becomes:

$$z = \frac{x - \mu_x}{\sigma_x} = \frac{\bar{x} - \bar{\bar{x}}}{\sigma_{\bar{x}}} \qquad\qquad 6(\text{iii})$$

Substituting from equations 6(i), 6(ii), and 6(iii) the standard equation then becomes

$$z = \frac{\bar{x} - \bar{\bar{x}}}{\sigma_{\bar{x}}} = \frac{\bar{x} - \mu_x}{\sigma_{\bar{x}}} = \frac{\bar{x} - \mu_x}{\frac{\sigma_x}{\sqrt{n}}} \qquad \qquad 6(\text{iv})$$

This is then the transformation relationship for samples. The Excel functions for the normal distribution can be used for samples except that now the mean value of the sample mean, $\bar{x}$, replaces the random variable, x, of the population distribution, and the standard error of the sampling distribution $\sigma_x/\sqrt{(n)}$ replaces the standard deviation of the population.

The following application case-exercise, *Safety valves* illustrates sampling of the means for an infinite, or very large population.

Application case-exercise: Safety valves

Situation

A German manufacturer produces safety pressure valves that are used on domestic water heaters. In the production process, the valves are automatically preset so that they open and release a flow of water when the upstream pressure in a heater exceeds 7 bars (7 kg/cm^2). In the manufacturing process there is a tolerance in the setting of the valves and the release pressure of the valves follows a normal distribution with a standard deviation of 0.30 bars.

Required

1   What proportion of randomly selected valves has a release pressure between 6.8 and 7.1 bars?
2   If many random samples of size eight were taken, what proportion of sample means would have a release pressure between 6.8 and 7.1 bars?
3   If many random samples of size 20 were taken, what proportion of sample means would have a release pressure between 6.8 and 7.1 bars?
4   If many random samples of size 50 were taken, what proportion of sample means would have a release pressure between 6.8 and 7.1 bars?

Worked out solutions

1   Here we are only considering a single valve, or a sample of size 1, from the population between 6.8 and 7.1 bars or either side of the mean. From Equation 5(ii) when $x = 6.8$ bars,

$$z = \frac{x - \mu_x}{\sigma_x} = \frac{6.8 - 7.0}{0.3} = -\frac{0.2}{0.3} = -0.6667$$

From **NORMSDIST** (Excel 2007) or **NORM.S.DIST** (Excel 2010 and later) in Excel this gives an area from the left end of the curve of 25.25%.

From Equation 5(ii) when $x = 7.1$ bars:

$$z = \frac{x - \mu_x}{\sigma_x} = \frac{7.1 - 7.0}{0.3} = \frac{0.1}{0.3} = 0.3333$$

From **NORMSDIST** (Excel 2007) or **NORM.S.DIST** (Excel 2010 and later) in Excel this gives a value from the left end of the curve of 63.06%. Thus the probability that a randomly selected valve has a release pressure between 6.8 and 7.1 bars is $63.06 - 25.25 = 37.81\%$.

2   Now we are sampling from the normal population with a sample size of eight. Using Equation 6(ii) the standard error is:

$$\sigma_{\bar{x}} = \frac{\sigma_x}{\sqrt{n}} = \frac{0.3}{\sqrt{8}} = \frac{0.3}{2.8284} = 0.1061$$

Using this value in Equation 6(iv) when $\bar{x} = 6.8$ bars:

$$z = \frac{\bar{x} - \mu_x}{\frac{\sigma_x}{\sqrt{n}}} = \frac{6.8 - 7.0}{0.1061} = -\frac{0.2}{0.1061} = -1.8850$$

From **NORMSDIST** (Excel 2007) or **NORM.S.DIST** (Excel 2010 and later) in Excel, using the standard error in place of the standard deviation, gives the area under the curve from the left of 2.97%.

Again from Equation 6(iv) when $\bar{x} = 7.1$ bars:

$$z = \frac{\bar{x} - \mu_x}{\frac{\sigma_x}{\sqrt{n}}} = \frac{7.1 - 7.0}{0.1061} = \frac{0.1}{0.1061} = 0.9425$$

From **NORMSDIST** (Excel 2007) or **NORM.S.DIST** (Excel 2010 and later) in Excel, using the standard error in place of the standard deviation, gives the area under the curve from the left of 82.71%. Thus the proportion of sample means that would have a release pressure between 6.8 and 7.1 bars is $82.71 - 2.97 = 79.74\%$.

3   Now we are sampling from the population with a sample size of 20. Using Equation 6(ii) the standard error is:

$$\sigma_{\bar{x}} = \frac{\sigma_x}{\sqrt{n}} = \frac{0.3}{\sqrt{20}} = \frac{0.3}{4.4721} = 0.0671$$

Using this value in Equation 6(iv) when $\bar{x} = 6.8$ bars:

$$z = \frac{\bar{x} - \mu_x}{\frac{\sigma_x}{\sqrt{n}}} = \frac{6.8 - 7.0}{0.0671} = -\frac{0.2}{0.0671} = -2.9814$$

From **NORMSDIST** (Excel 2007) or **NORM.S.DIST** (Excel 2010 and later), using the standard error in place of the standard deviation, gives the area under the curve from the left of 0.14%.

Again from Equation 6(iv) when $\bar{x} = 7.1$ bars:

$$z = \frac{\bar{x} - \mu_x}{\frac{\sigma_x}{\sqrt{n}}} = \frac{7.1 - 7.0}{0.0671} = \frac{0.1}{0.0671} = 1.4903$$

From **NORMSDIST** (Excel 2007) or **NORM.S.DIST** (Excel 2010 and later), using the standard error in place of the standard deviation, gives the area under the curve from the left of 93.20%. Thus the proportion of sample means that would have a release pressure between 6.8 and 7.1 bars is $93.20 - 0.14 = 93.06\%$.

4   Now we are sampling from the population with a sample size of 50. Using Equation 6(ii) the standard error is:

$$\sigma_{\bar{x}} = \frac{\sigma_x}{\sqrt{n}} = \frac{0.3}{\sqrt{50}} = \frac{0.3}{7.0711} = 0.0424$$

Using this value in Equation 6(iv) when $\bar{x} = 6.8$ bars:

$$z = \frac{\bar{x} - \mu_x}{\frac{\sigma_{\bar{x}}}{\sqrt{n}}} = \frac{6.8 - 7.0}{0.0424} = -\frac{0.2}{0.0424} = -4.714$$

From **NORMSDIST** (Excel 2007) or **NORM.S.DIST** (Excel 2010 and later), using the standard error in place of the standard deviation, gives the area under the curve from the left of 0.00%.

Again from Equation 7(v) when $\bar{x} = 7.1$ bars,

$$z = \frac{\bar{x} - \mu_x}{\frac{\sigma_{\bar{x}}}{\sqrt{n}}} = \frac{7.1 - 7.0}{0.0424} = \frac{0.1}{0.0424} = 2.3585$$

From **NORMSDIST** (Excel 2007) or **NORM.S.DIST** (Excel 2010 and later), using the standard error in place of the standard deviation, gives the area under the curve from the left of 99.08%. Thus the proportion of sample means that would have a release pressure between 6.8 and 7.1 bars is $99.08 - 0.00 = 99.08\%$.

The summarized results are in Table 6.11 and the concept is illustrated by the distributions of Figure 6.11.

Table 6.11

Sample size, n	1	8	20	50
$\sqrt{(n)}$	1.0000	2.8284	4.4721	7.0711
Standard error, $\sigma/\sqrt{(n)}$	0.3000	0.1061	0.0671	0.0424
Proportion between 6,8 and 7,1 bars	37.81%	79.74%	93.05%	99.08%

What happens is that not only does the standard error decrease as the sample size increases but also there is a larger proportion between 6.8 and 7.1 bars. That is, a larger cluster around the mean or the target value of 7.0 bars. Alternatively, as the sample size increases there is a smaller dispersion of the values. In the case of a sample size of one there is 37.81% of the data clustered around the values of 6.8 and 7.1 bars, which means that there is 62.19% (100%−37.81%) not clustered around the mean. In the case of a sample size of 50 there is 99.08% clustered around the mean and only 0.92% (100%−99.08%) not clustered around the mean. Note that in applying these calculations the assumption is that the sampling distributions of the mean follow a normal distribution, and the relation of the central limit theorem applies. As in all calculations using the normal distribution, we can avoid calculating the value of z, by using the **NORMSDIST** (Excel 2007) or **NORM.S.DIST** (Excel 2010 and later).

Finite population

A finite population is a collection of data that has a stated, limited, or small size. It implies that if one piece of the data from the population is destroyed, or removed, there would be an important

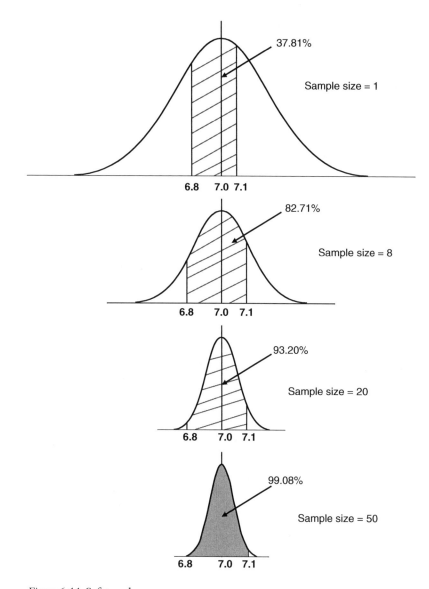

Figure 6.11 Safety valves.

impact on the data that remains. For example the probability of drawing a certain card from a pack is 1/52. If the card is not replaced the probability next draw of obtaining the certain card is 1/51; on the third draw it is 1/50, etc. That is, the probability or chance increases. If the population is considered finite, and we sample **with replacement** (after each item is sampled it is put back into the population), then we can use the equation for the standard error already presented in Equation 6(ii):

$$\sigma_{\bar{x}} = \frac{\sigma_x}{\sqrt{n}}$$

However, if we sample *without replacement*, when we are not replacing the sample in the population, then the standard error of the mean is modified by the relationship,

$$\sigma_{\bar{x}} = \frac{\sigma_x}{\sqrt{n}} \sqrt{\frac{N-n}{N-1}} \qquad\qquad 6(\text{v})$$

Here the term

$$\sqrt{\frac{N-n}{N-1}} \qquad\qquad 6(\text{vi})$$

is the **finite population multiplier**, where N is the population size, and n is the size of the sample. This correction is applied when the ratio of n/N is greater than 5%, meaning that the sample size is large relative to the population. In this case Equation 6(iv) now becomes:

$$z = \frac{\bar{x} - \mu_x}{\frac{\sigma_x}{\sqrt{n}}} = \frac{\bar{x} - \mu_x}{\frac{\sigma_x}{\sqrt{n}} \sqrt{\frac{N-n}{N-1}}} \qquad\qquad 6(\text{vii})$$

The use of the finite population multiplier is illustrated by the following application case-exercise, *Work week*.

Application case-exercise: **Work week**

Situation

A computer firm in France has 290 employees and records that they work an average of 35 hours per week with a standard deviation of 8 hours per week.

Required

1   What is the probability that an employee selected at random will be working within $\pm$ 2 hours per week of the population mean?
2   If a sample size of 19 employees is taken, what is the probability that the sample means lies within $\pm$ 2 hours per week of the population mean?

Worked out solutions

1   In this case again we have a single unit (an employee) taken from the population where the standard deviation σ_x is 8 hour/week. Thus $n = 1$, and $N = 290$.

   $n/N = 1/290 = 0.34\%$ or less than 5% and so the population multiplier is not needed.

   We know that the difference between the random variable and the population, $(x - \mu_x)$, is equal to $\pm$ 2. Thus, assuming that the population follows a normal distribution, then from Equation 6 (iii) for a value of $(x - \mu_x) = +2$:

$$z = \frac{x - \mu_x}{\sigma_x} = +\frac{2}{8} = 0.2500$$

From **NORMSDIST** (Excel 2007) or **NORM.S.DIST** (Excel 2010 and later) in Excel, the area under the curve from the left to a value of z of 0.2500 is 59.87%.

For a value of $(x - \mu_x) = -2$ we have again from Equation 6(ii):

$$z = \frac{x - \mu_x}{\sigma_x} = -\frac{2}{8} = -0.2500$$

Or we could have simply concluded that z is -0.2500 since the assumption is that the curve follows a normal distribution, and a normal distribution is by definition, symmetrical.

From **NORMSDIST** (Excel 2007) or **NORM.S.DIST** (Excel 2010 and later) in Excel, the area under the curve from the left to a value of z of -0.2500 is 40.13%.

Thus the probability that an employee selected at random will be working between ± 2 hours/week is:

$$59.87 \ - \ 40.13 \ = \ 19.74\%.$$

2   In this case again we have a sample, n, of size 19 from a population, N, of size 290. The ratio n/N is

$$\frac{n}{N} = \frac{19}{290} = 0.0655 \text{ or } 6.55\% \text{ of the population}$$

This ratio is greater than 5% and so we use the finite population multiplier in order to calculate the standard error. From Equation 6(vi):

$$\sqrt{\frac{N-n}{N-1}} = \sqrt{\frac{(290-19)}{(290-1)}} = \sqrt{\frac{271}{289}} = \sqrt{0.9377} = 0.9684$$

From Equation 6(v) the corrected standard error of the distribution of the mean is:

$$\sigma_{\bar{x}} = \frac{\sigma_x}{\sqrt{n}} \cdot \sqrt{\frac{N-n}{N-1}} = \frac{8}{\sqrt{19}} * 0.9684 = \frac{8 * 0.9684}{4.3589} = 1.7773$$

From Equation 6(vii) where now $\bar{x} - \mu_x = \pm 2$. First, for $\bar{x} - \mu_x = +2$ we have:

$$z = \frac{\bar{x} - \mu_x}{\frac{\sigma_x}{\sqrt{n}} \cdot \sqrt{\frac{N-n}{N-1}}} = \frac{2}{1.7773} = 1.1253$$

From **NORMSDIST** (Excel 2007) or **NORM.S.DIST** (Excel 2010 and later) in Excel, the area under the curve from the left to a value of z of 1.1253 is 86.98%.

Second, from Equation 6(vii) for $\bar{x} - \mu_x = -2$ we have:

$$z = \frac{\bar{x} - \mu_x}{\frac{\sigma_x}{\sqrt{n}} \cdot \sqrt{\frac{N-n}{N-1}}} = \frac{2}{1.7773} = -1.1253$$

From **NORMSDIST** (Excel 2007) or **NORM.S.DIST** (Excel 2010 and later) in Excel, the area under the curve from the left to a value of z of -1.1253 is 13.02%.

Thus the probability that the sample means lie between ± 2 hours/week is:

$$86.98 - 13.02 = \ 40.13 = 73.96\%.$$

Note that 73.96% is greater than 19.74%, obtained for a sample of size 1, because as we increase the sample size, the sampling distribution of the means is clustered around the population mean. This concept is illustrated in Figure 6.12.

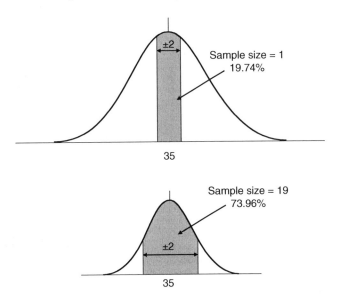

Figure 6.12 Work week.

Sampling for the proportion

In sampling we may not be interested in an absolute value of the population but in a proportion of the population. For example, what proportion of the population will vote Republican in the next US presidential elections? What proportion of the population in London has an annual salary of more than £100,000? What proportion of the houses in Los Angeles County, USA has a market value greater than $1,000,000? In these cases we have established a binomial situation. In the US elections either a person votes Republican or they do not. In London either an individual earns an annual salary of more than £100,000, or they do not. In Los Angeles County, either the houses have a market value greater than $1,000,000 or they do not. In these types of situations we use sampling for proportions.

Measuring the sample proportion

When we are interested in the proportion of the population the procedure is to sample from the population and then again use inferential statistics to draw conclusions about the population proportion. The sample proportion, $\bar{p}$, is the ratio of that quantity, x, taken from the sample having the desired characteristic divided by the sample size, n, or:

$$\bar{p} = \frac{x}{n}$$

6(viii)

For example, assume we are interested in peoples' opinion of gun control. We sample 2,000 people from the State of California and 1,450 say that they are for gun control. The proportion in the sample that says they are for gun control is thus 72.50% (1,450/2,000). We might extend this sample experiment further and say that 72.50% of the population of California is for gun control or even go further and conclude that 72.50% of the US population is for gun control. However, these would be very uncertain conclusions since the 2,000 sample size may be neither representative of California, nor, probably, of the USA. This experiment is binomial because either a person is for gun control, or is not. Thus the proportion in the sample that is against gun control is 27.50% (100% − 72.50%).

Sampling distribution of the proportion

In our sampling process for the proportion, assume that we take a random sample and measure the proportion having the desired characteristic and this is $\bar{p}_1$. We then take another sample from the population and we have a new value $\bar{p}_2$. If we repeat this process then we probably will have different values of $\bar{p}$. The probability distribution of all possible values of the sample proportion, $\bar{p}$, is the **sampling distribution of the proportion**. This is analogous the sampling distribution of the means, $\bar{x}$, of the previous section.

If in the sampling process there are only two possibilities in an outcome then this is binomial. In the binomial distribution the mean number of successes, μ, for a sample size, n, with a characteristic probability of *success, p*, is given by the relationship presented in Equation 4(xv) of *Chapter 4*:

$$\mu = n.p$$

Dividing both sides of this equation by the sample size, n, we have,

$$\frac{\mu}{n} = \frac{n.p}{n} = p \qquad\qquad 6(\text{ix})$$

The ratio μ/n is now the **mean proportion of successes**, written $\mu_{\bar{p}}$. Thus:

$$\mu_{\bar{p}} = p \qquad\qquad 6(\text{x})$$

Using Equation 4(x) from *Chapter 4*, the standard deviation of binomial distribution is given by the relationship,

$$\sigma = \sqrt{n.p.q} = \sqrt{n.p.(1-p)}$$

where the value q = 1 − p.
And again dividing by n:

$$\frac{\sigma}{n} = \frac{\sqrt{p.q.n}}{n} = \sqrt{\frac{p.q.n}{n^2}} = \sqrt{\frac{p.q}{n}} = \sqrt{\frac{p.(1-p)}{n}} \qquad\qquad 6(\text{xi})$$

Where the ratio σ/n is the **standard error of the proportion**, $\sigma_{\bar{p}}$, and thus:

$$\sigma_{\bar{p}} = \sqrt{\frac{p.q}{n}} = \sqrt{\frac{p.(1-p)}{n}} \qquad\qquad 6(\text{xii})$$

The sampling distribution of the proportion showing the standard error is given in Figure 6.13.

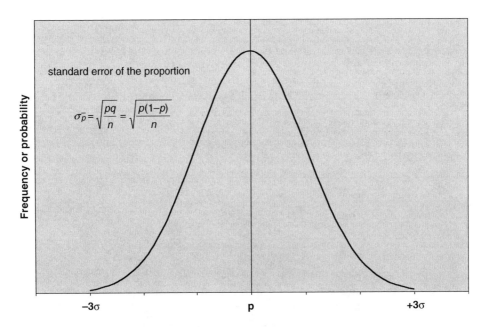

Figure 6.13 Sampling distribution of the proportion.

From Equation 6(iv) we have the relationship:

$$z = \frac{\bar{x} - \bar{\bar{x}}}{\sigma_{\bar{x}}} = \frac{\bar{x} - \mu_x}{\sigma_{\bar{x}}}$$

From Equations 5(iv) and 5(v) of *Chapter 5,* we indicated that we can use the normal distribution to approximate the binomial distribution when the following two conditions apply:

$$n.p \geq 5$$

$$n.(1 - p) \geq 5$$

That is, the product of $n.p$ and $n.(1 - p)$ are both greater or equal to five. Thus if these criteria apply then by substituting in Equation 6(iv) as follows:

$\bar{x}$, the sample mean, by the average sample proportion, $\bar{p}$

μ_x, the population mean, by the population proportion, p

$\sigma_{\bar{x}}$, the standard error of the sample means, by $\sigma_{\bar{p}}$ the standard error of the proportion.

and using the relationship developed in Equation 6(iii), we have:

$$z = \frac{\bar{x} - \bar{x}}{\sigma_{\bar{x}}} = \frac{\bar{x} - \mu_x}{\sigma_{\bar{x}}} = \frac{\bar{p} - p}{\sigma_{\bar{p}}} \qquad\qquad 6(\text{xiii})$$

Since from Equation 6(xii):

$$\sigma_{\bar{p}} = \sqrt{\frac{p.q}{n}} = \sqrt{\frac{p.(1 - p)}{n}}$$

then:

$$z = \frac{\bar{p} - p}{\sqrt{\frac{p.(1 - p)}{n}}}$$

6(xiv)

Alternatively we can say that the difference between the sample proportion, $\bar{p}$, and the population proportion p is,

$$\bar{p} - p = z.\sqrt{\frac{p.(1 - p)}{n}}$$

6(xv)

The application of this relationship is illustrated by the following application case-exercise, *Part-time workers*.

Application case–exercise: Part-time workers

Situation

Many countries in Europe and the USA use part-time workers, who work around 20 hours per week depending on the regulation in the country. Assume that in one particular country part-time employment accounts for 33% of all jobs.

1 If a sample of 100 people from the workforce were taken in that country what proportion, between 25% and 35% in the sample, would be part-time workers?
2 If a sample of 200 people from the workforce were taken in the country what proportion, between 25% and 35% in the sample, would be part-time workers?

Worked out solution

1 Now, the sample size is 100 and so we need to test whether we can use the normal probability assumption by using Equations 5(iv) and 5(v). Here p is 33%, or 0.33, and n is 100, thus from Equation 5(iv):

$n.p = 100 * 0.33 = 33$ or greater than 5.

From Equation 5(v):

$n.(1 - p) = 100(1 - 0.33) = 67$ or again greater than 5.

Thus we can apply the normal probability assumption.
The population proportion p is 33%, or 0.33, and thus from Equation 6(xii) the standard error of the proportion is:

$$\sigma_{\bar{p}} = \sqrt{\frac{0.33(1 - 0.33)}{100}} = \sqrt{\frac{0.33 * 0.67}{100}} = \sqrt{0.0022} = 0.0469.$$

The lower sample proportion, $\bar{p}$, is 25%, or 0.25, and thus from Equation 6(xiii):

$$z = \frac{\bar{p} - p}{\sigma_{\bar{p}}} = \frac{0.25 - 0.33}{0.0469} = -\frac{0.0800}{0.0469} = -1.7058.$$

From **NORMSDIST** (Excel 2007) or **NORM.S.DIST** (Excel 2010 and later) in Excel, the area under the curve from the left to a value of z of -1.7058 is 4.44%.

The upper sample proportion, $\bar{p}$, is 35%, or 0.35, and thus from Equation 6(xiii):

$$z = \frac{\bar{p} - p}{\sigma_{\bar{p}}} = \frac{0.35 - 0.33}{0.0469} = -\frac{0.02}{0.0469} = 0.4264.$$

From **NORMSDIST** (Excel 2007) or **NORM.S.DIST** (Excel 2010 and later) in Excel, the area under the curve from the left to a value of z of 0.4264 is 66.47%.

Thus the proportion between 25% and 35%, in the sample, that would be part-time workers is:

$66.47 - 4.44 = 62.03\%$ or 0.6203

2   First we need to test whether we can use the normal probability assumption by using equations 5(iv) and 5(v). Here p is 33%, or 0.33, and n is 200, thus from Equation 5(iv):

$n.p = 200*0.33 = 66$ or greater than 5.

From Equation 5(v):

$n.(1 - p) = 200(1 - 0.33) = 134$ or again greater than 5

Thus we can apply the normal probability assumption.

The population proportion p is 33%, or 0.33, and thus from Equation 6(xii) the standard error of the proportion is,

$$\sigma_{\bar{p}} = \sqrt{\frac{0.33(1 - 0.33)}{200}} = \sqrt{\frac{0.33 * 0.67}{200}} = \sqrt{0.0011} = 0.0332.$$

The lower sample proportion, $\bar{p}$, is 25%, or 0.25, and thus from Equation 6(xiii):

$$z = \frac{\bar{p} - p}{\sigma_{\bar{p}}} = \frac{0.25 - 0.33}{0.0332} = -\frac{0.0800}{0.0332} = -2.4061.$$

From **NORMSDIST** (Excel 2007) or **NORM.S.DIST** (Excel 2010 and later) in Excel the area under the curve from the left to a value of z of -2.4061 is 0.81%.
The upper sample proportion, $\bar{p}$, is 35%, or 0.35, and thus from Equation 6(xiii):

$$z = \frac{\bar{p} - p}{\sigma_{\bar{p}}} = \frac{0.35 - 0.33}{0.0332} = -\frac{0.02}{0.0332} = 0.6015.$$

From **NORMSDIST** (Excel 2007) or **NORM.S.DIST** (Excel 2010 and later) in Excel, the area under the curve from the left to a value of z of 0.6015 is 72.63%.
Thus the proportion between 25% and 35%, in a sample size of 200 that would be part-time workers is:

$72.63 - 0.81 = 71.82\%$ or 0.7182

Note that again this value is larger than in the first situation since the sample size was 100 rather than 200. As for the mean, as the sample size increases the values will cluster around the mean value of the population. Here the mean value of the proportion for the population is 33% and the sample proportions tested were 25% and 35%, on opposite sides of the mean value of the proportion. This concept is illustrated in Figure 6.14.

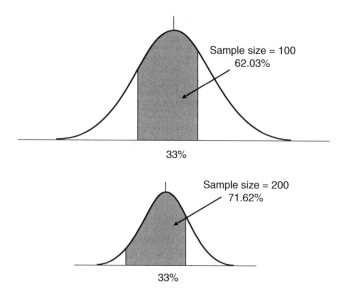

Figure 6.14 Part-time workers.

Moving on

This chapter has covered methods and theory of statistical sampling since, to find out, we must test. The following chapter considers how to use the sampling activity to make reliable estimates of population characteristics.

Notes

1 *Chicago Daily Tribune*, November 3, 1948.
2 Freidel, Frank and Brinkley, Alain, "America in the Twentieth Century," 5th edition, McGraw Hill, New York, 1982, pp. 371-2.

7 Estimating population characteristics

ICEBREAKER: Turkey and the margin of error

As of 2016 the European Union (EU) is made up of 28 countries: Austria; Belgium; Bulgaria; Croatia; Republic of Cyprus; Czech Republic; Denmark; Estonia; Finland; France; Germany; Greece; Hungary; Ireland; Italy; Latvia; Lithuania; Luxembourg; Malta; Netherlands; Poland; Portugal; Romania; Slovakia; Slovenia; Spain; Sweden and the UK. There are talks of admitting other countries (Serbia, Albania, Kosovo) and even of countries leaving (Greece, UK). A big contentious issue for the EU is the admission of Turkey.

Arguments in favor of admitting Turkey are that Istanbul is a great European city that lies at the economic and cultural heart of Turkey, and the country is an invaluable bridge between Europe and Asia. As a member, it would re-invigorate Europe's relations with fast-evolving regions like the energy-rich Caucasus, Central Asia, and the Middle East. Turkey's unique geo-strategic position, plus the strength of NATO's second-largest army, would greatly add to European security. Arguments against Turkey's membership are that Turkey is not a European

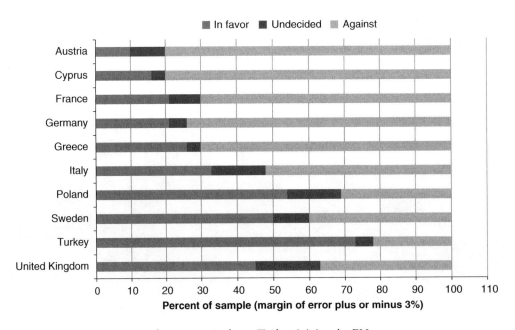

Figure 7.0 Survey response of country attitudes to Turkey joining the EU.

Figure 7.0.1 Turkey, Blue Mosque.

Source: Phil Moynihan, California (with permission).

country—about 97% of its territory lies in Asia—and the EU does not need shared borders with Syria, Iran, and Iraq. Agreeing to one non-European member would open the door for candidates from Cape Verde to Kazakhstan. Turkey is too big for the EU to absorb. With a population predicted to reach 91 million by 2050. This would make it the dominant EU member.

The EU, after a very heated debate, agreed in early October 2005 to open membership talks to admit Turkey, a Muslim country with, at that time, a population of 70 million people. This agreement came only after a tense night and day discussion with Austria, one of the 25 member states, who strongly opposed Turkey's membership. Austria had not forgotten fighting back the invading Ottoman armies in the sixteenth and seventeenth centuries. Reservations to Turkey's membership was also very strong in other countries, as shown in Figure 7.0: an estimated 70% or more of the population in each of Austria, Cyprus, Germany, France and Greece were opposed to membership.

This information was based on a survey response of a sample of about 1,000 people in each of the ten indicated countries conducted in the period May–June 2005 to estimate the feelings of each country's population. The indicated margin of error of this survey is ± 3 percentage points.[1] As of 2016 discussions are ongoing regarding the admission of Turkey to the EU. Estimating from sampling and the margin of error are the essence of this chapter.

Chapter subjects

✓ **Estimating the mean value** • Point estimates • Interval estimates • Confidence level and reliability • Confidence interval for an infinite population • Application case-exercise: *Paper* • Magnitude of sample size • Application case-exercise: *Jars of coffee* • Confidence interval for a finite population • Application case-exercise: *Printing*

✓ **Student's *t* distribution** • Concept • Degrees of freedom • Profile • Confidence intervals • Excel and the Student's *t* distribution • Application case-exercise: *Kiwi fruit* • Sample size • Re-look at the application case-exercise *Kiwi fruit* using the normal distribution

✓ **Estimating and auditing** • Estimating the population amount • Application case-exercise for large populations: *Tee-shirts* • Application case-exercise for a finite population: *Paperback books*

✓ **Estimating the proportion** • Large samples • Sample size for large samples • Application case-exercise: *Circuit boards*

✓ **Margin of error and confidence levels** • Margin of error • Confidence levels

---

*In Chapter 6, Theory and methods of statistical sampling, we discussed statistical sampling for the purpose of obtaining information about a population. This chapter expands upon this to use sampling to estimate, or infer, population parameters based entirely on the sample data. By its very nature, estimating is **probabilistic** as there is no certainty of the result. However, if the sample experiment is correctly designed then there should be a reasonable confidence about conclusions that are made. Thus from samples we might, with confidence, estimate the mean weight of airplane passengers for fuel-loading purposes, the proportion of the population expected to vote Republican, or the mean value of inventory in a distribution center.*

Estimating the mean value

The mean, or average value, of data is the sum of all the data taken divided by the number of measurements taken. The units of measurement can be financial units, length, volume, weight, etc.

Point estimates

In estimating, we could use a single value to estimate the true population mean. For example, if the grade point average of a random sample of students is 3.75 then we might estimate that the population average of all students is also 3.75. Or, we might select at random 20 items of inventory from a distribution center and calculate that their average value is £25.45. In this case we would estimate that the population average of the entire inventory is £25.45. Here we have used the sample mean $\bar{x}$ as a **point estimate** or an **unbiased estimate** of the true population mean, μ_x. The problem with one value or a point estimate is that they are presented as being exact and that unless we have a *super crystal ball*, the probability of them being precisely the right value is low. Point estimates are often inadequate since they are just a single value and are either right, or wrong. In practice it is more meaningful to have an interval estimate and to quantify these intervals by probability levels that give an estimate of the error in the measurement.

Interval estimates

With an **interval estimate** we might describe situations like one of the following examples. The estimate for the project cost is between $11.8 and $12.9 million and I am 95% confident of these figures. The estimate for the sales of the new products is between 22,000 and 24,500 units in the first year and I am 90% confidence of these figures. The estimate of the price of a certain stock is between $75 and $90 but I am only 50% confident of this information. The estimate of class enrolment for Business Statistics next academic year is between 220 and 260 students though I am not too confident about these figures. Thus the interval estimate is a range within which the population parameter is likely to fall.

Confidence level and reliability

Suppose Subcontractor A makes refrigerator compressors for Client B who assembles the final refrigerators. In order to establish the terms of the final customer warranty, the client needs information about the life of compressors since the compressor is the principal working component of the refrigerator. Assume that a random sample of 144 compressors is tested and that the mean life of the compressors, $\bar{x}$, is determined to be 6 years or 72 months. Using the concept of point estimates we could say that the mean life of all the compressors manufactured is 72 months. Here $\bar{x}$ is the **estimator** of the population mean, μ_x and 72 months is the **estimate** of the population mean obtained from the sample. However, this information says nothing about the reliability or confidence that we have in the estimate.

The subcontractor has been making these compressors for a long time and knows from past data that the standard deviation of the working life of compressors is 15 months. Then since our sample size of 144 is large enough the standard error of the mean can be calculated by using from *Chapter 6*, Equation 6(ii) from the central limit theorem:

$$\sigma_{\bar{x}} = \frac{\sigma_x}{\sqrt{n}} = \frac{15}{\sqrt{144}} = \frac{15}{12} = 1.25 \text{ months}$$

This value of 1.25 months is one standard error of the mean, or it means that $z = \pm 1.00$, for the sampling distribution. If we assume that the life of a compressor follows a normal distribution then we know from *Chapter 5* that 68.26% of all values in the distribution lie within ± 1 standard deviations from the mean. From Equation, 6(iv):

$$\pm z = \frac{\bar{x} - \mu_x}{\frac{\sigma_x}{\sqrt{n}}}$$

or:

$$\pm 1 = \frac{\bar{x} - 72}{1.25}$$

When $z = -1$ then the lower limit of the compressor life is:

$$\bar{x} = 72 - 1.25 = 70.75 \text{ months}$$

When $z = +1$ then the upper limit is:

$$\bar{x} = 72 + 1.25 = 73.25 \text{ months}$$

Thus we can say that the mean life of the compressors is about 72 months and there is a 68.26% (about 68%) probability that the mean value will be between 70.75 and 73.25 months.

Two standard errors of the mean, or when $z = \pm 2$, is 2*1.25 or 2.50 months. Again from *Chapter 5,* if we assume a normal distribution, 95.44% of all values in the distribution lie within ± 2 standard deviations from the mean.

When $z = -2$ then using Equation 6(iv), the lower limit of the compressor life is:

$$\bar{x} = 72 - 2*1.25 = 69.50 \, \text{months}$$

When $z = +2$ then the upper limit is:

$$\bar{x} = 72 + 2*1.25 = 74.50 \, \text{months}$$

Thus we can say that the mean life of the compressor is about 72 months and there is a 95.44% (about 95%) probability that the mean value will be between 69.50 and 74.50 months.

Finally, three standard errors of the mean is 3*1.25 or 3.75 months and again from *Chapter 5,* assuming a normal distribution, 99.73% of all values in the distribution lie within ± 3 standard deviations from the mean.

When $z = -3$ then using Equation 6(iv), the lower limit of compressor life is:

$$\bar{x} = 72 - 3*1.25 = 68.25 \, \text{months}$$

When $z = +3$ then the upper limit is:

$$\bar{x} = 72 + 3*1.25 = 75.75 \, \text{months}$$

Thus we can say that the mean life of the compressor is about 72 months and there is almost a 99.73%, (about 100 %) probability that the mean value will be between 68.25 and 75.75 months.

Thus in summary we can say that moving from a low to a high level of confidence:

- The best estimate is that the mean compressor life is 72 months and the manufacturer is about *68%* confident that the compressor life is in the range 70.75 to 73.25 months. Here the confidence interval is between 70.75 and 73.25 months, or a range of 2.50 months.
- The best estimate is that the mean compressor life is 72 months and the manufacturer is about *95%* confident that the compressor life is in the range 69.50 to 74.50 months. Here the confidence interval is between 69.50 and 74.50 months, or a range of 5.00 months.
- The best estimate is that the mean compressor life is 72 months and the manufacturer is *about 100%* confident that the compressor life is in the range 68.25 to 75.75 months. Here the confidence interval is between s are 68.25 and 75.75 months, or a range of 7.50 months.

It is important to note that as our confidence level increases, from 68% to 100%, the confidence interval expands, from a range of 2.50 to 7.50 months. This is to be expected since as we become more confident of our estimate, we give a broader range to cover uncertainties.

Confidence interval for an infinite population

The confidence interval is the range of the estimate being made. From the above compressor example, considering the $\pm 2\sigma$ confidence intervals, we have 69.50 and 74.50 months as the respective lower and upper limits. Between these limits this is equivalent to 95.44% of the area under the normal curve, or about 95%. A 95% confidence interval estimate implies that if all possible samples were taken, about 95% of them would include the true population mean, μ, somewhere within their interval, whereas about 5% of them would not. This concept is illustrated in Figure 7.1 for six different samples. The $\pm 2\sigma$ intervals for sample numbers 1, 2, 4, and 5 contain the population mean, μ, whereas the $\pm 2\sigma$ intervals for samples 3 and 6 do not contain the population mean, μ, within their interval.

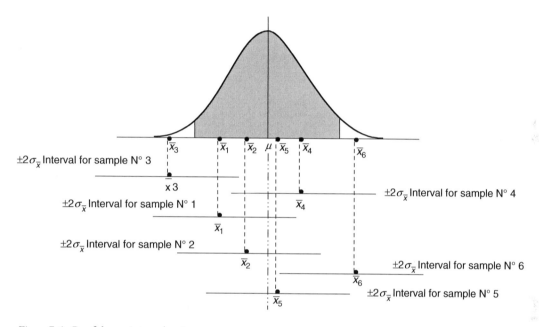

Figure 7.1 Confidence interval estimate.

The **level of confidence** is $(1 - \alpha)$, where α is the total proportion in the tails of the distribution outside of the confidence interval. Since the distribution is symmetrical the area in each tail is $\alpha/2$, as shown in Figure 7.2.

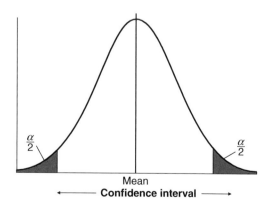

Figure 7.2 Confidence interval, and the area in the tails.

As we have shown in the compressor situation, the confidence intervals for the population estimate for the mean value are thus:

$$\text{Confidence intervals are, } \bar{x} \pm z.\sigma_{\bar{x}} = \bar{x} \pm z \cdot \frac{\sigma_x}{\sqrt{n}} \qquad\qquad 7(\text{i})$$

This implies that the population mean lies in the range given by the relationship:

$$\bar{x} - z \cdot \frac{\sigma_x}{\sqrt{n}} \leq \mu_x \leq \bar{x} + z \cdot \frac{\sigma_x}{\sqrt{n}} \qquad\qquad 7(\text{ii})$$

Estimation and confidence limits are illustrated by the following application case-exercise, *Paper*.

Application case-exercise: **Paper**

Situation

Inacopia, the Portuguese manufacturer of A4 paper commonly used in computer printers, wants to be sure that its cutting machine is operating correctly. The width of A4 paper is expected to be 21.00 cm and it is known that the standard deviation of the cutting machine is 0.0100 cm. The quality control inspector pulls a random sample of 60 sheets from the production line and the average width of this sample is 20.9986 cm.

Required

1   Determine the 95% confidence intervals of the mean width of all the A4 paper coming off the production line.
2   Determine the 99% confidence intervals of the mean width of all the A4 paper coming off the production line.

Worked out solutions

1   We have the following information:

Sample size, *n,* is 60

Sample mean, $\bar{x}$, is 20.9986 cm

Population standard deviation of the operation, σ, is 0.0100
Standard error of the mean is:

$$\frac{\sigma}{\sqrt{n}} = \frac{0.0100}{\sqrt{60}} = 0.0013$$

The area in the each tail for a 95% confidence limit is 2.5%. Using **NORMSINV** (Excel 2007) or **NORM.S.INV** (Excel 2010 and later) in Excel for a value *P(x)* of 2.5% gives a lower value of *z* of −1.9600. Since the distribution is symmetrical, the upper value is numerically the same at +1.9600. (Note: an alternative way of finding the upper value of

z is to enter in **NORMSINV** (Excel 2007) or **NORM.S.INV** (Excel 2010 and later) the value of 97.50% (2.50% + 95%), which is the area of the curve from the left to the upper value of z).

From Equation 7(i) the confidence limits are:

$$20.9986 \pm 1.9600 * 0.0013 = 20.9961 \text{ and } 21.0011 \text{ cm}$$

Thus we would say that our best estimate of the width of the computer paper is 20.9986 cm and we are 95% confident that the width is in the range 20.9961 and 21.0011. Since this interval contains the population expected mean value of 21.0000 cm we can conclude that there seems to be no problem with the cutting machine.

2.   The area in each tail for a 99% confidence limit is 0.5%. Using **NORMSINV** (Excel 2007) or **NORM.S.INV** (Excel 2010 and later) in Excel for a value $P(x)$ of 0.5% gives a lower value of z of −2.5758. Since the distribution is symmetrical, the upper value is +2.5758. (Note: an alternative way of finding the upper value of z is to enter in **NORMSINV** (Excel 2007) or **NORM.S.INV** (Excel 2010 and later) the value of 99.50% (0.50% + 99%), which is the area of the curve from the left to the upper value of z).

From Equation 7(i) the confidence limits are:

$$20.9986 \pm 2.5758 * 0.0013 = 20.9953 \text{ and } 21.0019 \text{ cm}$$

Thus we would say that our best estimate of the width of the computer paper is 20.9986 cm and we are 99% confident that the width is in the range 20.9953 and 21.0019. Again, since this interval contains the expected mean value of 21.0000 cm we can conclude that there seems to be no problem with the cutting machine.

Note that the limits in Question 2 are wider than those in Question 1 since we have a higher confidence level.

Magnitude of sample size

In sampling it is useful to know the size of the sample to take in order to estimate the population parameter for a given confidence level. We have to accept that unless the whole population is analyzed there will always be a sampling error. If the sample size is small, the chances are that the error will be high. If the sample size is large there may be only a marginal gain in reliability in the estimate of our population mean but what is certain that the analytical experiment will be more expensive. Thus, what is an appropriate sample size, n, to take for a given confidence level?

The confidence limits are related the sample size, n, by Equation 6(iv):

$$\pm z = \frac{\bar{x} - \mu_x}{\frac{\sigma_x}{\sqrt{n}}}$$

The range from the population mean, on the left side of the distribution is when z is negative is $-(\bar{x} - \mu_x)$ or $\mu_x - \bar{x}$ on the left side of the distribution, and $\bar{x} - \mu_x$ on the right side of the distribution curve. Reorganizing Equation 6(iv) by making the sample size, n, the subject gives:

$$n = \left(\frac{z\sigma_x}{\bar{x} - \mu_x}\right)^2 \qquad\qquad\qquad 7(\text{iii})$$

The term $\bar{x} - \mu_x$ is the sample error and if we denote this by e, then the sample size is given by:

$$n = \left(\frac{z\sigma_x}{e}\right)^2 \qquad\qquad 7(iv)$$

Thus for a given confidence level, which then gives the value of z, and a given confidence limit the required sample size can be determined. Note in Equation 7(iv) since n is given by squared value it does not matter if we use a negative or positive value for z. The following application case-exercise, *Jars of coffee*, illustrates the concept of confidence intervals and sample size for an infinite population.

Application case-exercise: Jars of coffee

Situation

The quality control inspector of a filling machine for coffee wants to estimate the mean weight of coffee in its 200 g jars to within ± 0.50 g. It is known that the standard deviation of the coffee filling machine is 2.00 g.

Required

1 What sample size should the inspector take to be 95% confidence of the estimate?

Worked out solution

1 Using Equation 7(iv):

$$n = \left(\frac{z\sigma_x}{e}\right)^2$$

The area in each tail for a 95% confidence limit is 2.5%. Using **NORMSINV** (Excel 2007) or **NORM.S.INV** (Excel 2010 and later) in Excel for a value $P(x)$ of 2.5% gives a lower value of z of -1.9600. Since the distribution is symmetrical, the upper value is numerically the same at $+1.9600$. (Note: an alternative way of finding the upper value of z is to enter in **NORMSINV** (Excel 2007) or **NORM.S.INV** (Excel 2010 and later) the value of 97.50% (2.50% + 95%), which is the area of the curve from the left to the upper value of z).
For the example:

- z is 1.9600 (it does not matter whether we use plus or minus since we square the value)
- σ_x is 2.00 g
- e is ±0.50 g

$$n = \left(\frac{1.9600 * 2.00}{0.50}\right)^2 = 61.463 = 62 \, (\textit{rounded up})$$

Thus the quality control inspector should take a sample size of 62(61 would be just slightly too small).

Confidence interval for finite population

As discussed in *Chapter 6* (Equation 6(vi)), if the population is considered finite, that is the ratio n/N is greater than 5%, then the standard error should be modified by the finite population multiplier according to the expression:

$$\sigma_{\bar{x}} = \frac{\sigma_x}{\sqrt{n}} \cdot \sqrt{\frac{N-n}{N-1}}$$

In this case the confidence limits for the population estimation from Equation 7(i) are modified as follows:

$$\bar{x} \pm z \cdot \sigma_{\bar{x}} = \bar{x} \pm z \cdot \frac{\sigma_x}{\sqrt{n}} \cdot \sqrt{\frac{(N-n)}{(N-1)}} \qquad\qquad 7(\text{v})$$

The confidence interval for a finite population is illustrated by the following case-exercise, *Printing*.

Application case-exercise: **Printing**

Situation

A printing firm runs off the first edition of a textbook of 496 pages. After the book is printed, the quality control inspector looks at 45 random pages selected from the book and finds that the average number of errors in these pages is 2.70. These include printing errors of color and alignment, but also typing errors which originate from the author and the editor. The inspector knows that based on past contracts for a first edition of a book the standard deviation of the number of errors per page is 0.5.

Required

1   What is a 95% confidence interval for the mean number of errors in the book?

Worked out solution

1   We have the following information:
Sample size, n, is 45
Population size, N, is 496
Sample mean, $\bar{x}$, errors per page is 2.70
Population standard deviation, σ, is 0.5
Ratio of n/N is $45/496 = 9.07\%$
This value is greater than 5%; thus we must use the finite population multiplier.

$$\sqrt{\frac{N-n}{N-1}} = \sqrt{\frac{496-45}{496-1}} = \sqrt{\frac{451}{495}} = 0.9545$$

Uncorrected standard error of the mean is $\frac{\sigma_x}{\sqrt{n}} = \frac{0.5}{\sqrt{45}} = 0.0745$

Corrected standard error of the mean; $\sigma_{\bar{x}} = \frac{\sigma_x}{\sqrt{n}} \cdot \sqrt{\frac{N-n}{N-1}} = 0.0745*0.9545 = 0.0711$.

Confidence level is 95%; thus area in each tail is 2.5%.

Using **NORMSINV** (Excel 2007) or **NORM.S.INV** (Excel 2010 and later) in Excel for a value *P(x)* of 2.5% gives a lower value of *z* of -1.9600. Since the distribution is symmetrical, the upper value is numerically the same at $+1.9600$.

Thus from Equation 7(v) the lower confidence limit is:

$$2.70 - 1.9600*0.0711 = 2.56$$

Thus from Equation 7(v) the upper confidence limit is:

$$2.70 + 1.9600*0.0711 = 2.84$$

Thus we could say that the best estimate of the number of errors in the book is 2.70 per page and that we are 95% confident that the number of errors lies between 2.56 and 2.84 per page.

Student's *t* distribution

There may be situations in estimating when we do not know the population standard deviation and the sample size is small. In this case an alternative distribution that we apply is called the Student's *t* distribution, or more simply the *t*-distribution.

Concept

In *Chapter 6,* in the paragraph entitled, "Sample size and shape of the sampling distribution of the means," we indicated that the sample size taken has an influence on the shape of the sampling distributions of the means. If we sample from population distributions that are normal, such that we know the standard deviation, σ, any sample sizes will give a sampling distribution of the means that are approximately normal. However, if we sample from populations that are not normal, we are obliged to increase our sampling size to at least 30 units in order that the sampling distribution of the means will be approximately normally distributed. Thus, what do we do when we have small sample sizes that are less than 30 units? To be statistically correct, we should use a **Student's *t* distribution**.

The Student's *t* distribution is for small amounts of data and, like the normal distribution, is a continuous distribution. It was developed by William Gossett of the Guinness Brewery, in Dublin, Ireland in 1908 (presumably when he had time between beer production!) and published under the pseudonym "Student" as the Guinness company would not allow him to put his own name to the development. The Student's *t* distributions are a family of distributions, each one a different shape and characterized by a parameter called the degrees of freedom. The density function from which the Student's *t* distribution is drawn is as follows:

$$f(t) = \frac{[(v-1)/2]!}{\sqrt{v \cdot \pi}[(v-2)/2]}\left[1 + \frac{t^2}{v}\right]^{-(v+1)/2} \tag{7(vi)}$$

Here, *y* is the degree of freedom, π is the value of 3.1416, and *t* is the Student's *t* value on the *x*-axis, similar to the *z* value of a normal distribution.

Degrees of freedom

The degrees of freedom are the choices you have regarding taking certain actions. For example, what is the degree of freedom that you have in maneuvering your car into a parking place? What is the degree of freedom that you have in contract negotiations? What is the degree of freedom that you have in negotiating a black run on the ski slopes? In the context of statistics the degrees of freedom in a Student's *t* distribution are given by $(n-1)$, where *n* is the sample size. This then implies that there is a degree of freedom for every sample size. To understand quantitatively the degrees of freedom consider the following.

There are five variables *v*, *w*, *x*, *y*, and *z* that are related by the following equation:

$$\frac{v + w + x + y + z}{5} = 13 \qquad\qquad 7\text{(vii)}$$

Since there are five variables we have a choice, or the degree of freedom, to select four of the five. After that, the value of the fifth variable is automatically fixed by the equation total amount. Assume that we give *v*, *w*, *x*, and *y* the values 14, 16, 12, and 18 respectively. Then from Equation 7(vii) we have:

$$\frac{14 + 16 + 12 + 18 + z}{5} = 13, \text{ and so}$$

$$z = 5*13 - (14 + 16 + 12 + 18) = 65 - 60 = 5$$

Thus automatically the fifth variable, *z*, is fixed at a value of 5 in order to retain the validity of the equation. Here we had five variables to give a degree of freedom of four. In general terms, for a sample size of *n* units, the degree of freedom is the value determined by $(n-1)$.

Profile

Three Student's *t* distributions, for sample size *n* of 6, 12, and 22, or sample sizes less than 30, are illustrated in Figure 7.3.

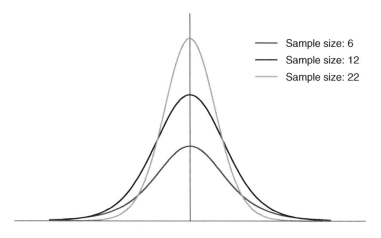

Figure 7.3 Three Student's *t* distributions for different sample sizes.

The degrees of freedom for these curves, using $(n-1)$ are respectively 5, 11, and 21. These three curves have a profile similar to the normal distribution but if we superimposed a normal distribution on a Student's t distribution as shown in Figure 7.4, we see that the normal distribution is higher at the peak and the tails are closer to the x-axis, compared to the Student's t distribution.

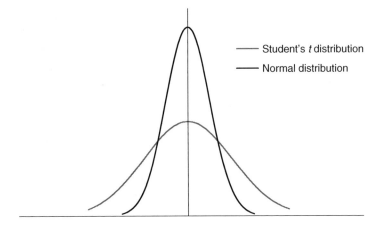

——— Student's t distribution

——— Normal distribution

Figure 7.4 Normal and Student's t distributions.

The Student's t distribution is flatter and you have to go further out on either side of the mean value before you are close to the x-axis, indicating greater variability in the sample data. This is the penalty you pay for small sample sizes and where the sampling is taken from a non-normal population. As the sample size increases the profile of the Student's t distribution approaches that of the normal distribution: for example, in Figure 7.3, the curve for a sample size of 22 has a smaller variation and is higher at the peak.

Confidence intervals

When we have a normal distribution the confidence intervals of estimating the mean value of the population are as given in Equation 7(i):

$$\bar{x} \pm z \cdot \frac{\sigma_x}{\sqrt{n}}$$

When we are using a Student's t distribution, Equation 7(i) is modified to give the following.

$$\bar{x} \pm t \cdot \frac{\widehat{\sigma}_x}{\sqrt{n}} \qquad\qquad 7(\text{viii})$$

Here the value of t has replaced z, and $\widehat{\sigma}$ has replaced σ, the population standard deviation. This new term, $\widehat{\sigma}$, is an estimate of the population standard deviation. Numerically it is equal to s, the sample standard deviation, by the relationship:

$$\widehat{\sigma} = s = \sqrt{\frac{\Sigma(x - \bar{x})^2}{(n-1)}} \qquad\qquad 7(\text{vix})$$

We could avoid writing $\widehat{\sigma}$, as some texts do, and simply write s since they are numerically the same. However, by putting $\widehat{\sigma}$ it is clear that our only alternative to estimate our confidence limits is to use an estimate of the population standard deviation as measured from the sample value.

Excel and the Student's t distribution

Two functions in Excel for applying the Student's *t* distribution are **TDIST** (Excel 2007) or **T.DIST** (Excel 2010 or later) and **TINV** (Excel 2007) or **T.INV.2T** (Excel 2010 or later).

- The **TDIST** (Excel 2007) or **T.DIST** (Excel 2010 or later) gives the area in the tails, or the area outside the distribution, by entering the value of the Student's *t* the degrees of freedom; and whether the situation is unilateral (one tail) or bilateral (two tails).
- The **TINV** (Excel 2007) or **T.INV.2T** (Excel 2010 or later) gives the two-tailed value of the Student's *t*, by entering the **total** area in the tails, α, or that area outside the distribution. There are also left- and right-handed versions in more recent versions of Excel.

Note that the logic of using the Student's *t* distribution is different than for the normal distribution, where you enter the area of the curve from the extreme left to a value on the *x*-axis.

The following application case-exercise, *Kiwi fruit* illustrates estimating using the Student's *t*, distribution.

Application case-exercise: Kiwi fruit

Situation

Sheila Hope, an agricultural inspector at Los Angeles, California wants to know, in mg, the level of vitamin C in a boat load of kiwi fruits imported from New Zealand in order to compare this information with kiwi fruits grown in the Central Valley, California. Sheila took a random sample of 25 kiwis from the ship's hold and measured their vitamin C content. Table 7.1 gives the results in milligrams per kiwi fruit sampled.

Table 7.1

109	88	91	136	93
101	89	97	115	92
114	106	94	109	110
97	89	117	105	92
83	79	107	100	93

Required

1   Estimate the average level of vitamin C in the imported kiwi fruits and give a 95% confidence level of this estimate. How would you express your findings?

Worked out solution

1   Since we have no information about the population standard deviation, and the sample size of 25 is less than 30, to be statistically correct we use a Student's *t*, distribution.
Using **AVERAGE**, mean value of the sample, $\bar{x}$ is 100.24.
Using **STDEV** (Excel 2007) or **STDEV.S** (EXCEL 2010 or later), standard deviation of the sample, *s*, is 12.6731.

Sample size, *n,* is 25.

Using **SQRT**, square root of the sample size, $\sqrt{n}$, is 5.00.

Estimate of the population standard deviation, $\widehat{\sigma} = s$, = 12.6731 of the sample distribution, $\frac{\widehat{\sigma}_x}{\sqrt{n}} = \frac{12.6731}{5.00} = 2.5346$.

Required confidence level (given) is 95%.

Area outside of confidence interval, α, $(100 - 95\%)$ is 5%.

Degrees of freedom, $(n - 1)$ or 24 $(25 - 1)$.

Using **TINV** (Excel 2007) or **T.INV.2T** (Excel 2010 or later), Student's *t* value is 2.0639

From Equation 7(viii):

Lower confidence level: $\bar{x} - t \cdot \frac{\widehat{\sigma}_x}{\sqrt{n}} = 100.24 - 2.0639*2.5346 = 100.24 - 5.2312 = 95.01$

Upper confidence level: $\bar{x} + t \cdot \frac{\widehat{\sigma}_x}{\sqrt{n}} = 100.24 + 2.0639*2.5346 = 100.24 + 5.2312 = 105.47$

The range of the confidence level is 10.46 mg $(105.47 - 95.01)$.

These results are illustrated on the Student's *t* distribution of Figure 7.5.

Thus the situation is expressed by saying that, "I estimate the average level of vitamin C in all of these imported kiwis is 100.24 mg and I am 95% confident that the lower level of the estimate is 95.01 mg and the upper level 105.47 mg."

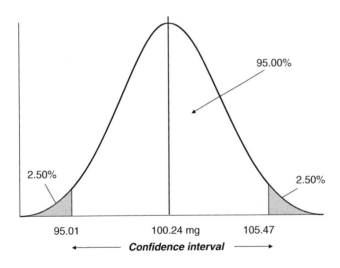

Figure 7.5 Confidence intervals for kiwi fruit.

Sample size

We have said that the Student's *t* distribution should be used when the sample size is less than 30 and the population standard deviation is unknown. Some statisticians are more rigid and use a sample size of 120 as the cut-off point. What should we use, a sample size of 30; a sample size of 120; or some other value? The movement of the value of *t* relative to the value of *z* is illustrated by the data in Table 7.2. Here we have the Student's *t* value for a confidence level of 95% for sample sizes ranging from 5 to 200. For the given confidence level in a Student's *t* distribution there is 2.50% in each tail. For the normal distribution the left-hand value of *z* is 2.50% and the right-hand value is at 97.50% (2.50 + 95.00) The value of *z* is constant at the 95% confidence level since *z* is not a function of sample size. The Student's *t* value in contrast declines with sample size as shown. The column $(t - z)/z$ gives the difference of the Student's *t* relative to the value *z*. For a sample size of 30 the difference between *t* and *z* is 4.35%. For a sample size of

Table 7.2

Sample size, n	Upper Student's t	Upper value z	(t − z)/z	Sample size, n	Upper Student's t	Upper value z	(t − z)/z
5	2.7764	1.95996	41.66%	130	1.9785	1.95996	0.95%
10	2.2622	1.95996	15.42%	135	1.9778	1.95996	0.91%
15	2.1448	1.95996	9.43%	140	1.9772	1.95996	0.88%
20	2.0930	1.95996	6.79%	145	1.9766	1.95996	0.85%
25	2.0639	1.95996	5.30%	150	1.9760	1.95996	0.82%
30	2.0452	1.95996	4.35%	155	1.9755	1.95996	0.79%
35	2.0322	1.95996	3.69%	160	1.9750	1.95996	0.77%
40	2.0227	1.95996	3.20%	165	1.9745	1.95996	0.74%
45	2.0154	1.95996	2.83%	170	1.9741	1.95996	0.72%
50	2.0096	1.95996	2.53%	175	1.9737	1.95996	0.70%
55	2.0049	1.95996	2.29%	180	1.9733	1.95996	0.68%
60	2.0010	1.95996	2.09%	185	1.9729	1.95996	0.66%
65	1.9977	1.95996	1.93%	190	1.9726	1.95996	0.64%
70	1.9949	1.95996	1.78%	195	1.9723	1.95996	0.63%
75	1.9925	1.95996	1.66%	200	1.9720	1.95996	0.61%
80	1.9905	1.95996	1.56%	205	1.9717	1.95996	0.60%
85	1.9886	1.95996	1.46%	210	1.9714	1.95996	0.58%
90	1.9870	1.95996	1.38%	215	1.9711	1.95996	0.57%
95	1.9855	1.95996	1.30%	220	1.9709	1.95996	0.56%
100	1.9842	1.95996	1.24%	225	1.9706	1.95996	0.54%
105	1.9830	1.95996	1.18%	230	1.9704	1.95996	0.53%
110	1.9820	1.95996	1.12%	235	1.9702	1.95996	0.52%
115	1.9810	1.95996	1.07%	240	1.9699	1.95996	0.51%
120	1.9801	1.95996	1.03%	245	1.9697	1.95996	0.50%
125	1.9793	1.95996	0.99%	250	1.9695	1.95996	0.49%

120 the difference is just 1.03%. Is this difference significant? It depends on what you are sampling. In the medical field small differences may be important but in the business world perhaps less so.

Figure 7.6 shows Table 7.2 graphically. It illustrates the exponential decline of how the values of the Student's *t* decline with sample size to approach the normal *z* value that is a horizontal straight line.

The following is a re-look at the application case-exercise *Kiwi fruit* using the normal *z* application.

Application case-exercise: **Kiwi fruit** *(now using the normal distribution)*

Required

1   Estimate the average level of vitamin C in the imported kiwi fruits and give a 95% confidence level of this estimate on the basis that a normal distribution can be used.

Worked out solution

1   Here all the provided data and the calculations are the same as previously, but now we will normal distribution for our analysis.

Required confidence level (given) is 95%.

Area outside of confidence interval, α, (100 − 95%) is 5%, which means that there is an area of 2.5% in both tails for a symmetrical distribution. Using **NORMSINV** (Excel 2007) or **NORM.**

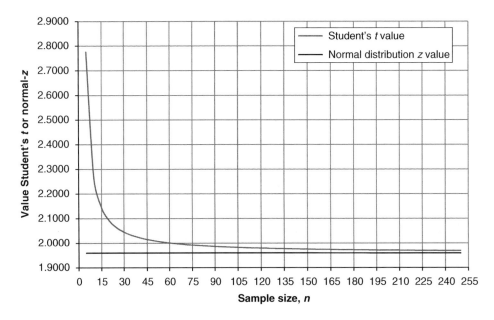

Figure 7.6 Comparing values of Student's *t* and normal *z* according to sample size.

S.INV (Excel 2010 and later) in Excel for a value *P(x)* of 2.5% gives the value of *z* as $\pm$ 1.9600. From Equation 7(i):

Lower confidence level: $\bar{x} - z \cdot \dfrac{\hat{\sigma}_x}{\sqrt{n}} = 100.24 - 1.9600 \times 2.5346 = 100.24 - 4.9678 = 95.27$

Upper confidence level: $\bar{x} + z \cdot \dfrac{\hat{\sigma}_x}{\sqrt{n}} = 100.24 + 1.9600 \times 2.5346 = 100.24 + 4.9678 = 105.21$

This gives a confidence range of 9.94 mg (105.21 − 95.27). This compares to a confidence range of 10.46 obtained using a Student's *t* distribution.

The corresponding values that we obtained by using the Student's *t* distribution were 105.47 and 95.01, or a difference of only some 0.3% in the values of the confidence levels. Since in reality we would probability report our confidence for the vitamin level of kiwis between 95 and 105 mg the difference between using *z* and *t* in this case is insignificant.

Estimating and auditing

Auditing is the methodical examination of financial accounts, inventory items, or operating processes and other activities to verify that they conform with standard practices, specifications, or target budget levels.

Estimating the population amount

We can use the concepts of estimating developed in this chapter to assess the total value of goods such as inventory held in a distribution center when for example it is difficult or very time-consuming to make an audit of the population. In this case we first take a random and representative sample and determine the mean financial value $\bar{x}$. If *N* is the total number of units, then the point estimate for the population total is the size of the population, *N*, multiplied by the sample mean, or:

Total $= N \cdot \bar{x}$ 7(x)

It is unlikely that we would know the standard deviation of the large population of inventory and so we would estimate the value from the sample. If the sample size is less than 30 we use the Student's t distribution. The confidence intervals are determined as follows, by multiplying both terms in Equation 7(viii) to give:

$$\text{Confidence intervals, } N \cdot \bar{x} \pm N \cdot t \frac{\hat{\sigma}}{\sqrt{n}} \qquad \qquad 7(\text{xi})$$

Alternatively, if the population is considered finite, that is the ratio of $n/N \geq 5\%$, then the standard error has to be modified by the estimated finite population multiplier to give:

$$\text{Estimated standard error} = \frac{\hat{\sigma}}{\sqrt{n}} \sqrt{\frac{N - n}{N - 1}} \qquad \qquad 7(\text{xii})$$

Thus the confidence intervals when the standard deviation is unknown, the sample size is less than 30, and the population is finite, are:

$$\text{Confidence intervals} = N \cdot \bar{x} \pm N \cdot t \frac{\hat{\sigma}}{\sqrt{n}} \sqrt{\frac{N - n}{N - 1}} \qquad \qquad 7(\text{xiii})$$

The following application case-exercise *Tee-shirts* illustrates auditing for large populations.

Application case-exercise for large populations: Tee-shirts

Situation

A store on Duval Street in Key West, Florida, wishes to estimate the total retail value of its tee-shirts, tank tops and sweaters that it has in its store. The inventory records indicate that there are 4,500 of these clothing articles on the shelves. The owner takes a random sample of 29 items and Table 7.3 shows the prices in $US indicated on the articles.

Required

1   Estimate the total retail value of the clothing items within a 99% confidence limit.

Worked out solution

1   From the data in Table 7.3 and using in Excel **AVERAGE** the sample mean value, $\bar{x}$ is $25.31. The sample size, n, is 29.
    Population size, N, is 4,500. Thus the estimated total retail value is $N \cdot \bar{x} = 4,500*25.31$ or $113,896.55

Table 7.3

16.50	25.00	25.50	42.00	37.00	22.00
21.00	20.00	21.00	9.50	24.50	11.50
52.50	15.50	32.50	18.00	18.50	19.00
29.50	16.00	21.00	44.00	17.50	50.50
27.00	29.50	12.50	32.00	23.00	

Sample size, *n,* is 29; ratio n/N is 29/4,500 or 0.64%. Since this value is less than 5% we do not need to use the finite population multiplier.

Sample standard deviation, *s,* is $11.0836 and since we have no value for the population we estimate the population standard deviation, $\widehat{\sigma}$, as also $11.0836.

Estimated standard error of the sample distribution: $\frac{\widehat{\sigma_x}}{\sqrt{n}} = \frac{11.0836}{\sqrt{29}}$ is 2.0582.

Since we do not know the population standard deviation, and the sample size is less than 30, we use the Student's *t* distribution.

Degrees of freedom $(n - 1)$ or 28.

Using Excel **TINV** (Excel 2007) or **T.INV.2T** (Excel 2010 or later) for a 99% confidence level, the Student's *t* value is 2.7633.

From Equation 7(xi) the lower confidence limit for the total value is:

$$N \cdot \bar{x} - N \cdot t \frac{\widehat{\sigma}}{\sqrt{n}} = \$113,896.55 - 4,500*2.7633*2.0582 \text{ or } 88,303.78$$

From Equation 7(xi) the upper confidence limit is,

$$N \cdot \bar{x} + N \cdot t \frac{\widehat{\sigma}}{\sqrt{n}} = \$113,896.55 + 4,500*2.7633*2.0582 \text{ or } \$139,489.33$$

Thus the owner estimates the average, or point estimate, of the total retail value of the clothing items in his Key West store as $113,897 (rounded) and he is 99% confident that the value lies between $88,303.78 (say $88,304 rounded) and $139,489.33 (say $139,489 rounded).

The following application case-exercise, *Paperback books* illustrates auditing for small populations.

Application case-exercise for a finite population: Paperback books

Situation

A newspaper and bookstore at Waterloo Station wants to estimate the value of paperback books it has in its store. The owner takes a random sample of 28 books and determines that the average retail value is £4.57 with a sample standard deviation of 53 pence. There are 12 shelves of books and the owner estimates that there are 45 books per shelf.

Required

1   Estimate the total retail value of the books within a 95% confidence limit.

Worked out solution

1   The estimated population number of books, N, is 12*45 or, 540.
    Mean retail value of books is £4.57. Thus the estimated total retail value is $N \cdot \bar{x} = 540*4.57$, or £2,467.80.
    Sample size, *n,* is 28; ratio *n/N* is 28/540 or 5.19%. Since this value is greater than 5% we use the finite population multiplier:

    Finite population multiplier $\sqrt{\frac{N-n}{N-1}} = \sqrt{\frac{540-28}{540-1}} = \sqrt{\frac{512}{539}} = 0.9746$

Sample standard deviation, s, is £0.53. Since we have no other information the estimated population standard deviation, $\hat{\sigma}$ or £0.53.

From Equation 7(xii) the estimated standard error is:

$$\frac{\hat{\sigma}}{\sqrt{n}}\sqrt{\frac{N-n}{N-1}} = \frac{0.53}{\sqrt{28}}*0.9746 = 0.0976$$

Degrees of freedom $(n-1)$ or 27.

Using Excel **TINV** (Excel 2007) or **T.INV.2T** (Excel 2010 or later) for a 95% confidence level, Student's t value is 2.0518.

From Equation 7(xiv) the lower confidence limit is:

$$N\cdot\bar{x} - N\cdot t\frac{\hat{\sigma}}{\sqrt{n}}\sqrt{\frac{N-n}{N-1}} = £2,467.80 - 540*2.0518*0.0976 = £2,359.64$$

From Equation 7(xiv) the upper confidence limit is:

$$N\cdot\bar{x} + N\cdot t\frac{\hat{\sigma}}{\sqrt{n}}\sqrt{\frac{N-n}{N-1}} = £2,467.80 + 540*2.0518*0.0976 = £2,575.96$$

Thus the owner estimates the average, or point estimate, of the total retail value of the paperback books in the store as £2,467.80 (£2,468 rounded) and she is 95% confident that the value lies between £2,359.64 (say £2,360 rounded) and £2,575.96 (say £2,576 rounded).

Estimating the proportion

Rather than making an estimate of the mean value of the population, we might be interested in estimating the proportion in the population. For example, we take a sample and say that our point estimate of the proportion expected to vote conservative in the next UK election is 37% and that we are 90% confident that the proportion will be in the range of 34% to 40%. When dealing with proportions then the sample proportion, $\bar{p}$, is a point estimate of the population proportion p. The value $\bar{p}$ is determined by taking a sample of size n and measuring the proportion of successes.

Large samples

When analyzing the proportions of a population in *Chapter 6* we developed the following Equation 6(xi) for the standard error of the proportion, $\sigma_{\bar{p}}$.

$$\sigma_{\bar{p}} = \sqrt{\frac{pq}{n}} = \sqrt{\frac{p(1-p)}{n}}$$

where n, is the sample size and p, is the population proportion of *successes* and q is the population proportion of *failures* equal to $(1-p)$. Further, from Equation 6(xiv),

$$z = \pm\frac{\bar{p} - p}{\sqrt{\frac{p\cdot(1-p)}{n}}}$$

Reorganizing this equation we have the following expression for the confidence intervals for the estimate of the population proportion as follows:

$$\bar{p} = p \pm z \sqrt{\frac{p(1-p)}{n}}$$ 7(xiv)

Thus, analogous to the estimation for the means, this implies that the confidence intervals for an estimate of the population proportion lie in the range given by the following expression:

$$\bar{p} - z \cdot \sqrt{\frac{p(1-p)}{n}} \leq p \leq \bar{p} + z \cdot \sqrt{\frac{p(1-p)}{n}}$$ 7(xv)

If we do not know the population proportion, p, then the standard error of the proportion can be estimated from the following equation formed by replacing p with $\bar{p}$:

$$\widehat{\sigma}_{\bar{p}} = \sqrt{\frac{\bar{p}(1-\bar{p})}{n}}$$ 7(xvi)

In this case, $\widehat{\sigma}_{\bar{p}}$, is the **estimated standard error of the proportion** and $\bar{p}$ is the sample proportion of successes. If we do this then Equation 7(xv) is modified to give the expression:

$$\bar{p} - z \cdot \sqrt{\frac{\bar{p}(1-\bar{p})}{n}} \leq p \leq \bar{p} + z \cdot \sqrt{\frac{\bar{p}(1-\bar{p})}{n}}$$ 7(xvii)

Sample size for large samples

In a similar way as for the mean, we can determine the sample size to take in order to estimate the population proportion for a given confidence level. From the relationship of 7(xiv) the intervals for the estimate of the population proportion are:

$$\bar{p} - p = \pm z \cdot \sqrt{\frac{p(1-p)}{n}}$$ 7(xviii)

Squaring both sides of the equation we have:

$$(\bar{p} - p)^2 = z^2 \cdot \frac{p(1-p)}{n}$$

Making n, the sample size, the subject of the equation gives:

$$n = z^2 \cdot \frac{p(1-p)}{(\bar{p}-p)^2}$$ 7(xix)

If we denote the sample error, $(\bar{p}-p)$, by e then the sample size is given by the relationship:

$$n = z^2 \cdot \frac{p \cdot (1-p)}{e^2}$$ 7(xx)

In using this equation for sample size, one question is what value to use for the true population proportion, p, when this is actually the value that we are trying to estimate! Two possible approaches are to use the value of $\bar{p}$ if this is available, or alternatively we can use a value of p equal to 0.5 or 50% as this will give the most conservative or largest sample size. This is because for a given value of the confidence level, say 95%, which defines z and the required sample error, e, then a value of p of 0.5 gives the maximum possible value of 0.25 in the numerator of Equation 7(xx). This is shown in Table 7.4 and illustrated by the graph in Figure 7.7.

Table 7.4

p	$(1 - p)$	$p.(1 - p)$
0.00	1.00	0.0000
0.05	0.95	0.0475
0.10	0.90	0.0900
0.15	0.85	0.1275
0.20	0.80	0.1600
0.25	0.75	0.1875
0.30	0.70	0.2100
0.35	0.65	0.2275
0.40	0.60	0.2400
0.45	0.55	0.2475
0.50	0.50	0.2500
0.55	0.45	0.2475
0.60	0.40	0.2400
0.65	0.35	0.2275
0.70	0.30	0.2100
0.75	0.25	0.1875
0.80	0.20	0.1600
0.85	0.15	0.1275
0.90	0.10	0.0900
0.95	0.05	0.0475
1.00	0.00	0.0000

The following is an application case-exercise, *Circuit boards* of the estimation for proportions including an estimation of the sample size.

Application case-exercise: **Circuit boards**

Situation

A firm in China manufactures electronic circuit boards for a variety of customers in Europe and the USA. One of the quality control inspectors, working for her US client, takes a random sample of 500 from the production line. Of these, 15 are found to be defective.

Required

1   What is a 90% confidence interval for the proportion of all the defective circuit boards produced in this manufacturing process?
2   If we required our estimate of the proportion of all the defective manufactured circuit boards to be within a margin of error of ± 0.01 at a 98% confidence level, what size of sample should we take?

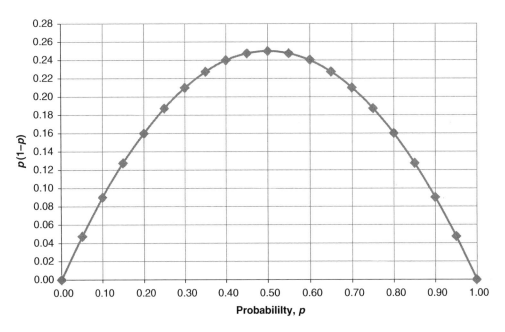

Figure 7.7 Variation of the product $p*(1-p)$ with p.

Worked out solutions

1   The proportion of defective circuit boards, $\bar{p}$ is $15/500 = 0.030$.
    Proportion that is good is $1 - 0.030$ or also $\frac{500-15}{500} = 0.97$.
    From Equation 7(xvi) the estimate of the standard error of the proportion is:

$$\hat{\sigma}_{\bar{p}} = \sqrt{\frac{\bar{p}(1-\bar{p})}{n}} = \sqrt{\frac{0.03*0.97}{500}} = \sqrt{\frac{0.0291}{500}} = 0.0076$$

When we have a 90% confidence interval, and assuming a normal distribution, then the area of the distribution up to the lower confidence level is $(100\% - 90\%)/2 = 5\%$ and the area of the curve up to the upper confidence level is $5\% + 90\% = 95\%$.
From Excel **NORMSINV** (Excel 2007) or **NORM.S.INV** (Excel 2010 and later), the value of z at the area of 5% is -1.6449.
From Excel **NORMSINV** (Excel 2007) or **NORM.S.INV** (Excel 2010 and later), the value of z at the area of 95% is $+1.6449$
From Equation 7(xvii) the lower confidence limit is:

$$\bar{p} - z \cdot \hat{\sigma}_{\bar{p}} = 0.03 - 1.6449*0.0076 = 0.03 - 0.0125 = 0.0175$$

From Equation 7(xvii) the upper confidence limit is:

$$\bar{p} + z \cdot \hat{\sigma}_{\bar{p}} = 0.03 + 1.6449*0.0076 = 0.03 + 0.0125 = 0.0425$$

Thus we can say that from our analysis the proportion of all the manufactured circuit boards which are defective is 0.03 or 3%. Further, we are 90% confident that this proportion lies in the range of 0.0175 or 1.75% and 0.0425 or 4.25%.

2   When we have a 98% confidence interval, and assuming a normal distribution, then the area of the distribution up to the lower confidence level is $(100\% - 98\%)/2 = 1\%$; the area of the curve up to the upper confidence level is $1\% + 98\% = 99\%$. From the Excel normal distribution function we have:
From Excel **NORMSINV** (Excel 2007) or **NORM.S.INV** (Excel 2010 and later), the value of z at the area of 1% is −2.3263.
From Excel **NORMSINV** (Excel 2007) or **NORM.S.INV** (Excel 2010 and later), the value of z at the area of 99% is +2.3263.
The margin of error, or sample error e, is ± 0.01.
If we use the sample proportion $\bar{p}$ for the population proportion p, or 0.03, then from Equation 7(xx):

$$n = z^2 \cdot \frac{p \cdot (1 - p)}{e^2} = 2.3263*2.3263*\frac{0.03*0.97}{0.01*0.01} = \frac{0.1575}{0.0001} = 1,575$$

(It does not matter which value of z we use, −2.3263 or +2.3263, since we are squaring z and the negative value becomes positive.)
Thus the sample size to estimate the population proportion of the number of defective circuits within an error of margin of error of ± 0.01 from the true proportion is 1,575.
An alternative, more conservative approach, is to use a value of $p = 0.5$. In this case the sample size is:

$$n = z^2 \cdot \frac{p \cdot (1 - p)}{e^2} = 2.3263*2.3263*\frac{0.50*0.50}{0.01*0.01} = \frac{0.2500}{0.0001} = 2,500$$

This value of 2,500 is almost 1,000 units higher than 1,575. This would add to the cost of the sampling experiment with not necessarily a significant gain in the accuracy of the results.

Margin of error and confidence levels

When we make estimates the question arises (or at least it should): "How good is your estimate?". That is to say, what is the margin of error? In addition, we might ask, "Why don't we always use a high confidence level of say 99% as this would signify a high degree of accuracy?" These two issues are related and are discussed below.

Margin of error

When we analyze our sample we are trying to estimate the population parameter, either the mean value, or the proportion. When we do this, there will be a **margin of error**. This is not to say that we have made a calculation error, although this can occur, but the margin of error measures the maximum amount that our estimate is expected to differ from the actual population parameter. The margin of error is a plus or minus value added to the sample result that tells us how good our estimate is. For example, if we are estimating the mean value:

Margin of error is $\pm z \cdot \dfrac{\sigma_x}{\sqrt{n}}$ 7(xxi)

This is the same as the confidence limits from Equation 7(i). In the worked case-example, *Paper*, at a confidence level of 95%, the margin of error is ± 1.9600*0.0013 or ± 0.0025 cm. Thus, another way of reporting our results is to say that we estimate that the width of all the computer paper from the production line is 20.9986 cm and we have a margin of error of ± 0.0025 cm at a 95% confidence. Now if we look at Equation 7(xxi), when we have a given standard deviation and a given confidence level the only term that can change is the sample size n. Thus we might say let us analyze a bigger sample in order to obtain a smaller margin of error. This is true, but as can be seen from Figure 7.8, which gives the ratio of $1/\sqrt(n)$ as a percentage according the sample size in units, there is a diminishing return. Increasing the sample size does reduce the margin of error but at a decreasing rate. If we double the sample size from 60 to 120 units the ratio of $1/\sqrt(n)$ changes from 12.91% to 9.13% or a difference of 3.78%. From a sample size of 120 to 180 the value of $1/\sqrt(n)$ changes from 9.13% to 7.45% or a difference of 1.68%. or, if we go from a sample size of 360 to 420 units the value of $1/\sqrt(n)$ goes from 5.27% to 4.88% or a difference of only 0.39%. With increasing the sample size the cost of testing increases and so there has to be a balance between the size of the sample and the cost.

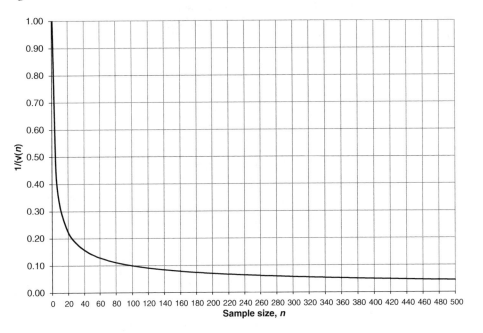

Figure 7.8 Change of $1/\sqrt(n)$ with sample size n.

If we are estimating for proportions then the margin of error is from Equation 7(xvii) the value:

$$\pm z \cdot \sqrt{\frac{\bar{p}(1-\bar{p})}{n}} \qquad\qquad 7(\text{xxii})$$

Since for proportions we are estimating the percentage for a situation then the margin of error is again a plus or minus percentage. In the circuit boards worked example the margin of error at a 90% level of confidence is:

$$\hat{\sigma}_{\bar{p}} = \pm z \cdot \sqrt{\frac{\bar{p}(1-\bar{p})}{n}} = \pm 1.6449 * \sqrt{\frac{0.03*0.97}{500}} = \pm 0.0125 = \pm 1.25\%$$

This means that our estimate could be 1.25% more or 1.25% less than our estimated proportion, a range of 2.50%. The margin of error quoted in a sampling situation is important as it can give uncertainty to our conclusions. If we look at Figure 7.0 of the Icebreaker regarding Turkey's admission for example we see that 52% of the Italian population is against Turkey joining the European Union. Based on just this information we might conclude that the majority of Italians are against Turkey's membership. However, if we then bring in the ± 3% margin of error then this means that we can have 49% against Turkey joining the Union (52 − 3), which is not now the majority of the population. Our conclusions are reversed and in cases like these we might hear the term for the media "the results are too close to call". Thus, the margin of error must be taken into account when surveys are made because the result could change. If the margin of error was included in the survey result of the Dewey/Truman election race, as presented in the Icebreaker of *Chapter 6,* the *Chicago Tribune* might not have been so quick to publish their morning paper! A recent pronouncement of the margin of error occurred during a survey analysis of who the leading Republican candidate for the 2016 presidential elections was. According to a poll, Ben Carson, a retired neurosurgeon, was the choice of 26% of Republican voters while Donald Trump, a real estate magnate, won support from 22%. However, the media qualified their announcement by indicating that the difference lies within the margin of sampling error.[2]

Confidence levels

If we have a confidence level that is high, say at 99%, the immediate impression is to think that we have a high accuracy in our sampling and estimating process. However, this is not the case since in order to have high confidence levels we need to have large confidence intervals or a large margin of error. In this case the large intervals give very broad or fuzzy estimates. This can be illustrated qualitatively as follows.

Assume that you have contracted a new house to be built of 170 m^2 living space on 2,500 m^2 of land. You are concerned about the time taken to complete the project since you have to make steps to move from your existing residence. You ask the contractor various questions concerning the timeframe. These are given in the first column of Table 7.5. Possible indicated responses to these are given in the second column and the third and fourth columns respectively give the implied confidence interval and the implied confidence level.

Table 7.5

Your question	Contractors response	Implied confidence interval	Implied confidence level
Will my house be finished in 10 years?	I am certain	> 99%	10 years
Will my house be finished in 5 years?	I am pretty sure	> 95%	5 years
Will my house be finished in 2 years?	I think so	> 80%	2 years
Will my house be finished in 18 months?	Possibly	About 50%	1.5 years
Will my house be finished in 6 months?	Probably not	About 1%	0.50 years

Thus, for a house to be finished in 10 years the contractor is almost certain because this is an inordinate amount of time and so we have put a very high confidence level of 99%. Again, to ask the question for five years the confidence level is still high at 95%. At two years there is a confidence level of 80% if everything goes better than planned. At 18 months there is a 50% confidence if there are, for example, ways to expedite the work. At six months we are essentially saying it is not possible. (The time to completely construct a house varies with location but some 18 months to 2 years to build and completely finish all the landscaping is a reasonable time frame.)

Moving on

This chapter has expanded on the sampling activity to use the concepts for estimating population characteristics. (It is usually the population which is of more interest.) The following two chapters extend both the ideas of sampling to validate hypothesized assumptions.

Notes

1 Champion, Marc, and Karnitschnig, Mathew, "Turkey gains EU approval to begin membership talks," *Wall Street Journal Europe*, October 4, 2005, pp. 1 and 14.
2 *International New York Times,* "Survey puts Carson in the lead nationally, displacing Trump," October 28, 2015, p. 5.

8 Hypothesis testing from a single population

ICEBREAKER: Radiation and exposure: you need to be objective

There are many familiar forms of radiation. We use light, heat, and microwaves daily. Doctors use X-rays in medical examinations. Radiation is naturally present in our environment. The sun and stars send a constant stream of cosmic radiation to Earth, rather like a steady drizzle of rain. The Earth itself is a source of terrestrial radiation. Radioactive materials, including uranium, thorium, and radium exist naturally in soil and rock. Air contains radon. Water contains small amounts of dissolved uranium and thorium and all organic matter, both plant and animal, contains radioactive carbon and potassium. People have internal radiation, mainly from radioactive potassium-40 and carbon-14 inside their bodies from birth and are thus sources of exposure to others. People are also exposed to radiation from the nuclear fuel cycle; from uranium mining and milling; from disposal of used or spent fuel. In addition, the public receives

Figure 8.0 A nuclear power plant adjacent to residential areas.

Source: Bugey nuclear power plant, 34 km from Lyon, France (permission to publish under CC BY-SA license).

minimal exposure from the transportation of radioactive materials and fallout from nuclear weapons testing and reactor accidents such as Three Mile Island, USA (1979), Chernobyl, Ukraine (1986) and Fukushima, Japan (2011).

Nuclear power plants sometimes release radioactive gases and liquids into the environment under controlled, monitored conditions to ensure that they pose no danger to the public or the environment. These releases dissipate into the atmosphere or a large water source and, therefore, are diluted to the point where it becomes difficult to measure any radioactivity. By contrast, most of an operating nuclear power plant's direct radiation is blocked by the plant's steel and concrete structures. The remainder dissipates in an area of controlled, uninhabited space around the plant, ensuring that it does not affect members of the public. Nuclear plant workers are no more radioactive than anyone else. Except in unusual circumstances, such as an accident at the plant, workers receive only minimal doses of radiation and rarely become contaminated with radiation.

For the same reason that we need to be careful around open flames, toxic chemicals, knives, or guns, the same applies to radiation. A knife is used to prepare and eat food. Misused it can cause injury and perhaps death. Similarly, when handled correctly, radioactive materials have many beneficial uses in medicine. Misused, it can pose a significant danger. Ionization can cause damage within a human cell that could eventually lead to cancer or a mutation in genetic material. Radiation will kill you if you are exposed to enough of it, by doing so much damage to your internal systems that your body can no longer function. The government regulatory system for radioactive materials is designed to prevent the possibility that anyone could receive an exposure even close to the levels that might inflict damage. The simplest preventions against harmful radiation exposure are time, distance and shielding. Limit the time of exposure to the radioactive source; increase the distance between you and the source; and shield yourself by placing objects between you and the source. These concepts form the basis of nuclear regulation so that we may enjoy the beneficial uses of radioactive materials while minimizing the risk to public health and the environment ... [1]

> ... *The government in a certain country says that radiation levels in the area surrounding a nuclear power plant are well below levels considered harmful. Three people in the area die of leukemia. The media and the local people immediately put the blame on radioactive fallout from the plant* ...

Does the death of three people make us assume that the government is wrong with its information and that we make the assumption, or hypothesis, that radiation levels in the area are abnormally high? Alternatively, do we accept that the deaths from leukemia are random and are not related to the nuclear power facility? You should not accept, or reject, a hypothesis about a population parameter—in this case radiation levels in the surrounding area of the nuclear power plant—simply by intuition. You need to be objective in decision making. For this situation an appropriate action would be to take representative samples of the incidence of leukemia cases over a reasonable time period and use these to test the hypothesis. This is the purpose of this chapter (and the following chapter)—to find out how to use hypothesis testing to determine whether a claim is valid. Sadly, there are many instances when published claims are not backed up by solid statistical evidence!

Chapter subjects

✓ **Concept of hypothesis testing** • Significance level • Null and alternative hypotheses
✓ **Presenting hypothesis testing for the mean** • Two-tailed test • One-tailed, right-hand test • One-tailed, left-hand test
✓ **Testing hypothesis for the mean** • Acceptance or rejection • Test statistics normal z, and Student's t • Probability value (p-value) • Confidence intervals • Application case-exercise for large samples: *Beer filling machine* • Application case-exercise for small samples: *Taxes*
✓ **Hypothesis testing for proportions** • Large samples • Application case-exercise: *Seaworthiness of ships*.
✓ **Considerations in hypothesis testing** • Interpretation of the probability value (p-value) • Errors in hypothesis testing • Type I and Type II errors • Cost of making an error • Power of the test.

A hypothesis is a judgment about a situation, outcome, or population parameter based simply on an assumption or intuition or gut feeling with no concrete backup information.

Concept of hypothesis testing

Hypothesis testing is taking sample data and making an objective decision based on the results of the test within an appropriate significance level. Thus, like estimating, hypothesis testing is an application of sampling presented in *Chapter 6*.

Significance level

When we make quantitative judgments, or hypotheses, about situations we are either right, or wrong. However, if we are wrong we may not be far from the real figure or our judgment is not significantly different. Thus our hypothesis may be acceptable. Consider the following:

• A contractor says that it will take nine months to construct a house for a client. The house is finished in nine months and one week. The completion time is not nine months; however, it is not *significantly* different from the estimated construction period of nine months.
• The local authorities estimate that there are 20,000 people at an open air rock concert. Ticket receipts indicate there are 42,000 attendees. This number of 42,000 is *significantly* different from 20,000.
• A financial advisor estimates that a client will make $15,000 on a certain investment. The client realizes $14,900. The number $14,900 is not $15,000 but it is not *significantly* different from $15,000 and the client really does not have a strong reason to complain. However, if the client made only $8,500 he would probably say that this is *significantly* different from the estimated $15,000 and has a justified reason to say that he was given bad advice.
• India and Pakistan were created by the British partition of the Indian subcontinent in 1947. India, principally Hindu, was the largest. Pakistan was Muslim but a country divided: in the northwest were Punjabis, Pashtuns, Baluchis, and Sindhis; far away in the east were Bengalis. In 1971 the Pakistani army slaughtered thousands of Bengalis after they won a legitimate election. Millions of refugees fled into neighboring India. Indira Ghandi, the prime minister of India, called it a

catastrophic situation. Nixon, the US President, and Henry Kissinger, the National Security Advisor, considered the slaughter of Bengalis and the refugee problem, *not significant* enough to warrant action on the part of the USA.[2]

Thus in hypothesis testing, we need to decide what we consider is the **significance level** or the level of importance in our evaluation. This significance level gives a ceiling level, usually in terms of percentages such as 1%, 5%, 10%, etc. To a certain extent this is the subjective part of hypothesis testing since one person might have a different criterion than another individual on what is considered significant. However, in accepting, or rejecting a hypothesis in decision making, we have to agree on the level of significance. This significance value, which is denoted as alpha, α, gives us the **critical value** for testing.

Null and alternative hypotheses

In hypothesis testing there are two defining statements premised on the binomial concept. One is the **null hypothesis,** which is that value considered correct within the given level of significance. The other is the **alternative hypothesis,** which is the hypothesized value that is not correct at the given level of significance. The alternative hypothesis as a value is also known as the **research hypothesis** since it is a value that has been obtained from a sampling experiment.

For example, the hypothesis is that the average age of the population in a certain country is 35. This value is the null hypothesis. The alternative to the null hypothesis is that the average age of the population is not 35 but is some other value. In hypothesis testing there are **three** possibilities. The first is that there is evidence that the value is significantly **different** from the hypothesized value. The second is that there is evidence that the value is significantly **greater** than the hypothesized value. The third is that there is evidence that the value is significantly **less** than the hypothesized value. Note that in these sentences we say *there is evidence* because as always in statistics there is no guarantee of the result as we are basing our analysis of the population on sampling alone and of course our sample experiment may not yield the correct result. These three possibilities lead to using a **two-tailed hypothesis test**, a **right-tailed hypothesis test**, and a **left-tailed hypothesis test** as explained in the next section.

Presenting hypothesis testing for the mean

In hypothesis testing for the mean an assumption is made about the mean or average value of the population. Then we take a sample from this population, determine the sample mean value, and measure the difference between this sample mean and the hypothesized population value. If the difference between the sample mean and the hypothesized population mean is small, then the probability is higher that our hypothesized population mean value is correct. If the difference is large then the probability is lower that our hypothesized value is correct.

Two-tailed test

A two-tailed test is used when we are testing to see if a value is **significantly different** from our hypothesized value. For example, in the above population situation the null hypothesis is that the average age of the population is 35 years and this is written as follows:

Null hypothesis : $H_0: \mu_x = 35$ 8(i)

In the two-tailed test we are asking: "Is there evidence of a difference?" In this case the alternative to the null hypothesis is that the average age is not 35 years. This is written as:

Alternative hypothesis : $H_1: \mu_x \neq 35$ 8(ii)

The two-tailed test is illustrated in Figure 8.1 where we have given a 10% level of significance and in this case for a two-tailed test there is 5% in each tail.

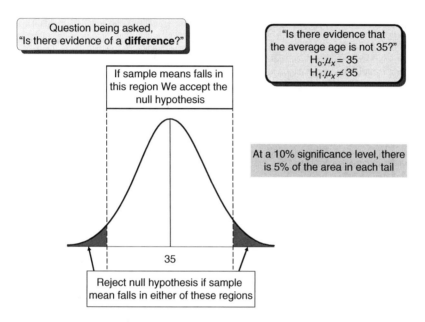

Figure 8.1 Two-tailed hypothesis test.

When we ask the question whether there evidence of a difference this means that the alternative value can be significantly lower or higher than the hypothesized value. For example, if we took a sample from our population and the average age of the sample was 36.2 years we might say that the average age of the population is not significantly different from 35 years. In this case we would accept the null hypothesis as being correct. However, if in our sample the average age was 52.7 years then we may conclude that the average age of the population is significantly different from 35 years since it is much higher. Alternatively, if in our sample the average age was 21.2 years then we may also conclude that the average age of the population is significantly different from 35 years since it is much lower. In both of these cases we would reject the null hypothesis and accept the alternative hypothesis. Since this is a binomial concept, when we reject the null hypothesis we accept the alternative hypothesis.

One-tailed, right-hand test

A one-tailed right-hand test is used to test if there is evidence that the value is **significantly greater** than our hypothesized value. For example in the above population situation, the null hypothesis is that the average age of the population is equal to or less than 35 years and this is written as follows:

Null hypothesis : $H_0: \mu_x \leq 35$ 8(iii)

The alternative hypothesis is that the average age is greater than 35 years and this is written:

Alternative hypothesis : $H_1: \mu_x > 35$ 8(iv)

Thus, if we took a sample from our population and the average age of the sample was say 36.2 years we would probably say that the average age of the population is not significantly greater than 35 years and we would accept the null hypothesis. Alternatively, if in our sample the average age was 21.2 years then although this is significantly less than 35, it is not greater than 35. Again we would accept the null hypothesis. However, if in our sample the average age was 52.7 years then we may conclude that the average age of the population is significantly greater than 35 years and we would reject the null hypothesis and accept the alternative hypothesis. Note that for this situation we are not concerned with values that are significantly less than the hypothesized value but only those that are significantly greater. Again, since this is a binomial concept, when we reject the null hypothesis we accept the alternative hypothesis. Conceptually the one-tailed right-hand test is illustrated in Figure 8.2. Again we say that there is a 10% level of significance, but in this case for a one-tailed test, all the 10% area is in the right-hand tail.

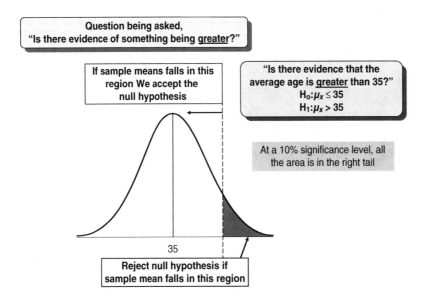

Figure 8.2 One-tailed hypothesis test (right hand).

One-tailed, left-hand test

A one-tailed left-hand test is used to test if there is evidence that the value is **significantly less** than our hypothesized value. For example again let us consider the above population situation. The null hypothesis, $H_o:\mu_x$, is that the average age of the population is equal to or more than 35 years and this is written as follows:

$H_0: \mu_x \geq 35$ 8(v)

The alternative hypothesis, $H_1:\mu_x$, is that the average age is less than 35 years. This is written:

$$H_1: \mu_x < 35 \qquad\qquad\qquad 8(\text{vi})$$

Thus, if we took a sample from our population and the average age of the sample was say 36.2 years we would say that there is no evidence that the average age of the population is significantly less than 35 years and we would accept the null hypothesis. Or, if in our sample the average age was 52.7 years, then although this is significantly greater than 35 it is not less than 35 and we would accept the null hypothesis. However, if in our sample the average age was 21.2 years then we may conclude that the average age of the population is significantly less than 35 years and we would reject the null hypothesis and accept the alternative hypothesis. Note that for this situation we are not concerned with values that are significantly greater than the hypothesized value but only those that are significantly less than the hypothesized value. Again, since this is a binomial concept, when we reject the null hypothesis we accept the alternative hypothesis. Conceptually the one-tailed left-hand test is illustrated in Figure 8.3. With the 10% level of significance shown means that for this one-tailed test all the 10% area is in the left-hand tail.

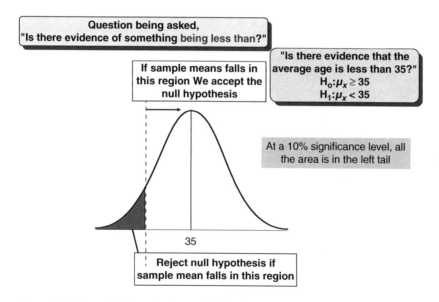

Figure 8.3 One-tailed hypothesis test (left hand).

Testing hypothesis for the mean

A hypothesis can be reasoned by comparing sample test values with the hypothesized benchmark. This might be a numerical statistic; the normal z or Student's t values; a percentage (probability) area under the normal distribution; or the positioning of a sample mean value relative to critical limits. This section explains.

Acceptance or rejection

The purpose of hypothesis testing is not to question the calculated value of the sample statistic, but to make objective judgments regarding the difference between the sample mean and the hypothesized population mean. Testing at the 10% significance level means that the null hypothesis would be rejected if the difference between the sample mean and the hypothesized population mean is so large that it, or a larger difference, occurs on average ten or fewer times for every 100 samples when the hypothesized population parameter is correct. Assuming the hypothesis is correct, then the significance level indicates the percentage of sample means that are outside certain limits. Even if a sample statistic does fall in the area of acceptance, this does not prove that the null hypothesis, H_0, is true. There simply is no statistical evidence to reject the null hypothesis. Acceptance or rejection is related to the values of the test statistic that are unlikely to occur if the null hypothesis is true. However, they are not so unlikely to occur if the null hypothesis is false.

Test statistics normal z, and Student's t

There are two relationships to use, analogous to those used in *Chapter 7*. If the **population standard deviation is known**, then from the central limit theorem for sampling, the test statistic, or the critical value is:

$$\text{Test statistic, } \pm z = \frac{\bar{x} - \mu_{H_0}}{\frac{\sigma_x}{\sqrt{n}}} \qquad \qquad 8(\text{vii})$$

- μ_{H_0} is the hypothesized population mean
- $\bar{x}$ is the sample mean
- the numerator, $\bar{x} - \mu_x$, measures how far the observed mean is from the hypothesized mean
- σ_x is the population standard deviation
- n is the sample size
- $\frac{\sigma_x}{\sqrt{n}}$, the denominator in the equation, is the standard error
- z is how many standard errors the observed sample mean is from the hypothesized mean. It can have a positive or negative value

For small sample sizes if the population standard deviation is unknown then the only standard deviation we can determine is the sample standard deviation, s. This value of s can be considered an estimate of the population standard deviation, sometimes written as $\hat{\sigma}_x$. If the sample size is less than 30 then we use the Student's t distribution, presented in *Chapter 7*, with $(n - 1)$ degrees of freedom, making the assumption that the population from which this sample is drawn is normally distributed. In this case, the test statistic is calculated by:

$$t = \frac{\bar{x} - \mu_{H_0}}{\frac{\hat{\sigma}_x}{\sqrt{n}}} \qquad \qquad 8(\text{viii})$$

- μ_{H_0} again the hypothesized population mean
- $\bar{x}$ is the sample mean
- the numerator, $\bar{x} - \mu_{H_0}$, measures how far the observed mean is from the hypothesized mean
- $\hat{\sigma}_x$ is the estimate of the population standard deviation and is equal to the sample standard deviation, s
- n is the sample size
- $\frac{\hat{\sigma}_x}{\sqrt{n}}$, the denominator in the equation, is the estimated standard error
- t is how many standard errors the observed sample mean is from the hypothesized mean

These methods of analysis select a significance level for the hypothesis, which then translates into a critical value of z or t, and then test to see whether the sample statistic lies within the boundaries of the critical values. If the test statistic falls within the boundaries then we accept the null hypothesis. If the test statistic falls outside, then we reject the null hypothesis and accept the alternative hypothesis. Here we have created a binomial "yes" or "no" situation by examining whether there is sufficient statistical evidence to accept or reject the null hypothesis.

Probability value (p-value)

Another approach to hypothesis testing is to ask: "What is the minimum probability level we will tolerate in order to accept the null hypothesis of the mean?" This level is called the p-value or the observed level of significance from the sample data. It answers the question: "If H_0 is true, what is the probability of obtaining a value of $\bar{x}$, (or $\bar{p}$, in the case of proportions) this far or more from the hypothesized value H_0?". If the p-value, determined from the sample, is greater than or equal to α the null hypothesis is accepted. If it is less, the null hypothesis is rejected and the alternative hypothesis is accepted.

Confidence intervals

This approach determines the confidence intervals established by the hypothesized criteria and then tests to establish whether the sample value lies within these intervals. If it does, the hypothesized value is accepted; if not, it is rejected. This is a similar approach to determining confidence intervals described in *Chapter 7*.

The following applications exercise, *Beer filling machine*, illustrates the three procedures for hypothesis testing for large populations.

Application exercise for large samples: **Beer filling machine**

Situation

A production line of a brewery is filling cans whose net volume, indicated on the label, is 50 cl. It is known that the standard deviation of the filling process is 3 cl. The quality control inspector performs an analysis on the line to test whether the process is operating according to specifications. If the volume of liquid in the cans is higher than the specification limits then this costs the firm too much money. If the volume is lower than the specifications then this can cause a problem with the external inspectors. A sample of 40 cans was taken at random from the filling line and their volume in cl is given in Table 8.1.

Table 8.1

52.67	55.47	46.92	53.62	47.95	45.23	48.97	48.42
46.34	50.92	48.45	53.56	53.56	50.65	52.23	53.68
53.28	49.83	53.12	55.98	52.25	45.98	48.23	48.95
51.92	53.27	51.69	49.69	50.28	50.78	48.56	48.23
52.34	51.78	51.89	51.78	48.97	48.89	53.89	55.57

Required

1   Using the test statistic, at a significance level, α, of 5% is there evidence that the volume of beer in the cans from this bottling line is significantly different than the target volume of 50 cl?
2   Using the p-value approach for a difference does your calculation corroborate the answer to Question 1?

3   In determining the critical limits for a difference, does the sample average lie within these limits and in which case also verify your answer to Question 1?
4   Using the test statistic, at a significance level, α, of 5% is there evidence that the volume of beer in the cans from this bottling line is significantly greater than the target volume of 50 cl?
5   Using the *p*-value approach testing for greater than, does your calculation corroborate the answer to Question 4?
6   In determining the critical limit does the sample average lie below this limit and in which case also verify your answer to Question 4?

Worked out solutions

1   Here we are asking the question whether there is there evidence of a difference so this means using a two-tailed test. The null and alternative hypotheses are written as follows:
Null hypothesis: $H_o : \mu_x = 50$ cl.
Alternative hypothesis: $H_1 : \mu_x \neq 50$ cl.
Since we know the population standard deviation we can use Equation 8(vii):

- hypothesized population mean H_0 is 50 cl
- sample mean $\bar{x}$ is 50.8948 using the Excel function **AVERAGE**
- $\bar{x} - \mu_{H_0}$, is $50.89 - 50.00 = 0.89$ cl
- σ_x is the population standard deviation, or 3.00 cl
- n is the sample size, or 40, using the Excel function **COUNT**
- $\sqrt{n} = 6.3246$
- thus, the standard error of the mean for the sample is $3.00/6.3246 = 0.4743$

The test statistic from Equation 8(vii) is $z = \frac{\bar{x} - \mu_{H_0}}{\frac{\sigma_x}{\sqrt{n}}} = \frac{0.8900}{0.4743} = 1.8863$

At a significance level of 5% for test of a difference there is 2.5% in each tail. Using Excel function **NORM.S.INV** gives a critical value of z of $\pm\ 1.9600$.

Since the value of the test statistic, 1.88863 is less than the critical value of 1.9600, or alternatively within the boundaries of $\pm\ 1.9600$, there is no statistical evidence that the volume of beer in the cans is significantly different than 50 cl. Thus we accept the null hypothesis. The situation is illustrated by the normal distribution in Figure 8.4.

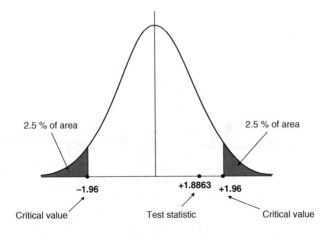

Figure 8.4 Beer filling machine, Question 1.

2 The test statistic is 1.8863 as determined in Question 1. From the Excel function **NORM.S. DIST** for this value of z the area from the left of the curve is 97.06%. Area to the right of the sample statistics is 2.96% (100 − 97.06). The value of 2.96% is greater than the critical probability of 2.50% so this corroborates the answer to Question 1. This situation is illustrated in Figure 8.5.

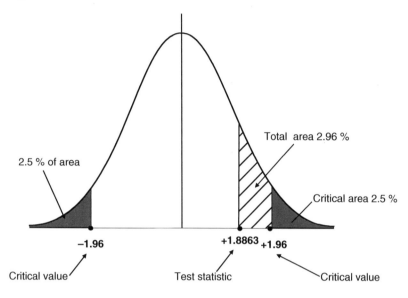

Figure 8.5 Beer filling machine, Question 2.

3 The critical limits are given by the relationship $\mu_{H_0} \pm z.(\sigma/\sqrt{n})$

- μ_{H_0} = 50 cl
- z is ± 1.9600
- $(\sigma/\sqrt{n})$ is the standard error of the mean = 0.4743
- critical limits are 50 ± 1.9600*0.4743 or 49.0703 on the lower side; 50.9297 on the upper side

The sample x value of 50.8948 cl. lies between these values so again this verifies the conclusion for Question 1.

4 Here we are asking the question whether there is evidence that the volume in the cans is significantly greater than the target value or the hypothesized amount. This is a one-tailed right-hand test. The null and alternative hypotheses are as follows:

- Null hypothesis: $H_0: \mu_x \leq 50$ cl
- Alternative hypothesis: $H_1: \mu_x > 50$. cl

Nothing has changed regarding the test statistic and it remains 1.8863 as calculated in Question 1. However, for a one-tailed test, at a significance level of 5% for the test there is 5% in the right tail. The area of the curve for the upper level is (100% − 5.0%) or 95.00%. Using Excel function **NORM.S.INV** gives a critical value of z of + 1.6449.

Since now the value of the test statistic of 1.8883 is greater than the critical value of 1.6449 then there is evidence that the volume of beer in all of the cans is significantly greater than 50 cl. The concept is illustrated in Figure 8.6.

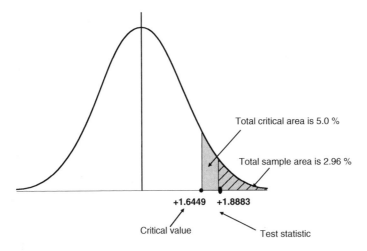

Figure 8.6 Beer filling machine, Questions 4 & 5.

5   For the *p*-value, the area of the distribution from the left to the sample statistic is 97.04%. Area to the right of the sample statistics is 2.96% (100% − 97.04%). The value of 2.96% is less than the critical value of 5.00% so this corroborates the answer to Question 4. This situation is illustrated in Figure 8.6.

6   The critical limits are given by the relationship $\mu_{H_0} \pm z.(\sigma/\sqrt{n})$

- $\mu_{H_0} = 50$ cl
- z is 1.6449
- $(\sigma/\sqrt{n})$ is the standard error = 0.4743
- critical limit is 50 + 1.6449*0.4743 = 50.7802

The sample *x* value of 50.8948 cl. is greater than this value, it lies to the right and so again this verifies the conclusion for Question 4.

In the previous application case-exercise, *Beer filling machine*, we considered when the population standard deviation was known, since we analyzed an operation that had been carried out innumerable times and the sample size was relatively large. There are circumstances when we may not know the standard deviation, the sample size is relatively small, and we have to make use of estimates. This condition is illustrated by the following application exercise, *Taxes*, where the Student's *t* distribution is used.

Application exercise for small samples: Taxes

Situation

A certain state in the USA has made its budget on the basis that the average individual tax payments for the year will be $30,000. The financial controller takes a random sample of annual tax returns and these amounts in US dollars are as in Table 8.2.

Table 8.2

34,000	12,000	16,000	10,000	24,000	15,000	19,000	12,000
2,000	39,000	7,000	72,000	23,000	14,000	6,000	43,000

Required

1   Using the test statistic, at a significance level, α, of 5%, is there evidence that the average tax returns of the state will be different than the budget level of $30,000 in this year?
2   Verify your answer to Question 1 using the *p*-value approach.
3   Validate your answer to Question 1 by comparing your sample value with the critical limits.
4   Using the test statistics, at a significance level, α, of 5%, is there evidence that the tax returns of the state will be lower than the budget level of $30,000 in this year?
5   Substantiate your answer to Question 4 using the *p*-value approach.
6   Prove your answer to Question 4 by comparing the sample value with the critical limits.

Worked out solutions

1   The null and alternative hypotheses are:

- Null hypothesis: $H_0 : \mu_x = \$30,000$
- Alternative hypothesis: $H_1 : \mu_x \neq \$30,000$

Since we have no information of the population standard deviation, and the sample size is less than 30, we use a Student's *t* distribution:

- Sample size, *n*, is 16
- Degrees of freedom, $(n-1)$, are 15.

Using **T.INV.2T** from Excel the Student's *t* value is ± 2.1314 and these are the critical values. Note that since this is a two-tailed test there is 2.5% of the area in each of the tails and *t* has a plus or minus value.
From Excel, using the function **AVERAGE**

- Mean value of this sample data, $\bar{x}$, is $21,750.00

$\bar{x} - \mu_x$ is $(21,750.00 - 30,000.00) = -\$8,250.00$

Using Excel function **STDEV.S**, the sample standard deviation, s, is $17,815.72 and this can be taken as an estimate of the population standard deviation, $\widehat{\sigma}_x$.
Estimate of the standard error of the mean is $\frac{\widehat{\sigma}_x}{\sqrt{n}} = \frac{17,815.72}{\sqrt{16}} = \$4,453.93$
From Equation 8(viii) the sample statistic is:

$$t = \frac{\bar{x} - \mu_{H_0}}{\frac{\widehat{\sigma}_x}{\sqrt{n}}} = -\frac{8,250}{4,453.93} = -1.8523$$

Since the sample statistic, -1.8523, is not less than the test statistic of -2.1314, there is no reason to reject the null hypothesis and we accept that there is no evidence that the average of all the tax receipts will be significantly different from $30,000. Note that in this situation, as the test statistic is negative we are on the left side of the curve, and so we only make an evaluation with the negative values of *t*. Another way of making the analysis, when we are looking to see if there is a difference, is to see whether the sample statistic of -1.8523 lies within the critical boundary values of $t = \pm 2.1314$. In this case it does. Figure 8.7 illustrates the concept.

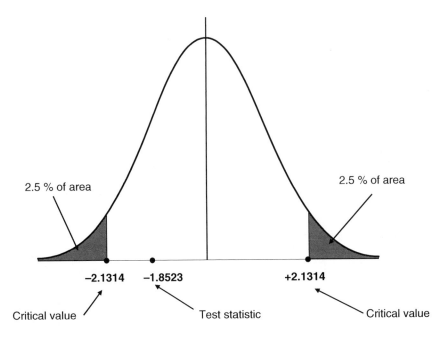

Figure 8.7 Taxes, Question 1.

2   The significance level, α, is 5%, the total area outside the distribution. In a test for a difference there is an area of 2.5% in each tail. The sample Student's t value from Question 1 is -1.8523. Using Excel function **T.DIST.2T** this gives an area of 8.48%. (We must use the absolute value of the Student's t if we have a negative value.) Since 8.48% is greater than the critical limit of 5.00% this confirms the conclusions from Question 1. Alternatively, since this is a two-tailed test, we can say that as 4.19% (8.18/2) is greater than 2.50% this again confirms that we have no reason to reject the null hypothesis.

3   The critical limits are given by the relationship $\mu_{H_0} - t.(\sigma/\sqrt{n})$

- $\mu_{H_0} = \$30,000$
- t is ± 2.1314
- $(\sigma/\sqrt{n})$ is the standard error of the mean $= \$4,453.93$
- critical limits are $30,000 \pm 2.1314*4,453.93$ or $\$20,506.67$ on the lower side and $\$39,493.33$ on the upper side

The sample x value of $\$21,750.00$ lies between these values so again this verifies the conclusion for Question 1.

4   This is a left-hand, one-tailed test. The null and alternative hypothese are as follows:

- Null hypothesis: $H_0:\mu_x \geq \$30,000$
- Alternative hypothesis: $H_1:\mu_x < \$30,000$

Again, since we have no information of the population standard deviation, and the sample size is less than 30, we use a Student's t distribution.

Sample size, n, is 16

Degrees of freedom, $(n-1)$, is 15.

Here we have a one-tailed test and thus all of the value of α, or 5%, lies in one tail. However, the Excel function for the Student's t value is based on input for a two-tailed test so in order to determine t we have to enter the area value of 10% (5% in one tail and 5% in the other tail). Using Excel function **TINV** (Excel 2007) or **T.INV** (Excel 2010 or later) gives a critical value of $t = -1.7531$.

The value of the sample statistic t remains unchanged at -1.8532 as calculated in Question 1. Since now the sample statistic, -1.8523, is less than the test statistic, -1.7531, there is reason to reject the null hypothesis and to accept the alternative hypothesis that there is evidence the average value of all the tax receipts is significantly less than $30,000. Note that in this situation we are on the left side of the curve and so we are only interested in the negative value of the Student's t. Figure 8.8 illustrates the concept.

5   The sample statistic gives a Student's t value equal to -1.8523 and from the Excel function **T. DIST** this sample statistic, for a one-tailed test, indicates a probability of 4.19%. Since 4.19% < 5.00% we reject the null hypothesis and conclude that there is evidence to indicate that the average tax receipts are significantly less than $30,000. This is the same conclusion as for Question 4. The situation is noted in Figure 8.8.

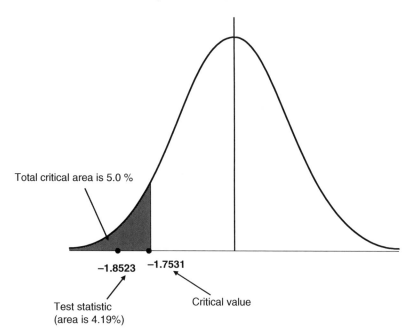

Figure 8.8 Taxes, Questions 4 and 5.

6   The critical limit is given by the relationship $\mu_{H_0} - t.(\sigma/\sqrt{n})$

- $\mu_{H_0} = \$30,000$
- t is -1.7531
- $(\sigma/\sqrt{n})$ is the standard error of the mean = $4,453.93
- critical limit is $30,000 - 1.7531*4,453.93$ or = $22,192.04

The sample x value of $21,750.00 is lower than $22,192.04 so again this verifies the conclusion for Question 4.

Hypothesis testing for proportions

In hypothesis testing for the proportion we test the assumption about the value of the population proportion. In the same way as for the mean, we take a sample from this population, determine the sample proportion, and measure the difference between this proportion and the hypothesized population value. If the difference between the sample proportion and the hypothesized population proportion is small, then the higher is the probability that our hypothesized population proportion value is correct. If the difference is significantly large then the probability that our hypothesized value is correct is low.

Large samples

In *Chapter 6,* we developed the relationship from the binomial distribution between the population proportion, p, and the sample proportion $\bar{p}$. On the assumption that we can use the normal distribution as our test reference then from Equation 6(xii) we have the value of z as follows.

$$\pm z = \frac{\bar{p} - p}{\sigma_{\bar{p}}} = \frac{\bar{p} - p}{\sqrt{\frac{p.(1-p)}{n}}}$$

In hypothesis testing for proportions we use an analogy as for the mean where p is now the hypothesized value of the proportion and may be written as p_{H_0} Thus, Equation 6(xii) becomes:

$$\pm z = \frac{\bar{p} - p_{H_0}}{\sigma_{\bar{p}}} = \frac{\bar{p} - p_{H_0}}{\sqrt{\frac{p_{H_0}\left(1 - p_{H_0}\right)}{n}}} \qquad\qquad 8(\text{ix})$$

Hypothesis testing for proportions is illustrated by the application case-exercise *Seaworthiness of ships.*

Application exercise: **Seaworthiness of ships**

Situation

On a worldwide basis, governments say that 0.80, or 80%, of merchant ships are seaworthy. Greenpeace, the environmental group, takes a random sample of 150 ships and their analysis indicates that from this sample 111 ships prove to be seaworthy.

Required

1   Using the test statistic at a 5% significance level, is there evidence to suggest that the seaworthiness of ships is different than the hypothesized 80% value?
2   Using the *p*-value approach at a 5% significance level verify your conclusion to Question 1.
3   Validate your answer to Question 1 by comparing your sample value with the critical limits.
4   At a 5% significance level is there evidence to suggest that the seaworthiness of ships is less than the 80% indicated?
5   Using the *p*-value approach at a 5% significance level, verify your conclusion to Question 4.
6   Validate your answer to Question 4 by comparing your sample value with the critical limit.

Worked out solution

1   Since we are asking the question: "Is there a difference?", this is a two-tailed test with 2.5% of the area in the left tail, and 2.5% in the right tail (5% divided by 2).

From the Excel function **NORM.S.INV** the value of z, or the critical value when the tail area is 2.5% is ± 1.9600.

The hypothesis test is written as follows:

- $H_o{:}p = 0.80$. The proportion of ships that are seaworthy is equal to 0.80 or 80%
- $H_1{:}p \neq 0.80$. The proportion of ships that are not seaworthy is different from 0.80 or 80%
- Sample size n is 150
- Sample proportion $\bar{p}$ that is seaworthy is $111/150 = 0.74$ or 74.00%
- From the sample, the number of ships that are not seaworthy is 39 $(150 - 111)$
- Sample proportion $\bar{q} = (1-\bar{p})$ that is not seaworthy is $39/150 = 0.26$ or 26.00%

The standard error of the proportion, or the denominator in Equation 8(ix), is:

$$\sigma_{\bar{p}} = \sqrt{\frac{p_{H_0}(1 - p_{H_0})}{n}} = \sqrt{\frac{0.80 * 0.20}{150}} = \sqrt{\frac{0.16}{150}} = 0.0327$$

$$\bar{p} - p_{H_0} = 0.74 - 0.80 = -0.06$$

Thus the sample test statistic from Equation 6(xii) is:

$$z = \frac{\bar{p} - p}{\sigma_{\bar{p}}} = -\frac{0.06}{0.0327} = -1.8349$$

Since the test statistic of -1.8349 is not less than -1.9600 then we accept the null hypothesis and say that at a 5% significance level there is no evidence of a significance difference between the 80% of seaworthy ships postulated. Conceptually this situation is shown in Figure 8.9.

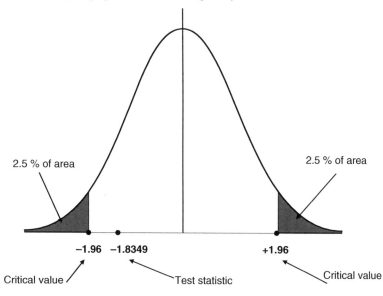

2.5 % of area   2.5 % of area

−1.96   −1.8349   +1.96

Critical value   Test statistic   Critical value

Figure 8.9 Seaworthiness of ships, Question 1.

2 This is a two-tailed test with an area of 2.5% in each tail.

The sample test z value is -1.8349

From the Excel function **NORM.S.DIST** this sample statistic, for a two–tailed test, indicates a probability of 3.31%. Since 3.31% > 2.50% we accept the null hypothesis and conclude that there is no evidence to indicate that the seaworthiness of ships is different from the hypothesized value of 80%.

3 The critical limits are given by the relationship: $p_{H_0} \pm z.\sigma_{\bar{p}}$ (reorganizing Equation 8(ix))

- $p_{H_0} = 0.80$
- z is ± 1.9600
- $\sigma_{\bar{p}}$ the standard error of the proportion is 0.0327
- critical limits are $0.80 \pm 1.9600*0.03$ or 0.7360 (73.60%) on the left side; 0.8640 (86.40%) on the right side

The sample proportion of 0.74 or 74.00% lies between these limits so this again confirms our conclusion arising from Question 1.

4 This is now a one-tailed left-hand test where we are asking whether there is evidence that the proportion is less than 0.80 or 80%. The hypothesis test is written:

- $H_0:p \geq 0.80$. The proportion of ships unseaworthy is not less than 0.80 or 80%
- $H_1:p < 0.80$. The proportion of ships unseaworthy is less than 0.80, or 80%.

In this situation the value of the sample statistics remains unchanged at -1.8349, but the critical value of z is different. From the Excel function **NORM.S.INV** the value of z, or the critical value when the tail area is 5%, is $z = -1.6449$.

Now we reject the null hypothesis because the value of the test statistic, -1.8349 is less than the critical value of -1.6449. Thus our conclusion is that there is evidence that the proportion of ships that are not seaworthy is significantly less than 0.80 or 80%. Conceptually this situation is shown in Figure 8.10.

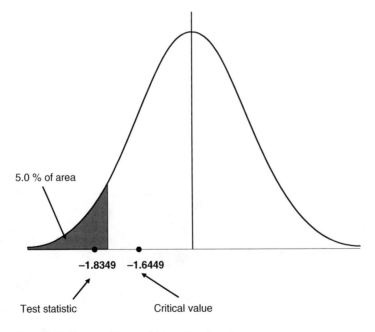

5.0 % of area

−1.8349   **−1.6449**

Test statistic   Critical value

Figure 8.10 Seaworthiness of ships, Question 4.

5 This is a one-tailed left-hand test with 5% of the area in the tail.

The p-value for the sample experiment, 3.31% (from the answer to Question 2), is less than the critical probability of 5.00%. So we reject the null hypothesis and conclude there is evidence that the seaworthiness of ships is less than the hypothesized value of 80%; same conclusion as for Question 4.

6 The critical limit is given by the relationship: $p_{H_0} - z.\sigma_{\bar{p}}$ (reorganizing Equation 8(ix))

- $P_{H_0} = 0.80$
- z is -1.6449
- $\sigma_{\bar{p}}$ the standard error of the proportion is 0.0327
- critical limit is $0.80 \pm 1.6449*0.0327$ or 0.7463 (74.63%)

The sample proportion of 0.74 or 74.00% is less than the value of 0.7463 (74.63%) so this again verifies the conclusion arising from Question 4.

Considerations in hypothesis testing

In sampling for hypothesis testing, assumptions are drawn about the population parameter. These are just suppositions but you have organized your experimental work such that you hope you are right.

Interpretation of the probability value (p-value)

In hypothesis testing we make inferences about a population based only on sampling. The sampling distribution permits us to make probability statements about a sample statistic on the basis of the knowledge of the population parameter. In the case of the application case-exercise, *Beer filling machine*, where we are asking whether there is evidence that the volume of beer in the can is greater than 50 cl, the sample size obtained is 50.8948 cl. The probability of obtaining a sample mean of 50.8948 cl from a population whose mean is 50 cl is 2.96% or quite small. Thus we have observed an unlikely event or an event so unlikely that we should doubt our assumptions about the population mean in the first place. Note that in order to calculate the value of the test statistic we assumed that the null hypothesis is true and thus in this case we have reason to reject the null hypothesis and accept the alternative.

The p-value provides useful information as it measures the amount of statistical evidence that supports the alternative hypothesis. Consider Table 8.3 that gives values of the sample mean, the value of the test statistic, and the corresponding p-value for the application case-exercise *Beer filling machine*. As the sample mean gets larger, or moves further away from the hypothesized population mean of 50 cl, the smaller is the p-value. Values of the sample mean, $\bar{x}$, far above 50 cl tend to indicate that the alternative hypothesis is true or the smaller the p-value, the more the statistical evidence there is to support the alternative hypothesis.

Table 8.3

Sample mean, cl	z	p-value
50.00	0.0000	50.00%
50.40	0.4000	34.46%
50.80	0.8000	21.19%
51.20	1.2000	11.51%
51.60	1.6000	5.48%
52.00	2.0000	2.28%
52.40	2.4000	0.82%

The *p*-value is not to be interpreted by saying that it is the probability that the null hypothesis is true. You cannot make a probability assumption about the population parameter 50 cl as this is not a random variable.

Errors in hypothesis testing

The higher the value of the significance level, α, (the area outside the distribution) that is used for hypothesis testing then the higher is the percentage of the distribution in the tails. In this case, when α is high, the greater is the probability of rejecting a null hypothesis. Since the null hypothesis is true, or is not true, then as α increases there is a greater probability of rejecting the null hypothesis when in fact it is true. Looking at it another way, with a high significance level, that is a high value of α, it is unlikely we would accept a null hypothesis when it is in fact not true, or false. This relationship is illustrated in the normal distributions of Figure 8.11. At the 1% significance level, the probability of accepting the hypothesis when it is false is greater than at a significance level of 50%. Alternatively, the risk of rejecting a null hypothesis when it is in fact true is greater at a 50% significance level than at a 1% significance level. These errors in hypothesis testing are referred to as Type I and Type II errors.

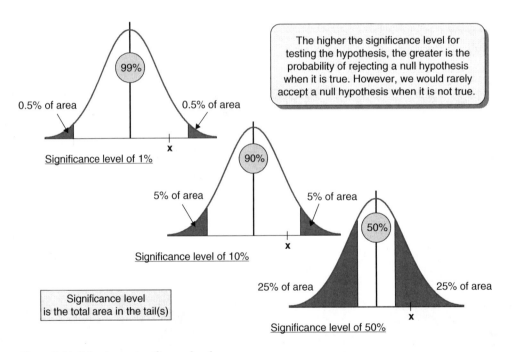

Figure 8.11 Selecting a significance level.

Type I and Type II errors

A Type I error occurs if the null hypothesis is rejected when in fact it is true. The probability of a Type I error is called α, where α is also the level of significance. A Type II error is accepting a null hypothesis when it is not true, or false. The probability of a Type II error is called β. When the acceptance region is small, or α is large, it is unlikely we would accept a null hypothesis when it is false.

However, at a risk of being this sure, we will often reject a null hypothesis when it is in fact true. The level of significance to use depends on the risk or "cost" of the error as illustrated as follows.

Cost of making an error

Consider a pharmaceutical firm that makes a certain drug. A quality inspector tests a sample of the product from the reaction vessel where the drug is being made. He makes a Type I error in his analysis. That is, he rejects a null hypothesis when it is true or concludes from the sample that the drug does not conform to quality specifications when in fact it really does. As a result, all the production quantity in the reaction vessel is dumped and the firm starts the production all over again. In reality the batch was good and could have been accepted. In this case, the firm incurs all the additional costs of repeating the production operation. Alternatively, suppose the quality inspector makes a Type II error, or accepts a null hypothesis when it is in fact false. In this case the produced pharmaceutical product is accepted and commercialized but it does not conform to quality specifications. This may mean that users of the drug could become sick, or at worst die. The "cost" of this error would be very high. In this situation, a pharmaceutical firm would prefer to make a Type I error, destroying the production lot, rather than take the risk of poisoning the users. This implies having a high value of α such as 50% as illustrated in Figure 8.11. A very sad, related situation occurred in the mid-1950s in several Western countries. Thalidomide was a drug developed in Germany that became popular with pregnant women to prevent morning sickness. It was hypothesized that the drug was safe and was marketed by several pharmaceutical companies in Germany, Canada, USA, and Britain with limited testing. As it turned out the drug caused phocomelia, or malformation of the limbs in the fetus of the child in the mother's womb. Experts estimated that the drug thalidomide led to the death of approximately 2,000 children and serious birth defects in more than 10,000, about 5,000 of them in the, then, West Germany. A total of 17 children with thalidomide-induced malformations were born in the USA. Those afflicted with the diseases were crippled for life and died prematurely. The drug was withdrawn from the market in the 1960s.[3,4]

Suppose that in another situation a manufacturing firm makes mechanical components used in the assembly of washing machines. An inspector takes a sample of this component from the production line and measures the appropriate properties. He makes a Type I error in the analysis. That is, he rejects the null hypothesis that the component conforms to specifications. In fact the null hypothesis is true and the components conform. To correct this (wrong) conclusion would involve an expensive disassembly and repair procedures of many components in the factory that have already been produced. On the other hand if the inspector had made a Type II error, accepting a null hypothesis when it is in fact false, this might involve a less expensive warranty repairs by the dealers when the washing machines are commercialized. In this second case, the cost of the error is probably lower than modifications in the factory. In this case a manufacturer perhaps is more likely to prefer a Type II error. In this case, the manufacturer will set low levels for α, such as 10% as illustrated in Figure 8.11. However, with the cost of litigation and the adverse marketing impact this may not always be a reliable option. Classic illustrations are the Bridgestone/Firestone tire company who in 2000 recalled some 6.5 million of its products as problems with treads with their tires were linked to hundreds of accidents and at least 46 deaths,[5] Toyota in 2013 and 2014 recalling some 2.92 million vehicles in the USA as faulty airbags were linked to some fatal accidents,[6] and Volkswagen in 2015 who hypothesized they could escape detection from authorities on the high levels of emissions from their diesel engines.[7]

The cost of an error in some situations might be infinite and irreparable. Consider for example a murder trial. Under Anglo-Saxon law the null hypothesis is that a person, if charged with murder, is considered innocent of the crime and the court has to prove guilt. In this case, the jury (we presume) would prefer to commit a Type II error, accepting a null hypothesis that the person is innocent, when it is in fact not true. Thus the guilty person is set free. The alternative would be to accept a Type I error, rejecting the null hypothesis that the person is innocent, when it is in fact true. In this case the person would be found guilty and risk the death penalty (at least in the USA) for a crime that they did not commit. Unfortunately there have been several instances when innocent people were executed for crimes that it was subsequently proved that they did not commit.

Power of a test

In all analytical work we want the chance of making an error to be small. In hypothesis testing we want the probability of making a Type I error, α, (rejecting the null hypothesis when it is in fact true) to be small. In addition we want the Type II error, β, (accepting the null hypothesis when it is false) also to be low. If a null hypothesis is false then we want the test to reject the conclusion every time. However, hypothesis tests are not perfect and when a null hypothesis is false, a test may not reject it and consequently a Type II error, β, is made, accepting a null hypothesis when it is false.

When the null hypothesis is false it implies that the true population value does not equal the hypothesized population value but instead equals some other value. For each possible value for which the alternative hypothesis is true, or the null hypothesis is false, there is a different probability, β, of accepting the null hypothesis when it is false. We would like this value of β to be as small as possible. Alternatively, we would like $(1 - \beta)$, the probability of rejecting a null hypothesis when it is false, to be as large as possible. Rejecting a null hypothesis when it is false is exactly what a good hypothesis test ought to do. A high value of $(1 - \beta)$ approaching 1.0 means that the test is working well. Alternatively, a low value of $(1 - \beta)$ approaching zero means that the test is not working well and the test is not rejecting the null hypothesis when it is false. The value of $(1 - \beta)$, the measure of how well the test is doing, is called the **power of the test**. In summary, Table 8.4 gives the possibilities in a situation when the null hypothesis is correct; Table 8.5 gives the possibilities in circumstances when the null hypothesis H_0 is false.

Table 8.4

Fact	Null hypothesis, H_0 is correct
Analysis No.1	Test statistic falls in region $(1 - \alpha)$
Decision after analysis No.1	Null hypothesis, H_0 is accepted
Situation	No error is made
Analysis No.2	Test statistic falls in region α
Decision after analysis No.2	Null hypothesis, H_0 is rejected
Situation	A Type I error α is made

Table 8.5

Fact	Null hypothesis, H_0 is false
Analysis No.3	Test statistic falls in region $(1 - \alpha)$
Decision after analysis No.3	Null hypothesis, H_0 is accepted
Situation	A Type II error β is made
Analysis No.4	Test statistic falls in region α
Decision after analysis No.4	Null hypothesis, H_0 is rejected
Situation	No error is made. Power of test is $(1 - \beta)$

Again, as in all statistical work, in order to avoid errors in hypothesis testing, utmost care must be made to ensure that the extracted random sample fairly represents the population from where the sample is taken.

Moving on

This chapter has considered hypothesis testing for sampling for the mean and the proportion from a single population. The next chapter covers hypothesis testing from more than one population.

Notes

1 Adapted from: www.nrc.gov/about-nrc/radiation (accessed July 31, 2015).
2 Bass, Gary J, "*The Blood Telegram: Nixon, Kissinger, and a forgotten genocide,*" Vintage Books, USA, July 2014.
3 *Dictionary of English Language and Culture*, Addison Wesley Longman, 1998, p. 1394.
4 Available online at: https://en.wikipedia.org/wiki/Thalidomide (accessed October 29, 2015).
5 Available at: www.history.com/this-day-in-history/bridgestonefirestone-announces-massive-tire-recall (accessed October 29, 2015).
6 Available at: www.cnbc.com/2015/06/16/toyota-expanding-us-airbag-related-recalls-by-137m-vehicles.html (accessed October 29, 2015).
7 "The Volkswagen scandal: A mucky business," *The Economist*, September 26, 2015, p. 19.

9 Hypothesis testing from different populations

ICEBREAKER: Overweight, obesity and weight loss programs

In the USA, UK, Australia, and some countries of the Middle East and Latin America there is a high incidence of people being overweight or obese. Overweight refers to an excess amount of body weight that comes from muscles, bone, fat, and water; obesity refers to an excess amount of body fat. Data from the USA indicates that more than 2 in 3 adults are considered to be overweight or obese; more than 1 in 3 are described as obese; more than 1 in 20 are considered extremely obese. About one-third of children and adolescents aged 6–19 are considered to be overweight or obese and more than 1 in 6 children and adolescents aged 6–19 are considered to be obese.

The body mass index (BMI) is a standard measure for determining if a person is overweight or obese. It is calculated by the ratio of the (weight, kg)/(height, m)^2. For example if a person is 1.70 m tall and weighs 60 kg then that persons BMI is $60/(1.70*1.70) = 20.76$. Classifications according to the numerical value of BMI for adults over 20 are given in Table 9.0.

Figure 9.0 Overweight and obese (participants on a walk against diabetes and for general fitness around Nauru airport).

Source: Lorrie Graham/AUS AID, Wikimedia Commons, the free media repository (licensed under Creative Commons Attribution 2.0 Generic).

Table 9.0

BMI	18.5 to 25	>25 ≤ 30	>30 ≤ 40	>40
Classification	Normal weight	Overweight	Obese	Extremely obese

Overweight and obesity result from an energy imbalance. The body needs a certain amount of energy or calories for basic life functions. Body weight tends to remain stable when the number of calories eaten equals the number the body uses or "burns." When people eat and drink more calories than they burn, the energy imbalance leads toward weight gain, overweight, and obesity. Overweight and obesity are considered risk factors for Type 2 diabetes, heart disease, high blood pressure, nonalcoholic fatty liver disease, osteoarthritis, some types of cancer, and strokes.[1]

Overweight conditions have given rise to innumerable weight loss and diet programs giving claims to their effectiveness. *Weight Watchers* states that you can drop up to two pounds a week (0.90 kg) using their PointsPlus program and group support. The key is choosing nutritionally dense foods that have a healthy ratio of fiber, fat, protein, and carbohydrates. Fresh fruits and vegetables are highly encouraged and carry zero points, so people can eat their fill of these foods. Candy bars and processed foods are higher in points and allowed in smaller portions. The *Health Management Resources* (*HMR*) program is a lifestyle. The theory is that reducing calorie intake, increasing physical activity, and personal accountability helps people lose weight and keep it off. It's divided into two phases: Phase I is that meal replacements including low-calorie shakes and meals, nutrition bars, and multi-grain hot cereal. Phase II is maintenance, including telephone coaching sessions with dietitians and exercise physiologists, where a person begins to incorporate other healthy food options into their meal replacements and focuses on lifestyle changes. The *Biggest Loser Diet* proposes eating lots of fruits and vegetables, choosing lean protein sources over red meat, keeping a food journal, and exercising at least two and a half hours per week, for example by walking or jogging, along with a couple of days of muscle-strengthening activities. The *Jenny Craig Diet* claims that devotees will drop up to two pounds (0.9 kg) each week by following their restricted calorie diet. Prepackaged meals limit calories and satisfy a sweet tooth without worrying about overindulging. This plan also includes weekly supportive counseling sessions to help members through their weight loss journeys right from the start. One downside with prepackaged meals: dining out can be tricky. The *Raw Food Diet* is for people who are looking to shed some pounds, improve their health and support the environment. It relies heavily on vegetables and completely eliminates food that has been cooked, processed, micro-waved, irradiated, genetically engineered, or exposed to pesticides or herbicides. The claim is that this program guarantees weight loss because those who follow it consume only half the calories they would eat on a diet that included cooked and processed foods.[2]

These and other weight loss programs claim (hypothesize) that if the regime is followed a certain amount of weight will be lost within a specified time period. However, for this type of statement to have validity a sampling program needs to be carried out before the program is started and then afterwards, in order to justify the hypothesis or claim. This is the purpose of this chapter.

Chapter subjects

✓   **Difference between the mean of two independent populations** • Large samples
• Test statistic for large sample • Application case-exercise: *Salaries of men and women*
• Small samples • Application case-exercise, *Production output*
✓   **Differences of the means between dependent or paired populations** • Application
case-exercise, *Health spa and weight loss*
✓   **Difference between the proportions of two populations with large samples**
• Standard error of the difference between two proportion • Application case-exercise:
Commuting time
✓   **Chi-squared test for dependency** • Chi-squared distribution • Contingency table and
observed frequency, f_o • Hypothesis test and expected frequency, f_e • Degrees of freedom
• Application case-exercise: *Work schedule preference* • Excel and chi-squared functions

---

In this chapter we look at hypothesis testing when there is more than one population involved in the analysis. It can be used in a before and after situation as for example in weight loss programs discussed in the Icebreaker to prove a claim.

Difference between the mean of two independent populations

The difference between the mean of two independent populations is a hypothesis test to sample to see if there is a significant difference between the parameters of two independent populations, for example:

- A human resource manager wants to know if there is a significant difference between the salaries of men and the salaries of women in his multi-national firm.
- A professor of Business Statistics is interested to know if there is a significant difference between the grade level of students in her morning class and in a similar class in the afternoon.
- A company wants to know if there is a significant difference in the productivity of the employees in one country and another country.
- A firm wishes to know if there is a difference in the absentee rate of employees in the morning shift and the night shift.
- A company wishes to know if the sales volumes of a certain product in one store are different from a store in a different location.

In these cases, we are not necessarily interested in the specific value of a population parameter but more to understand something about the relation between the two parameters from the populations. That is, are they essentially the same, or is there a significant difference?

Large samples

The hypothesis testing concept between two population means is illustrated in Figure 9.1. The figure on the left gives the normal distribution for Population 1 and the figure on the right gives the normal distribution for Population 2. Underneath the respective distributions are the sampling distributions of the means taken from that population. From the data another distribution can be constructed that is then the **difference** between the values of sample means taken from the respective populations.

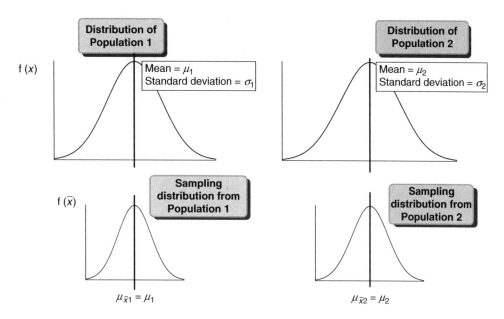

Figure 9.1 Two independent populations.

Assume for example that we take a random sample from Population 1, which gives a sample mean of $\bar{x}_1$. Similarly we take a random sample from Population 2 and this gives a sample mean of $\bar{x}_2$. The difference between the values of the sample means is then:

$$\bar{x}_1 - \bar{x}_2 \qquad\qquad 9(\text{i})$$

When the value of $\bar{x}_1$ is greater than $\bar{x}_2$ then the result of Equation 9(i) is positive. When the value of $\bar{x}_1$ is less than $\bar{x}_2$ then the result of Equation 9(i) is negative

If we construct a distribution of the difference of the entire sample means then we will obtain a sampling distribution of the differences of all the possible sample means, as shown in Figure 9.2.

The mean of the sample distribution of the differences of the mean is written:

$$\mu_{\bar{x}_1 - \bar{x}_2} = \mu_{\bar{x}_1} - \mu_{\bar{x}_2} \qquad\qquad 9(\text{ii})$$

When the mean of the two populations are equal then: $\mu_{\bar{x}_1} - \mu_{\bar{x}_2} = 0$.

From *Chapter 6*, Equation 6(ii), using the central limit theorem we developed the following relationship for the standard error of the sample mean:

$$\sigma_{\bar{x}} = \frac{\sigma_x}{\sqrt{n}}$$

Extending this relationship for sampling from two populations, the **standard deviation of the distribution of the difference between the sample means**, as given in Figure 9.2, is determined from the relationship:

$$\sigma_{\bar{x}_1 - \bar{x}_2} = \sqrt{\frac{\sigma_1^2}{n_1} + \frac{\sigma_2^2}{n_2}} \qquad\qquad 9(\text{iii})$$

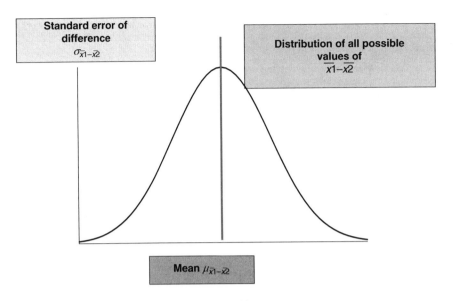

Figure 9.2 Distribution of all possible values of between two means.

where σ_1^2 and σ_2^2 are respectively the variance of Population 1 and Population 2; σ_1 and σ_2 are their standard deviations and n_1 and n_2 are the sample sizes taken from these two populations. This relationship is also the **standard error of the difference between two means**. If we do not know the population standard deviations, then we use the sample standard deviation, s, as an estimate of the population standard deviation $\hat{\sigma}$. In this case the **estimated standard deviation of the distribution of the difference between the sample means** is:

$$\hat{\sigma}_{\bar{x}_1 - \bar{x}_2} = \sqrt{\frac{\hat{\sigma}_1^2}{n_1} + \frac{\hat{\sigma}_2^2}{n_2}}$$

9(iv)

Test statistic for large samples

From *Chapter 6*, Equation 6(iv), for a single population, the test statistic z for large samples (greater than 30), is:

$$z = \frac{\bar{x} - \mu_x}{\frac{\sigma_x}{\sqrt{n}}}$$

When we test a hypothesis for the difference between the means of two populations then the equation for the test statistic becomes:

$$z = \frac{(\bar{x}_1 - \bar{x}_2) - (\mu_1 - \mu_2)_{H_0}}{\sqrt{\frac{\sigma_1^2}{n_1} + \frac{\sigma_2^2}{n_2}}}$$

9(v)

Here the subscript H_0 refers to the hypothesized difference of the means. Alternatively, if we do not know the population standard deviation, then Equation 9(v) becomes as follows, which is numerically the same as given by Equation 9(v):

$$z = \frac{(\bar{x}_1 - \bar{x}_2) - (\mu_1 - \mu_2)_{H_0}}{\sqrt{\frac{\hat{\sigma}_1^2}{n_1} + \frac{\hat{\sigma}_2^2}{n_2}}} \qquad\qquad 9(vi)$$

In these equations, $(\bar{x}_1 - \bar{x}_2)$ is the difference between the sample means taken from the population and $(\mu_1 - \mu_2)_{H_0}$ is the difference of the hypothesized means of the population. This concept is illustrated by the following application case-exercise, *Salaries of men and women*.

Application case-exercise: Salaries of men and women

Situation

A large firm in the USA wants to know the relationship between the salaries of men and women employed at its company. The hypothesis is that there is no significant difference. Sampling the employees gave the information in $US in Table 9.1.

Table 9.1

Statistic	Sample mean $\bar{x}$	Sample standard deviation	Sample size n
Sample 1, women	$28.65	$2.40	130
Sample 2, men	$29.15	$1.90	140

Required

1　Present appropriate hypothesis relationships for this situation.
2　Using z values, at a 10% significance level, is there evidence of a difference between the salaries of men and women?
3　Corroborate your conclusions using the p-value approach.
4　Would your conclusions change if you used a significance level of 5% for the test of a difference? Use both the z values and the p-value approach.
5　Would your conclusions change if you used a significance level of 5% for the test to see if there is evidence that the salaries of women are less than those of men? Write the hypothesis relationships and quantify your response.
6　What are your comments related to this analysis?

Worked out solutions

1　The null and alternative hypotheses are:

- Null hypothesis, $H_0: \mu_1 = \mu_2$ is that there is no significant difference in the salaries
- Alternative hypothesis, $H_1: \mu_1 \neq \mu_2$ is that there is a significant difference in the salaries

2　At a 10% significance level we are asking the question: "Is there a difference?" That is, can be greater or less than. This is a two-tailed test with 5.0% in each of the tails. Using in Excel the function **NORMSINV** or **NORM.S.INV** the critical value of z is ± 1.6449.

Since we have only a measure of the sample standard deviation, s, and not the population standard deviation σ then the test or sample statistic z is calculated from:

$$z = \frac{(\bar{x}_1 - \bar{x}_2) - (\mu_1 - \mu_2)_{H_0}}{\sqrt{\frac{\hat{\sigma}_1^2}{n_1} + \frac{\hat{\sigma}_2^2}{n_2}}}$$

Here, $\bar{x}_1 - \bar{x}_2 = 28.65 - 29.15 = -0.50$.

$(\mu_1 - \mu_2)_{H_0} = 0$ since the null hypothesis is that there is no difference between the population means.

The standard error of the difference between the means is given by:

$$\hat{\sigma}_{\bar{x}_1 - \bar{x}_2} = \sqrt{\frac{\hat{\sigma}_1^2}{n_1} + \frac{\hat{\sigma}_2^2}{n_2}} = \sqrt{\frac{2.40^2}{130} + \frac{1.90^2}{140}} = 0.2648$$

Thus $z = -\frac{0.50}{0.2648} = -1.8886$.

Since the sample, or test statistic, of -1.8886 is less than the critical value of -1.6449 we reject the null hypothesis and conclude that there is evidence to indicate that the wages of women are significantly different from those of men.

3   With a sample value of $z = -1.8886$, and using the function **NORMSDIST** or **NORM.S. DIST**, gives an area in the tail of 2.95%. Since 2.95% is less than 5% we reject the null hypothesis. This is the same conclusion as in Question 2. The representation of the p-value and the z values are given in Figure 9.3.

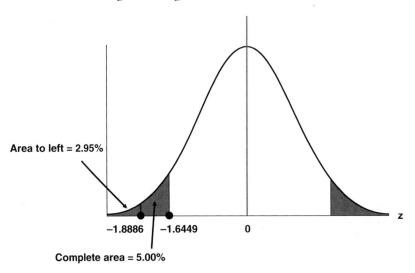

Figure 9.3 Salary differences between men and women; signicance level of 10%.

4   At a 5% significance level this is still a two-tailed test with now 2.5% in each of the tails. For this area the critical value of z is ± 1.9600. The sample value is still -1.8886 and since numerically the sample value is less than the critical value we accept the null hypothesis and say that there is no significant different between the salaries of men and women. Alternatively we might say the critical value is less than the sample test value The p-value of the sample statistic is still 2.95%, that

is, greater than 2.5% and so this confirms our conclusion in Question 4. The representation of the *p*-value and the *z* values are given in Figure 9.4.

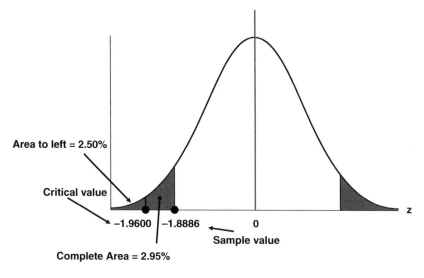

Figure 9.4 Salary differences between men and women: significance level 5%.

5   In this case for a test of "less than":

- Null hypothesis is $H_0:\mu_1 \geq \mu_2$: Women's salaries are not significantly less than men's salaries
- Alternative hypothesis is $H_1:\mu_1 < \mu_2$: Women's salaries are significantly less than men's salaries

    The sample *z* value of the statistic remains the same at -1.8886. This is now a left-tailed test with the area in the tail being 5%. This gives a critical value of *z* of -1.6449. Since the absolute value of the sample statistic is greater than the absolute value of the critical value than we reject the null hypothesis and conclude that there is evidence that the salaries of women are less than those of men. (Alternatively we can say that the value of -1.8886 is less than the value -1.6449.) The *p*-value of the sample is 2.95% and this is less than the critical value of 5% and this corroborates our conclusion. The representation of these results is given in Figure 9.5.

6   The conclusions drawn depend on the level of significance and in what manner the hypothesis is stipulated. In this case it can lead to a conflict between say the unions and management.

Small samples

When the sample size is small, or less than 30 units, to be correct we must use the Student's *t* distribution. Using the Student's *t* distribution the population standard deviation is unknown. Thus to estimate the standard error of the difference between the two means we use Equation 9(iv) introduced previously:

$$\widehat{\sigma}_{\bar{x}_1 - \bar{x}_2} = \sqrt{\frac{\widehat{\sigma}_1^2}{n_1} + \frac{\widehat{\sigma}_2^2}{n_2}}$$

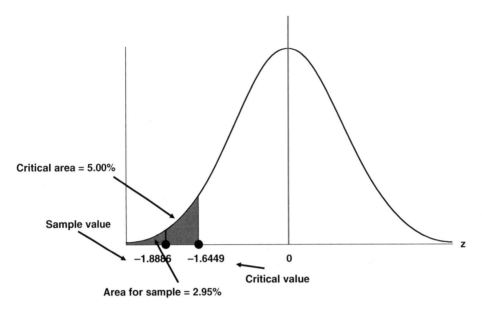

Figure 9.5 Are women's salaries less than men's?—signifiance level 5%.

However, a difference from the hypothesis testing of large samples is that here we make the assumption that the variance of Population 1, $\sigma_1{}^2$, is equal to the variance of Population 2, $\sigma_2{}^2$, or $\sigma_1{}^2 = \sigma_2{}^2$. This then enables us to use a pooled variance such that the sample variance, s_1^2, taken from Population 1 can be pooled, or combined, with s_2^2, to give a value s_p^2. This value of the pooled estimate s_p^2 is given by:

$$s_p^2 = \frac{(n_1 - 1) \cdot s_1^2 + (n_2 - 1) \cdot s_2^2}{(n_1 - 1) + (n_2 - 1)}$$ \hfill 9(vii)

This value of s_p^2 is now the best estimate of the variance common to both populations, σ^2, on the assumption that the two population variances are equal. Note that the denominator in Equation 9(vii), can be rewritten:

$$(n_1 - 1) + (n_2 - 1) = (n_1 + n_2 - 2)$$ \hfill 9(viii)

This is so because we now have two samples and thus two degrees of freedom. (Note that in *Chapter 8,* with one sample of size n in order to use the Student's t distribution we had $(n-1)$ degrees of freedom.)

Combining equations 9(iv) and 9(vii), relationship for the estimated standard error of the difference between two sample means, when there are small samples and on the assumption that the population variances are equal, is given by:

$$\widehat{\sigma}_{\bar{x}_1 - \bar{x}_2} = s_p \sqrt{\frac{1}{n_1} + \frac{1}{n_2}}$$ \hfill 9(ix)

Then by analogy to Equation 9(vi) the value of the Student's t distribution is given by:

$$t = \frac{(\bar{x}_1 - \bar{x}_2) - (\mu_1 - \mu_2)_{H_0}}{\sqrt{s_p^2\left(\frac{1}{n_1} + \frac{1}{n_2}\right)}}$$

9(x)

If we take samples of equal size from each population then since $n_1 = n_2$, Equation 9(vii) becomes:

$$s_p^2 = \frac{(n_1 - 1)\cdot s_1^2 + (n_2 - 1)\cdot s_2^2}{(n_1 - 1) + (n_2 - 1)} = \frac{(n_1 - 1)\cdot s_1^2 + (n_1 - 1)\cdot s_2^2}{(n_1 - 1) + (n_1 - 1)} = \frac{(n_1 - 1)(s_1^2 + s_2^2)}{(n_1 - 1)(1 + 1)} = \frac{(s_1^2 + s_2^2)}{2}$$

9(xi)

Further, the relationship in the denominator of Equation 9(x) can be rewritten.

$$\left(\frac{1}{n_1} + \frac{1}{n_2}\right) = \left(\frac{1}{n_1} + \frac{1}{n_1}\right) = \frac{2}{n_1}$$

9(xii)

Thus Equation 9(x) can be rewritten as:

$$t = \frac{(\bar{x}_1 - \bar{x}_2) - (\mu_1 - \mu_2)_{H_0}}{\sqrt{\left(\frac{s_1^2 + s_2^2}{n_1}\right)}}$$

9(xiii)

The use of the Student's t distribution for small samples is illustrated by the following application case-exercise, *Production output*.

Application case-exercise: **Production output**

Situation

One part of a car production firm is the assembly line of the automobile engines. In this area of the plant, the firm employs three shifts: morning 07:00–15:00, evening 15:00–23:00, and the night shift 23:00–07:00. The manager of the assembly line believes that the production output on the morning shift is greater than that of the night shift, implying that productivity of the employees is better in the morning than in the evening. Before the manager takes any action he first records the output in production units for 16 days for the morning shift, and 13 days for the night shift. This information is given in Table 9.2.

Table 9.2

Morning shift (1)	29	24	28	29	31	27	29	28	26	23	25	28	27	27	30	23
Night shift (2)	22	23	21	25	31	22	28	30	20	22	23	25	26			

Required

1   Develop the hypothesis relationships for this situation.
2   Using t values, at a 1% significance level, is there evidence that the output of engines on the morning shift is greater than on the evening shift?
3   Confirm your answer to Question 2 using the p-value approach.
4   Using t values, at a 5% significance level, is there evidence that the output of engines on the morning shift is greater than on the evening shift?
5   Confirm your answer to Question 4 using the p-value approach.
6   What are your comments regarding this hypothesis test?

Worked out solutions

1   The null hypothesis is that the output on the morning shift is not greater than the output on the night shift, $H_0 : \mu_M \leq \mu_N$. The alternative hypothesis is that the output on the morning shift is greater than that on the night shift, $H_1 : \mu_M > \mu_N$.
2   At a 1% significance level we are asking the question: "Is there evidence of the output on the morning shift output being greater than the output on the night shift?" This is then a one-tailed test with 1% in the upper tail. Degrees of freedom are $(16 + 13 - 2) = 27$. Using function **TINV** (Excel 2007) or **T.INV.2T** (Excel 2010 or later) gives a critical value of Student's t of 2.4727. From the sample data using in Excel the standard deviation and the mean we have the statistics in Table 9.3.

Table 9.3

n_1 (Morning shift)	16
n_2 (Night shift)	13
s_1	2.3910
s_2	3.4548
$\bar{x}_1$	27.1250
$\bar{x}_2$	24.4615
$\bar{x}_1 - \bar{x}_2$	2.6635

The value of the pooled variance is:

$$s_p^2 = \frac{(n_1 - 1) \cdot s_1^2 + (n_2 - 1) \cdot s_2^2}{(n_1 - 1) + (n_2 - 1)} = \frac{(16 - 1)*2.3910^2 + (13 - 1)*3.4548^2}{(16 - 1) + (13 - 1)} = 8.4808$$

The sample or test value of the Student's t value is:

$$t = \frac{(\bar{x}_1 - \bar{x}_2) - (\mu_1 - \mu_2)_{H_0}}{\sqrt{s_p^2 \left(\frac{1}{n_1} + \frac{1}{n_2}\right)}} = \frac{27.1250 - 24.4615 - 0}{\sqrt{8.4808 \left(\frac{1}{16} + \frac{1}{13}\right)}} = 2.4494$$

Since the sample test value of t of 2.4494 is less than the critical value t of 2.4727 we conclude that there is no significant difference between the production output in the morning and night shift.

3   If we use the p-value approach for this hypothesis test then using function **TDIST** (Excel 207) or **T.DIST.RT** (Excel 2010 or later) in Excel for a one-tailed test then the area in the tail for the

sample information is 1.05%. This is greater than 1.00% and so our conclusion is the same in that we accept the null hypothesis. The concept for Questions 1 and 2 is illustrated in Figure 9.6.

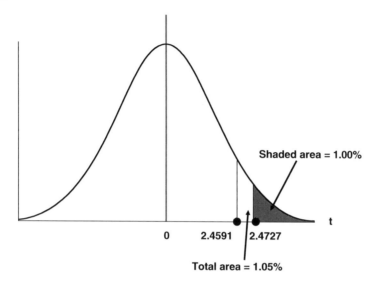

Figure 9.6 Production output between the morning and night shift (1% level of significance).

4   In this situation nothing happens to the sample or test value of the Student's *t* which remains at 2.4494. However, now we have 5% in the upper tail and using function **TINV** (Excel 2007) or **T.INV.2T** (Excel 2010 or later) gives a critical value of Student's *t* = 1.7033. Since 2.4494 is greater than 1.7033 we conclude that at a 5% level the production output in the morning shift is significantly greater than for the night shift. We reject the null hypothesis.

5   If we use the *p*-value approach for this hypothesis test then using **TDIST** (Excel 2007) or **T.DIST.RT** (Excel 2010 or later) in Excel for a one-tailed test then the area in the tail for the sample is still 1.05%. This is less than 5.00% and so our conclusion is the same in that we reject the null hypothesis. The concept for Questions 3 and 4 is illustrated on Figure 9.7.

6   Conclusions are sensitive to the data collected and also the level of significance used.

Differences of the means between dependent or paired populations

In the previous section we discussed analysis of populations that were essentially independent from each other. In the wage example we chose samples from a population of men and a population of women. In the production output example we looked at the population of the night shift and the morning shift. Sometimes in sampling experiments we are interested in the differences of **paired samples** or those that are dependent or related, often in a before and after situation. Examples might be weight loss of individuals after a diet program; productivity improvement after an employee training program; or sales increases of a certain product after an advertising campaign. The purpose of these tests is to see if improvements have been achieved as a result of a new action. When we make this type of analysis we remove the effect of other variables or extraneous factors. The analytical procedure is to consider statistical analysis on the **difference** of the values since there is a direct relationship rather than the values before and after. The use of this concept is illustrated by the following application case-exercise, *Health spa and weight loss*.

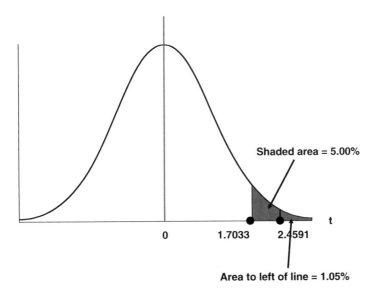

Figure 9.7 Production output between the morning and night shift (5% level of significance).

Application case-exercise: **Health spa and weight loss**

Situation

A health spa connected to a hotel in the center of Brussels, Belgium advertises a combined fitness and diet program where it guarantees that participants who are overweight will lose at least 10 kg in six months if they scrupulously follow the course. The weights of all participants in the program are recorded each time they come to the spa. The authorities are somewhat skeptical of the advertising claim so they select, at random, 13 of the regular participants and their recorded weights in kilograms before and after six months in the program are in Table 9.4.

Required

1   Define appropriate hypothesis relationships.
2   At a 5% significance level, is there evidence that the weight loss of participants in this program is greater than 10 kg?
3   Confirm your conclusions of Question 2 using the *p*-value approach.
4   Would the conclusions change if a 10% significance level was used?
5   Confirm your conclusions of Question 4 using the *p*-value approach.
6   What are your observations on this statistical experiment?

Worked out solution

1   Null hypothesis is that the weight loss is not more than 10 kg or $H_0{:}\mu \le 10$ kg; alternative hypothesis is that the weight loss is more than 10 kg, or $H_1{:}\mu > 10$ kg.
2   We are interested not in the absolute weights before and after but in the difference of the weights before and after. We extend Table 9.4 to give Table 9.5. The test is now very similar to hypothesis testing for a single population since we are making our analysis just on the difference.
     At a significance level of 5% all of the area lies in the right-hand tail. Using function **TINV**

Table 9.4

Before, kg	120	95	118	92	132	102	87	92	115	98	109	110	95
After, kg	101	87	97	82	121	87	74	84	109	87	100	101	82

Table 9.5

Before, kg	120	95	118	92	132	102	87	92	115	98	109	110	95
After, kg	101	87	97	82	121	87	74	84	109	87	100	101	82
Difference, kg	19	8	21	10	11	15	13	8	6	11	9	9	13

(Excel 2007) or **T.INV.2T** (Excel 2010 or later) gives a critical value of Student's $t = 1.7823$. Using Excel the statistics for Table 9.5 are:

$\bar{x}$ (Difference) $= 11.7692$ kg; $s = \hat{\sigma} = 4.3999$; n the sample size is 13. Estimated standard error of the mean is $\bar{\sigma}_x = \frac{\hat{\sigma}}{\sqrt{n}} = \frac{4.3999}{\sqrt{13}} = 1.2203$.

Sample, or test value of Student's t is:

$$t = \frac{\bar{x} - \mu_{H_0}}{\hat{\sigma}/\sqrt{n}} = \frac{11.7692 - 10}{1.2203} = \frac{1.7692}{1.2203} = 1.4498$$

Since this sample value of t of 1.4498 is less than the critical value of t of 1.7823 we accept the null hypothesis and conclude that based on our sampling experiment the weight loss in this program over a 6-month period is not more than 10 kg.

3 If we use the p-value approach for this hypothesis test then using function **TDIST** (Excel 2007) or **T.DIST.RT** (Excel 2010 or later) in Excel for a one-tailed test then the area in the tail for sample information is 8.64%. This is greater than 5.00% and so our conclusion is the same in that we accept the null hypothesis. Figure 9.8 shows the concepts for the p-value and the Student's t distributions.

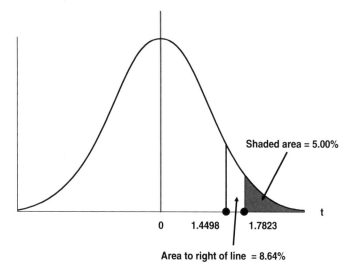

Figure 9.8 Health spa and weight loss (5% significance).

4   In this case at a significance level of 10% all of the area lies in the right-hand tail and using **TINV** (Excel 2007) or **T.INV.2T** (Excel 2010 or later) gives a critical value of Student's $t = 1.3562$. The sample or test value of the Student's t remains unchanged at 1.4498. Now, $1.4498 > 1.3562$ and thus we reject the null hypothesis and conclude that the publicity for the program is correct and that the average weight loss is greater than 10 kg.

5   Using the p-value approach for this test then with function **TDIST** (Excel 2007) or **T.DIST. RT** (Excel 2010 or later) in Excel for a one-tailed test, the area in the tail is still 8.64%. This is less than 10.00% and so our conclusion is the same in that we reject the null hypothesis. Figure 9.9 shows the concepts for the p-value and the Student's t distributions.

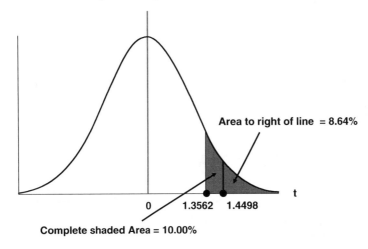

Figure 9.9 Health spa and weight loss (10% significance).

6   Again, as in all hypothesis testing, remember that the conclusions are sensitive to the level of significance used in the test.

Differences between the proportions of two populations with large samples

There are situations where we might be interested to know if there is a significant difference between the proportion or percentage of some criterion of two different populations. For example, is there a significant difference between the percentage output of two of a firm's production sites? Is there a difference between the health of British people and Americans? (The answer is yes, according to a study in the *Journal of the American Medical Association*.[3]) Is there a significant difference between the percentage effectiveness of one drug and another drug for the same ailment? In these situations we take samples from each of the two groups and test for the percentage or proportional difference in the two populations. The procedure behind the test work is similar to the testing of the differences in means except rather than looking at the difference in numerical values we have the differences in percentages.

Standard error of the difference between two proportions

In *Chapter 6*, Equation 6(xi), we developed the following equation for the standard error of the proportion, $\sigma_{\bar{p}}$.

$$\sigma_{\bar{p}} = \sqrt{\frac{p \cdot q}{n}} = \sqrt{\frac{p \cdot (1 - p)}{n}}$$

where is the sample size, p is the population proportion of *successes* and q is the population proportion of *failures* equal to $(1 - p)$. Then by analogy to Equation 9(iii) for the difference in the standard error for the means we have the equation for the **standard error of the difference between two proportions** as:

$$\sigma_{\bar{p}_1 - \bar{p}_2} = \sqrt{\frac{p_1 \cdot q_1}{n_1} + \frac{p_2 \cdot q_2}{n_2}}$$

9(xiv)

where p_1, q_1 are respectively the proportion of *success* and *failure* and n_1 is the sample size taken from Population 1; p_2, q_2, and n_2 are the corresponding values for Population 2. If we do not know the population proportions then the **estimated standard error of the difference between two proportions** is:

$$\widehat{\sigma}_{\bar{p}_1 - \bar{p}_2} = \sqrt{\frac{\bar{p}_1 \cdot \bar{q}_1}{n_1} + \frac{\bar{p}_2 \cdot \bar{q}_2}{n_2}}$$

9(xv)

Here $\bar{p}_1, \bar{q}_1, \bar{p}_2, \bar{q}_2,$ are the values of the proportion of *successes* and *failures* taken from the sample.

Now in *Chapter 8*, Equation 8(ix), we developed that the number of standard deviations, z, in hypothesizing for a single population proportion was as follows:

$$z = \frac{\bar{p} - p_{H_0}}{\sigma_{\bar{p}}}$$

By analogy, the value of z for the difference in the hypothesis for two population proportions is:

$$z = \frac{(\bar{p}_1 - \bar{p}_2) - (p_1 - p_2)_{H_0}}{\widehat{\sigma}_{\bar{p}_1 - \bar{p}_2}}$$

9(xvi)

The use of this concept is illustrated by the following application case-exercise, *Commuting time*.

Application case-exercise: **Commuting time**

Situation

A study was made to see if there was a significance difference between the commuting time of people working in downtown Los Angeles in Southern California and the commuting time of people working in downtown San Francisco in Northern California. The benchmark for commuting time was at least 2 hr per day. A random sample of 302 people was selected from Los Angeles and 178 said that they had a daily commute of at least 2 hr. A random sample of 250 people was selected in San Francisco and 127 replied that they had a commute of at least 2 hr.

Required

1 Develop the relationships for this hypothesis test.
2 Using a normal z value approach, at a 5% significance level, is there evidence to suggest that the commuting time of people in Los Angeles is different from those of San Francisco?
3 Confirm your conclusion to Question 2 using the p-value approach.

4   Using a normal z value approach, at a 5% significance level, is there evidence to suggest that the commuting time for people in Los Angeles is greater than for those working in San Francisco?
5   Confirm your conclusion to Question 4 using the p-value approach.
6   What are your comments on this case-exercise?

Worked out solution

1   Sample proportion for Los Angeles is p_1 and sample proportion for San Francisco is p_2. Null hypothesis is that there is no difference, or $H_0:p_1 = p_2$. The alternative hypothesis is that there is a difference, or $H_1:p_1 \neq p_2$
2   Sample proportion, p_1 of people commuting at least 2 hr to Los Angeles is:

$$\bar{p}_1 = 178/302 = 0.5894 \text{ and } \bar{q}_1 = 1 - 0.5894 = 0.4106$$

Sample proportion of people, p_2, commuting at least 2 hr to San Francisco is:

$$\bar{p}_1 = 127/250 = 0.5080 \text{ and } \bar{q}_1 = 1 - 0.5080 = 0.4920$$

This is a two-tailed test since we are asking the question: "Is there a difference?"

- Null hypothesis is that there is no difference or $H_0:p_1 = p_2$
- Alternative hypothesis is that there is a difference or, $H_1:p_1 \neq p_2$
  The estimated standard error of the difference between two proportions is:

$$\hat{\sigma}_{\bar{p}_1 - \bar{p}_2} = \sqrt{\frac{\bar{p}_1 \cdot \bar{q}_1}{n_1} + \frac{\bar{p}_2 \cdot \bar{q}_2}{n_2}} = \sqrt{\frac{0.5894 * 0.4106}{302} + \frac{0.5050 * 0.4920}{250}} = 0.0424$$

The sample value of z is:

$$z = \frac{(\bar{p}_1 - \bar{p}_2) - (p_1 - p_2)_{H_0}}{\hat{\sigma}_{\bar{p}_1 - \bar{p}_2}} = \frac{(0.5894 - 0.5080) - 0}{0.0424} = 1.9181$$

$(p_1 - p_2)_{H_0} = 0$ because the null hypothesis is that there is no difference.
This is a two-tailed test at 5% significance so there is 2.50% in each tail. Using function **NORMSINV** (Excel 2007) or **NORM.S.INV** (Excel 2010 or later) gives a critical value of z of $\pm$ 1.9600. Since 1.9181 < 1.9600 we accept the null hypothesis and conclude that there is no significant difference between commuting times in Los Angeles and San Francisco.
3   We obtain the same conclusion when we use the p-value for making the hypothesis test. Using function **NORMSDIST** (Excel 2007) or **NORM.S.DIST** (Excel 2010 or later) for a sample value z of 1.9181 the area in the upper tail is 2.75%. This area of 2.75% is greater than the critical value of 2.50%, and so again we accept the null hypothesis. The concepts for the p-value and the z values are illustrated in Figure 9.10.
4   At a 5% significance level, is there evidence to suggest that the commuting time in Los Angeles is greater than for those working in San Francisco?
This is a one-tailed test since we are asking the question: "Is the commuting time of one population greater than the other?" Here all the 5% is in the upper tail.

- Null hypothesis is that the commuting time is not greater, or $H_0:p_1 \leq p_2$
- Alternative hypothesis is that commuting time is greater, or $H_1:p_1 > p_2$.

Here we use $\leq$ since less than or equal is not greater than and so thus satisfies the null hypothesis. The sample test value of z remains unchanged at 1.9181. However, using function

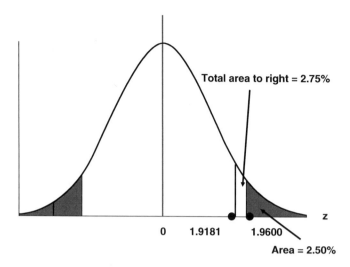

Figure 9.10 Commuting time (5% significance two-tailed).

NORMSINV (Excel 2007) or **NORM.S.INV** (Excel 2010 or later) the 5% in the upper tail corresponds to a critical z value of 1.6449. Since the value of 1.9181 > 1.6449 we reject the null hypothesis and conclude that there is statistical evidence that the commuting time for Los Angeles people is significantly greater than for those persons in San Francisco.

5   Using the p-value approach, the area in the upper tail corresponding to a sample test value of 1.9181 is still 2.75%. Now this value is less than the 5.00% significant value and so the conclusion is the same: there is evidence to suggest that the computing time for those in Los Angeles is greater than for those in San Francisco. This new situation is illustrated in Figure 9.11.

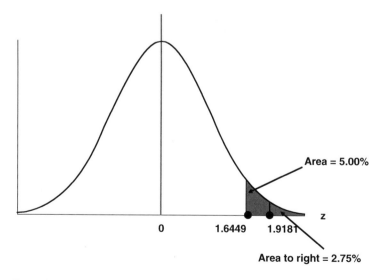

Figure 9.11 Commuting time (5% significance one-tailed).

Chi-squared test for dependency

In testing samples from two different populations we examined the difference between either two means, or alternatively two proportions. If we have sample data that gives proportions from more than two populations then a **chi-squared distribution** can be used to draw conclusions. From this distribution a chi-squared test is used that enables us to decide whether the differences among several sample proportions is significant, or that the difference is only due to chance. Suppose for example that a sample survey on the proportion of people in certain states of the USA who exercise regularly was found to be 51% in California, 34% in Ohio, 45% in New York and 29% in South Dakota. If these differences are considered significant then a conclusion may be that location affects the way people behave. If they are not significant, then the differences are just due to chance. Thus, assuming a firm is considering marketing a new type of jogging shoe then, if there is a significant difference between states, its marketing efforts should be weighted more to the state with a higher level of physical fitness.

Chi-squared distribution

The chi-squared distribution is a continuous probability distribution and, like the Student's t distribution, there is a different curve for each degree of freedom, y. The x-axis is the value of chi-squared, written x^2 where the symbol χ is the Greek letter c. Since we are dealing with x^2, or x to the power of two, the values on the x-axis are always positive and extend from zero to infinity. The y-axis is the frequency of occurrence, $f(x^2)$ where this probability density function is given by:

$$f(\chi^2) = \frac{1}{[(\upsilon/2 - 1)]!} \cdot \frac{1}{2^{\upsilon/2}} \cdot (\chi^2)^{(\upsilon/2 - 1) \cdot e^{-x^2/2}}$$

9(xvii)

Figure 9.12 gives three chi-squared distributions for degrees of freedom, y, of 4, 8, and 12. For small values of y the curves are positively or right skewed. As the value of y increases the curve takes on a form similar to a normal distribution. The mode or the peak of the curve is equal to the degrees of freedom less two. For example, for the three curves illustrated, the peak of each curve is for values of x^2 equal to 2, 6, and 10 respectively.

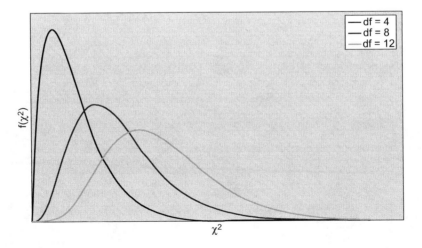

Figure 9.12 Chi-square distribution for three different degrees of freedom.

To use the chi-squared test we need to develop a **contingency table** and determine the **degrees of freedom** associated with this table.

Contingency table and observed frequency, f_o

A contingency or cross-classification table, discussed in *Chapter 2,* shows information by cross classifying variables according to certain criteria of interest. In this way the cross-classification accounts for all contingencies in the sampling data. Assume a survey is taken regarding the preference for wine and beer in Spain, Italy, Norway, and England. The sample data is in Table 9.6. In this contingency table, the columns give preference according to country and the rows give preference according to the type of beverage. These sample values are the observed frequencies of occurrence, denoted f_o. This is a four by four contingency table as there are four rows and four columns. Neither the row totals nor the column totals are considered in determining the dimension of the table.

Hypothesis test and expected frequency, f_e

From Table 9.6 the null hypothesis for the chi-squared test is that the choice of beverage is independent of country and the null hypothesis of the proportions is written: $H_0 : p_S = p_I = p_N = p_E$. The alternative hypothesis is $H_1 : p_S \neq p_I \neq p_N \neq p_E$ where the indices refer to the first letter of the country. On the basis that the null hypothesis is correct then we develop an expected value of frequencies f_e as shown in Table 9.7. One way of determining these values is to use the relationship:

$$f_e = \frac{(Total\ of\ the\ row)*(Total\ of\ the\ column)}{Sample\ size,\ n}$$

Table 9.6

Preference	Spain	Italy	Norway	England	Total
White wine	220	115	185	125	645
Rosé wine	150	90	32	101	373
Red wine	125	250	97	245	717
Beer	110	69	197	220	596
Total	605	524	511	691	2,331

Table 9.7

Preference	Spain	Italy	Norway	England	Total
White wine	167.41	144.99	141.40	191.20	645.00
Rosé wine	96.81	83.85	81.77	110.57	373.00
Red wine	186.09	161.18	157.18	212.55	717.00
Beer	154.69	133.98	130.65	176.68	596.00
Total	605.00	524.00	511.00	691.00	2 331.00

For example, if we consider the preference of white wine for Spain, the total of the row is 645; total of the column is 605; sample size is 2,331. The expected value for Spain is then $(645*605)/2{,}331 = 167.41$.

- Alternatively, this can be expressed by saying that the sample size for Spain is 605.
- Proportion of people in the sample who prefer white wine is 645/2331 = 27.67%.
- Thus expected value of the people in Spain who prefer white wine is 605*27.67% = 167.41.

The other values are similarly calculated.

Degrees of freedom

In order to use the chi-squared test we need to know the degrees of freedom in the contingency table. The degrees of freedom in a cross–classification table are calculated by the relationship:

$$(\text{Number of rows} - 1)*(\text{Number of columns} - 1) \qquad 9(\text{xviii})$$

Consider Table 9.8 which is a 3 by 4 contingency table as there are three rows and four columns. The value of the row totals and the column totals are fixed. The "yes" or "no" in the cells indicate whether or not we have freedom to choose a value in this cell. For example, in Column 1 we have only the freedom to choose two values; the third value is automatically fixed by the total of that column. The same logic applies to the rows. In this table we have the freedom to choose only six values or the same as determined from Equation 9(xviii):

$$\text{Degrees of freedom} = (3 - 1)*(4 - 1) = 2*3 = 6$$

Table 9.8

	Column 1	Column 2	Column 3	Column 4	Total
Row 1	Yes	Yes	Yes	No	Total Row 1
Row 2	Yes	Yes	Yes	No	Total Row 1
Row 3	No	No	No	No	Total Row 1
Total	Total Column 1	Total Column 2	Total Column 3	Total Column 4	TOTAL

In Table 9.6 the degrees of freedom are $(4-1)*(4-1) = 9$. These concepts serve as the basis for the development of hypothesis testing using chi-squared. A detailed example is illustrated by the following application exercise, *Work schedule preference*.

Application exercise: **Work schedule preference**

Situation

An international company with activities in the USA, Germany, Italy, and England is interested in knowing the employees' preference between the current 8 hr per day, 5 days per week work schedule and a proposed 10 hr per day, 4 days per week. Samples were taken and the number of people responding from each country, giving their preference to the proposals, is given in Table 9.9.

Table 9.9

Preference	USA	Germany	Italy	England
8 hr/day	227	213	158	218
10 hr/day	93	102	97	92

Required

1   What are the null and alternative hypotheses for this situation?
2   For the chi-squared test, what are the degrees of freedom?
3   Using the chi-squared values, at a 5% level of significance is there evidence of a relationship between the preference of the work week and the country or are the differences merely chance?
4   Using the *p*-value approach confirm your conclusions to Question 3 at a 5% significance level.
5   Using the chi-squared values, at a 10% level of significance is there evidence of a relationship between the preference of the work week and the country or are the differences merely chance?
6   Using the *p*-value approach confirm your conclusions to Question 5 at a 10% significance level.

Worked out solutions

1   Let's say p_U is the proportion in the USA who prefer the present work schedule; p_G is the proportion in Germany who prefer the present work schedule; p_I is the proportion in the Italy who prefer the present work schedule; and p_E is the proportion in England who prefer the present work schedule. Then the null hypothesis, H_0, is that the population proportion favoring the current work schedule is not significantly different from country to country and thus we write the null hypothesis as follows:

$$H_0 : p_U = p_G = p_I = p_E$$

This is also saying that for the null hypothesis of the preference of work schedule is independent of the country of work. Thus, the chi-squared test is also known as a **test of independence**. The alternative hypothesis is that population proportions are not the same and that the preference for the work schedule is dependent on the country of work. In this case, the alternative hypothesis H_1 is written:

$$H_1 : p_U \neq p_G \neq p_I \neq p_E$$

Thus in hypothesis testing using the chi-squared distribution we try to determine if the population proportions are independent, or dependent, according to a certain **criterion**, in this case the country of employment. This test determines frequency values as follows.

2   From Table 9.9 not including the total columns and rows, the degrees of freedom are given by:

$$(\text{Number of rows} - 1) * (\text{Number of columns} - 1) = (2 - 1) * (4 - 1) = 3.$$

3   From the observed data of Table 9.9, we calculate the totals of the rows and columns. This is in Table 9.10. These are the absolute values of the observed frequencies, f_o
    From Table 9.10 if the null hypothesis is true that there is no difference in the preference for the work schedule, then:

   •   Population proportion who prefer the 8 hr/day schedule is $816/1,200 = 0.6800$ or 68%
   •   Population proportion who prefer the 10 hr/day schedule is $384/1,200 = 0.3200$, or 32%

   We then use these proportions of the sample data to **estimate** the population proportion that prefer the 8 hr/day or the 10 hr/day schedule. These values are shown in Table 9.11.

Table 9.10

Preference	USA	Germany	Italy	England	Total
8 hr/day	227	213	158	218	816
10 hr/day	93	102	97	92	384
Total	320	315	255	310	1,200

Table 9.11

Preference	USA	Germany	Italy	England	Total
8 hr/day	217.60	214.20	173.40	210.80	816.00
10 hr/day	102.40	100.80	81.60	99.20	384.00
Total	320.00	315.00	255.00	310.00	1,200.00

For example, the sample size for the USA is 320 and so assuming the null hypothesis, the estimated number that prefer the 8 hr/day schedule is 0.6800*320 = 217.60. The estimated number that prefer the 10 hr/day schedule is 0.3200*320 = 102.40. This value is also given by 320 − 217.60 = 102.40 since the choice is one schedule or the other. Thus the complete expected data, on the assumption that the null hypothesis is correct, is as in Table 9.11. These are then the expected frequencies, f_e. Another way of calculating the expected frequency is from the relationship:

$$f_e = \frac{(Total\ of\ the\ row) * (Total\ of\ the\ column)}{n}$$

For example, from Table 9.10 consider the cell that gives the observed frequency for Germany for a preference of an 8 hr/day schedule. The total in the row for the 8hr/day schedule is 816; total of the column for Germany is 315; the size of the sample taken is 1,200. Thus the expected frequency:

$$f_e = \frac{(Total\ of\ the\ row) * (Total\ of\ the\ column)}{n} = 816*315/1,200 = 214.20$$

The value of chi-squared, x^2, is given by: $\chi^2 = \sum \frac{(f_o - f_e)^2}{f_e}$ and these terms in the numerator and denominator are developed in Table 9.12.

Table 9.12

Schedule	f_o	f_e	$f_o - f_e$	$(f_o - f_e)^2$	$(f_o - f_e)^2/f_e$
USA: 8 hr	227	217.60	9.40	88.3600	0.4061
Germany: 8 hr	213	214.20	−1.20	1.4400	0.0067
Italy: 8 hr	158	173.40	−15.40	237.1600	1.3677
England: 8 hr	218	210.80	7.20	51.8400	0.2459
USA: 10 hr	93	102.40	−9.40	88.3600	0.8629
Germany: 10 hr	102	100.80	1.20	1.4400	0.0143
Italy: 10hr	97	81.60	15.40	237.1600	2.9064
England: 10 hr	92	99.20	−7.20	51.8400	0.5226
Total	1,200.00	1,200.00	0.0	757.60	6.3325

Thus from this information in Table 9.12 the value of the sample chi-squared as shown is:

$$\chi^2 = \sum \frac{(f_o - f_e)^2}{f_e} = 6.3325$$

Note that in order to verify that calculations are correct, the total amount in the f_o column must equal to total in the f_e column; and also the total $(f_o - f_e)$ must be equal to zero, as is shown. This is a long way to determine the chi-squared value. A quicker way is to use the Excel function **CHITEST** or **CHISQ.TEST** and enter the observed data from Table 9.10 and the expected values from Table 9.11. This gives 9.65%, the area under the chi-squared. Then using the function **CHIINV** (Excel 2007) or **CHISQ.INV.RT** (Excel 2010 or later), enter the degrees of freedom, 3, and the area 9.65% to give the chi-squared value of 6.3325.

The level of significance is 5%. Using Excel function **CHIINV** (Excel 2007) or **CHISQ. INV.RT** (Excel 2010 or later) for 3 degrees of freedom and a significance level of 5% gives a critical chi-squared value of 7.8147. Since the value of the sample chi-squared statistic, 6.3325, is less than the critical value of 7.8147 at the 5% significance level, we accept the null hypothesis and say there is no statistical evidence to conclude that the preference for the work schedule is significantly different from country to country. These results are shown on the chi-squared distribution of Figure 9.13.

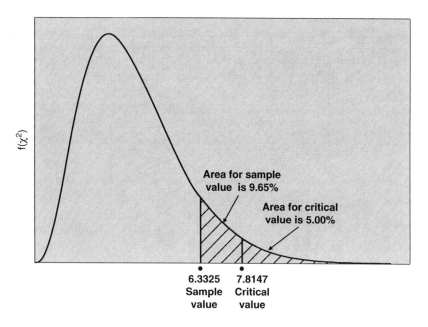

Figure 9.13 Chi-squared distribution for work preferences: 5% significance.

4   Using the function **CHIDIST** (Excel 2007) or **CHISQ.DIST.RT** (Excel 2010 or later) and entering the sample chi-squared value of 6.3325 and degrees of freedom of 3 gives the sample area of 9.65%. Alternatively using the function **CHITEST** (Excel 2007) or **CHISQ.TEST** (Excel 2010 or later) and entering the observed data from Table 9.10 and the expected values from Table 9.11 gives the same result. Since 9.65% is greater than the significant level of 5% this verifies our conclusion of Question 3. The situation for the 5% significance level for the chi-squared values and the areas is also shown in Figure 9.13.

5   At the 10% significance level the sample value of the chi-squared is still 6.3325. Using Excel function **CHIINV** (Excel 2007) or **CHISQ.INV.RT** (Excel 2010 or later) for 3 degrees of freedom and a significance level of 10% gives a critical chi-squared value of 6.2514. Since now the value of the sample chi-squared statistic, 6.3325, is greater than the critical value of 6.2514, we reject the null hypothesis and say there is evidence to conclude that the preference for the work schedule is perhaps contingent on the country.

6   The area or the probability value for the sample rests unchanged at 9.65%. This probability value is now less than 10% and so this confirms our rejection of the null hypothesis. The situation for the 10% significance level for the chi-squared values and the areas is shown in Figure 9.14.

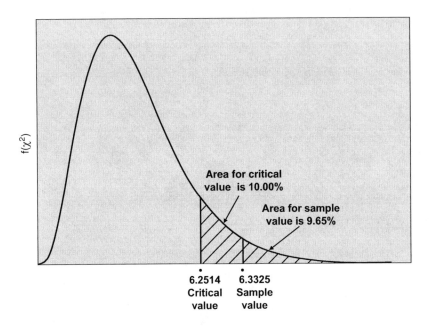

Figure 9.14 Chi-squared distribution for work preferences: 10% significance.

Excel and chi-squared functions

In Microsoft Excel there are three functions for chi-squared testing:

- **CHIDIST** (Excel 2007) or **CHISQ.DIST.RT** (Excel 2010 or later)
  This generates the area in the chi-distribution when you enter the chi-squared value and the degrees of freedom of the contingency table.
- **CHIINV** (Excel 2007) or **CHISQ.INV.RT** (Excel 2010 or later)
  This generates the chi-squared value when you enter the area in the chi-squared distribution and the degrees of freedom of the contingency table.
- **CHITEST** (Excel 2007) or **CHISQ.TEST** (Excel 2010 or later)
  This generates the area in the chi-squared distribution when you enter the observed frequency and the expected frequency values assuming the null hypothesis.

Moving on

This chapter has extended hypothesis assumptions by comparing different populations. The following chapter examines statistical forecasting methods, a requirement that triggers all business functions.

Notes

1 "Clinical guidelines on the identification, evaluation, and treatment of overweight and obesity in adults," *National Institutes of Health*, www.nhlbi.nih.gov/health-pro/guidelines/archive/clinical-guidelines-obesity-adults-evidence-report [accessed May 16, 2016].
2 http://health.usnews.com\best-diet\best-weight-loss-diets [accessed July 31,2015].
3 "Compared with Americans, the British are the picture of health," *International Herald Tribune*, May 22, 2006, p. 7.

10 Forecasting from correlated data

ICEBREAKER: The Euro

The Euro originated in the European Monetary System in 1979 by the creation of the European Currency Unit. Originally 12 of the 15 EU countries: Germany, France, Austria, Spain, Portugal, Italy, Belgium, Luxembourg, Netherlands, Finland, Greece and Ireland; were members of this so-called "Eurozone." The Euro was first launched as an accounting currency on January 1, 1999 after the acceptance of the Maastricht treaty. The participating countries fixed domestic currencies to the Euro and were not allowed to let their domestic currencies fluctuate against the Euro or each other's currencies. The Euro as a cash currency became legal tender on January 1, 2002. The German Mark, French Franc, Italian Lira and other local currencies ceased to exist on July 1, 2002. This date was the beginning of centralized monetary policies for the countries that adopted the Euro who were joined later by Slovenia (2007), Cyprus (2008), Malta (2008), Slovakia (2009), Estonia (2011), Latvia (2014), and Lithuania (2015) making 19 European countries members of the Eurozone out of the 28 countries in the European Union.

The goal of the Euro was to create a more stable European economy. Economic growth in the Eurozone improved; offered more integration among financial markets; the Euro currency

Figure 10.0 Euro coins and banknotes.

Source: author.

Figure 10.0.1 Exchange rate of the Euro to the $US.

strengthened European presence globally through being a reserve currency; and helped ease exchange rate volatility among different European nations. On the downside all members of the Eurozone have to have relatively similar interest rates to avoid interest rate arbitrage. This has created problems for some economies since the Euro has taken away interest rates as a tool of fiscal policy. If a country's economy slows, that country's government cannot lower interest rates to stimulate growth.

Europe, which represents 28 countries, lives on exports and imports, a lot of which is denominated in $US. The exchange rate of the Euro to the $US has fluctuated enormously since the Euro's inception, as shown by the time-series data in Figure 10.0.1. This fluctuation for Eurozone countries poses headaches in forecasting future sales.

In France in 2007 it was difficult for luxury companies like LVMH exporting to the USA when the exchange rate was above €1.50 per $US. However, in 2015 forecast revenues were up when the exchange rate was close to parity[1]. The reverse was true for US companies exporting to Europe. Forecasting is the substance of this chapter.

Chapter subjects

✓ **Time series and correlation** • Scatter diagram • Application case-exercise: *Snowboard sales, Part I* • Coding time-series data • Application case-exercise: *Snowboard sales, Part II* • Correlation coefficients • Application case-exercise: *Snowboard sales, Part III* • Linear regression line • Application case-exercise: *Snowboard sales Part IV* • Variability of the estimate • Application case-exercise: *Snowboard sales, Part V* • Forecasting and confidence limits • Application case-exercise: *Snowboard sales, Part VI*

✓ **Linear regression and causal forecasting** • Application case-exercise: *Surface area and home prices, Part I*

✓ **Forecasting using multiple regression** • Multiple independent variables • Standard error of the estimate • Application case-exercise: *Supermarket and coffee sales*

✓ **Forecasting using non-linear regression** • Polynomial function • Exponential function • Application case-exercise: *Surface area and home prices, Part II*

✓ **Seasonal patterns in forecasting** • Application case-exercise: *Spanish soft drinks sales*

✓ **Considerations in statistical forecasting** • Time horizons • Product life cycle • Collected data • Coefficient of variation • Market changes • Models are dynamic • Model accuracy • Selecting the best model

---

*A useful part of statistical analysis is **correlation**, or the measurement of the strength of a relationship between collected variables from historical data. If there is a reasonable correlation, then **regression** analysis is a mathematical technique to develop an equation that describes the relationship between the variables in question. The practical use of this part of statistical analysis is that on the basis that collected data mimics future conditions, correlation and regression can be used for forecasting. The function, forecasting, is that activity that kicks off most of the other business functions—strategic considerations, operations, employment changes, budgeting, etc.*

Time series and correlation

A time series is past data presented in regular time intervals such as weeks, months, or years to illustrate the movement of variables. Financial data such as revenues, profits, or costs can be presented in a time series. Operating data, for example, customer service level, capacity utilization of a tourist resort, or quality levels can be similarly shown. Macro-economic data such as gross national product, consumer price index, or wage levels are typically illustrated by a time series. In a time series we are presenting a variable, such as revenues, against another variable, time, and this is called **bivariate data.**

Scatter diagram

A scatter diagram is the presentation of the time-series data by dots on an *x-y* graph to see if there is a correlation between the two variables. The time, or independent variable, is presented on the *x*-axis or abscissa and the variable of interest, on the *y*-axis, or the ordinate. The variable on the *y*-axis is considered the dependent variable since it is "dependent," or a function, of the time. Time is always shown on the *x*-axis and considered the independent variable since whatever happens today—an earthquake, a flood, or a stock market crash, tomorrow will always come! The scatter diagram is the starting point of regression forecasting and the following application case-exercise, *Snowboard sales, Part I*, illustrates the procedure and subsequent calculations.

Application case-exercise: **Snowboard sales, Part I**

Situation

Mammoth Mountain in the Sierra Nevadas is a major winter resort area serving principally the Los Angeles population in Southern California and the San Francisco residents in the north. There are many stores renting and selling winter sports equipment including skis, snowboards, etc. A particular location in Mammoth recorded the historical sales of snowboards since 1999 and this information is shown in Table 10.1.

Table 10.1

Year	Sales (units)	Year	Sales (units)
1999	60	2007	580
2000	90	2008	985
2001	110	2009	1,600
2002	210	2010	2,100
2003	250	2011	1,875
2004	650	2012	2,500
2005	400	2013	2,100
2006	800	2014	1,475

Required

1 Plot the data in Table 10.1 as a scatter diagram putting the year on the *x*-axis and the sales in units on the *y*-axis. What are your observations?

Worked out solutions

1 Using in Excel the chart type **XY(scatter)**, with the *x*-value the year, and the *y*-value the sales units gives the scatter diagram in Figure 10.1. This is a visual display correlating sales with time. Observations are that there is an increase of snowboard sales over the years though the data is somewhat scattered in the later period

Coding time-series data

We were able to plot the scatter diagram in Figure 10.1 because both the *x*-axis and *y*-axis data are numerical values. However, if the time period on the *x*-axis was in months; January, February, March, etc. then with Excel we cannot develop easily a correlated graph with a mix of numerical values and letters. To overcome this we give the time period sequentially a code number: January = 1, February = 2, March = 3, etc. We might wish to do this even with the numerical values of the years since they are large numbers and are not convenient in performing calculations. Coding information is illustrated by of the application case-exercise, *Snowboard sales Part II*.

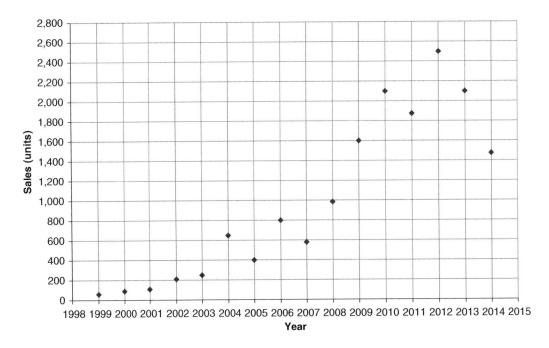

Figure 10.1 Snowboard sales: scatter diagram of historical sales.

Application case-exercise: Snowboard sales, Part II

Required

1   Give the years of the data in Table 10.1 a code with 1999 = 1; 2000 = 2; 2001 = 3; etc and plot this data as a scatter diagram where now the code value is the *x*-axis and the *y*-axis is still the sales. What are your observations?

Worked-out-solutions

1   The code values for the years are given in Table 10.2. Using in Excel the chart type **XY(scatter)**, with the *x*-value now the code, and the *y*-value the units of sales gives the scatter diagram in Figure 10.2. We have developed a visual display of the correlation of sales with a coded value of time. The profile of Figure 10.2 is identical to Figure 10.1.

Table 10.2

Year	Code value	Sales (units)	Year	Code value	Sales (units)
1999	1	60	2007	9	580
2000	2	90	2008	10	985
2001	3	110	2009	11	1,600
2002	4	210	2010	12	2,100
2003	5	250	2011	13	1,875
2004	6	650	2012	14	2,500
2005	7	400	2013	15	2,100
2006	8	800	2014	16	1,475

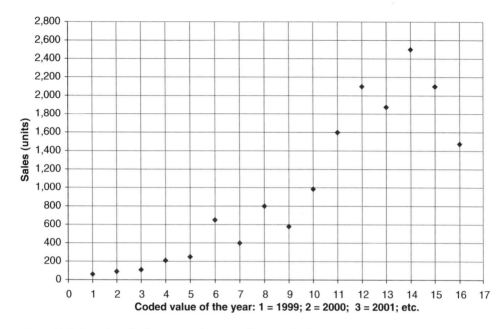

Figure 10.2 Snowboard sales: scatter diagram of historical sales.

Correlation coefficients

Visually in both of the scatter diagrams of our *Snowboard sales* application exercise it appears that there is a reasonable relationship between sales and time. How can we quantify this relationship? There are two possibilities: the **coefficient of correlation, r**; or the **coefficient of determination, r^2**. Numerically the coefficient of correlation, r, is positive if the data increases with time; negative if the data is decreasing with time. The coefficient of determination, r^2, is always positive since it is the square of r and regardless of whether r is positive or negative the squared value will always be positive. The values of these coefficients can be determined from the following rather complex looking equations. Here x is the variable on the horizontal axis; y is the variable on the vertical axis; n is the number of paired values, (x, y). Microsoft Excel performs these calculations easily.

$$\text{Coefficient of correlation: } r = \frac{n\sum xy - \sum x \sum y}{\sqrt{\left[n\sum x^2 - (\sum x)^2\right]\left[n\sum y^2 - (\sum y)^2\right]}} \qquad \text{10(i)}$$

$$\text{Coefficient of determination: } r^2 = \frac{(n\sum xy - \sum x \sum y)^2}{\left[n\sum x^2 - (\sum x)^2\right]\left[n\sum y^2 - (\sum y)^2\right]} \qquad \text{10(ii)}$$

Both coefficients measure the strength of the relationship of y with x. Numerically r^2 can have a value of from 0 to 1; r can take a value from 0 to $+1$ (data increasing with time) or 0 to -1 (data decreasing with time). For either the coefficient of correlation or the coefficient of determination, a numerical value of 1.00 means there is a perfect correlation; a numerical value of 0.00 means there is no correlation. I use a threshold value of 0.80 for the value of the coefficient of determination above which suggests there is a reasonable strength of the relationship between the x and y data. (Numerically this r^2

value of 0.80 is equal to a value of r of 0.89.) The use of the correlation coefficients is illustrated in the application case-exercise, *Snowboard sales, Part III*.

Application case-exercise: **Snowboard sales, Part III**

Required

1   Determine the coefficient of correlation and the coefficient of determination for the historical sales data given in Table 10.1. What are your observations?
2   Determine the coefficient of correlation and the coefficient of determination for the historical sales data given in Table 10.2 using the coded values for the year. What are your observations? How do these values compare to those from Question 1?

Worked out solution

1   Using the function **COEFF** and entering all the given 16 values of x and y gives a value for the coefficient of correlation, r, of 0.9066. Alternatively, the function **PEARSON** can be used and this gives also the value 0.9066.
    Using the function **COEFF**, entering all the given 16 values of x and y, and squaring the result gives a value for the coefficient of determination, r^2, of 0.8220 (when calculated with Excel). Since the value of 0.8220 is greater than 0.80 we can say that there is a reasonable correlation between the sales of snowboards and time.
2   Performing the same calculation using the coded values for the year identical values are obtained. This confirms our conclusion that the profile of the two scatter diagrams is the same. For quantifying the correlation it makes no difference which data is used.

Linear regression line

Once we have developed a scatter diagram for time-series data, and have concluded that the strength of the relationship between the dependent variable, y, and the independent time variable, x, is reasonably strong, then we develop a linear regression equation to define the relationship. Afterwards we can use this equation to forecast beyond the time periods given. The linear regression line is the best straight line that minimizes the error between the data points on the regression line and the corresponding actual data from which the regression line is developed. The following equation represents the regression line:

$$\hat{y} = a + bx \qquad\qquad 10(\text{iii})$$

- a is a constant value and equal to the intercept on the y-axis
- b is a constant value and equal to the slope of the regression line
- x is the time and the independent variable value
- $\hat{y}$ is the predicted, or forecast value, of the actual dependent variable, y

The values of the constants, a and b, can be calculated by the **least squares method** using the following:

$$a = \frac{\sum x^2 \sum y - \sum x \sum xy}{n \sum x^2 - \left(\sum x\right)^2} \qquad\qquad 10(\text{iv})$$

$$b = \frac{n\sum xy - \sum x \sum y}{n\sum x^2 - (\sum x)^2} \qquad\qquad 10(\text{v})$$

Another approach is to calculate b and a using the average value of x or $\bar{x}$, and the average value of y or $\bar{y}$ using the two equations below. It does not matter which we use as the result is the same.

$$b = \frac{\sum xy - n\bar{x}\bar{y}}{\sum x^2 - n(\bar{x})^2} \qquad\qquad 10(\text{vi})$$

$$a = \bar{y} - b\bar{x} \qquad\qquad 10(\text{vii})$$

Again it is not necessary to perform these calculations because all the relationships can be developed from Microsoft Excel as illustrated by the application exercise, *Snowboard sales, Part IV.*

Application case-exercise: Snowboard sales, Part IV

Required

1   Using Excel, on the scatter diagram developed from the data in Table 10.1, show the regression line on the graph together with the linear regression equation and the coefficient of determination.
2   Using Excel, on the scatter diagram developed from the data in Table 10.2 that gives the code values for x, show the regression line on the graph together with the linear regression equation, and the coefficient of determination. How does this compare with the answer of Question 1?

Worked-out-solution

1   Select the data points on the scatter diagram and then:
   - In the Excel menu select, *Chart*
   - Select, *Add trend line*
   - Select, *Type*
   - Select, *Linear*
   - Select *Options* and, check "*Display equation on chart*" and "*Display R-squared value on chart*"

   The regression line on the scatter diagram is given in Figure 10.3.
2   Repeat the procedure for the Question 1 now using the scatter diagram that has the code values and this gives the regression line as in Figure 10.4. The diagram is the same as for Question 1 as is the value of the coefficient of regression but the linear equation is different. For Question 1 it is:

$$y = 158.6250x - 317,294.5000$$

and for Question 2 where the code values of x have been used, it is:

$$y = 158.6250x - 361.7500$$

The slope of the regression line, given by $158.6250x$ is the same. However, the intercepts on the y-axis given by $-317,294.5000$ for Question 1; and -361.7500 for Question 2 are different.

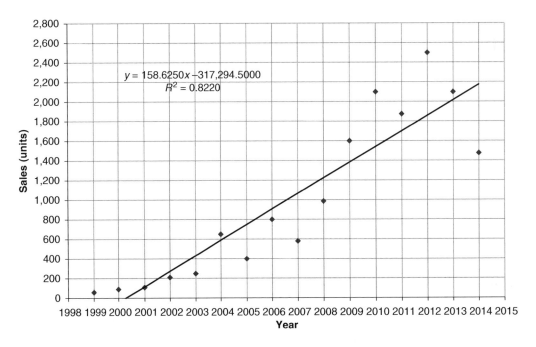

Figure 10.3 Snowboard sales: scatter and regression of historical sales.

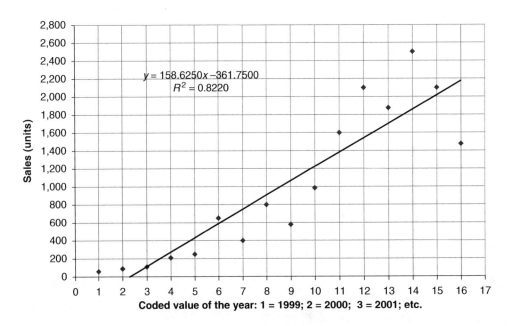

Figure 10.4 Snowboard sales: scatter and regression of historical sales.

This is because the units of the *x*-axis are different. Note also the format of the regression line given by Excel: $y = bx + a$ is the reverse of the form: $\hat{y} = a + bx$ and Excel uses *y* rather than the regression term, $\hat{y}$. Further, Excel uses R^2 rather than r^2.

Variability of the estimate

In Equation 1(vii) of *Chapter 1,* we presented the sample standard deviation, *s*, of data by the equation:

$$s = \sqrt{s^2} = \sqrt{\frac{\sum (x - \bar{x})^2}{(n - 1)}}$$

The standard deviation is a measure of the variability around the sample mean, $\bar{x}$, for each random variable *x*, in a given sample size, *n*. And, the deviation of all the observations, *x*, about the mean value, $\bar{x}$, is zero, or from Equation 1(viii):

$$\sum (x - \bar{x}) = 0$$

Similarly, a measure of the variability around the regression line is the **standard error of the estimate, s_e,** given by:

$$\text{Standard error of the estimate, } s_e = \sqrt{\frac{\sum (y - \hat{y})^2}{n - 2}} \qquad \text{10(viii)}$$

Here *n* is the number of bivariate data points (*x, y*). The value of s_e has the same units of the dependent variable, *y*. The denominator in this equation is (*n* − 2) or the degrees of freedom, rather than (*n* − 1) in Equation 1(vii). In Equation 10(xiii) two degrees of freedom are lost because two statistics, "a" and "b," are used in regression to compute the standard error of the estimate. Like the standard deviation, the closer to zero the value of the standard error the less the scatter or deviation around the regression line. This translates into saying that the linear regression model is a good fit of the observed data. Then we should have reasonable confidence in the forecast estimate made.

The regression equation is determined so that the vertical distance between the observed, or data values, *y*, and the predicted values, $\hat{y}$, balance out when all data are considered. Thus, analogous to Equation 1(viii) this means that:

$$\sum (y - \hat{y}) = 0 \qquad \text{10(ix)}$$

Again, we do not have to go through a stepwise calculation but the standard error of the estimate, together with other statistical information, can be determined by using Excel. This is illustrated by the following application case-exercise, *Snowboard sales, Part V.*

Application case-exercise: Snowboard sales, Part V

Required

1 Using Excel for the data in Table 10.1 determine the standard error of the regression together with the complete statistical characteristics for this information.
2 From the appropriate statistical data, calculate the regression data values that are indicated by the regression line of Figure 10.4.
3 Using Excel from the data in Table 10.2 using the code values of *x*, determine the standard error of the regression and the complete statistical characteristics for this information.

4   From the appropriate statistical data, calculate the regression data values that are indicated by the regression line of Figure 10.5.

Worked-out-solution

1   Select a virgin block of cells of dimension 2 columns by 5 rows. Select function **LINEST** and in the window enter the given x and y from Table 10.1. Input "1" both times for the constant data. Like the frequency distribution execute this function by pressing simultaneously on control-shift-enter keys [Ctrl-↑-↵]. (That is for a PC.) This gives the information as shown in Table 10.3. In order to clarify these terms columns on the left and right of the calculated statistical values gives their meaning. It includes some of terms that have already been calculated: slope "b"; intercept "a" and the coefficient of determination, r^2. It also calculates the degree of freedom as shown in the fourth line.
     The standard error the estimate, s_e is 363.7938 as indicated in the third line of Table 10.3.

Table 10.3

Slope of the line, b	b	158.6250	−317,294.5000	a	Intercept on the y-axis
Standard error for the slope	s_{e_b}	19.7295	39,587.3252	s_{e_a}	Standard error for intercept, a
Coefficient of determination	r^2	0.8220	363.7938	s_e	Standard error of estimate
F-ratio for analysis of variance	F	64.6415	14	df	Degrees of freedom $(n − 2)$
Sum of squares due to regression (Explained variation)	SS_{reg}	8,555,042.8125	1,852,843.1250	SS_{resid}	Sum of squares or residual (Unexplained variation)

2   Using the constant values of "a" and "b" from Table 10.3 and multiplying these by the corresponding year, gives for example:

   - 1999: forecast sales $\hat{y} = −317,294.50000 + 158.6250 * 1999 = −203.13$
   - 2000: forecast sales $\hat{y} = −317,294.50000 + 158.6250 * 2000 = −44.50$
   - 2001: forecast sales $\hat{y} = −317,294.50000 + 158.6250 * 2001 = 114.13$
   - Etc.

     The complete data is given in Column 3 of Table 10.4. For comparison the actual sales data is shown in the second column. If we add this data to the line graph of Figure 10.4 it would be superimposed on the regression line developed graphically from Excel. These lines would be indistinguishable.

3   Repeat the procedure for Question 1 using now the data from Table 10.2 with the coded values. This gives the information in Table 10.5 together with an explanation of these values.
     The standard error the estimate, s_e is 363.7938 as indicated in the third line of Table 10.3. This is the same as if we use the year rather than the code value. Obtaining the same value is to be expected since the profile of the two regression lines is the same. The only difference between Table 10.5 and 10.3 is "a," the numerical value of the intercept on the y-axis and s_{e_a} the standard error of the estimate for "a". This is because the values of the x-axis are different.

4   Repeat the procedure for Question 1 using now the data from Table 10.5 with the coded values. The calculation procedure is the same as previously except the code values for the year are used and the intercept on the y-axis is different. As an illustration:

   - 1999: forecast sales $\hat{y} = −361.75 + 158.6250 * 1 = −203.13$
   - 2000: forecast sales $\hat{y} = −361.75 + 158.6250 * 2 = −44.50$

Table 10.4

Year x	Actual sales y	Forecast sales ŷ
1999	60	−203.13
2000	90	−44.50
2001	110	114.13
2002	210	272.75
2003	250	431.38
2004	650	590.00
2005	400	748.63
2006	800	907.25
2007	580	1,065.88
2008	985	1,224.50
2009	1,600	1,383.13
2010	2,100	1,541.75
2011	1,875	1,700.38
2012	2,500	1,859.00
2013	2,100	2,017.63
2014	1,475	2,176.25

Table 10.5

Slope of the line, b	b	158.6250	−361.7500	a	Intercept on the y-axis
Standard error for the slope	s_{e_b}	19.7295	190.7751	s_{e_a}	Standard error for intercept, a
Coefficient of determination	r^2	0.8220	363.7938	s_e	Standard error of estimate
F-ratio for analysis of variance	F	64.6415	14	df	Degrees of freedom (n − 2)
Sum of squares due to regression (Explained variation)	SS_{reg}	8,555,042.8125	1,852,843.1250	SS_{resid}	Sum of squares or residual (Unexplained variation)

- 2001: forecast sales $\hat{y} = -361.75 + 158.6250 * 3 = 114.13$
- Etc.

The complete forecast information is given in Table 10.6. It is identical to Table 10.4. Again superimposed on the regression line of Figure 10.5 it would be indistinguishable from the original line.

Forecasting and confidence limits

Once a regression line has been developed, and there is assurance between the strength of the related variables, we can use the regression line to forecast beyond the last date of the historical information. From Equation 10(iii) equation of the regression line is:

$$\hat{y} = a + bx$$

Knowing the values of the constants "a" and "b" then x will represent the period for which we are making a forecast. If we have numerical data for time, such as years, then x will be the year in question.

Table 10.6

Code for year x	Actual sales y	Forecast sales $\hat{y}$
1	60	−203.13
2	90	−44.50
3	110	114.13
4	210	272.75
5	250	431.38
6	650	590.00
7	400	748.63
8	800	907.25
9	580	1,065.88
10	985	1,224.50
11	1,600	1,383.13
12	2,100	1,541.75
13	1,875	1,700.38
14	2,500	1,859.00
15	2,100	2,017.63
16	1,475	2,176.25

If not, it will be the code value for that period. Once a forecast has been developed then the confidence limits can be determined as follows. For a sample size greater than 30 then the confidence limits of the forecast are:

$$\hat{y} \pm z.s_e \hspace{8cm} 10(x)$$

With sample sizes no more than 30, we use a Student's *t* relationship and the confidence limits of the forecast are given by:

$$\hat{y} \pm t.s_e \hspace{8cm} 10(xi)$$

where s_e is the standard error of the regression, z is the number of standard deviations for a normal distribution, and *t* is the value of the Student's *t* distribution. These confidence limits are analogous to the confidence limits developed for univariate data in *Chapter 7*.

Application case-exercise: **Snowboard sales, Part VI**

Required

1   Using the regression data develop a forecast for snowboard sales in 2016 and 2020 using the years for the required value *x*.
2   Using the regression data develop a forecast for snowboard sales in 2016 and 2020 using the coded values for the required value *x*.
3   What are the 85%, 90%, and 95% confidence limits for the forecasts developed in Question 1 and 2?
4   What are your comments on the information developed?

Worked out solutions

1   Use the Excel function **FORECAST** (Excel 2007) or **FORECAST.LINEAR** (Excel 2010 or later) that requires inserting the forecast year in question, actual values of *y* (sales), and the actual values of *x* (the years). This then gives the data in Table 10.7.

Table 10.7

Year	Forecast sales
2016	2,493.50
2020	3,128.00

These values can be determined manually using the regression equation as follows by substituting the values of a and b from Table 10.3:

$$\hat{y} = a + bx = -317,294.5000 + 158.6250x$$

When $x = 2016$: $\hat{y} = a+bx = -317,294.5000+158.6250*2016 = 2,493.50$
When $x = 2020$: $\hat{y} = a+bx = -317,294.5000+158.6250*2020 = 3,128.00$

2   Use the Excel function **FORECAST** (Excel 2007) or **FORECAST.LINEAR** (Excel 2010 or later). This requires inserting the code value for the forecast year in question, actual values of y (sales), and the actual values of x, the code values of the years. This then gives the data in Table 10.8.

Table 10.8

Year	Code value	Forecast sales
2016	18	2,493.50
2020	22	3,128.00

Again, these values can be determined manually using the regression equation as follows by substituting the values of a and b from Table 10.3:

$$\hat{y} = a + bx = -361.75 + 158.6250x$$

When $x = 2016$: $\hat{y} = a+bx = -361.75+158.6250*18 = 2,493.50$
When $x = 2020$: $\hat{y} = a+bx = -361.75+158.6250*22 = 3,128.00$

Note that, as mentioned before, we deal with smaller numbers when calculating with code values as opposed to the year.

3   Since the sample size is 16, or below 30, to be statistically correct we should use the Student's t distribution to determine the confidence limits. For this we need the degrees of freedom and this is 14 (16 − 2), or as already determined in row 4 of either Table 10.3 or Table 10.5. Table 10.9 gives the complete step–by–step measures to determine the 85%, 90%, and 95% confidence levels for the 2016 and 2020 forecasts.

4   The 2016 forecast should be more accurate than the 2020 forecast since it is closer to 2014, the last year for which actual data is available. The confidence levels, as illustrated by the range, are broader at a 90% confidence than at an 85% confidence. There is a risk associated with giving confidence limits, so the better the confidence level, the broader the limits. Remember a forecast is just that: a forecast. You assume that the historical pattern will be repeated in the future. You can be close to the real value or wide of the mark. In this example that considers snowboards, sales could be impacted by demographic changes, decline in the popularity of winter sports, or the

Table 10.9

Measure	Value	Value	Value	Source
Confidence limit, CL	85.00%	90.00%	95.00%	Required
Area outside distribution	15.00%	10.00%	5.00%	(100% − CL)
Degrees of freedom	14	14	14	Table 10.3 or 10.4
Student's t	1.5231	1.7613	2.1448	Excel [Function Student's t]
Standard error, s_e	363.7938	363.7938	363.7938	Table 10.3 or 10.4
$t * s_e$	554.0926	640.7537	780.2601	
Forecast 2016	2,493.50	2,493.50	2,493.50	Table 10.7 or 10.8
Lower limit of 2016 forecast, LL	1,939.41	1,852.75	1,713.24	Equation 10(xi) $(\hat{y} - t.s_e)$
Upper limit of 2016 forecast, UL	3,047.59	3,134.25	3,273.76	Equation 10(xi) $(\hat{y} + t.s_e)$
Range limits for 2016 forecast	1,108.19	1,281.51	1,560.52	UL − LL
Forecast 2020	3,128.00	3,128.00	3,128.00	Table 10.7 or 10.8
Lower limit of 2020 forecast	2,573.91	2,487.25	2,347.74	Equation 10(xi) $(\hat{y} - t.s_e)$
Upper limit of 2020 forecast	3,682.09	3,768.75	3,908.26	Equation 10(xi) $(\hat{y} + t.s_e)$
Range limits for 2020 forecast	1,108.19	1,281.51	1,560.52	UL − LL

level of the snowpack. (California is in a drought situation right now.) To offset these even-
tualities the store might modify inventory levels or stock up on equipment on a just in time basis.

Linear regression and causal forecasting

In the previous sections we discussed correlation and how a dependent variable changed with time.
Another type of correlation is when one variable is dependent on, or a function of, not time, but some
other variable. For example, the sale of household appliances is in part a function of the sale of new
homes; the demand for medical services increases with an aging population; or for many products,
the quantity sold is a function of price. The price of wine can be determined by causal regression
where data is plotted according to the weather in a particular year, the vineyard's history of medallion
winning, and other factors. From the regression line for a particular vintage an estimation of the price
can be predicted.[2] In these situations we say that the movement of the dependent variable, y, is caused
by the change of the dependent variable, x, and the correlation can be used for **causal forecasting**.
The analytical approach is similar to time-series linear regression but time is replaced by the causal
variable. This approach is illustrated by the application case-exercise, *Surface area and home prices, Part I*.

Application case-exercise: **Surface area and home prices, Part I**

Situation

In a certain community in Southern France, a real estate agent has recorded the past sale of houses
according to sales price and the square meters of living space. This information is in Table 10.10.

Required

1   Develop a scatter diagram for this information. Does there appear to be a reasonable correlation
    between the price of homes and the square meters of living space?
2   Show the regression line and the coefficient of determination on the scatter diagram. Compute
    the coefficient of correlation. What can you say about the coefficients of determination and
    correlation? What is the slope of the regression line and how is it interpreted?

Table 10.10

Square meters (m²)	Price €	Square meters (m²)	Price €
100	260,000	120	282,500
180	1,225,000	370	2,945,500
190	600,000	280	1,252,500
250	921,000	450	5,280,250
360	2,200,000	425	3,652,000
200	2,760,500	390	3,825,240
195	880,250	60	140,250
110	690,250	125	280,125

3   Develop the complete statistical information for this relationship. What are the degrees of freedom and the standard error of the estimate, s_e?

4   If a house on the market has a living space of 310 m², what would be a reasonable estimate of the price? Give the 85% confidence intervals for this price.

5   If a house was on the market and had a living space of 600 m², what is a reasonable estimate for the sales price of this house? What are your comments about this figure?

Worked out solutions

1   The scatter diagram is in Figure 10.5. There is a reasonable correlation between the price of homes and the square meters of living space.

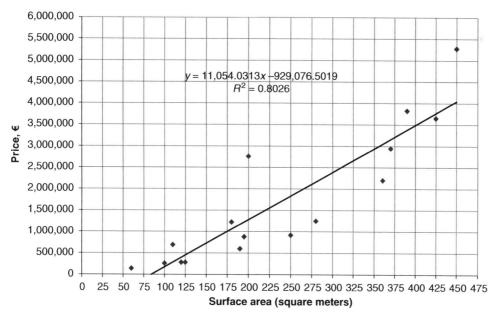

Figure 10.5 Surface area and home prices.

2   The regression line and coefficient of determination are also shown on Figure 10.5.

- The coefficient of determination is 0.8026.
- Using function **CORREL** gives of value of 0.8959.

The coefficient of determination is greater than 0.80, and the coefficient of correlation is 0.90 (rounded) so that there is a reasonable strength between surface area and home price. The slope of the line, b, is 11,054.0313. This means that the house price increases by about €11,000 for every additional square meter of surface area.

3   Using the function **LINEST** the complete statistical information is given in Table 10.11. The sample size is 16. So the degrees of freedom are 14 (16 − 2). This number is given in the second column, fourth row of Table 10.11. The standard error of the estimate, s_e, is 717,702.2901. This is given in the second column, third row.

Table 10.11

b	11,054.0313	−929,076.5019	af
se_b	1,465.0084	391,885.4774	se_a
r^2	0.8026	717,702.2901	s_e
F	56.9326	14	df
SS_{reg}	2.93258E+13	7.21135E+12	SS_{resid}

4   The regression equation is: $11,054.0313x − 929,076.5019$. Thus for a surface area of 310 m^2 the estimated home price is $11,054.0313 * 310 − 929,076.5019 = $ €2,497,673 say €2.5 million. This value can also be obtained in Excel by using the function **FORECAST** (Excel 2007) or **FORECAST.LINEAR** (Excel 2010 or later).
Since the sample size is less than 30, we use a Student's *t* distribution. At an 85% confidence level the total area in the tails is 15%. This gives a Student's *t* value of 1.5231.

- Lower confidence level is: $\hat{y} − t.s_e = $ €2, 497, 673 − 1.5231 * 717, 702.2901 = €1, 404, 544
- Upper confidence level is: $\hat{y} − t.s_e = $ €2, 497, 673 − 1.5231 * 717, 702.2901 = €3, 590, 802

Thus you would say the estimated home price for 310 m^2 is €2.5 million and I am 85% confident that the price is between €1.4 million and €3.6 million.

5   The regression equation is: $11,054.0313x − 929,076.5019$. Thus for a surface area of 600 m^2 the estimated home price is $11,054.0313 * 600 − 929,076.5019 = $ €5,703,342 say €5.7 million. This value can also be obtained in Excel by using function **FORECAST** (Excel 2007) or **FORECAST.LINEAR** (Excel 2010 or later).
The danger with making this estimate is that 600 m^2 is outside of the limits of our observed data (which ranges from 60 to 450 m^2). Thus the assumption that the linear regression equation is still valid for a living space area of 600 m^2 may be erroneous. Thus you must be careful in using causal forecasting beyond the range of data collected.

Forecasting using multiple regression

In the previous section on causal forecasting we considered the relationship between just one dependent variable and one independent variable. Multiple regression takes into account the relationship of a dependent variable with more than one independent variable. For example, obesity, the dependent variable, is a function of food intake and the amount of exercise. Automobile accidents are a function of driving speed, road conditions, and levels of alcohol and/or drugs in the blood. In business, sales revenues are a function of advertising expenditures, number of sales staff, number of branch offices, unit prices, number of competing products on the market, etc. For these types of situations, the forecast estimate is again a causal regression equation but this time containing more than one independent variable.

Multiple independent variables

The multiple regression model is represented by the equation:

$$\hat{y} = a + b_1 x_1 + b_2 x_2 + b_3 x_3 + \ldots + b_k x_k \qquad\qquad 10(\text{xii})$$

- a is a constant and the intercept on the y-plane
- $x_1, x_2, x_3,$ and x_k are the independent variables
- b_1, b_2, b_3 and b_k are constants and slopes of the line corresponding to $x_1, x_2, x_3, \ldots x_k$
- $\hat{y}$ is the forecast or predicted value given by the best fit for the actual data
- k is a value equal to the number of independent variables in the model.

Since there are now more than just an x and y variable a function as given by Equation 10(xii) cannot be represented on a two-dimensional graph. Further, the more the number of independent variables in the relationship the more complex is the model, and possibly more uncertainty in the predicted value $\hat{y}$.

Standard error of the estimate

As in linear regression, there is a standard error of the estimate, s_e, that measures the degree of dispersion around the multiple regression plane. It is expressed as follows:

$$s_e = \sqrt{\frac{\sum (y - \hat{y})^2}{n - k - 1}} \qquad\qquad 10(\text{xiii})$$

- y = actual value of the dependent variable
- $\hat{y}$ = corresponding predicted value of dependent variable from the regression equation
- n = number of bivariate data points
- k = number of independent variables

This is similar to Equation 10(viii) for linear regression except now there is a term k in the denominator where $(n-k-1)$ are the degrees of freedom. To illustrate, if the number of bivariate data points or sample size, n, is 16, and there are four independent variables the degrees of freedom are 11 $(16-4-1)$. In linear regression, with the same sample size of 16, the number of independent variables, k, is 1. Thus the degrees of freedom are 12 $(14-1-1)$ as demonstrated by the application case-exercise *Snowboards*. Again, the degrees of freedom are automatically determined in Excel when the function **LINEST** is used. As before, the smaller the value of the standard error of the estimate, the better is the fit of the regression equation. And, similar to linear regression, there is a coefficient of multiple determination, r^2, that measures the strength of the relationship between all the independent variables and the dependent variable.

The procedure for using multiple regression is illustrated by the following application case-exercise, *Supermarket and coffee sales*.

Application case-exercise: **Supermarket and coffee sales**

Situation

Brands of instant coffee include Nestlé, Maxwell House and private label brands. A distributor of Nestlé coffee sold in supermarkets and hypermarkets in Scandinavia visits the stores periodically to meet with the store manager to negotiate shelf space, to discuss pricing, and other sales-related activities. For one particular store in Helsinki, Finland, the Nestlé distributor had collected the historical data in Table 10.12 regarding the annual unit sales for the 200 g jar format of instant coffee; the number of visits made by the distributor to the store; the total shelf space allotted to the product; and the unit price of a 200 g jar.

Table 10.12

Annual unit sales y	Store visits per months x_1	Shelf space, m^2 x_2	Price, €/jar x_3
90,150	7	3.50	3.25
58,750	4	1.75	5.75
71,250	6	2.33	4.75
63,750	5	1.83	5.65
39,425	3	1.83	5.90
55,487	6	2.67	5.50
76,975	7	2.92	5.70
74,313	6	2.92	4.65
71,813	8	2.33	5.20
33,125	3	1.33	6.90

Required

1   From the information in Table 10.12, develop the complete statistical data for three separate linear regression cases: unit sales versus store visits per month; unit sales versus store shelf space; and unit sales versus price. What are your comments?
2   From the information in Table 10.12, develop the complete statistical data for the multiple regression model of unit sales as a function of the three independent variables. How does this information compare with that from Question 1?
3   Define the equation that represents this multiple regression model.
4   Using the multiple regression model, what is a reasonable forecast of sales if 8 visits per month were made to the supermarket, allotted shelf space was 3.00 m^2, and the coffee price was €5.00? What are the 85% confidence levels for this estimate?

Worked out solutions

1   Using the Excel function **LINEST**, and inserting the appropriate values for x and y, Table 10.13 gives the statistics according to store visits; Table 10.14 according to shelf space; and Table 10.15 according to coffee price.

Table 10.13 Store visits

b	8,655.4151	15,898.9170	a
s_{e_b}	1,870.4103	10,728.3969	s_{e_a}
r^2	0.7280	9,628.5263	s_e
F	21.4142	8	df
SS_{reg}	1,985,279,577	741,668,143	SS_{resid}

Table 10.14 Shelf space

b	21,953.5251	12,095.8620	a
s_{e_b}	5,013.5327	12,159.8985	s_{e_a}
r^2	0.7056	10,017.4864	s_e
F	19.1744	8	df
SS_{reg}	1,924,147,445	802,800,274	SS_{resid}

Table 10.15 Coffee price

b	−15,288.8106	144,916.6166	a
s_{e_b}	3,387.6499	18,303.7363	s_{e_a}
r^2	0.7180	9,804.4524	s_e
F	20.3681	8	df
SS_{reg}	1,957,929,426	769,018,293	SS_{resid}

In all three cases, the value of the coefficient of determination r^2 is less than the threshold value of 0.80 so we would conclude that unit sales according to individual variables are not strong.

2   Using the Excel function **LINEST**, and inserting the values for y and all the x-value matrix as one entry gives the statistical data for the multiple regression model in Table 10.16. Table 10.17 identifies the meaning of these numbers. (This can also be found in Appendix III.) The terms b_1, b_2 and b_3 are the slopes according to variables x_1, x_2, and x_3 and $s_{e_{b1}}$, $s_{e_{b2}}$ and $s_{e_{b3}}$ are the standard errors for the slopes. The other terms have been identified previously. Note here that the degrees of freedom are 6 $(10 - 3 - 1)$ where $k = 3$. For the individual relationships the degrees of freedom was 8.

Table 10.16

-8,731.8789	1,166.0060	5,150.2375	78,944.2515
4,583.4385	8,028.6804	2,257.4927	38,434.5551
0.8807	7,362.2856	#N/A	#N/A
14.7699	6	#N/A	#N/A
2,401,728,226	325,219,494	#N/A	#N/A

Table 10.17

b_3	b_2	b_1	a
$s_{e_{b3}}$	$s_{e_{b2}}$	$s_{e_{b1}}$	s_{e_a}
r^2	s_e	#N/A	#N/A
F	df	#N/A	#N/A
SS_{reg}	SS_{resid}	#N/A	#N/A

Table 10.18 summarizes the data for the coefficient of determination, r^2, and the standard error of the estimate, s_e, using individual variables and all the variables together showing:

When all the variables are taken into account as a multiple regression model, the value of the coefficient of determination is 0.8807 or above the threshold value of 0.80. Also the standard error is lower than for the other three cases, underscoring that the multiple regression model is better.

Table 10.18

Variable	s_e	r^2
Visits	9,628.5263	0.7280
Shelf space	10,017.4864	0.7056
Price	9,804.4524	0.7180
All	7,362.2856	0.8807

3   The multiple regression model equation is:

$$\hat{y} = 78,944.2515 + 5,150.2375x_1 + 1,166.0060x_2 - 8,731.8789x_3$$

Note that the variable x_3 refers to the price. The slope is negative, meaning that as the price increases the sales decrease. Not surprising; this is the basis of market economics.

4   With 8 visits per month; shelf space of 3.00 m^2 and a price of €5.00, forecast unit sales are:

$$\hat{y} = 78,944.2515 + 5,150.2375 * 8 + 1,166.0060 * 3.00 - 8,731.8789 * 5.00$$

$$= 79,984.7751 \;(\text{say } 79,985)$$

For an 85% confidence limit there is an area of 15% outside the distribution. Using Excel function **TINV** (Excel 2007) or **T.INV.2T** (Excel 2010 or later) gives a value of 1.6502.

Confidence limits are given by $\hat{y} \pm t * s_e$

Lower limit is 79,984.7751 − 1.6502 * 7,362.2856 = 67,835.7291 (say 67,836)
Upper limit is 79,984.7751 + 1.6502 * 7,362.2856 = 92,133.8211 (say 92,134)

Thus you could say that if there are 8 visits per month, the allotted shelf space is 3.00 m^2, and the price of a 200 g jar of coffee is €5.00 the best estimate of annual unit sales is 79,985 and I am 85% confident that sales will lie between 67,836 units and 92,134 units.

Forecasting using non-linear regression

Up to this point we have considered that the dependent variable is a linear function of one or several independent variables. For some situations the relationship of the dependent variable, y, to x, the independent variable may be non-linear but instead a **curvilinear function.** Examples include the sales of mobile phones from about 1995 to 2000; the increase of HIV contamination in Africa; the increase in the sale of DVD players; growth of algae in the sea; growth of the rabbit population; etc. In essence the curvilinear growth is more rapid than a linear change. A curvilinear function may have a **polynomial** or **exponential** relationship.

Polynomial function

The general form of a polynomial relationship is:

$$y = a + bx + cx^2 + dx^3 + \ldots kx^n \qquad\qquad 10(\text{xiv})$$

where y is the dependent variable; x is the independent variable; and a, b, c, d ... k are constants. Since there are only two variables, x and y, a scatter diagram can be developed for the relationship. From the scatter diagram Microsoft Excel can be used to develop the regression line. To do this we first select the data points on the graph and then from the [Menu chart] proceed sequentially as follows:

- Add trend line
- Type polynomial power
- Options
- Display equation on chart and Display R-squared value on chart.

In Microsoft Excel there are options of a polynomial function with the powers of x ranging from two to six.

Exponential function

The general form of an exponential relationship is:

$$y = ae^{bx} \hspace{4cm} 10(x)$$

where x and y are the independent and dependent variables respectively and a and b are constants. Again, since there are only two variables, x and y, a scatter diagram can be developed and from Excel the exponential curve can be drawn using the same procedure as in the previous paragraph except that the exponential option is chosen. (*See Chapter 5*, section *Exponential distributions*.)

The polynomial and exponential options are illustrated by looking at the application case-exercise, *Surface area and home prices, Part II*.

Application case-exercise: **Surface area and home prices, Part II**

Required

1 Using the data for surface area and home prices, develop a graph for a polynomial relationship for a power of x of 2. Show on the graph the regression equation and the coefficient of determination. What is an estimate of a home price when the surface area is 310 m^2; 600 m^2?
2 Using the data for surface area and home prices, develop a graph for a polynomial relationship when the power of x is 6. Show on the graph the regression equation and the coefficient of determination. What is an estimate of a home price when the surface area is 310 m^2; 600 m^2?
3 Using the data for surface area and home prices, develop a graph for the exponential relationship and indicate the exponential function. Show on the graph the regression equation and the coefficient of determination. What is an estimate of a home price when the surface area is 310 m^2; 600 m^2?
4 For the polynomial relationships when $x = 2$ and $x = 6$; the exponential function; and include as a comparison the linear regression from *Surface area and home prices, Part I* determine forecasts when the surface area is 310 m^2 and 600 m^2. What are your comments?

Worked-out-solutions

1 The graph for the polynomial regression relationship when x has the power of 2 is given in Figure 10.6. The equation for the function is:

$$\hat{y} = 26.6424x^2 - 2,729.5309x + 442,435.1650.$$

The coefficient of determination, r^2 is 0.8475

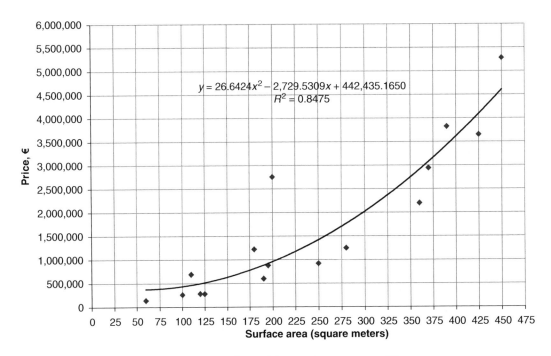

Figure 10.6 Surface area and home prices: polynomial regression to power of 2.

2 The graph for the polynomial regression relationship when x has the power of 6 is given in Figure 10.7 The equation for the function is:

$$\hat{y} = 0.00000010x^6 - 0.00016645x^5 + 0.10647985x^4 - 33.70387171x^3$$
$$+ 5,448,22679517x^2 - 409,213.41142900x + 11,160,023.23240020.$$

The coefficient of determination, r^2 is 0.8475.

3 For Question 3 the graph for the exponential regression relationship is given in Figure 10.8. The equation for the function is:

$$\hat{y} = 158\,427,5350e^{0,0079x}$$

The coefficient of determination, r^2 is 0.8173.

4 A summary of the coefficients of determination and the forecast home prices for 310 m^2 and 600 m^2 for the four regression relationships are given in Table 10.19.

Compared to the linear regression relationship, the coefficient of determination is higher for all of the other cases, indicating a better correlation. The forecast values at 310 m^2 surface area follow the corresponding curves and since this area is within the data collected, we might conclude that the forecast for the polynomial relationship when x has a power of 6 is the best. However, the curve traces the given points and this is perhaps distorting the forecast. At 600 m^2 surface area the forecasts differ widely. This surface area is beyond the data collected and the justification of such rapid growth for the polynomial and exponential functions cannot be demonstrated. Further, these values are very sensitive to the coefficient of x particularly when the power of x is high.

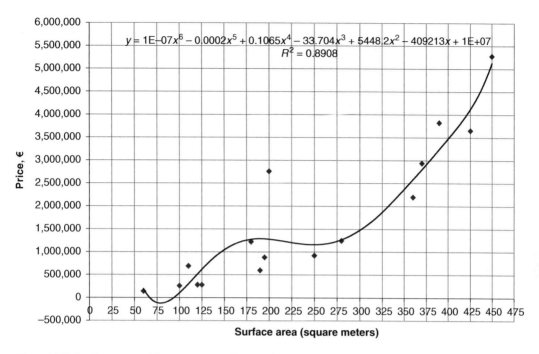

Figure 10.7 Surface area and home prices: polynomial regression to the power of 6.

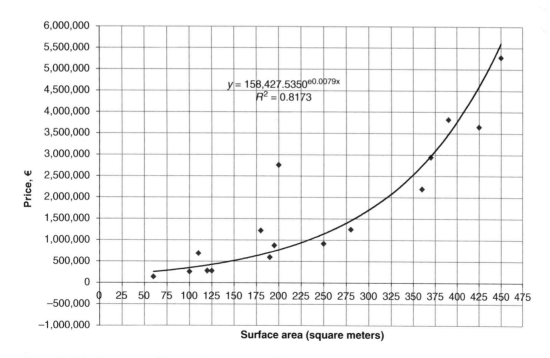

Figure 10.8 Surface area and home prices: exponential.

Table 10.19

	Linear: power x = 1	Polynomial: power x = 2	Polynomial: power x = 6	Exponential
Coefficient of determination, r^2/	0.8026	0.8475	0.8908	0.8173
Forecast home price for area = 310 m^2 (€)	2,497,673	2,156,615	1,620,103	1,834,078
Forecast home price for area = 600 m^2(€)	5,703,342	8,395,981	86,389,571	18,129,528

Seasonal patterns in forecasting

In business, particularly when selling is involved, seasonal patterns often exist. In the northern hemisphere the sale of swimwear is higher in the spring and summer; the demand for heating oil is higher in the fall and winter; the sale of cold beverages is higher in the summer. Linear regression analysis for a time-series analysis, presented earlier, can be modified to take into consideration seasonal effects. The following application case-exercise *Spanish soft drink sales,* illustrates one approach.

Application case-exercise: **Spanish soft drink sales**

Situation

Table 10.20 gives the past data for the number of pallets of soft drinks that have been shipped from a distribution center in Spain to various retail outlets on the Mediterranean coast.

Table 10.20

Year	Quarter	Actual sales (pallets)	Year	Quarter	Actual sales (pallets)
2009	Winter	14,844	2012	Winter	18,226
	Spring	15,730		Spring	19,295
	Summer	16,665		Summer	20,028
	Fall	15,443		Fall	17,769
2010	Winter	15,823	2013	Winter	18,909
	Spring	16,688		Spring	20,064
	Summer	17,948		Summer	20,965
	Fall	16,595		Fall	18,503
2011	Winter	16,480	2014	Winter	19,577
	Spring	17,683		Spring	20,342
	Summer	18,707		Summer	21,856
	Fall	17,081		Fall	19,031

Required

1   Plot the actual data and see if a seasonal pattern exists.
2   From the historical data remove the seasonal effects and develop a linear regression forecast model.
3   From the forecast model developed in Question 2 develop a forecast for 2015 by quarter.

Worked out solutions

1   The complete calculations, as explained later, are in Table 10.21. The plotted actual data, using a code value for the quarter, Column 3 of Table 10.21, is given in Figure 10.9. Clearly the data is seasonal.

Table 10.21

Col. 1	2	3	4	5	6	7	8	Col. 9
Year	Quarter	Code value	Actual sales (pallets)	CMA	Sales/ CMA	Seasonal index SI	Sales/SI	Regression line
2009	Winter	1	14,844			0.9740	15,240.97	15,438.30
	Spring	2	15,730			1.0173	15,462.69	15,669.15
	Summer	3	16,665	15,792.88	1.0552	1.0601	15,719.93	15,899.99
	Fall	4	15,443	16,035.00	0.9631	0.9486	16,279.60	16,130.84
2010	Winter	5	15,823	16,315.13	0.9698	0.9740	16,246.15	16,361.68
	Spring	6	16,688	16,619.50	1.0041	1.0173	16,404.41	16,592.53
	Summer	7	17,948	16,845.63	1.0654	1.0601	16,930.17	16,823.37
	Fall	8	16,595	17,052.13	0.9732	0.9486	17,494.00	17,054.22
2011	Winter	9	16,480	17,271.38	0.9542	0.9740	16,920.72	17,285.06
	Spring	10	17,683	17,427.00	1.0147	1.0173	17,382.50	17,515.91
	Summer	11	18,707	17,706.00	1.0565	1.0601	17,646.12	17,746.75
	Fall	12	17,081	18,125.75	0.9424	0.9486	18,006.33	17,977.60
2012	Winter	13	18,226	18,492.38	0.9856	0.9740	18,713.42	18,208.44
	Spring	14	19,295	18,743.50	1.0294	1.0173	18,967.10	18,439.29
	Summer	15	20,028	18,914.88	1.0588	1.0601	18,892.21	18,670.13
	Fall	16	17,769	19,096.38	0.9305	0.9486	18,731.60	18,900.98
2013	Winter	17	18,909	19,309.63	0.9793	0.9740	19,414.68	19,131.82
	Spring	18	20,064	19,518.50	1.0279	1.0173	19,723.03	19,362.67
	Summer	19	20,965	19,693.75	1.0646	1.0601	19,776.07	19,593.51
	Fall	20	18,503	19,812.00	0.9339	0.9486	19,505.37	19,824.36
2014	Winter	21	19,577	19,958.13	0.9809	0.9740	20,100.55	20,055.20
	Spring	22	20,342	20,135.50	1.0103	1.0173	19,996.31	20,286.05
	Summer	23	21,856			1.0601	20,616.55	20,516.89
	Fall	24	19,031			0.9486	20,061.97	20,747.74
2015	Winter	25	20,432			0.9740		20,978.58
	Spring	26	21,576			1.0173		21,209.43
	Summer	27	22,729			1.0601		21,440.27
	Fall	28	20,557			0.9486		21,671.12

2   The development of the regression line is a stepwise approach as explained as follows:

Step 1: Determine a centered moving average

A centered moving average (CMA) is the average value around a designated center point. We determine the average value around a particular season for a 12-month period, or four quarters. The following relationship indicates the calculation of the CMA around the summer quarter (usually 15 August) for the current year, calling it "n":

$$\frac{0.5 * \text{winter}(n) + 1.0 * \text{spring}(n) + 1.0 * \text{summer}(n) + 1.0 * \text{autumn}(n) + 0.5 * \text{winter}(n+1)}{4}$$

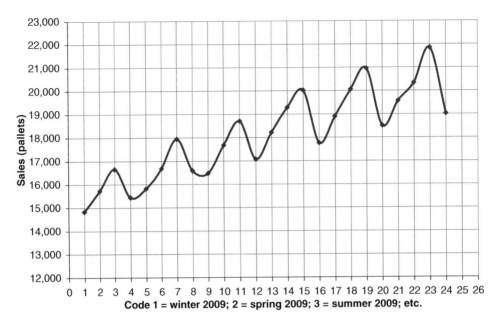

Figure 10.9 Spanish soft drink sales.

Here the numerator has a total of four quarters and so dividing by 4 gives one quarter.

For example, considering the center period for summer 2009, the CMA around this quarter using the actual data from Table 10.21 is:

$$\frac{0.5 * 14,844 + 1.0 * 15,730 + 1.0 * 16,665 + 1.0 * 15,443 + 0.5 * 15,823}{4} = 15,792.88$$

We are determining a CMA and so the next center period is fall 2009. For this quarter we drop the data for winter 2009 and add spring 2010 and thus the CMA around fall 2009 is:

$$\frac{0.50 * 15,730 + 1.0 * 16,665 + 1.0 * 15,443 + 1.0 * 15,823 + 0.5 * 16,688}{4} = 16,035.00$$

Step 2: Divide the actual sales by the moving average to give a period seasonal index, SI_p

The period seasonal index is:

$$SI_p = (\text{Actual recorded sales in a period})/(\text{CMA for the same period})$$

This data is in Column 6 of Table 10.21. What we have done is compared actual sales to the average for a 12-month period. It gives a specific seasonal index for each month. For example considering 2011 we have the ratios as in Table 10.22. These are interpreted by saying that sales in the winter 2011 are about 4.6% below the average for the year $(1 - 0.9542)$, in the spring they are

Table 10.22

Season	Index
Winter	0.9542
Spring	1.0147
Summer	1.0565
Fall	0.9424

Table 10.23

Winter	0.9740
Spring	1.0173
Summer	1.0601
Fall	0.9486

about 1.5% above the average for the year $(1.0147 - 1)$; 5.7% above the year for the summer $(1.0565 - 1)$, and about 5.8% below the year for fall 2011 $(1 - 0.94240)$.

Step 3: Determine an average seasonal index, SI, for the four quarter for the complete data

This is determined by taking the average of all the ratios, SI_p for similar seasons. For example, the seasonal index for the summer is:

$$\frac{1.0552 + 1.0654 + 1.0565 + 1.0588 + 1.0646}{5} = 1.0601$$

The seasonal indices for the four quarters are in Table 10.23.

The complete seasonal indices for the four seasons are in Column 7 of Table 10.21. Each corresponding quarter has the same value. Note that the average value of these indices must be very close to unity since they represent the movement for one year. In this case, they are equal to unity as the following calculation shows:

$$\frac{0.9740 + 1.0173 + 1.0601 + 0.9486}{4} = 1.0000$$

Step 4: Divide the actual sales by the seasonal index, SI

This data is shown in Column 8 of Table 10.21. What we have done is remove the seasonal effect of the sales, to just show the trend without any contribution from the season. Another way to look at it is to say is that the sales are de-seasonalized. The line graph for these de-seasonalized sales is in Figure 10.10.

Step 5: Develop the regression line for the de-seasonalized sales

The regression line for the de-seasonalized sales with the forecast equation and the coefficient of regression direct from the graph is also shown in Figure 10.10. In order to calculate the values for each quarter we use the Excel function **LINEST** entering the data "y" from Column 8 and the "x" data from Column 3. This gives the statistical information in Table 10.24.

The calculated data for the regression line in Column 9 is determined from the regression equation:

$$\hat{y} = 15,207.4554 + 230.8451x$$

For example:

- Winter quarter 2009: Code 1: $\hat{y} = 15,207.4554 + 230.8451 * 1 = 15,538.30$
- Summer quarter 2010: Code 7: $\hat{y} = 15,207.4554 + 230.8451 * 7 = 16,823.37$

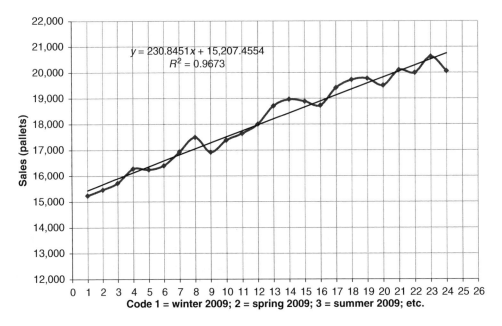

Figure 10.10 Spanish soft drink sales: de-seasonalized.

Table 10.24

b	230.8451	15,207.4554	a
s_{e_b}	9.0539	129.3687	s_{e_a}
r^2	0.9673	307.0335	e
F	650.0810	22.0000	df
SS_{reg}	61,282,860.7242	2,073,930.8417	SS_{resid}

Step 6: To forecast the seasonal values for 2015

Determine the values for the four quarters in 2015. These are calculated the same way as in Step 5 using the appropriate code values for these quarters:

- Winter quarter 2015: Code 25: $\hat{y} = 15,207.4554 + 230.8451 * 25 = 20,978.58$
- Spring quarter 2015: Code 26: $\hat{y} = 15,207.4554 + 230.8451 * 26 = 21,209.43$
- Summer quarter 2015: Code 27: $\hat{y} = 15,207.4554 + 230.8451 * 27 = 21,440.27$
- Fall quarter 2015: Code 28: $\hat{y} = 15,207.4554 + 230.8451 * 28 = 21,671.12$

Then multiply these values by the corresponding seasonal index to give a forecast according to the season. This is the reverse of the procedure for determine de-seasonalized sales.

- Winter quarter 2015 is 20,978.58 * 0.9740 = 20,432
- Spring quarter 2015 is 21,209.43 * 1.0173 = 21,576
- Summer quarter 2015 is 21,440.27 * 1.0601 = 22,729
- Fall quarter 2015 is 21,671.12 * 0.9486 = 20,557

These are the forecast values by season and are shown in the last four rows of Table 10.21. The actual and forecast sales are shown in Figure 10.11.

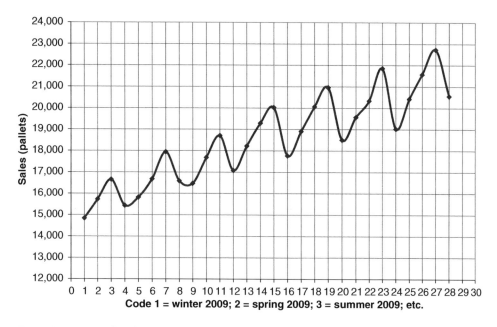

Figure 10.11 Spanish soft drink sales, actual and forecast sales.

Although the calculation procedure may seem laborious, it can be very quickly executed using Excel and its appropriate functions.

Considerations in statistical forecasting

We must remember that a forecast is just that: a forecast. Thus using statistical analysis to forecast future patterns we have to exercise caution when we interpret the results. The following are considerations.

Time horizons

Often in business, managers would like a forecast to extend as far into the future as possible. However, the longer the time period the more uncertain is the model because of the changing environment; new technologies; demographic changes—India, African, and China's population is expected to increase, Europe's to decrease; how will interest rates move? An approach to recognize this is to develop forecast models for different time periods—say short, medium, and long-term. The forecast model for the shorter time period would provide the most reliable information.

Product life cycle

When a forecast for sales is being made understand where you are in the product life cycle as shown in Figure 10.12.

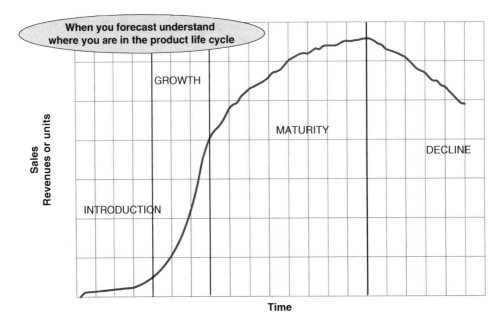

Figure 10.12 Product life cycle.

In the introduction of growth phase, a polynomial or exponential model may be appropriate. However, in the maturity and decline phases linear models may be best. Again as indicated in the application case-exercise, *Surface area and house prices*, caution must be made in using curvilinear functions, as the predicted value $\hat{y}$ changes rapidly with x. Even though the actual collected data may exhibit a curvilinear relationship, an exponential or polynomial growth often cannot be sustained in the future often because of economic, market, or demographic reasons.

Collected data

Quantitative forecast models use collected or historical data to estimate future outcomes. In collecting data it is better to have detailed rather than aggregate information, as the latter might camouflage situations. For example, assume that you want to forecast sales of a certain product of which there are six different models. You could develop a model of revenues for all of the six models. However, revenues can be distorted by market changes, price increases, or exchange rates if exporting or importing is involved. You would be better first to develop a time-series model on a unit basis according to product range. This base model would be useful for tracking inventory movements. It can then be extended to revenues simply by multiplying the data by unit price.

Coefficient of variation

When past data is collected to make a forecast, the coefficient of variation of the data, or the ratio of the standard deviation to the mean (α/μ), is an indicator of how reliable is a forecast model. For example, consider the time-series data in Table 10.25. For Product A the coefficient of variation is low,

Table 10.25

Period	Product A	Product B
January	1100	700
February	1024	65
March	1080	564
April	1257	120
May	1320	16
June	1425	456
July	1370	56
August	1502	98
September	1254	854
October	1120	820
November	1420	95
December	1250	102
Mean, μ	1 260,17	328,83
std, s	153,90	326,30
s/μ	12,21%	99,23%

meaning that the dispersion of the data relative to its mean is small. In this case a forecast model should be quite reliable. On the other hand, for Product B the coefficient of variation is high. Here a forecast model would be less reliable. In situations like this perhaps there is a seasonal activity of the product and this should be taken into account in the selected forecast model. In using the coefficient of variation as a guide care should be taken as, if there is a trend in the data, this will of course impact the coefficient. As already discussed in the chapter, plotting the data on a scatter diagram would be a visual indicator of how good is the past data for forecasting purposes. Note that in determining the coefficient of variation we have used the sample standard deviation, s, as an estimate of the population standard deviation, $\hat{\sigma}$.

Market changes

Market changes should be anticipated in forecasting. For example, in the past, steel requirements might be correlated with the forecast sale of automobiles. However, plastic and composite materials have rapidly replaced steel, so this factor would distort the forecast demand for steel if the old forecasting approach were used. Alternatively, more and more uses are being found for plastics, so this element would need to be incorporated into a forecast for the demand for plastics. Labor requirements have declined as many production processes have been replaced by information systems, computer operations, and robots. These types of events may not affect short-term planning but certainly are important in long-range forecasting when capital appropriation for plant and equipment is a consideration.

Models are dynamic

A forecast model must be a dynamic working tool with the flexibility to be updated or modified as soon as new data becomes available that might impact the outcome of the forecast. For example, an economic model for the German economy had to be modified with the fall of the Berlin Wall in 1989 and the fusion of the two Germanys. Similarly, models for the European economy have been modified to take into account the impact of the Euro single currency.

Model accuracy

All managers want an accurate model. The accuracy of the model, whether it is estimated at 10, 20 or, say, 50% can only be within a range bounded by the error in the collected data. Further, accuracy must be judged in light of the control a firm has over resources and external events. Besides accuracy, also of interest in a forecast is when turning points in situations might be expected, such as a marked increase (or decrease) in sales, so that the firm can take advantage of the opportunities or be prepared for the threats.

Selecting the best model

It is difficult to give hard and fast rules to select the best forecasting model. The activity may be a trial and error process selecting a model and testing it against actual data or opinions. If a quantitative forecast model is used there needs to be consideration of subjective input, and vice-versa.

Models can be complex. In the 1980s, in a marketing function in the USA, I worked on developing a forecast model for world crude oil prices. This model was needed to estimate financial returns from future oil exploration, drilling, refinery, and chemical plant operation. The model basis was a combined multiple regression and curvilinear relationships incorporating variables in the US economy such as changes in the GNP, interest rates, energy consumption, chemical production and forecast chemical use, demographic changes, taxation, capital expenditure, seasonal effects, and country political risk. Throughout the development, the model was tested against known situations. The model worked. It proved to be a reasonable forecast of future oil prices.

A series of forecast models have been developed by a group of political scientists who study the US elections. These models use combined factors such as public opinions in the preceding summer, the strength of the economy, and the public's assessment of its economic well-being. The models have been used in all the US elections since 1948 and have proved highly accurate.[3] In 2007 the world economy suffered a severe decline as a result of bank loans to low income homeowners. Jim Melcher, a money manager based in New York, using complex derivative models, forecast this downturn and pulled out of this risky market and saved his clients $USmillions. [4]

Moving on

This chapter has examined the process of forecasting. There are risks in forecasting. The next chapter looks at just this, "Decision and Risk at the strategic an operating level."

Notes

1 "Euro hides luxury firms' woes," *The Wall Street Journal*, July 27, 2015, p. 15.
2 "Quants and quaffs," *The Economist*, August 8, 2015, p. 58.
3 "Mathematically, Gore is a winner," *International Herald Tribune*, September 1, 2000.
4 "Warnings were missed in U.S. loan meltdown," *International Herald Tribune*, August 20, 2007.

11 Business decisions and risk

ICEBREAKER: Everest, decisions, risk ... and death

Many of the British young men who survived the bloody trenches and mutilation of the Somme, Passchendaele, and Ypres in the 1914–18 war were disillusioned with life. They felt adrift. For many, travel became a source of irrational happiness, a moving celebration of the sheer joy of just being alive. For those men England only offered a memory of lost youth; betrayal, and lies; the residue of four years of repression; casualty lists, and mass murder. It was in this world-weary climate that discussions began about climbing Mount Everest. The feeling was that if a British team succeeded in reaching the summit this would boost British morale and atone for Captain Scott's disastrous 1912 Antarctic expedition. On January 11, 1921 Tibet sanctioned an Everest expedition. However, the enormity and *risks* of the challenge were brought into focus by a leading Himalayan mountaineering specialist, Sir Martin Conway, who announced that: no **statistical** data was known of the mountain; no European had ever reached its base; the immediate approaches remained unchartered; little was understood of the snow *characteristics*, ice, topography, and nature of the rock formation. And he added that medical data on man's ability to climb at altitudes around 27,000

Figure 11.0 Mountain climbing and the inherent risks.

Source: Michel Wintenberger, France (with permission).

feet (8,223 m) was unknown. The only information was based on a nineteenth century experiment by the French meteorologist Gaston Tissandier who ascended rapidly in a balloon. He passed out at 26,500 feet (8,077 m). On regaining consciousness, he was deaf. His two companions were dead. All this missing **statistical information** presented considerable **risk** to the adventure.

Regardless of the absence of **data**, a decision was made for a climb in mid-1921. Harold Raeburn, an irascible Scot, was selected to lead the climbing party. Although he had considerable climbing experience, at 56, he was old. A serious fall on a Scottish face had robbed him of courage and he constantly suffered from abdominal pain, most likely from an undiagnosed ulcer. Charles Howard-Bury, who fought at the Somme in 1917, was named the expedition leader. A prominent member of the team was George Mallory, a wounded war-veteran, with climbing experience in the Alps. The climbing party set off from Darjeeling, India in two groups. The going was tough with high winds and difficulty walking at altitudes of 17,000 feet. Alexander Kellas, the doctor, became sick and at Kampa Dzong died of a massive heart attack. Base camp was established some four miles south of the Rongbuk glacier. On July 6 the monsoon broke blanketing the camp and leaving the climbers snowbound for two days. The teams reached the North Col of Everest in September but the men were in bad shape from frostbite, altitude sickness, and dehydration. An assault on the summit was abandoned.

In England there was pressure for another assault on Everest in 1922. Mallory was at first reticent, but later agreed to be on the team. The party took the same route as in 1921, arriving at the Rongbuk glacier on May 1, 1922. Climatic conditions and the men's physical ability were no better than before. On May 29, Tom Longstaff, the medical doctor, feeling the margin of safety for an assault was slim, gave the men a physical exam. Only one, Howard Somervell, was deemed fit. Mallory had frostbitten fingers and a detectable heart murmur. Longstaff was of the opinion that the expedition had shot its bolt and it was time to go home. Mallory was not about to give up. He was a **risk taker**. He left Base Camp for Camp III with three others on June 3. One was later forced to return. Snow began to fall and the climbing party of three huddled into a Whymper tent to wait out the opening squalls of the monsoon. The plan was to move up to the North Col on June 7 for a final assault on Everest. They set out with porters at 8:00 a.m. climbing at some 22,500 feet. The going was brutal and excruciatingly slow. At about 1:30 p.m. they were still 600 feet below the col. They pushed ahead some 100 feet up a slope when according to Mallory they heard an explosion like gunpowder. In an instant the entire slope gave way. Mallory rode the surge and, though first buried, was able to swim and struggle until mercifully, he broke surface. His two companions were equally lucky. The Sherpa porters were not. Seven of them were never seen again. The expedition was finished. Mallory had deep regrets about his assault decision and the subsequent loss of the Sherpas.

Another Everest expedition set out in 1924. Mallory was again part of the team, as was a newcomer, Sandy Irvine, just 22, with little climbing experience. Again the same severe conditions affected the climbers. One team member became severely snow-blinded and could not be moved. Irving was so badly sunburned that skin peeled of his face making the wearing of sun goggles painful. Mallory did not want another failure and decided to take a **risk**. He set off with Irvine on June 6 for an assault on the summit. They were never seen again. Whether Mallory and Irvine ever reached the summit is pure speculation. On May 1, 1999 the body of Mallory was found. The right leg was broken and the position of the body suggested that he was alive when he came to rest. [1]

... *This chapter is about decision and calculated risk.*

Chapter subjects

✓ **Decision environment** • Risk • Businesses are systems • Industry insight: *Burger King's wrong decision* • Magnitude of the business decision • Steps in the decision-making process • Segmenting the decision environment

✓ **Decision making under certainty** • Time horizons • Production and sales

✓ **Decision making under uncertainty** • You have not been there before • Optimistic, pessimistic, middle of the road, and regret analysis • Application case-exercise: *Chemical plant, Rangoon*

✓ **Decision making under risk with assigned probabilities** • Expected values (EV) • Application case-exercise: *Boutique hotels Africa, Part I* • Decision trees • Application case-exercise: *Boutique hotels Africa, Part II* • Expected value of perfect information • Application case-exercise: *Boutique hotels Africa, Part III* • Sensitivity • Application case-exercise: *Boutique hotels Africa, Part IV*

✓ **Marginal analysis** • Marginal profit and marginal loss • Application case-exercise: *Black Forest cake, Part I* • Marginal analysis and the normal distribution • Payoff tables • Application case-exercise: *Black Forest cake, Part II*

✓ **Utility** • Different personalities • Gambling • Insurance industry

---

In the business environment decisions may be at a **strategic level,** long term and involving large capital investment. What products should we produce for the market? Which geographic markets to pursue? When to invest in the new technology? Should we build a new facility in India? The decision may be at the **operating level.** How much food and beverages to purchase? When should we start to prepare the banquet? When should we install the new equipment? How many new people should we hire? How much inventory should we carry? Should we use temporary labor?

Decision environment

Making decisions is part of life. Should I marry Joan? Is Bill the right person for me? Should I take the job in Hanoi? Should I buy the house in London? What are the risks? Perhaps the marriage won't work out. Perhaps it will be stressful working in Hanoi. The investment required for the London house is high. Making these types of decisions always involves future events. We do not have a crystal ball. We simply do not know with any certainty. These are personal decisions. In business it is the same. There is no guarantee of the future.

Risk

Whenever a decision is made, there is always an element of risk. Plans may not proceed as desired; markets alter; technology changes; material costs increase unexpectedly; the economy turns sour. As such, the more systematic and the more logic used in analyzing variables involved in a decision, the less likely is the **downside-risk**. A decision is usually made from a combination of **quantitative analysis**: costs, prices, historical statistical data, economic factors, etc. plus **subjective reasoning** based on the experience of persons involved in the decision-making process. Marriot, for example, is opening new facilities in Africa—Kigali, Rwanda, and Accra, Ghana—and announced that it plans to double the number of its hotels in Africa and the Middle East to 100 over the next five years.[2] A big financial risk! An expensive decision gone wrong is illustrated by Industry insight 11.1.

Businesses are systems

A business can be considered a **system** comprised principally of three subsystems, Marketing, Operations or Production, and Finance. Decisions made for one subsystem will impact other sub-systems. A decision that seems logical for one subsystem may not be so for another. For example, the Marketing Department wishes to introduce another product line. However, there is a risk this would saturate the capacity of the Production Department and a risk that it would also strain company finances. In arriving at the final decision, the system has to be considered as a complete unit and not only at the subsystem level. In decision making, **trade-offs** occur, that is, making a choice between one decision and another. With both there is a risk and both have advantages and disadvantages. The Finance Department wants Production to keep inventories low in order to reduce the financial investment and working capital. Production wants to keep a safety stock of inventory because of uncertainty of supplier deliveries, and fluctuations in final product demand. There is a trade-off between what levels of inventories to hold. In **quality control** there are trade-offs between internal product inspections, cost, and the number of defects that slip through the operating process—an external cost. Thus there is a trade-off of levels of quality control. Managers are continually confronted with decisions. One task is to sort decisions in order that **priority** is given to the most important. Should the problem of quality control in the hotel in Europe be dealt with now, or should the project plans for the new hotel in Japan be developed? Should the purchase decision of raw materials be made now, in order to get a preferential price or should the sales negotiations be finalized first? Should we start hiring now to meet the expected market demand, or should we wait and see if the market expands as forecast?

Industry insight 11.1 Burger King's wrong decision[3]

Burger King is the world's No.2 hamburger chain with some 11,400 restaurants worldwide including 8,400 in the USA. McDonald's, the undisputable No.1 in the fast food restaurant business, has close to 16,000 restaurants worldwide with some 12,400 in the USA. For years Burger King beat out McDonald's in taste tests for its hamburgers but McDonald's always came out ahead regarding the taste of its French fries. As such, Burger King made a decision that it was also going to be first in the category of fries. French fries are an important product in the fast food market as the profit margin is up to a hefty 80 cents on the dollar, making them the most lucrative food on the menu. Further, since 1996, American per capita consumption of frozen French fries, the type served at most fast food restaurants, increased by almost 30% to about 12 kg per year.

　　Market research showed that people wanted a fry that was crispy and remained hot. Burger King seized on the idea that crunchiness would distinguish its fry from McDonald's popular version and thus inspire a mass loyal following. Burger King began developing a new French fry in 1996 which was a potato stick coated with a layer of starch designed to help retain heat and add crunch to the product. It developed a 19-page French fry specification for the product which included that the degree of crispiness should be determined by an audible crunch that must be present for seven or more chews and loud enough to be audible to an evaluator. This seven audible-crunch mandate was considered way outside the norms of the fry business as evaluators had to eat the product and grade it by listening and hearing crunches. The other element in the specification was the starch coating to retain the heat. Burger King assembled a French fry team of 100 marketing executives, food scientists, franchises, potato suppliers and others to test a coated fry. With help from the potato processing company Lamb Weston, a division of ConAgra

Foods Inc., Burger King developed a clear-coated fry that required a complicated preparation formula. Further, the product had to meet staple requirements of length, color, tenderness, mouth feel, "toothpack", or the degree to which a fry sticks to the surface of the tooth, and the avoidance of "marriage" that is two full fry units bonding together by one-third of their surface. In addition to all this, there were numerous suppliers handling different ingredients and preparatory steps. These many players and complex specifications made quality control difficult.

Before Burger King commercialized its fry, the product went through numerous tests. In the testing stage, it was agreed that the fries, cooked carefully under test-kitchen conditions, were a tremendous success. For the product launching, each Burger King fry supplier was outfitted with new equipment and about 300,000 restaurant managers and crew were "certified" in the new operations including mastering new frying procedures and different salting techniques. On Friday January 2, 1998 it launched its "Free Fryday" by giving away 15 million orders of French fries throughout the United States. Governors in three states officially welcomed the new product and Burger King promoted children's meal toys in the likeness of Mr. Potato Head, the new fries official emblem or "spokespud." The company booked expensive TV-commercial time during the 1998 Super Bowl and as part of its "Decision 98: Try the Fry America" it had a 15 m advertising trailer crisscrossing the country. Over $70 million was spent on marketing the product. In 1999 Burger King repeated the "Free Fryday" to promote its new product in London, UK.

After its introduction, consumers for the first time ranked Burger King's fries better than McDonald's. In an independent national test 57% preferred Burger King's fries compared to 35% for McDonald's, with 8% with no opinion. In the six months following the US "Free Fryday", Burger King had sold 150 million more orders of fries than in the same period a year before. At the headquarters in Miami, Burger King management basked in its assault on McDonald's. However, less than one year later consumers started avoiding the new fry in droves and franchises and suppliers complained of drastically falling sales. By the end of 1999 Burger King confirmed that sales of fries had fallen 14% from the year before. Consumers and franchises said the fry's taste was subject to change; they often seemed under salted, they clumped easily, and when the potatoes cooled they became tough and bitter. Fast food thrives on consistency and this new fry was proving too complicated to get right under anything less than ideal conditions. It was unforgiving in the cooking process and its flavor and consistency varied widely if it was not thawed and cooked exactly to the letter. The crunchy fry was also getting crunchier. As sales started to drop, suppliers, mindful of the "seven crunch minimum," started to add more batter to give more crunch. However, as the crunch increased, flavor suffered and with little to hold in the heat, the fry grew colder much faster. Then the fry was so brittle that it snapped like a potato chip!

Finger pointing for the poor sales extended from the Burger King Company, the franchises, and the potato suppliers. Burger King blamed the cooking procedures of the franchises and even said that a poor potato crop in 1998 may have been a factor. At this comment, a Burger King franchises operator said that in this case McDonald's must have got its potatoes from the planet Mars! In the summer of 2000, some 30 months after the launch of the new French fry, Burger King's executives conceded they had a big problem and something needed to be done. They agreed that the new fry would be replaced by one with less coating, more potato taste and further, the supply chain would be simplified and product specifications reassessed. In the annals of consumer product flops, Burger King's 1996 decision turned out to be a whopping and very costly mistake.

Magnitude of the business decision

The depth of analysis for making decisions depends on its complexity, costs involved, and time horizons. A decision to purchase a new quantity of raw material may be relatively **short-term** and simple to make, and, under a **just-in-time** operation, the raw material should be consumed within a few days. Deciding on the purchase of a new oven for the hotel kitchen may take three months to finalize with a long-term impact on the outcome since there will probably be five to ten years use of the equipment. The decision for a French company to build a new hotel in the USA has **long-term** consequences. A decision might be **intuitive**. Consider the sales manager of a Paris fashion house who receives a telephone call from a prospective client in the USA. He decides to take a night flight to New York to meet the client even though it means cancelling other appointments. He goes because, intuition tells him, that as he knows this particular customer, he is likely to obtain an order. The client gives him a large order for ladies' apparel, at an asking price lower than normal. The manager is not immediately clear of the profit margin and on the flight back to Paris he does some **quick and dirty,** or "**back of the envelope**" calculations. On arrival at the office he obtains complete material and labor costs and makes additional **computations,** after talking with design and production people before getting back to the New York client.

If the decision is big, a **financial model** may be developed. A model represents a version of the real thing and is a tool to aid decision making. A hotel company is deciding on whether to build a hotel-spa complex in Kenya. Before a final decision is made financial models are developed analyzing costs, preparing pro-forma financial statements, analyzing expected future revenues, and market demand. A food company is considering acquiring a chocolate manufacturer. Before a decision on the acquisition is made, financial models are made to analyze the expected revenues from the acquisition, market forecasts, and additional debt on net returns. A construction company develops a proposal for constructing a new hotel in Las Vegas, USA. Three dimensional **computer models** are developed as an aid to the client to make a decision on the final design. A boat builder is developing a new yacht as an entrant to the America's Cup. A model, either physical or three dimensional, is developed to decide on the final aerodynamic design.

Another approach in decision making is to use a **task force** to analyze first-hand variables involved in the decision. For example, a US-based construction company was deciding whether to build a large petrochemical facility in Kuwait. A team comprising a project manager, a market analyst, an estimator, a scheduler and two engineers was sent over to Kuwait for five weeks to analyze the proposition. This task force then returned to the home office with a recommendation based on their work in Kuwait City. (In this real illustration, the author was the market analyst!)

Steps in the decision-making process

Steps in the decision-making process, as shown schematically in Figure 11.1, depend on the magnitude of the decision and the enormity of the risk should a wrong decision be made.

The initial consideration is what is the **reason** for needing a decision? Perhaps markets have changed; competition has increased; or new technology has been developed. Technological changes have a big impact on just about every sector of business—the medical field, retailing, teaching, manufacturing, finance, tourism, etc. A next step is to **define** the current situation, for example, in terms of cost, prices, markets, and quality. It might involve performing research, interviewing personnel, benchmarking, competitive analysis. etc. Once the situation has been defined the **objectives** of proposed changes are established according to established criteria: reduced costs; low risk; increased profits; enhanced company image; increased return on investment; increased productivity; or improved quality. There may be **alternatives** in the decision

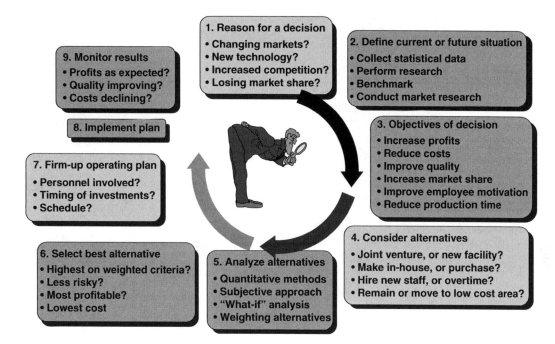

Figure 11.1 Decision-making process.

made that perhaps cost less. For example a firm has a sudden increase in customer demand for its product. The alternatives to meet this requirement might be to put the factory on overtime; hire additional workers; rent additional factory space; subcontract the work; or purchase semi-finished units from elsewhere and then finish them in the firm's factory. These alternative decisions need to be **analyzed** quantitatively before the best alternative is **selected**. In parallel, or after the decision has been made, a **plan for implementation** needs to be made. The plan might include: deciding where the work will be performed; deciding on the personnel involved; the equipment to be employed; timing of investments; and the schedule. Once the decision has been made and implemented, results should be **monitored** to be sure results are according to expectations. If not, then why not? Perhaps this involves more decision making. Elements to monitor might include: Are profits as expected? Is quality improving? Are costs declining? Is productivity improving? Has delivery time been reduced?

Segmenting the decision environment

The environment for business decision making can be segmented into three segments: **decision making under certainty; decision making under uncertainty**; and **decision making under risk.** In each division, decision theory can be applied in order to select the best alternative for a **future** outcome. Remember the future is unknown. If a decision depends on market demand this might be high, flat, or low. If a decision is a function of interest rates these might increase, decrease, or remain flat. If a decision concerns unemployment levels they may decline, increase, or remain unchanged. If a decision depends on a competitor activity they may introduce new products, or they may not. We simply do not know these future conditions are: they are called **states of nature**. They are external events over which the decision maker usually has no control. **Probability** values may be assigned to these states of nature. Selection of the best alternative is usually based on some **financial**

payoff such as maximum profit, highest revenue, or lowest cost taking into consideration the state of nature and probabilities if assigned.

Decision making under certainty

Decision making under certainty implies that all relevant parameters in the decision-making process—costs, capacity, labor rates, market demand, price, style, interest rates—are known and fixed during the decision-making period. These values are not expected to change. In decision making under certainty it is known pretty much for certain which of the future conditions will actually happen. The states of nature are clearly defined and the decision is relatively straightforward in that the choice is the alternative with the highest payoff under that given state of nature.

Time horizons

In decision making under certainty the time horizon is very often fixed thus providing the environment of certainty. For example labor contracts fix labor rates at $18.50 for one year. During the year you decide to have your employees spend 6,000 hours on a job. Thus you know that total labor costs will be $111,000 (18.50 * 6,000). A supplier of raw materials indicates that prices offered are €15.00/unit and are good for three months. Thus within three months you decide to purchase 500 units and thus you know that your purchase costs will equal €7,500 (15 * 500). A teacher/consultant fixes her rates at £150/hour and specifies that they will remain unchanged for one year. You give the consultant 120 hours of work within the year. Thus you know that costs will be £18,000 (120 * 150). If US government certificate of deposits are currently at an annual yield of 7.50% then $200,000 invested for one year will give a return of $15,000 ($200,000 * 0.075).

Production and sales

Other decisions under certainty may not directly involve a time period. For example a firm requires a decision on supplying 500 engine parts. If the units are made in-house, the production costs per unit will be $20.00. Thus for certainty the total cost will be $10,000 (500 * $20.00). If the parts are purchased outside at $19.50 per unit then the price paid will be $9,750 (500 * $19.50). A marketing decision is required concerning 750 office desks that a manufacturer has in stock that cost $960 per unit to produce. One client wishes to purchase all these office desks and has agreed to pay $1,100 per unit. Thus, the profit to be realized for the manufacturer is $140 per unit (1,100 − 960) or a total of $105,000 (140 * 750). Another client wishes to purchase 700 of these desks at a price of $1,117 per unit. This would yield a profit of $109,900 [700 * (1,117 − 960)]. All the parameters are known, and the decision most likely to be taken is the one that maximizes the final outcome.

Decision making under uncertainty

Decision making under uncertainty is a concept that implies the activity, or similar activities, have not before been undertaken. As such there is no foundation for logical decisions. Although there is risk, it is extremely difficult, if not impossible, to apply probabilities to assess outcomes.

You have not been there before

If you are a 22-year-old bachelor and you decide to get married, what is the chance of you staying married for life? Having healthy children? Always having a stable job? You do not know since you have never been there before—you have never been married. Who knows your situation when you are 10 years older! You could consult statistical data on divorce rates, family size, education and employment, but does the data apply to your particular situation? This is a personal decision but business decisions are equally uncertain. They are often long-term, complex, involve many players, and may be political. Certainly if government and politics are involved the outcome is sure to be unpredictable! In ten years will US firms realize reasonable profits by doing business in Myanmar (Burma) now that, thanks to Aung San Suu Kyi, the door has been opened to democracy?[4] However the military is still in control.[5] If new investments are made in nuclear technology, will this industry be profitable the next 30 years, especially as some countries are moving toward renewable resources? Will Hong Kong, under Chinese rule, maintain the same spectacular growth? In fifteen years time will the income from the investment in Google stock be more attractive than in Apple stock, or Facebook, or Amazon? What will the relationship of Europe and the USA with Russia be in 20 years time? These examples involve decision making under uncertainty.

Optimistic, pessimistic, middle of the road, and regret analysis

In decision making under uncertainty states of nature are unclear and assigning probabilities is wild. Thus they cannot be considered. Nevertheless there are always business situations where management requires a decision even in the absence of concrete information. Here methods called Maximax, Minimax, Average, and Minimax regret can be used. **Maximax** means choosing the maximum of the best return. This is an optimistic approach. Perhaps the decision makers are young, gung-ho and **risk-takers** (rather like Mallory on Everest in the Icebreaker). In the business environment we often see that it is the relatively young who are prepared to take risks and are optimistic. Look at Jeff Bezos of Amazon; Larry Page, at Google; Mark Zuckerberg with Facebook; and the late Steve Jobs of Apple, who was young when he established his company.[6] **Minimax** is choosing the best of the worst returns. It is pessimistic and usually, but not always, an older group of the workforce. In this case they look at the worst possible financial outcome and select "the best of the worst." **Middle of the road** is in between and it is the average value that is selected. **Minimax regret** means selecting the choice that gives the least disappointment. These methods are illustrated by the following application case-exercise, *Chemical plant, Rangoon*.

Application case-exercise: **Chemical plant, Rangoon**

Situation

A US construction firm is considering building a chemical plant near Rangoon, Myanmar. The decision will be based on three options: to build a completely new grassroots facility; to update and expand an existing war-damaged plant; or to build a new facility in a joint venture with an Asian firm. These are management decisions and thus there is certain control. Table 11.1 gives the estimated net income in $millions over five years after completion of the project according to the states of nature regarding marketing of the chemical products. Since Myanmar is a new territory for the firm, it is considered inappropriate to assign probabilities to the project.

Table 11.1

Project decision	Low market ($millions)	Stable market ($millions)	High market ($millions)
Grassroots	−100	60	200
Expansion	−50	100	150
Joint venture	20	60	90

The financial risk to the US firm is higher for the grassroots plant if things go badly but the rewards are high if things go well. The financial risk is lower for the joint-venture operation but financial returns are also lower since this is a shared activity. The firm has a strong history of building chemical plants so that cost data such as equipment, local and expatriate labor is reasonably reliable though there is uncertainty on the productivity of labor and this influences cost. More uncertainty lies with revenues since data on market demand for the chemical products in Myanmar is non-existent. However, if the politics stabilize, the market could grow. Thus it is the market demand, or the external states of nature, that govern the success of the project and over which management has no control. It is this uncertainty that impacts revenues and consequently net income. (Remember: Net income = Revenues − Cost.)

Required

1 Assume that management is optimistic, or risk-taking; what would be the preferred decision?
2 Assume that management is pessimistic, or risk averse; what would be the preferred decision?
3 Assume that management is middle-of-the road by nature; what would be the preferred decision?
4 What would be the preferred decision if the Minimax Regret method was used?

Worked out solutions

1 Table 11.2 gives the solution. The best of the maximum outcome is $200 million. Thus the decision would be to build a grassroots facility. Here the assumption is that market demand for the product will be high. This is the Maximax and optimistic approach.

[*In Excel apply function* **MAX** *on the data for each project decision to give the first three values in the last column. Then apply function* **MAX** *on these to give the final result.*]

Table 11.2

Project decision	Low market ($millions)	Stable market ($millions)	High market ($millions)	Best of decision ($millions)
Grassroots	−100	60	200	200
Expansion	−50	100	150	150
Joint venture	20	60	90	90
Maximum				**200**

2 Table 11.3 gives the solution. The best of the minimum outcome is $20 million. Thus the decision would be to build a joint-venture facility with an Asian firm. Here the assumption is that market demand for the product will be low. This is the Maximin and pessimistic approach.

[*In Excel apply function* **MIN** *on the data for each project decision to give the first three values in the last column. Then apply function* **MAX** *on these to give the final result.*]

3 Table 11.4 gives the solution. The best of the average decisions is €66.67 million. Thus the decision would be to expand the existing facility. Here this is a middle-of-the road or equally likely approach.

Table 11.3

Project decision	Low market ($millions)	Stable market ($millions)	High market ($millions)	Worst of decision ($millions)
Grassroots	−100	60	200	−100
Expansion	−50	100	150	−50
Joint venture	20	60	90	20
Maximum				**20**

Table 11.4

Project decision	Low market ($millions)	Stable market ($millions)	High market ($millions)	Average of decision ($millions)
Grassroots	−100	60	200	53.33
Expansion	−50	100	150	66.67
Joint venture	20	60	90	56.67
Maximum				**66.67**

[*In Excel apply function* **AVERAGE** *on the data for each project decision to give the first three values in the last column. Then apply function* **MAX** *on these to give the final result.*]

4   Table 11.5 and Table 11.6 show the approach. The step in Table 11.5 is to determine the best outcome according to the market from the original data in Table 11.1. This is shown in the last line.

[*In Excel apply function* **MAX** *on the data for each of the market outcomes to give the last line.*]

Table 11.6 gives the "regret" or difference between the best according to market and the original data from Table 11.1. For example, the regret for a grassroots decision when the market is low is

Table 11.5

Project decision	Low market ($millions)	Stable market ($millions)	High market ($millions)
Grassroots	−100	60	200
Expansion	−50	100	150
Joint venture	20	60	90
Best according to market	20	100	200

Table 11.6

Project decision	"Regret" Low market ($millions)	"Regret" Stable market ($millions)	"Regret" High market ($millions)	Maximum regret
Grass roots	120	40	0	120
Expansion	70	0	50	70
Joint venture	0	40	110	110
Minimum of maximum regret				**70**

[20 − (−100) = 120]; the regret for a grassroots decision in a stable market is (100 − 60 = 40); etc. The last column gives the maximum regret according to the project decision. Then the last line is the minimum of the maximum regret. In this case the decision would be to expand the existing facility.

[*In Excel once the "regret" values have been determined, apply function* **MAX** *on the data for each project decision to give the first three values in the last column. Then apply function* **MIN** *on these to give the final result.*]

Note that in this approach to decision making developing a spreadsheet changes nothing regarding reliability or confidence to proposed outcomes. Decisions made depend on the feeling and characters of those responsible for making the decision. They may not agree with opinions of others. However, what the spreadsheet does is to **"lay the information on the table"** and serves as a **benchmark** for further discussion. In this way it is better than making a "blue sky" decision—looking up into the sky and saying "I believe . . . !"

Decision making under risk with assigned probabilities

In decision making under risk, probability estimates are assigned to the states of nature as a certain confidence can be attributed because similar activities have been performed in the past in a comparable external environment. For example, the selling price per unit of an item is $40. Based on previous experience, there is a 75% probability of selling all the 500 units made. A contractor is 70% certain that it will receive a building contract of value of $23 million. A software dealer is 90% certain that this year's profits will exceed the forecast. A hotel manager is 90% sure that the hotel will be fully booked during the London Olympics in July 2012. Here the probabilities are taken into account in order to try to maximize the benefits of expected outcomes. When probabilities are assigned they are mutually exclusive. For example, if we are looking at market demand it might be low, stable, or high. It cannot be all three because of the mutual exclusivity of the given probabilities. However, together they must add up to unity, or 100%.

Expected value (EV)

In decision making under risk, the idea of **expected value (EV)** is used. This is a result obtained by weighting the outcomes according to probabilities in a related manner described in *Chapter 4*. The decision that has the highest expected value in, say, profit, or lowest. in the case of cost, would then be the preferred decision. The EV is a number used for decision making though this quantity cannot be the actual payoff because of the mutual exclusivity of the states of nature as only one of the states of nature can occur. Again, the market demand will be high, low, or level but cannot be all three. This method is illustrated by the following case-exercise application, *Boutique hotels Africa, Part I*.

Application case-exercise: **Boutique hotels Africa, Part I**

Situation

A group of four graduates from the French Hotel Management Institute, Paul Bocuse have set up their own consulting business, principally covering Africa, as they believe this is the market of the future.[7] Their activity is to convert existing properties into boutique hotels. These will be unique facilities, with no more than 20 rooms, which provide a comfortable, friendly, and personalized service in a rustic atmosphere. The décor of each room will be different, fitting local style. Restaurant food is organic and locally grown. One project concerns facilities in Cape Town, South Africa; Mombasa, Kenya; Dakar, Senegal; and Abidjan, Ivory Coast. Due to financing limitations they can only embark on one facility at this time. Table 11.7 gives an estimate of net income from these four projects over a five-year period from completion. These numbers are based on renovation costs, operating expenses when the facility opens, and revenues. Probability values are given that take into consideration the reliability of the costs, market demand, and prices that can be charged.

Table 11.7

Project decision	Net income $millions according to market		
	Low market	Level market	High market
Cape Town, South Africa	25	25	60
Mombasa, Kenya	15	30	65
Dakar, Senegal	10	35	70
Abidjan, Ivory Coast	5	20	85
Probability	20%	35%	45%

Required

1   Using the concept of expected values determine the preferred choice.

Worked out solution

1   Table 11.8 shows the expected values in the last column using the probabilities. The highest is $46.25 (5*20% + 20*35% + 85*45%) for Abidjan, Ivory Coast.

[*In Excel use function* **SUMPRODUCT** *where the probability line is fixed and project decisions are variable values.*]

Table 11.8

Project decision	Net income $ according to market			Expected value (EV)
	Low	Level	High	
Cape Town, South Africa	25	25	60	40.75
Mombasa, Kenya	15	30	65	42.75
Dakar, Senegal	10	35	70	45.75
Abidjan, Ivory Coast	5	20	85	**46.25**
Probability	20%	35%	45%	100%

Decision trees

A decision tree, as shown in Figure 11.2, is a visual representation of alternatives in decision making, useful for decision under risk. By convention a decision tree has nodes, branches, and payoffs as illustrated. There are **square nodes** (1, 4, and 5), which denote a decision point where the decision maker has control. There are **circular nodes** (2 and 3) that denote an external chance event or the states of nature. Here the decision maker has little, if any, control. There are **branches**

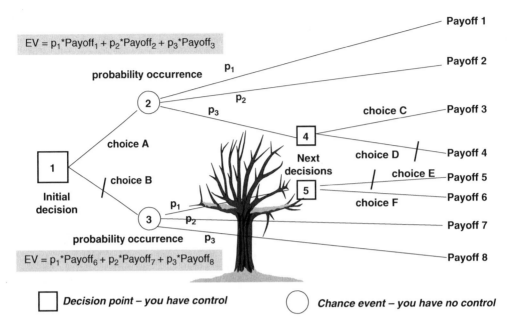

Figure 11.2 Decision tree layout.

(the straight lines in the figure) that indicate the direction the decision is being made, although the lengths have no bearing on the decision process. All the possible financial outcomes, or **payoffs**, in the decision-making process, are listed on the right of the tree for each alternative, and probability of occurrence.

From the decision tree, branches are severed by drawing a single or double line across the branch if this alternative is less attractive than an associated alternative.

- Consider the decision point at Node 4. If the Payoff 3, for choice C, is greater than Payoff 4 for choice D, then branch for choice D is severed as shown.
- Consider the decision point at Node 5. If the Payoff 6, for choice F, is greater than Payoff 5 for choice E, then branch for choice E is severed as shown.

The EV for each node is then calculated as follows:

$$EV \text{ (Node 2)} = p_1*Payoff_1 + p_2*Payoff_2 + p_3*Payoff_3$$
$$EV \text{ (Node 3)} = p_1*Payoff_6 + p_2*Payoff_7 + p_3*Payoff_8$$

Assume that the expected value at Node 2 is greater than the expected value at Node 3; then the branch for choice B is severed as shown. The preferred decision route is then choice A described as follows:

- If the probability turns out to be p_1, then the result would be Payoff 1
- If the probability turns out to be p_2, then the result would be Payoff 2
- If the probability turns out to be p_3, then the result would be Payoff 3.

Use of decision trees is illustrated by the application case-exercise, *Boutique hotels Africa, Part II.*

Application case-exercise: **Boutique hotels Africa, Part II**

Required

1   Using the given data from Table 11.7 show the condition on a decision tree with the preferred route to take.

Worked out solution

1   Figure 11.3 gives the decision tree for the boutique hotel using the data from Table 11.7. The choices for Cape Town, Mombasa, and Dakar have been severed leaving Abidjan as the preferred choice.

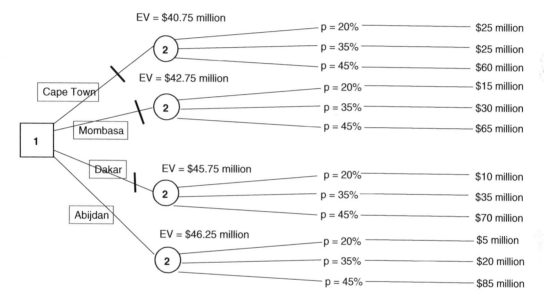

Figure 11.3 Decision tree: boutique hotel, Africa.

Expected value of perfect information

In decision theory under risk there is a concept called **expected value of perfect information** (EVPI). This is expected value under certainty (EVUC) less the expected value under risk (EVUR) or:

$$EVPI = EVUC - -EVUR \qquad\qquad 10(i)$$

The **expected value under certainty** is the weighted outcome, based on the probabilities, of all the best possibility outcomes. The **expected value under risk** is the expected value calculated by taking into account the probabilities. The purpose of determining the EVPI is that in decision making it does not make sense to pay more than the expected value of perfect information to be more certain of an alternative. For example, in deciding whether to launch a new product onto the market, the success of the product launch will be based on whether there is a high, low, or medium market demand. Conceivably one could spend more money on market research in order to be "more certain" of the outcome. However, it would not make sense to spend more than the EVPI value on the market research to be more certain of the outcome. You take the chance under risk. (A fraudulent illustration of EVPI concerns under-the-table payments. If you are one of the three bidders for a certain project, you would not give more than the EVPI in illegal payments; you would take the risk!) The concept of expected value of perfect information is demonstrated by the following application case-exercise, *Boutique hotels Africa, Part III*.

Application case-exercise: **Boutique hotels Africa, Part III**

Required

1   Determine the expected value of perfect information for the Boutique hotels Africa.

Worked out solution

1   Using the basic data from Table 11.7:

- If we knew for certainty in advance that the market demand would be low we would build a boutique hotel in Cape Town, South Africa that would give the best return of $25 million.
- If we knew for certainty in advance that the market demand would be level we would build a boutique hotel in Dakar, Senegal Town that would give the best return of $35 million.
- If we knew for certainty in advance that the market demand would be high we would build a boutique hotel in Abidjan, Ivory Coast that would give the best return of $85 million.

Since only one of these probability outcomes will occur, we need to consider probability, thus:

- Expected value under certainty, EVUC = 25*20% + 35*35% + 85*45% = $55.50 million
- Expected value under risk, EVUR = $46.25 million (already calculated from Table 11.8)
- Expected value of perfect information, EVPI = 55.50 − 46.25 = $9.25 million.

Thus the consulting group should not spend more than $9.25 million to do additional research and market evaluation to be sure of the result; otherwise they would use the data under risk.

Sensitivity

In decision making the final outcome is sensitive to assumed future external events and the question arises, **What if this happens?** or, **What if that occurs?** As an illustration, in July 2010 Boeing and Airbus received $24 billion in jet orders at the Farnborough, England air show on the basis of a rebound in air travel.[8] Good news for the companies, but this is future activity and what if the rebound does not occur as forecast bearing in mind that the building of new airplanes takes several years? Thus under decision making with conditions of risk, the choice of the decision is a function, or is **sensitive** to probability. Probabilities are difficult to quantify exactly, and thus there is always a risk in making a decision based on precise probabilities. A sensitivity analysis will illustrate how the final decision outcome might change with various probabilities. The concept is illustrated as follows in the application case-exercise, *Boutique hotels Africa, Part IV*.

Application case-exercise: **Boutique hotels Africa, Part IV**

Required

1   What would the decision be if the probability for a level market was 45% and for a high market 35%?, that is, reversed from Table 11.7?

Solution

1   The Table 11.9 shows the expected values using the new probabilities.
    The highest expected value in the last column is now $42.25 million (10*20% + 35*45% + 70*35%) for Dakar, Senegal. This illustrates the sensitivity of decision making to probability estimates.

Table 11.9

Project decision	Net income $ according to market			Expected value (EV)
	Low	Level	High	
Cape Town, South Africa	25	25	60	37.25
Mombasa, Kenya	15	30	65	39.25
Dakar, Senegal	10	35	70	42.25
Abidjan, Ivory Coast	5	20	85	39.75
Probability	20%	45%	35%	100%

Marginal analysis

Marginal analysis evaluates net financial benefits by calculating additional revenues and costs from increasing an activity by incremental amounts, often one unit.

Marginal profit and marginal loss

What is the marginal benefit if a baker makes an additional cake recognizing that the cake has to be sold the day of baking or it must be trashed? What is the marginal benefit of a retailer opening one hour earlier in the morning taking into account utility and labor costs and the fact that early sales might have occurred later? What is the marginal benefit of adding an additional employee to a production line taking into account labor costs and increased output? As marginal analysis considers small changes it is also referred to as **incremental analysis**. It is an alternative to looking at average, or total values. Consider a firm that makes standardized products, which go into inventory for subsequent sale.

- A Marginal Profit (MP) is realized if the product is made and sold.
- A Marginal Loss (ML) is realized if the product is made but not sold.
- Probability of selling the additional unit that is made is p.
- Probability of not selling the additional unit is $(1 - p)$. Selling is binomial; either it is sold or it is not.
- Expected marginal profit is the probability of sale multiplied by the marginal profit, or $p*MP$.
- Expected marginal loss is the probability of not selling multiplied by marginal loss, or $(1 - p)*ML$.

Remember nothing has happened yet, we are just anticipating future outcomes. Thus a logical business decision is to only make an additional unit if it will be subsequently sold. That is, the expected marginal profit will be greater, or as a minimum equal to the expected marginal loss. This relationship is written as:

$$p*MP \geq (1 - p)*ML$$

Reorganizing to make p the subject:

$$p*MP \geq (1 - p)*ML$$

$$p*MP \geq ML - p*ML$$

$$p*(MP + ML) \geq ML$$

Reorganizing:

$$p \geq \frac{ML}{MP + ML}$$

This implies that we should always go ahead and make an additional unit providing that the probability of sale is at least greater or equal to the ratio [ML/(MP + ML)]. This value of p is the minimum required probability of selling at least one additional unit of stock that justifies the making of that unit. It only makes sense to make additional units as long as the probability of selling at least an additional unit is greater than p. This technique is illustrated by the following case-exercise application, *Black Forest cake, Part I.*

Application case-exercise: **Black Forest cake, Part I**

Situation

Ron Piller is a baker in Normandy, France. One of Ron's specialties is a Black Forest chocolate cake that he sells in his bakery shop for €21.00. The cost of making the cake, including his time and the ingredients, is €10.80. Ron bakes his cakes in the morning before 07:00 and hopes to sell them that same day before he closes the store at 19:00. Any of these Black Forest cakes remaining at the end of the day are sold to the local retirement home for €4.00. There is never any problem selling his cakes to the retirement home. However, Ron wants to better optimize his operation. He plans to use historical sales data and use this to plan future activity. Table 11.10 gives historically the number of Black Forest chocolate cakes sold each day over the last six months, or 180 days. Never knowing the sales quantity, Ron had always baked more than he sold, so that there were times that he made a loss on some of the units.

Required

1   Determine the Marginal Profit (MP) and the Marginal Loss (ML) for this situation.
2   From the basic data for the last 180 days determine the level of cakes sold. Then using classical probability on this discrete data, calculate the probability of sale of each unit of chocolate cake.
3   Using marginal analysis, determine the minimum required probability in order to justify Ron making an additional cake.
4   At the probability level in Question 3 what would be the number of cakes to make?
5   Verify your answer with a table showing expected marginal loss, and expected marginal profit for each level of the cake demanded.
6   How would your decision change if Ron was unable to sell the cakes to the retirement home and had to trash them?

Table 11.10

27	26	21	28	24	24	27	25	26	28
20	21	27	24	26	26	28	26	27	21
24	25	26	27	25	28	25	27	21	25
27	26	24	26	26	28	23	24	25	27
25	23	23	27	25	24	26	24	28	25
27	26	26	20	22	30	26	29	26	26
20	26	26	23	27	25	28	21	25	24
23	27	22	22	24	24	28	27	28	29
28	27	28	28	27	27	27	25	29	25
26	27	22	25	24	25	25	21	21	28
23	26	28	30	27	26	27	27	27	26
25	28	27	22	26	25	21	25	25	24
21	22	23	24	27	24	23	26	23	22
24	22	25	23	23	27	28	24	23	27
27	26	27	25	24	26	29	27	27	27
25	25	27	22	26	27	28	24	23	23
27	26	24	24	22	27	26	26	24	20
26	28	23	22	27	26	23	22	25	26

Worked out solutions

1   Each cake is sold for €21.00 and costs €10.80 to make. Thus marginal profit is (€21.00 – 10.80) or €10.20. If cakes are sold to the retirement home for €4.00 then profit is (€4.00 – 10.80) or –€6.80, which is a marginal loss for each cake of €6.80.

2   From Table 11.10, the minimum and maximum number of cakes sold is determined and a frequency distribution is developed. This is in Table 11.11. The number of cakes sold is between 20 and 30 according to Column 1 of Table 11.11. The number of times this amount is sold is in Column 2. Thus the frequency or probability of sale is the amount sold divided by 180. For example, in the 180 days, 20 chocolate cakes are sold 4 times so the probability is 4/180 = 2.22%; there are 9 times that 21 chocolate cakes so probability is 5.00%, etc. The complete probability data is in Column 3 of Table 11.2.

Table 11.11

Column 1 No. of cakes sold, x	Column 2 No. of times x amount sold	Column 3 Frequency or probability of sale	Column 4 Probability, p, of "x" amount of more	Column 5 Expected marginal profit, $p*MP$	Column 6 Expected marginal loss $(1 - p)*ML$
20	4	2.22%	100.00%	10.20	0.00
21	9	5.00%	97.78%	9.97	0.15
22	12	6.67%	92.78%	9.46	0.49
23	16	8.89%	86.11%	8.78	0.94
24	22	12.22%	77.22%	7.88	1.55
25	25	13.89%	65.00%	6.63	2.38
26	32	17.78%	51.11%	5.21	3.32
27	36	20.00%	33.33%	3.40	4.53
28	18	10.00%	13.33%	1.36	5.89
29	4	2.22%	3.33%	0.34	6.57
30	2	1.11%	1.11%	0.11	6.72
	180	100.00%	0.00%	0.00	6.80

Table 11.12

Cakes made	Number of cakes demanded States of nature											EV
	20	21	22	23	24	25	26	27	28	29	30	
20	204.00	204.00	204.00	204.00	204.00	204.00	204.00	204.00	204.00	204.00	204.00	204.00
21	197.20	214.20	214.20	214.20	214.20	214.20	214.20	214.20	214.20	214.20	214.20	213.82
22	190.40	207.40	224.40	224.40	224.40	224.40	224.40	224.40	224.40	224.40	224.40	222.79
23	183.60	200.60	217.60	234.60	234.60	234.60	234.60	234.60	234.60	234.60	234.60	230.63
24	176.80	193.80	210.80	227.80	244.80	244.80	244.80	244.80	244.80	244.80	244.80	236.96
25	170.00	187.00	204.00	221.00	238.00	255.00	255.00	255.00	255.00	255.00	255.00	241.21
26	163.20	180.20	197.20	214.20	231.20	248.20	265.20	265.20	265.20	265.20	265.20	243.10
27	156.40	173.40	190.40	207.40	224.40	241.40	258.40	275.40	275.40	275.40	275.40	241.97
28	149.60	166.60	183.60	200.60	217.60	234.60	251.60	268.60	285.60	285.60	285.60	237.43
29	142.80	159.80	176.80	193.80	210.80	227.80	244.80	261.80	278.80	295.80	295.80	231.20
30	136.00	153.00	170.00	187.00	204.00	221.00	238.00	255.00	272.00	289.00	306.00	224.59
Prob. of sale	2.22%	5.00%	6.67%	8.89%	12.22%	13.89%	17.78%	20.00%	10.00%	2.22%	1.11%	

[In Excel create the first column "No. of cakes sold x". For Column 2, "No. of times x amount sold" apply function **COUNTIF** *to the dataset in Table 11.11 and where the criteria is the "No. of cakes sold x." For Column 3, divide the number of cakes sold by the total amount of 180.]*

3 The minimum probability is given by:

$$p \geq \frac{ML}{MP + ML} \text{ or } p = 6.80/(10.20 + 6.80) = 40.00\%$$

4 Column 4 of Table 11.11 gives the probability of more than a certain level of cakes being sold. If 26 is the number of cakes sold the probability is 51.11%; if 27 is the number the probability is 33.33%. The value of 40.00% lies between the two and thus the number of cakes to make is 26. If we selected 27 then *p* would be less than 40.00%. Note cakes are discrete values so we can only select whole cakes.

[In Excel the first entry in Column 4 "Probability p, of x amount or more" is the total amount or 100%. The second entry of 97.88% is (100 − 2.22). The third entry is 92.78% (97.78 − 5.00) etc.]

5 Columns 5 and 6 of Table 11.11 give the expected marginal profit and expected marginal loss. The expected marginal profit is always greater than the expected marginal loss just up to 26 cakes being sold. After this the expected marginal profit becomes less (at 27 cakes €3.40 versus €4.53). This corroborates the answer to Question 4.

[In Excel the first entry in Column 5 "Expected marginal profit" is 10.20 (100%* 10.20). The second entry is 9.97 (10.20*97.78%); the third entry is 9.46 (10.20* 92.78%) etc. In Column 6 "Expected marginal loss" the first entry is 0.00 (100%− 100%) * 6.80; the second entry is 0.15 (100%− 97.18)* 6.80; third entry is 0.49 (100 − 92.78) * 6.80 etc.]

6 If the cakes are trashed, the marginal loss is €10.80 or the cost of making the cake. In this case *p* = 10.80/(10.20 + 10.80) = 51.43%. Now 26 cakes, at a value of *p* = 51.11% is just too small so the number of cakes to make is 25. This confirmed by the following expected marginal values.

- At 25 cakes expected marginal profit is €6.63 as before.
  Expected marginal loss = (1 − 65.00%) * 10.80 = €3.78.
- For 26 cakes expected marginal profit is €5.21 as before.
  Expected marginal loss = (1 − 51.11%) * 10.80 = €5.28.

Marginal analysis and the normal distribution

If the assumption can be made that the distribution of the sale or stocking of a product follows a normal distribution, then marginal analysis can be used in conjunction with the normal distribution as illustrated in Figure 11.4. The normal distribution represents the probability of sale according the number of units *x*.

- Point x_1 is -3σ and so the probability of this number of units x_1 or more being sold is 100%, or the area to the right of x_1.
- Point x_2 is the mean value or where $\sigma = 0$. The probability at this point of selling x_2 units, or more, is 50%, the area of the curve to the right.
- Point x_4 is $+3\sigma$ the extreme right of the curve and so the probability at this point of selling x_4 units or more is 0%, the area of the curve to the right.

Thus as we move from left to right on the curve, the probability of selling a certain quantity of units, or more, declines from 100% to 0%. The point x_3 as shown in Figure 11.4 is the minimum probability of selling a unit and in marginal analysis is given by

$$p \geq \frac{ML}{(MP + ML)}$$

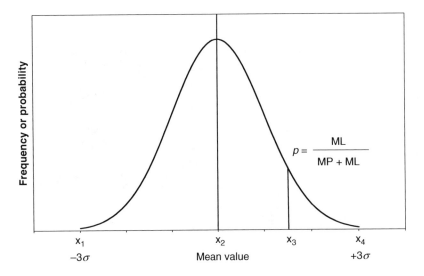

Figure 11.4 Normal distribution and marginal analysis.

Payoff tables

A payoff table is another way of using expected values to make a decision from the outcomes of various alternatives. A payoff table shows all the possible outcomes for options according to likely external events. These outcomes are then weighted using assigned probabilities to determine the expected values. The alternative selected is that which has the highest expected value if it is profit; lowest if it is cost. Again, since only one state of nature is possible the expected value is only a quantitative measure to determine the preferred alternative. The real outcome will be only one of the external events or states of nature, since the states of nature are mutually exclusive. When a situation is a recurring event, such as a inventory stocking, selling, labor levels it is unlikely the optimum result occurs each time. What the payoff table does is to optimize the best long-run alternative assuming that the probability data remains constant. Again, there is the question of sensitivity in that the chosen decision is based on the probability estimates. If these change, then the decision alternative may be modified. Payoff tables are illustrated as follows by the application case-exercise, *Black Forest cake, Part II*.

Application case-exercise: **Black Forest cake, Part II**

Required

1   Develop a payoff table for conditional profit using the states of nature selling from 20 to 30 cakes and the decisions being to produce from 20 to 30 cakes. Weight the outcome of each decision level according to probability levels developed for Part I of this application

case-exercise in Column 3 of Table 11.11. Opportunity costs are not a consideration. What are your conclusions?

Worked out solutions

1   Table 11.12 gives the conditional profit payoff spreadsheet. Column 1 from 20 to 30 is the number of cakes made. The baker is in control of this activity. The line at the top of table gives the number of cakes demanded by customers. Again this is from 20 to 30 and is the states of nature. The line at the bottom gives the probability of the states of nature and is those values derived from the historical data. The last column gives the expected value, EV.

The conditional profit is determined as follows for three situations.

• The baker makes 20 cakes. Assume that there is a demand for 24. He can only sell the 20 that he made to give a profit of €204.00 (20 * 10.2). He "loses" a sale of 4 cakes, which is an opportunity cost.
• The baker makes 24 cakes. Assume that there is a demand for 22. He sells 22 cakes as demanded for a profit of €224.40. For the 2 cakes that he does not sell he loses €13.60 (2 * 6.80). Thus the net profit is €210.80 (224.40 − 13.60).
• The baker makes 27 cakes. Assume that there is a demand for 27. His production matches demand. He sells all 27 cakes for a profit of €275.40 (27 * 10.2).

These are all conditional situations: the condition that the baker makes a certain quantity; and the condition that there is a certain sale. The expected values given in the last column are calculated by multiplying each conditional outcome by the probability. If the baker makes 26 cakes the expected value is:

$$2.22\% * 163.20 + 5.00\% * 180.20 + 6.67\% * 197.20 + 8.89\% * 214.20$$
$$+ 12.22\% * 231.20 + 13.89\% * 248.20 + 17.78\% * 265.20 + 20.00\% * 265.20$$
$$+ 10.00\% * 265.20 + 2.22\% * 265.20 + 1.11\% * 265.20$$
$$= €243.10.$$

This is the highest of the expected values and thus the baker should make 26 cakes to maximize his profit in the long term. This is the same conclusion as obtained using marginal analysis of Part I of this application case-exercise.

[In Excel use the function IF *in each cell according to the criteria "if" the number of cakes demanded is greater than the number made.]*

Utility

The concept of **utility** attempts to quantify human behavior that in itself is subjective or qualitative. The subject is important since managers make many business decisions and often in the work place we hear remarks like, "why did he make that decision?" or "that decision makes no sense" when numerical or statistical data would suggest an alternative.

Different personalities

In the paragraph on decision making under uncertainty we used the idea of optimistic, pessimistic, and middle-of-the road. In these circumstances one manager may be optimistic and another pessimistic. Why might this be? People are simply different! In using expected values we calculate the value of alternatives by multiplying probability by payoff. Logic says that the alternative that has the highest expected value should be selected. However because of fear, gut feeling, or risk-averse personalities, decisions are not always made according to the expected value. This is utility: the satisfaction, disappointment, or even nervousness, as the result of a certain outcomes. People have different personalities and conflicting ideas in making decisions. (In the Everest expedition of the Icebreaker members of the climbing team were diverse and these differences in part contributed to the fatal accidents.)

Different personalities can be quantified by using the concept of utility as illustrated in Figure 11.5 that illustrates utility and risk. Here, utility has been given an index of 0 to 10 (positive utility, or satisfaction) and 0 to −10 (negative utility, or disappointment).

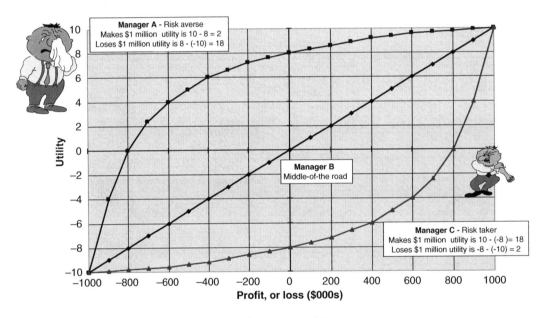

Figure 11.5 Utility curves for managers with different personalities.

The lower curve, Manager C, is the **risk taker**. For this person, taking a great risk is a big challenge. He gets a "high." Thus, his satisfaction of reaping great rewards such as an acquisition, succeeding with a new product, or making a stock investment, increases sharply. Thus, with high profits, the slope of his utility curve is high (the portion of the curve to the right of the y-axis). For example, if his profit increases by $1 million (from 0 to $1 million on the x-axis) then his utility increases by a factor of 18 (from −8 to +10 on the y-axis). In contrast, his disappointment, or loss, is relatively less intense as illustrated by the smaller slope to the left of the y-axis. For a loss of $1 million, (from 0 to − $1 million on the x-axis) his utility or disappointment only changes by a factor of 2 (from −8 to −10 on the y-axis).

The upper curve, Manager A, is for the **risk-averse** individual. Here the situation is reversed. The thought of a loss sharply decreases his utility (left of the y-axis). For example, for a loss of $1 million (from 0 to minus $1 million on the *x*-axis) his utility decreases by a factor of 18 (from +8 to −10 on the *y*-axis). The idea of a gain, although it increases his utility, is much less marked. For a gain of $1 million (from 0 to $1 million on, the *x*-axis) his utility only increases by a factor of 2 (from +8 to +10 on the *y*-axis). The center curve is for the **middle-of-the-road** manager, Manager B, where utility (satisfaction, or disappointment) is of equal value whether the outcome is a gain, or loss. Here for a gain of $1 million his utility increases ten-fold (from 0 to +10 on the curve). For a loss of $1 million his utility decreases ten-fold (from 0 to −10 on the curve).

Gambling

In *Chapter 3,* there were examples relating to gambling. Now consider a gambler at a casino playing blackjack where the probability of winning is 70%. Alternatively since the outcome is binomial, the probability of losing is 30% (100 − 70). Illustrated in Figure 11.6 are the expected gains for gambling when placing bets of $5, $50, $500, $5,000, $50,000, $100,000 and $1,000,000. In all cases, the "expected gain" is greater than the "expected loss," and greater than the level of bet. Thus, if expected value were the decision criteria used, the gambler would always play. However, in reality this is not the case. Placing a bet of $5 perhaps causes little concern, even if one loses. The payoff, if the gambler wins, is $500 with an expected outcome of $350 (500 * 70%). In the next situation, we might be more cautious of putting down $50 even though the potential winnings are $5,000 with an expected gain of $3,500. As the value of the bet placed becomes higher, a gambler (usually!) becomes more and more cautious about placing a bet, and may refuse to place bets at the $50,000 and $100,000 levels as this may require mortgaging his house, even though the expected gain is greater than the expected loss; or greater than the value of the bet. In this case, the prospect, or the utility

Level of bet, $	Potential payoff $	Probability of winning	Probability of losing	"Expected" gain, $	"Expected" loss, $
$1	$100	70%	30%	$70	$30
$5	$500	70%	30%	$350	$150
$50	$5,000	70%	30%	$3,500	$1,500
$500	$50,000	70%	30%	$35,000	$15,000
$5,000	$500,000	70%	30%	$350,000	$150,000
$50,000	$5,000,000	70%	30%	$3,500,000	$1,500,000
$100,000	$10,000,000	70%	30%	$7,000,000	$3,000,000
$1,000,000	$100,000,000	70%	30%	$70,000,000	$30,000,000

Always
Expected gain is > than expected loss and > than level of bet.
However, there is a point at which people will stop gambling.
This depends in the individual's utility.

Figure 11.6 When expected values are not the criteria for decision making.

or the **value** in the mind of the gambler of losing $50,000 or $100,000 outweighs the prospect of winning.

Insurance industry

The insurance industry makes its fortune on risk. (Insurance is a bad investment if there is never a claim from the insured.) Consider **industrial insurance** for a company that operates a small manufacturing operation producing steel and aluminum aircraft components. The facility is constructed of brick. The company is valued at £2 million. The probability of the company being destroyed by fire is 0.001% based on historical data of industrial accidents in England. The insurance premium is £2,500 per year. In this case the expected loss is £2,000,000 × 0.00001 = £20. Using expected value criteria, the expected loss of £20 is less than the insurance premium of £2,500 and based on this it does not make sense to insure. However, most, if not all, manufacturing concerns have insurance. The thought (utility) of losing £2 million in assets far outweighs the cost of the premium. Consider **transportation insurance** for a US manufacturer of compressors that sends a new installation to Germany for use in a pharmaceutical operation. The value of the equipment is $4 million. The probability of the equipment being damaged, or lost, is 1 in 10,000, or 0.01%. The expected loss is $4 × 0.0001 = $400. Insuring the equipment cost $2,000. Since $2,000 is greater than the $400, then using expected value criteria, it makes no sense to insure the merchandise. However, losing $4 million would be very heavy for the company. Thus it takes out insurance.

Some homes in Southern California, certainly in Beverly Hills, Brentwood, and San Marino, have market values in excess of $2 million. The possibility of a home being completely destroyed by an earthquake is small (though not remote according to geologists' reports on the San Andreas Fault). Assume the value of the home is $2 million and the probability of completely losing the home in a destructive earthquake is 0.02%. Then expected loss (EL) = $2,000,000 * 0.02% = $400. The insurance premium costs $4,000. The expected loss is less than the cost of the premium, thus logic says it makes no sense to insure. However, people insure as the thought (utility) of losing the home greatly outweighs the insurance cost. Some insurance companies have "do it yourself" insurance booths at major airports. Here, for example, for $10 one is able to take out an **accident insurance** of say $100,000 in the event the airplane crashes. The probability of the plane crashing is 1 in 500,000 or 0.0002%. The expected loss, if an accident, is 100,000 x 0.000002 = $0.20. This expected loss is less than the insurance premium, and under expected value criteria it does not make sense to take out insurance. However, many travelers do since if there were an accident, the loss of the family breadwinner would be disastrous. (As an engineer in the 1980s I traveled quite a bit with project managers. One particular manager, when we arrived at LA airport, would go the insurance booth and take out insurance to the benefit of his wife and children in case the airplane crashed. He retired and died in old age!) Overall **life insurance** is a bad investment since the probability of dying prematurely is low. However, with the concept of utility many people have life insurance for family protection. (With good returns to the insurance companies!!)

Moving on

This chapter has presented decision making and risk using giving qualitative steps to consider and quantitative methods based on statistical concepts. Various business climates have been discussed—certainty, uncertainty, and risk with assigned probabilities. The next chapter, covering statistical process control, demonstrates, using statistics, how process and quality risk can be minimized.

Notes

1 Davis, Wade, "Into the silence: The Great War, Mallory, and the Conquest of Everest," Random House, Canada, 2011.
2 "Large Hotels and Airline Expanding Operations in Africa," *International Herald Tribune*, March 19, 2012, p. 19.
3 Jordonez, Jennifer, "Burger King sought the perfect fry but got burned: Sales soon went limp for crispier potato stick," *The Wall Street Journal Europe*, January 17, 2001, p. 1.
4 "The President sets a precedent," *Time*, December 3, 2012, pp. 10–12.
5 "An election in Myanmar: Change in the air," *The Economist*, October 31, 2015, p. 49.
6 "Another game of thrones," *The Economist*, December 1, 2012, p. 25.
7 "Africa Rising," *Time*, December 3, 2012, p. 32.
8 "$24 billion in jet orders confirm air travel revival," *International Herald Tribune*, July 20, 2010, p. 1.

12 Statistical process control (SPC)

ICEBREAKER: Using SPC perhaps the *Titanic* would not have sunk!

The White Star Line was owned by the American financier J. Pierpont Morgan who dreamed of creating a monopoly against his bitter rival, Cunard, on the North Atlantic shipping routes. In the early 1900s the largest liner afloat, the *Titanic*, was conceived by White Star. It was a remarkable technical achievement but it was handled with a folly that turned it into the world's largest coffin. The *Titanic* had every conceivable luxury—cafés, squash courts, a swimming pool, Turkish baths, a barbershop, and three libraries. It was built in the shipyards of Harland & Wolff in Belfast, Northern Ireland and was protected by a double bottom and 16 watertight compartments, formed by 15 bulkheads across the ship. Doors in the bulkheads could be closed instantly by a switch on the bridge. Should any two of the largest compartments become flooded, the liner could remain afloat indefinitely. These safeguards led White Star to boast that the ship was practically unsinkable!

On 10 April 2012 at 12:15 the *Titanic* sets off for its maiden voyage from Southampton. None of the passengers or well-wishers realized that the preparations for the voyage were fraught with difficulties. A national coal strike had left the new ship without sufficient fuel for the long voyage to New York. Other liners, paralyzed by the same problem, gave up their meager

Figure 12.0 Sinking of the RMS *Titanic* (German artist Willy Stöwer (1864–1931), 1912).

Source: Wikipedia.

reserves to help. But transferring coal was a dirty business and before the liner could leave it had to be cleaned from stem to stern. As she set sail the *Titanic* would have collided with another liner, the *New York*, in Southampton Water near the Isle of Wight if it wasn't for the quick action of the Titanic's captain, Edward J; Smith, who ordered the port propeller reversed. This prompt reaction and the attention of the tugs prevented a nasty collision.

The year 1912 was exceptional for the number of icebergs seen in the Atlantic. The winter in the Arctic had been mild and ice flows drifted into the Gulf Stream further south than anyone could remember. On Sunday April 14, two days out from Southampton, the *Titanic* received many ice warnings from ships in the area. With no explanation that morning's life boat drill was cancelled. Although in 1912 there were no statutory requirements for drills or musters White Star did have its own rules. However, it chose to ignore them. The previous evening the wireless set had broken down and the two wireless operators were unable to send or receive messages until early Sunday morning when they concentrated on a backlog of passenger messages. At 13:40 their routine was disturbed by a communication from the Greek Liner *Athinai*: " . . . sighting of icebergs and large quantities of field ice in latitude 41.51N, longitude 49.11W . . . " Other messages from the steamships *La Touraine* and *Rappahannock* also reported sighting icebergs on the *Titanic's* route. The most important message was from the steamer *Mesaba* who telegraphed " . . . ice report in latitude 42N to 41.25N, longitude 49W to 50.3W . . . saw much heavy pack ice and a great number of large icebergs." This message was never sent to Captain Smith as the *Titanic's* wireless operators were still concentrating on the passenger's messages.

At 23:40 the *Titanic* struck an iceberg and by 02:20 on April 15 the ship had gone, taking with it 1,523 lives out of the 2,200 on board. The pressure exerted by the iceberg caused the plates to buckle, the rivets gave way under the strain, and eventually water flowed in to all 16 "water tight compartments." In any event some 94 years later, scientists concluded that the ship was doomed from the start because of poor quality rivets made of iron, rather than stronger steel, used in the hull assembly. The scientists, using process quality control tests and computer simulations on 48 rivets taken from the wreckage three kilometers deep in the Atlantic, found many contained high concentrations of slag. This is a glassy residue from iron smelting that can make rivets brittle and fracture easily. Studies of the *Titanic's* wreckage showed that, rather than a gigantic gash in the ship's hull as was first thought, there were six narrow slits where bow plates appeared to have parted. These openings were in the area where poor quality rivets were used. Naval experts suspected that rivets had popped along the seams, letting seawater rush in at high pressure. Further, from archives it was found that the ship's constructors, Harland & Wolff in Belfast, Ireland had difficulty in finding skilled and properly trained riveters, which possibly increased the accident risk. Statistical process and quality control, throughout the conception, construction, and voyage of the ship may have averted this terrible disaster! Statistical process control is the theme of this chapter.[1,2]

*Statistical process control (SPC), is the periodic sampling and analysis of elements from services or manufacturing to determine if a process is performing to some target level or conforms to specification. If sample data is within the specification limits then the process is considered to be meeting requirements, or more bluntly, is **in control**. Alternatively, if sample data falls outside the specification limits, it is not meeting conditions and is **out of control**. In business a goal is to always provide specification or perfect quality to customers within the constraints of operational objectives.*

Sampling in statistical process control

In manufacturing, sampling may be analyzing units from operations such as drilling, soldering, bottle filling, packing, molding, milling, painting, assembly, etc. In services, sampling may be from activities such as airline flight arrival/departure times; the excellence of a professor's teaching; the accuracy of bank statements or other financial records; the cleanliness of hotel rooms, or restaurant facilities in hospitality; delivery conditions for merchandize in distribution and transportation; or the quality of medical services perhaps measured by the accuracy of the diagnosis. If the sampling indicates that units or the activity are acceptable, no action is taken as the process is considered in control. If the output is not acceptable, then the process is stopped and the necessary corrective action is taken as the process is judged out of control.

Binomial characteristics

The test of quality might simply be that the product, process, or service works is acceptable or is not acceptable. This criterion is considered an **attribute** and is a binomial characteristic as there are only two possible outcomes: yes or no; pass or fail; open or closed; go or no-go; defective or non-defective; works or does not work; late or on time; present or absent; on or off; good or bad; fresh or not fresh; right or wrong; clean or dirty; conforming or non-conforming; sharp or dull (blades); etc. In production attribute classification is basically a product functions or it does not: a battery generates a current or does not; a bulb illuminates or does not; a lid closes or does not. In hospitality-related services the attribute classification is that the service is acceptable or is not: the tour bus arrives on time or is late; the hotel billing statement is correct or is wrong; the hotel room is clean or is not; the meal is tasty or is not; the blood analysis is negative or positive. Attributes are count data or integer values.

Six of the sample batteries were defective; three of the hotel rooms did not conform; five of the bills were incorrect; four of the meals were cold; nine of the dinner plates had cracks. We could not say 6½ of the sample batteries were defective; $3^3/_4$ of the hotel rooms did not conform; 5¼ of the bills were incorrect; 4½ of the meals were cold; 9¼ of the dinner plates had cracks. For attribute data, a discrete probability distribution such as the **binomial** or **Poisson** is the basis for statistical analysis. *Chapter 4* gives information on discrete distributions.

Acceptance range

An alternative acceptance of quality is that units are within a given **variable** range, measured on a continuous scale, resulting in a spectrum of acceptance to meet specifications. Examples might be color, weight, length, diameter, viscosity, volume, hardness, or time. For example the specification for the volume of wine in a bottle may be 75 cl $\pm$ 3 cl; the weight of coffee in a jar, 200 g $\pm$ 4 g; the time to make hamburger, 4.0 min $\pm$ 30 sec; the time to clean a hotel room, 15 minutes $\pm$ 2 min; the weight of a piece of steak, 200 g $\pm$ 10 g. When variables are involved it is almost impossible to have exactly the nominal amount specified. The volume of beer in a can is indicated on the label as 33 cl. However, it is highly unlikely that the volume will be 33.0000 cl but perhaps something like 32.9985 cl or 33.0204 cl. Both these volumes would probably be acceptable to the consumer or to a weights and measures inspector. For most work, the narrower the specification range, the better is the quality of the units produced and this is the concept of **Taguchi** and a requirement for **six-sigma** as presented in *Chapter 13*. For variables it is the **normal distribution** that serves as the basis for statistical analysis.

Target values and process variation

In a game of darts, often found in English pubs, the objective is to throw a dart into the bull's-eye. This is the target value and if obtained makes the player a winner. However some of the darts may not hit the bull's-eye but be dispersed around it or even miss the board altogether. This is the same in statistics in management: there are targets, but also variation. In statistical process control variation is measured around a target value. Variation might be **random** or also referred to as **common cause**, or **inherent** variation. In the filling operation of 33 cl cans of beverages the nominal or target value is 33 cl. However, because of differences—in the viscosity of the beverage; its foaming characteristics; the velocity of fluid in the feed pipe; the diameter of the filling nozzles; the rotational speed of the filling platform, etc—each can does not contain exactly 33.0000 cl of liquid. Some will have more than 33 cl, say between 33.0000 and 33.8250 and some less, say between 32.1750 and 33.0000. However, if the volume of a large quantity of cans of beverages were measured, most observations would lie near the target value of 33 cl with some being below this value and some above. The result would produce a symmetrical or normal distribution centered on the mean value of 33 cl as shown in Figure 12.1. As random variations are often a natural occurrence a spread of data would be acceptable, giving specifications for the volume of beverage in the can for example as 33 $\pm$ 0.8250 cl.

Figure 12.1 is just a static situation. In addition this distribution profile must remain constant over a suitable operating time period as shown in Figure 12.2. From this figure the conclusion is that the process is operating satisfactorily, or in control, since the profiles "yesterday," "today," and "tomorrow" are identical.

Similarly, in services such as a hotel, there are targets: an 80% room occupancy rate; rooms cleaned in 20 minutes; guests checked out in 5 minutes; daily revenues according to an established objective, etc. In a restaurant we may have a target that a bottle of wine is sold for every three customers; the time between ordering and serving the main meal is 15 minutes; the number of place settings sold is 90% of

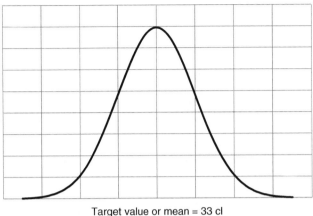

Figure 12.1 Normal distribution centered on a mean of 33 cl.

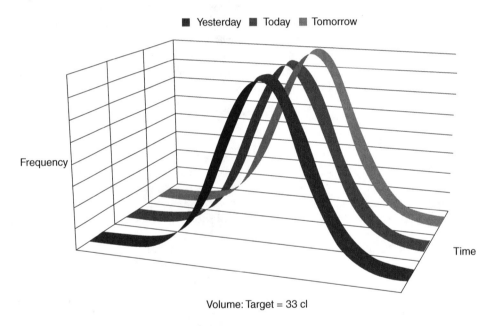

Figure 12.2 Distribution profile unchanged over time: process is in control.

capacity. It is the process that is being analyzed for conformity to meet the target value though often it is the measurement of the "product" that is used in the sampling process. We measure the "product," rooms sold, but the process includes forecasts, price levels or other elements of revenue management. In the sale of wine in the restaurant it is the inventory level that is counted though the process is the information and recommendation supplied by the sommelier.

In a production operation such as the beverage filling operation, over time the bearings on the rotating filling platform become sloppy; the diameters of the filling nozzles decrease slightly owing to deposits of sugar; wear on the feed valve causes the fluid velocity to change; etc. In this case, the average volume of beverage in the can may change from 33 cl to some upper or lower value as illustrated in Figure 12.3.

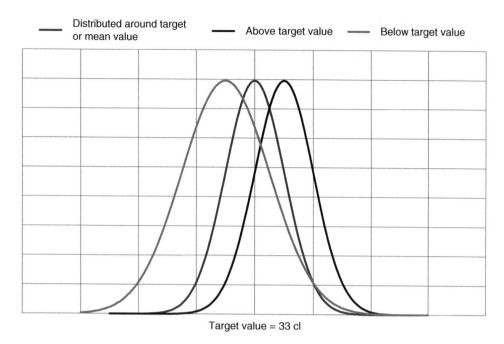

Figure 12.3 Average volume has changed: process is out of control.

Alternatively, the distribution, or spread, of the volumes in the can may change, but the average volume remains the same, as illustrated in Figure 12.4.

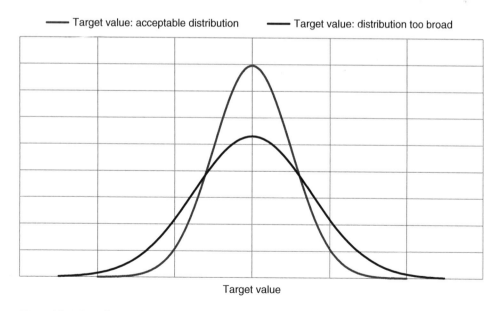

Figure 12.4 Distribution has changed: process is out of control.

Thus over time the distribution profiles obtained are different, as shown in Figure 12.5. In these cases there are **assignable** variations as we can detect or locate the reasons for the changes. Here the

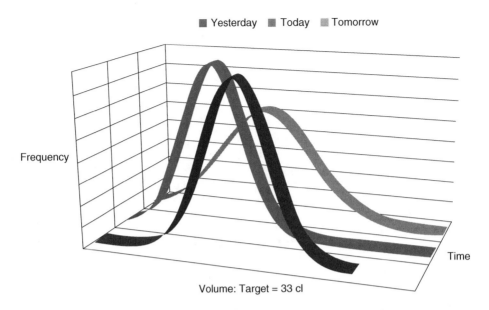

Figure 12.5 Distribution profile changes over time: process is out of control.

solution in the production operation would be to shut down the operation and take appropriate corrective action.

In services we have the time to perform the activity, revenue targets, costs, arrival and departure times in transportation, quantities delivered, billing statement errors, hotel quality, etc. Here there will be variation around the target level and distributions can be developed for these divergences. Ideally we would like no variation but when we have supply chain activities like the delivery of food and beverages, variation may be impacted by traffic congestion or weather. These would be assignable variations. When people perform certain services, there may be times when they have "off-days" and a small amount of variation may be tolerated, and this would be considered random variations. However, if the error problems continued, because of poor training, for example, these would be assignable variations. Steps would be taken to rectify the problem.

Control charts

The control chart is the analytical tool in statistical process control whether the purpose is to control, costs, absenteeism, accidents, temperature, weight, time, color, quality, blood levels, etc. There are three basic control charts: the percentage or **p-chart** and the count, or **c-chart**, both of which are for measuring discrete values. The other is the **x-bar and range chart** used together for measuring variable values.

Parameters of control charts

All control charts have three **benchmark** parameters; a **center line (CL)** that is the target or value expected; an **upper control limit (UCL)**; and a **lower control limit (LCL)**. The units of these depend on the process. It might be the arrival time (airplane, train, bus); number of errors (printing, financial statements); evaluation level (teaching, service level); absentee rate (personnel); cleanliness (hotel rooms, restaurant), financial (revenues, costs, profits). With tangible products it could be weight (chocolate bars, bread, slices of meat); color (paper, textile, linoleum); hardness (gear wheels, metal surfaces); viscosity (oil, ointments, blood); or temperature (food, chemical processes). The units of the upper and lower control limits are the same as the units of the center line or target value. When these limits are established, they represent the benchmarks to compare data. The logic of the control chart is that if sample data falls outside these control limits a process is considered out of control, meaning that performance is unacceptable. There may be situations when data falls below a lower control limit, such as the errors in billing, the blemishes on fabric, or the time to clean a room. Alternatively there may be occurrences when sample data falls above the upper control limit, such as hotel revenues, yield rate, or evaluation level. These would not be a cause for concern, though it this happens frequently it means performance is better than expected and control limits should be revised.

Analytical procedures

A typical control chart is given in Figure 12.6 where samples have been taken from a production lot and the percent defective units measured. The analytical procedure in statistical process control is to measure the variable of interest with a defined sampling program: number of samples; sampling time; and sampling period. We might measure the average weight of a sample of 25 chocolate bars from the production lot; the quality of a professor's teaching by the 30 students in the class; the absentee level out of the 600 employees; the correctness of 25 hotel rooms out of a population of 250, etc. This variable value is then compared to the defined standard. If the variable measurement falls within the control limits then no action is necessary. If the measurement falls outside the limit, then the situation should be investigated and appropriate action taken. This might include, in services, modification of the schedule (transportation); replacing the professor (teaching); motivation programs (personnel); replace the chef (restaurant). In manufacturing it could be replacement

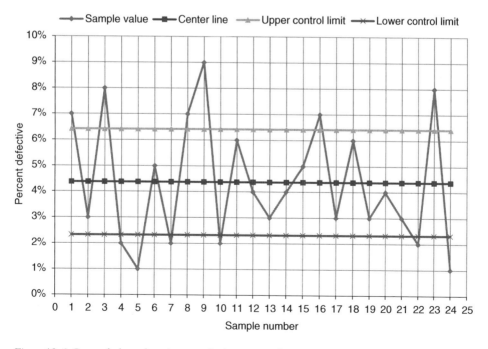

Figure 12.6 Control chart showing sample data, center line, and upper and lower limits.

of worn tools, adjustment of machines, retraining of operating staff, replacing equipment, etc. Figure 12.6 indicates that the process is out of control since samples 3, 8, 9, 16, and 23 are above the upper control limit. (Those below the lower control limit may not be a problem.)

In statistical process control the ultimate interest is to determine if the whole operation or the population is performing as expected. We are interested in the *whole* production line of chocolate; *all* of the professors teaching; the cleanliness of *all* the hotel rooms. We are using sample data as an estimate of the population since usually it is impossible or uneconomic to perform statistical process control on all the elements.

Percentage or p–chart

A p–chart indicates the percentage or fraction of defective or non–conforming units in a sample, n, where the value of p is given by:

$$p = \frac{\text{Number of defective units in sample}}{\text{Total number of units examined}}$$

A p–chart is used in the measurement of discrete data and the binomial distribution is the probability function that serves as the basis. However, for large sample sizes the normal distribution can be approximated for the binomial distribution. (See *Chapter 5* giving these conditions.) The control limits in the p–chart are:

Center line, $CL = p$
Upper control limit, $UCL = p + z\sigma_p$
Lower control limit, $LCL = p - z\sigma_p$

- p is the target or benchmark fraction or percentage "non-conforming" units. If there is no target level, then the value of p is equal to $\bar{p}$, the average fraction defective in a sample of size, n. For the binomial situation $p = (1 - q)$ where q is the fraction or percentage of "conforming" units.
- z is the number of standard deviations. The smaller the value of I, the more rigorous are control specifications.
- σ_p is the standard deviation of the sampling distribution, and is estimated from the relationship:

$$\sigma_p = \sqrt{\frac{\bar{p} \cdot (1 - \bar{p})}{n}} = \sqrt{\frac{\bar{p} \cdot \bar{q}}{n}}$$

Thus, in this case with a centerline or target value of, $\bar{p}$, the control limits are:

$$CL = \bar{p}$$

$$UCL = \bar{p} + z\sqrt{\frac{\bar{p} \cdot (1 - \bar{p})}{n}} = p + z\sqrt{\frac{\bar{p} \cdot \bar{q}}{n}}$$

$$LCL = \bar{p} - z\sqrt{\frac{\bar{p} \cdot (1 - \bar{p})}{n}} = p - z\sqrt{\frac{\bar{p} \cdot \bar{q}}{n}}$$

The development of a p–chart is illustrated by the following application case-exercise, *Record keeping*.

Application case-exercise: **Record keeping**

Situation

A government agency, responsible for keeping records of land ownership, is in the process of changing its computer database system. In doing this, it has secretarial staff manually entering account numbers from the old system into the new system. The work is done six days per week and every Saturday student helpers assist the secretaries. The agency is concerned about the accuracy of the creation of the new records and so the office manager carries out a quality control check. This quality control check is carried out over a five-week period, six days per week with samples taken every day starting on a Monday. Each day, samples of 200 records are taken and the accuracy of the records is verified. Accuracy means that the account number been correctly recorded. Table 12.1 gives the number of errors found during the five-week sampling period. Day 1 corresponds to the Monday when the quality control process was put into effect, Day 2 Tuesday, etc.

Table 12.1

Day sample was taken	No. of wrong account numbers	Day sample was taken	No. of wrong account numbers
1	6	16	2
2	5	17	1
3	3	18	13
4	6	19	6
5	2	20	2
6	15	21	6
7	1	22	8
8	0	23	9
9	2	24	10
10	2	25	3
11	4	26	1
12	14	27	4
13	4	28	5
14	3	29	7
15	0	30	10

Required

1 Construct a p-chart for this operation, using sigma limits, z, of $\pm\,3$. What would you conclude from this 3σ control chart?
2 What would happen if you used a z value of 1? This would be a more rigorous control.

Worked-out-solutions

1 The calculated information, for $z = 3$, from which the control chart is developed is in Table 12.2. The sample size, n, is 200:

- For first sample the fraction defective is 6/200 or 0.0300
- For second sample the fraction defective is 5/200 or 0.0250
- For third sample the fraction defective is 3/200 or 0.0150
- Etc.

Table 12.2

Day sample taken	Specific day	No. wrong accounts	Fraction accounts defective	Center line (CL)	Upper limit (UCL)	Lower limit (LCL)	Status relative to UCL
1	Monday	6	0.0300	0.0257	0.0592	0.0000	Below
2	Tuesday	5	0.0250	0.0257	0.0592	0.0000	Below
3	Wednesday	3	0.0150	0.0257	0.0592	0.0000	Below
4	Thursday	6	0.0300	0.0257	0.0592	0.0000	Below
5	Friday	2	0.0100	0.0257	0.0592	0.0000	Above
6	Saturday	15	0.0750	0.0257	0.0592	0.0000	Below
7	Monday	1	0.0050	0.0257	0.0592	0.0000	Below
8	Tuesday	0	0.0000	0.0257	0.0592	0.0000	Below
9	Wednesday	2	0.0100	0.0257	0.0592	0.0000	Below
10	Thursday	2	0.0100	0.0257	0.0592	0.0000	Below
11	Friday	4	0.0200	0.0257	0.0592	0.0000	Above
12	Saturday	14	0.0700	0.0257	0.0592	0.0000	Below
13	Monday	4	0.0200	0.0257	0.0592	0.0000	Below
14	Tuesday	3	0.0150	0.0257	0.0592	0.0000	Below
15	Wednesday	0	0.0000	0.0257	0.0592	0.0000	Below
16	Thursday	2	0.0100	0.0257	0.0592	0.0000	Below
17	Friday	1	0.0050	0.0257	0.0592	0.0000	Above
18	Saturday	13	0.0650	0.0257	0.0592	0.0000	Below
19	Monday	6	0.0300	0.0257	0.0592	0.0000	Below
20	Tuesday	2	0.0100	0.0257	0.0592	0.0000	Below
21	Wednesday	6	0.0300	0.0257	0.0592	0.0000	Below
22	Thursday	8	0.0400	0.0257	0.0592	0.0000	Below
23	Friday	9	0.0450	0.0257	0.0592	0.0000	Below
24	Saturday	10	0.0500	0.0257	0.0592	0.0000	Below
25	Monday	3	0.0150	0.0257	0.0592	0.0000	Below
26	Tuesday	1	0.0050	0.0257	0.0592	0.0000	Below
27	Wednesday	4	0.0200	0.0257	0.0592	0.0000	Below
28	Thursday	5	0.0250	0.0257	0.0592	0.0000	Below
29	Friday	7	0.0350	0.0257	0.0592	0.0000	Below
30	Saturday	10	0.0500	0.0257	0.0592	0.0000	Below
Mean			0.0257				

The control chart is in Figure 12.7.

Average fraction of samples defective, p, is 0.0257. This is the *CL*.

Average fraction of samples not defective, q, is 0.9743 (1 − 0.0257).

Standard deviation of sampling proportion is $\sqrt{(p \cdot q/n)} = \sqrt{(0.0257 * 0.9743/200)} = 0.0112$.

At 3σ limits, *LCL* is: $0.0257 − 3 * 0.0112 = −0.0079$. You cannot have negative control limits so the pragmatic value is 0.0.

At 3σ limits, *UCL* is: $0.0257 + 3 * 0.0112 = 0.0592$. (Slight difference between Excel and a calculator.)

As can be seen from the chart, or as given in the last column of Table 12.2, there are three points that are above the upper limit. Thus the process is out of control. The problem always occurs on a Saturday. It could be influenced by the student helpers who are not trained properly, though as time goes on the level of errors is falling.

2   At 1σ limits, *LCL* is: $0.0257 − 1 * 0.0112 = 0.0145$.

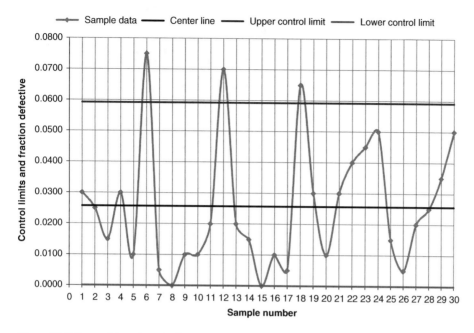

Figure 12.7 Record keeping: 3σ limits.

At 1σ limits, *UCL* is: $0.0257 + 1 * 0.0112 = 0.0369$.

This control chart with the revised lower and upper control limits is in Figure 12.8. Now there are seven data values above the upper control limit. Thus with this more rigorous control, there is additional evidence of an out of control situation.

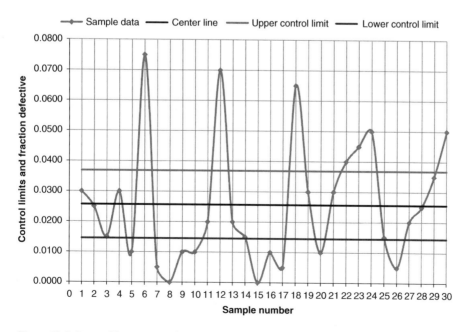

Figure 12.8 Record keeping: 1σ limits.

Count or c-chart

A c-chart is used where counting the number of errors is appropriate rather than a percentage value. For example, the number of imperfections on a 1 m^2 sample of fabric, the number of blemishes on given surface area of a bobbin of paper, the number of imperfections on a piece of wooden furniture, or the number of typing errors on one page of printing, etc. It is the Poisson distribution that is used as the basis for analysis.

For the count chart a sample n is taken and the number of errors in each sample is counted. The total number of errors is thus $\sum c$. The average number of errors is $\sum c^n$ or $\bar{c}$.

In the Poisson distribution the variance is equal to the mean value, $\bar{c}$, and since the standard deviation is always the square root of the variance then the standard deviation is $\sqrt{\bar{c}}$. The control limits in the c-chart are:

$$CL = \bar{c}$$

$$UCL = \bar{c} + z\sqrt{\bar{c}}$$

$$LCL = \bar{c} - z\sqrt{\bar{c}}$$

Here $\bar{c}$ the sample average value has been used. However, as in the p-chart we may have a target value, c, that is not directly related to the average value. The value z is the number of standard deviations. Again the smaller the value of z, the more rigorous are the specifications.

The development of a c-chart is illustrated by the following application case-exercise, *Curtain fabric*.

Application case-exercise: **Curtain fabric**

Situation

A distributor of curtain material, whose clients include owners of offices, hotels, restaurants, and other commercial establishments, currently buys fabric from Spain. It is now looking at a new supplier in China that has material less expensive than the fabric it currently uses. However, before it makes any decision on changing suppliers it wants to verify the quality of the fabric from the proposed Chinese supplier. The distributor asks a sourcing agent in China to perform a quality inspection on the fabric produced by the Chinese supplier. The sourcing agent selects at random 12 bobbins of fabric and from each of these 12 bobbins the agent inspects 10 m^2 of fabric and counts the number of blemishes. The imperfection criteria is a blemish of length of 1 mm. If a blemish is 2 mm in length then this would be considered two blemishes, a length of 3 mm three blemishes, etc. The sample information is given in Table 12.3 where the number of blemishes has been converted to the 1 mm criterion. You are to construct a c-control chart for this data where the x-axis is the number of the bobbin and the center line, benchmark, is the average number of 1 mm blemishes per m^2 based on the analysis of the 12 bobbins. The standard deviation for a c-chart is the square root of the average blemishes per m^2.

Required

1   To the nearest two decimal places, what is the value of the center line in equivalent 1 mm length blemishes per m^2?

Table 12.3

Bobbin No.	Equivalent 1 mm blemishes on 10 m² fabric	Bobbin No.	Equivalent 1 mm blemishes on 10 m² fabric	Bobbin No.	Equivalent 1 mm blemishes on 10 m² fabric
1	52	5	72	9	63
2	32	6	15	10	10
3	35	7	40	11	15
4	42	8	12	12	8

2   Determine the lower and upper limits for a 3σ control chart and plot the chart. How many of the bobbins have average blemishes above the upper control limit?

3   Develop the more rigorous 1σ control chart. What is the value of the upper control limit to the nearest two decimal places? From the 1σ control chart how many of the sample points are above the upper control limit? What are your conclusions?

Worked out solutions

1   The framework for when $z = 3$, from which the control is developed is in Table 12.4.

Table 12.4

Bobbin No.	Equivalent 1 mm blemishes on 10 m² fabric	Blemishes/m² on fabric	CL	UCL	LCL
1	52	5.2000	3.3000	8.7498	0.0000
2	32	3.2000	3.3000	8.7498	0.0000
3	35	3.5000	3.3000	8.7498	0.0000
4	42	4.2000	3.3000	8.7498	0.0000
5	72	7.2000	3.3000	8.7498	0.0000
6	15	1.5000	3.3000	8.7498	0.0000
7	40	4.0000	3.3000	8.7498	0.0000
8	12	1.2000	3.3000	8.7498	0.0000
9	63	6.3000	3.3000	8.7498	0.0000
10	10	1.0000	3.3000	8.7498	0.0000
11	15	1.5000	3.3000	8.7498	0.0000
12	8	0.8000	3.3000	8.7498	0.0000
Total	396	3.3000			

Total number of blemishes is 396. There are 12 bobbins so average blemish/bobbin is 396/12 = 33.0000. Surface area of fabric tested is 10 m². Thus average blemishes per m² is 33/10 = 3.3000. This is the center line of the control chart.

2   For the c-chart the standard deviation is $\sqrt{\text{(average value)}} = \sqrt{(3.3000)} = 1.8166$

- At 3σ limits, lower control limit = $3.3000 - 3 * 1.8166 = -2.1498$. Pragmatic lower level is 0 as you cannot have a negative limit.
- At 3σ limits, upper control limit = $3.3000 + 3 * 1.8166 = 8.7498$.

The control chart is in Figure 12.9. There are no data points above the upper control limit.

3   For the c-chart the standard deviation is $\sqrt{\text{(average value)}} = \sqrt{(3.3000)} = 1.8166$

- At 1σ limits, lower control limit = $3.3000 - 1 * 1.8166 = 1.4834$.
- At 1σ limits, upper control limit = $3.3000 + 1 * 1.8166 = 5.1166$.

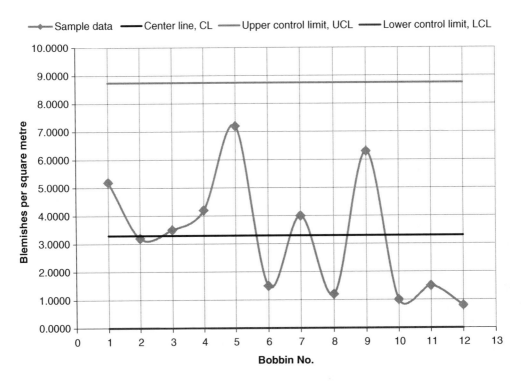

Figure 12.9 Curtain fabric: 3σ control chart (c-chart).

The control chart is in Figure 12.10. There are three data points above the upper control limit. Thus for this more rigorous requirement indications are that the process is out of control.

x-bar and range chart for variables

Variables are those characteristics that can take on a range of values such as kg, decibels, meters, hours, liters, km/hour, gallons, etc. In quality control for variables, random samples are withdrawn from populations of products and are described by the sample mean, $\bar{x}$, and the sample range, R. The sample mean is the average of all of the values in a specific sample and the sample range is the difference between the greatest and smallest value in the sample. For correct quality control of products, an $\bar{x}$ (x-bar) chart and an R-chart need to be used together. Consider a chocolate manufacturer who purchases sacks of sugar by truckload from a local supplier. The weight of each sack is indicated as 25 kg. From a supplied lot, a sample of ten sacks is removed and weighed. These weights in kg are given in Table 12.5.

The average weight of these ten samples, $\bar{x}$ is then:

$$\frac{21 + 24 + 23 + 29 + 25 + 27 + 28 + 25 + 22 + 26}{10} = 25.00 \text{ kg}$$

That is, according to specifications. However, the range is 8 kg (29 − 21) or 32% of the average value (8/25), which is high, thus indicating a problem. By having dual monitoring, with both an x-bar and range chart, this controls both the average values and the variation of values from their means. It

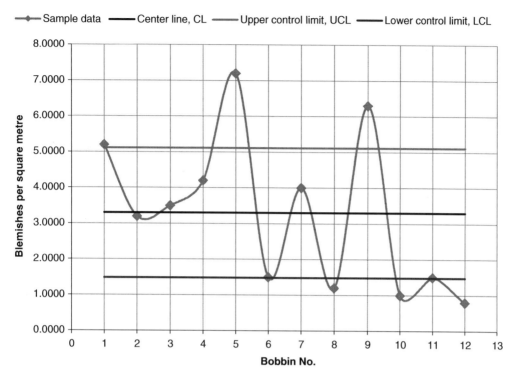

Figure 12.10 Curtain fabric: 1σ control chart (c–chart).

Table 12.5

Sack No.	1	2	3	4	5	6	7	8	9	10
Weight, kg	21	24	23	29	25	27	28	25	22	26

cannot be concluded that a process is in control by just monitoring sample means but the variation or range within a sample must also be measured.

The control limits in the *x*-bar chart are:

$CL = \bar{\bar{x}}$ or the average of all the sample averages

$UCL = \bar{\bar{x}} + z\sigma_{\bar{x}}$

$LCL = \bar{\bar{x}} - z\sigma_{\bar{x}}$

If the standard deviation is unknown, or difficult to determine, it can be replaced with the average range of values, $\bar{R}$ (R-bar) since the range is also a measure of deviation. In this case the control limits are:

$UCL = \bar{\bar{x}} + A\bar{R}$

LCr

Range or R–chart, the control limits are:

$$CL = \bar{R}$$
$$LCL = B\bar{R}$$
$$UCL = C\bar{R}$$

The terms A, B, and C are the control chart factors for variables and are given in Table 12.6.

Table 12.6

Sample size n	Factor A for sample means	Factor B for sample ranges	Factor C for sample ranges
2	1.880	0.000	3.269
3	1.023	0.000	2.574
4	0.729	0.000	2.282
5	0.577	0.000	2.114
6	0.483	0.000	2.004
7	0.419	0.076	1.924
8	0.373	0.136	1.864
9	0.337	0.184	1.816
10	0.308	0.223	1.777
11	0.285	0.256	1.744
12	0.266	0.283	1.717
13	0.249	0.308	1.692
14	0.235	0.328	1.672
15	0.223	0.347	1.653
16	0.212	0.363	1.637
17	0.203	0.378	1.622
18	0.194	0.391	1.609
19	0.187	0.403	1.597
20	0.180	0.414	1.586
21	0.173	0.425	1.575
22	0.167	0.434	1.566
23	0.162	0.443	1.557
24	0.157	0.452	1.548
25	0.153	0.460	1.540
>25	$0.751/\sqrt{n}$	$0.450 + 0.001n$	$1.550 - 0.001n$

For sample sizes greater than $n = 25$, values A, B, and C are linear approximations according to n.

They were determined by statistical work at the former Bell laboratory of AT & T.[3] The factors are a function of sample size, n. For example, in a process control experiment where the sample size is 15, the value of A is 0.223; B 0.347; and C 1.653.

The development of an x-bar chart and a range chart, using the control chart factors, is illustrated by the following application case-exercise, *Candy Company*.

Application case-exercise: Candy Company

Situation

The Candy Co. outside of Berlin, Germany manufactures a large variety of candies. One of its principal products is Gumbo Bears. (Jelly candies in the shape of the German Bear.) After production, the Gumbo Bears are cooled and packed in 250, 500, and 1,000 g plastic bags. The Gumbo Bears are fed automatically into the bags, which are then sealed by a hot press. Lately, the manager of production has noticed that there have been problems with the hot press sealer and he wonders if this has something to do with the filling operation. Particularly this has occurred with the 500 g bags. The company decided to investigate the operation of the automatic filling machine for the 500 g bags and asked Michael Grand to undertake the analysis and verify the settings. The filling machine was first shut down and checked to see if was operating properly. This was just a quick check, and no extensive adjustments were made. Next, Michael randomly sampled the bags of Gumbo Bears and weighed the contents. Sample sizes were 20 bags each, and in total 200 samples were taken. The average of the 200 sample means was 502.50 g. The average of the sample range was 24.20 g. Michael used this information to construct control charts for sample means and sample ranges. The following week Michael, together with his assistant, Anne Box, carried out a detailed analysis of the filling machine. Each hour, they took random samples of the 500 g bags of Gumbo Bears. Michael took samples for the first 8 hours, and Anne took samples for the next 8. Thus in total they took 16 samples. As before, there were 20 bags in each sample. The mean weight of the sample was determined, and also the heaviest and the lightest bag in each sample were recorded. This data is in Table 12.7.

Table 12.7

Sample No.	Sample Average	Maximum Weight	Minimum Weight	Sample No.	Sample Average	Maximum Weight	Minimum Weight
1	504.00	508.00	499.20	9	498.20	509.60	499.20
2	502.00	510.00	485.50	10	499.80	507.60	499.60
3	501.00	512.60	496.50	11	500.90	503.20	498.40
4	501.30	514.90	498.60	12	501.30	514.60	486.50
5	499.20	520.20	487.20	13	501.90	521.00	498.60
6	501.00	504.90	489.50	14	501.70	521.20	485.60
7	502.10	508.60	489.60	15	502.10	521.90	475.60
8	499.20	512.50	475.60	16	503.20	512.80	498.20

Required

1 Develop a control chart for the sample mean and the sample range. Plot the data for the 16 samples from Table 12.7 on the corresponding control charts.
2 What are your conclusions about the automatic filling machine?

Worked-out-solutions

1 The sample size, *n*, is 20. From Table 12.6, for control chart factors for variables, the constants A, B, and C are according to Table 12.8.

Table 12.8

Sample size n	Factor A For sample means	Factor B For sample ranges	Factor C For sample ranges
20	0.180	0.414	1.586

Using these values and the test value for the sample mean of 502.50 g and for the test value of 24.20 g for sample range we calculate control limits for the average and range charts.

For the mean x-bar chart:

$$UCL = \bar{\bar{x}} + A\bar{R}$$

$$LCL = \bar{\bar{x}} - A\bar{R}$$

- CL is the average value of 502.50 g.
- UCL is: $502.50 + 0.180 * 24.20 = 506.86$ g.
- LCL is: $502.50 - 0.180 * 24.20 = 498.14$ g.

For the range chart:

$$CL = \bar{R}$$

$$LCL = B\bar{R}$$

$$UCL = C\bar{R}$$

- CL is the average value of 24.20 g.
- LCL is: $0.414 * 24.20 = 10.02$ g.
- UCL is: $1.586 * 24.20 = 38.38$ g.

The complete data from which the control charts are developed is in Table 12.9. The x-bar chart is given in Figure 12.11 and the range chart in Figure 12.12.

Table 12.9

No.	Mean g	Max wt. g	Min wt. g	Range G	x-bar chart CL	x-bar chart LCL	x-bar chart UCL	Range chart CL	Range chart LCL	Range chart UCL
1	504.00	508.00	499.20	8.80	502.50	498.14	506.86	24.20	10.02	38.38
2	502.00	510.00	485.50	24.50	502.50	498.14	506.86	24.20	10.02	38.38
3	501.00	512.60	496.50	16.10	502.50	498.14	506.86	24.20	10.02	38.38
4	501.30	514.90	498.60	16.30	502.50	498.14	506.86	24.20	10.02	38.38
5	499.20	520.20	487.20	33.00	502.50	498.14	506.86	24.20	10.02	38.38
6	501.00	504.90	489.50	15.40	502.50	498.14	506.86	24.20	10.02	38.38
7	502.10	508.60	489.60	19.00	502.50	498.14	506.86	24.20	10.02	38.38
8	499.20	512.50	475.60	36.90	502.50	498.14	506.86	24.20	10.02	38.38
9	498.20	509.60	499.20	10.40	502.50	498.14	506.86	24.20	10.02	38.38
10	499.80	507.60	499.60	8.00	502.50	498.14	506.86	24.20	10.02	38.38
11	500.90	503.20	498.40	4.80	502.50	498.14	506.86	24.20	10.02	38.38
12	501.30	514.60	486.50	28.10	502.50	498.14	506.86	24.20	10.02	38.38
13	501.90	521.00	498.60	22.40	502.50	498.14	506.86	24.20	10.02	38.38
14	501.70	521.20	485.60	35.60	502.50	498.14	506.86	24.20	10.02	38.38
15	502.10	521.90	475.60	46.30	502.50	498.14	506.86	24.20	10.02	38.38
16	503.20	512.80	498.20	14.60	502.50	498.14	506.86	24.20	10.02	38.38

2   The x-bar chart data is in control though there is a downward trend in the first 8 hours and an upward trend in the second 8 hours. Also, much of the data is below the center line. This should be investigated as something has changed between the control experiment and test experiment. For the range chart it is essentially in control with only one sample above the limit. This may be a measurement error. Again there are over 50% of the samples below the center line and even three below the lower limit. This underscores that the test work should be investigated. It would

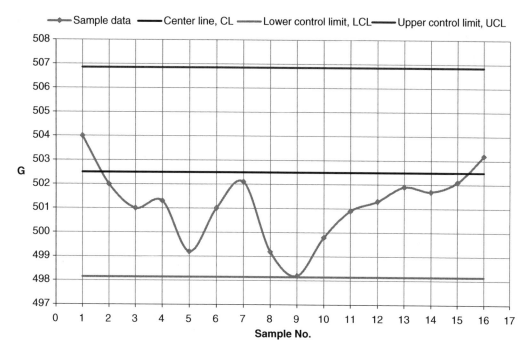

Figure 12.11 Candy Company: *x*-bar control chart.

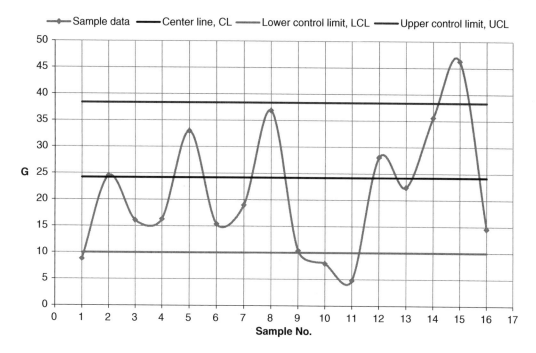

Figure 12.12 Candy Company: range chart.

be more expedient to have online sampling with the process control equipment connected directly into the operating process. Measurement would be instantaneous.

Sigma limits of control charts

In SPC the sample distribution for the variable being examined is considered to follow a normal distribution. Historically the control limits for the process have been at three-sigma (3σ) limits. However, as the application case-exercises have illustrated, other limits can be used such as two-sigma (2σ), one-sigma (1σ) or even non-integer values of sigma.

Normal distribution

The relationship between the value of control limits and the sampling distribution is illustrated in Figure 12.13. Here the sigma limits from -4σ to $+4\sigma$ are illustrated. Remember that for a normal distribution, as presented in *Chapter 5*:

- 68.26% of all data lies between $\pm 1\sigma$
- 95.44% of all data lies between $\pm 2\sigma$
- 99.74% of all data lies between $\pm 3\sigma$

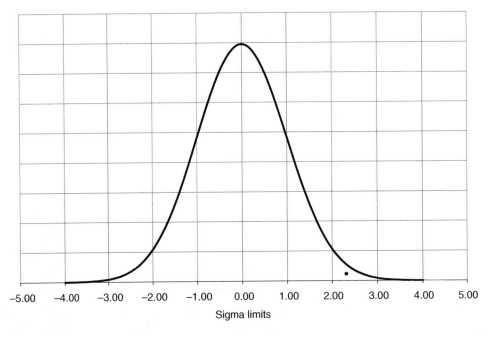

Figure 12.13 Control limits related to the normal distribution.

If for example $\pm 3\sigma$ limits are being used and all the collected sample data lies between these limits the process will be in control even though there are inherent or random variations occurring that explain the spread of data. If some sample data lies outside of this $\pm 3\sigma$ limit range, then the process would be exhibiting assignable causes of variation and is considered out of control. In interpreting sample data in control charts a Type I error or a Type II error might be encountered.

Type I error

A Type I error occurs when a process is considered out of control when in reality this is not the case. Assume that a given process is in control and that ± 3σ limits are used to set the boundaries. In Figure 12.13 a sample data point illustrated by ● is shown. Since this data point falls inside (to the left) of the +3σ limit the process would be considered, quite correctly, in control. However, now assume that ±2σ limits are used: then the data point is now outside (to the right) of the +2σ limit. Thus, the assumption would be that the process is out of control but it is known that this is not true. The conclusion is that the tighter the control limits, or the smaller the number of standard deviations, the higher is the probability, or **risk** of making a Type I error that a process is considered out of control when this is in fact not true.

Type II error

A Type II error occurs when a process is considered in control when this is not the case. Assume that a process is out of control and that again ± 3σ control limits are used, as illustrated in Figure 12.14. Consider the sample data point shown relative to the expected operating conditions shown by the left–hand normal distribution curve. This sample data point is at about a sigma value of 4.75 or beyond the 3σ limit indicating that the process is out of control, which is in fact the case. Now if the process is really out of control the distribution will move. Suppose it has shifted to the right as illustrated by the right–hand normal distribution curve. The "real" sigma limits of the right–hand distribution are indicated by the first row of Table 12.10. These correspond to the Figure 12.14 sigma limits by the second row of Table 12.10.

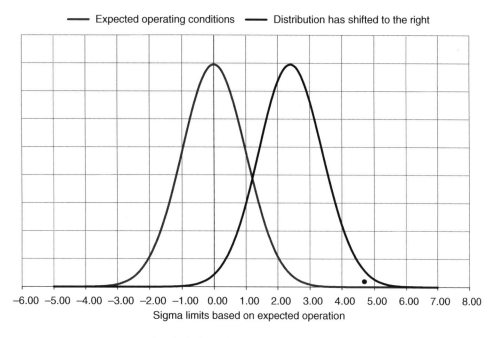

—— Expected operating conditions   —— Distribution has shifted to the right

Sigma limits based on expected operation

Figure 12.14 Target average has shifted: a risk of making a Type II error.

Table 12.10

Sigma limits with shift	−4.00	−3.00	−2.00	−1.00	0.00	+1.00	+2.00	+3.00	+4.00
z-limits shown		−1.50	−0.50	+0.50	+1.50	+2.50	+3.50	+4.50	+5.50	+6.50

Now the indicated data point, originally at z of 4.75, falls between $+2.00\sigma$ and $+3.00\sigma$ limits or within the 3σ limits of this new distribution curve. The assumption then would be that the process is in control when this is not in fact the case. However, if tighter 2σ control limits are used then the point originally at $z = 4.75$ is now beyond the 2σ limit. Thus the process would now be considered out of control, which is in fact true. Thus the conclusion is that the tighter the control limits, or the smaller the number of standard deviations, z, the lower is the probability or risk of making a Type II error or the risk of concluding a process is in control, when in fact this is not the case.

In the pharmaceutical industry, such as the production of vaccine, the control limits would be tight in order to avoid making a Type II error, as it would be too risky and the costs too high to have a poor quality vaccine on the market. A similar argument goes for food processing and other industries when toxicity is a consideration. The relationship with the control limits, the normal distribution, and out of control conditions are illustrated in Figure 12.15.

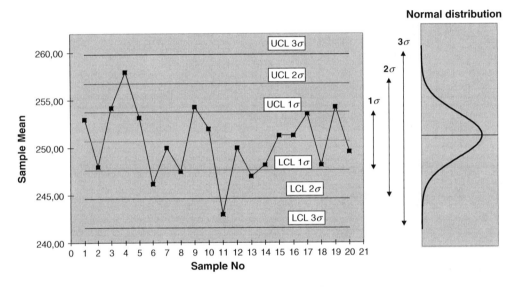

Figure 12.15 Out-of-control situation depends on sigma limits.

As in all analytical work using statistics there are no **guarantees**. Even if all observations are within limits, it does not guarantee that assignable variations are not present. That means that the process may be out of control even though the chart indicates otherwise. Even if some observations are outside the control limits, this does not guarantee that assignable variations are present. That means that the process may be in control even though the chart indicates otherwise. If 3σ limits are used, 99.74% of sample means should be within limits, and 0.26% outside limits when only random variations are present. If 2σ limits are used, 95.44% of sample means should be within limits and 4.56% outside limits when only random variations are present. If 1σ limits are used, 68.26% of sample means should be within limits and 31.74% outside limits when only random variations are present. Using 3σ limits will reduce the a risk of concluding that a process is out of control when only random variations account for points outside the control limits. However, wider limits make it more difficult to detect non-random variations when they are present.

In six-sigma management the goal is to hit the target level to arrive at no more than 3.4 defects per million (*Chapter 13*). This underscores the criteria of the Japanese Genichi **Taguchi**

(1924–2012) who believed that quality level should be measured by the deviation from the target level. This should be as small as possible and implies tight specification levels, as illustrated in Figure 12.16.

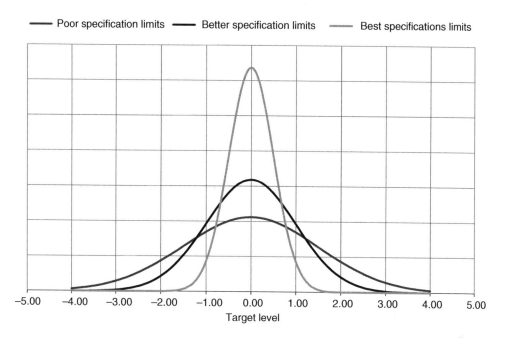

Figure 12.16 Taguchi criteria and control limits.

Here the slender, leptokurtic curve, with a sharp peak and a low distribution is the target distribution for six-sigma. The broad platykurtic curve with a large deviation is inappropriate. The mesokurtic curve within an intermediate standard deviation is better but still not a six-sigma requirement. Taguchi's concept is that the system should be correctly designed to make it insensitive to random variations. In order to achieve tighter limits the quality level must be built into the design of the process, product, or service and levels must not rely upon the old idea of inspection and rework. In the service industries this implies having personnel qualified for the work with the availability of continuous training programs.

Statistical process control in the medical field

Statistical process control is an important tool in the medical field to monitor the health of patients. Many medical problems cannot be solved instantaneously and the patient's conditions are monitored over time. Three such procedures are for measuring body mass index, blood pressure, and blood fluidity.

Body mass index

The body mass index (BMI) is a measure of a person's body size and calculated by a weight in kilograms divided by the square of height in meters. A person who is 1.80 m and weighs 117 kg would have a body mass index of 36.1 ($117/1.80^2$). At this level the person would be considered severely obese according to Table 12.11[4]. Obesity is usually a result of a sedentary lifestyle, insufficient exercise and/or a high caloric consumption. see *Icebreaker, Chapter 9*.

Table 12.11

Category	BMI range (kg/m²)
Very severely underweight (VSU)	Less than 15.0
Severely underweight (SU)	15.0 to 16.0
Underweight (U)	16.0 to 18.5
Normal, healthy weight (N)	18.5 to 25.0
Overweight (O)	25.0 to 30.0
Obese Class I, moderately overweight (MO)	30.0 to 35.0
Obese Class II, severely overweight (SO)	35.0 to 40.0
Obese Class III, very severely overweight (VSO)	Greater than 40

BMI has taken an importance in medicine because being severely overweight, or obese, leads to diabetes, heart, and other vascular problems. In the USA and Saudi Arabia, 33% of the population is obese, with the UK and Australia not far behind with 27% of the population dangerously fat. Nutritionists warn that obesity-related diseases like heart disease and diabetes are now the world's biggest killers.[5] Depending on the severity of the overweight condition, treatment consists of dieting, an exercise regime, or even perhaps surgery to reduce the food intake to the stomach.

Figure 12.17 gives a statistical process control chart of three hypothetical individuals, John, 1.70 m; Susan, 1.80 m; and Thomas, 1.90 m; going through a weight reduction program over a 26-week period (6 months).

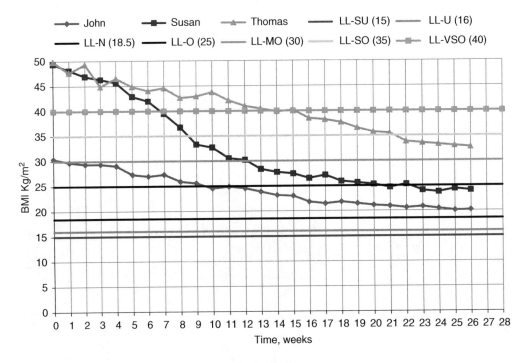

Figure 12.17 Body mass index values for three individuals.

The following are the control limits according to the category and BMI ranges of Table 12.11.

- (LL–SU) is the lower control limit for severely underweight: 15.0 kg/m^2
- (LL–U) is the lower control limit for underweight: 16.0 kg/m^2. It is also the upper control limit for severely underweight (SU).
- (LL–N) is the lower control limit for normal healthy weight: 18.5 kg/m^2 It is also the upper control limit for underweight (U).
- (LL–O) is the lower control limit for overweight: 25.0 kg/m^2. It is also the upper control limit for normal healthy weight (N).
- (LL–MO) is the lower control limit for obese Class I: 30.0 kg/m^2. It is also the upper control limit for overweight (O).
- (LL–SO) is the lower control limit for obese Class II: 35.0 kg/m^2. It is also the upper control limit for obese Class I (MO).
- (LL–VSO) is the lower control limit for obese Class III: 40.0 kg/m^2.

At the beginning John is on the borderline of being overweight to moderately obese and is able to bring his BMI down to a normal level at the end of 6 months; Susan, who is very severely obese at the beginning, is also able to bring her BMI level to normal at the end of 6 months; Thomas, however, who at the start is also very severely obese, is only able to bring his BMI down to a level of moderately obese. On this basis he should continue with his weight reducing program.

Blood pressure and heart rate

Blood pressure is the force of the blood against the walls of the arteries. It is stated by two numbers as for example 120/80. The first number, the higher or systolic pressure, is the measure while the heart contracts to pump blood to the body. The lower, or diastolic pressure is a measure when the heart relaxes between beats. According to guidelines by the World Health Organization (WHO) Table 12.12 gives qualifications of blood pressure levels in mm mercury. Blood pressure levels are important as high blood pressure can lead to coronary or other vascular problems. Low blood pressure can lead to dizziness or fainting.

Table 12.12 Blood pressure levels

Category	Systolic pressure (mm mercury)	Diastolic pressure (mm mercury)
Normal	120	80
Borderline	130	85
Mild hypertension (Level I)	140	90
Moderate hypertension (Level II)	160	100
Severe hypertension (Level III)	180	130

Related to blood pressure is the heart pulse rate. The normal pulse rate is around 70 beats/minute though it is lower for athletic individuals. Figure 12.18 gives a statistical process control chart for blood pressure and heart pulse rate of a patient in early 2014.

Compared to the normal control limits of 120 mm mercury for the systolic pressure and 80 mm mercury for the diastolic blood pressures indications are that the patient is borderline of a mild hypotension, Level I. The pulse rate is relatively low suggesting the patient is physically active. Blood pressure levels can vary according to the time of day the measurements are taken and to such factors

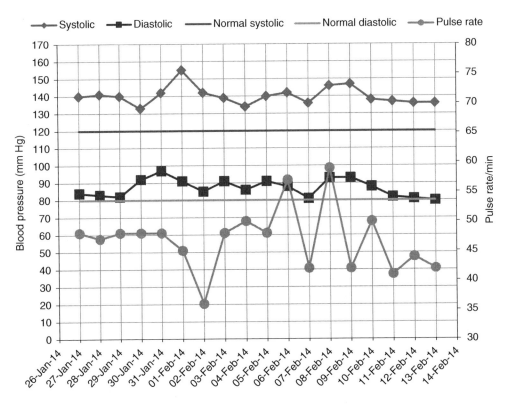

Figure 12.18 Blood pressure and pulse rate.

as whether the person has been exercising, eating, smoking, under stress, drinking alcohol, and other. Like all statistical process control studies measurements should be taken over a sufficient time period to be meaningful and to indicate whether there are trends.

Blood fluidity levels and the International Normalized Ratio

Figure 12.19 is a statistical process control chart used to test the fluidity of blood in a patient based on the International Normalized Ratio, INR, also in early 2014.

The more fluid in the blood, the less is the risk of having thrombosis or blood clots, but the greater is the risk of external or internal hemorrhaging. The reverse is true the less fluid the blood. The control limits for the INR are between 2.0 and 3.0 as indicated in Figure 12.19 and this range is obtained by a patient taking anticoagulants such as warfarine, fluindione, or other appropriate medicines. The anticoagulants act by reducing the level of vitamin K which is a blood clotting agent. Vitamin K is present in vegetables such as broccoli, asparagus, spinach, cabbage, cauliflower, and Brussels sprouts. Thus a patient should limit the consumption of these and also drink alcohol in moderation.

Sample control charts

Even though sample data may fall within the prescribe specification limits, indicating an in control situation, the format or trend of the curves should also be considered. The following illustrates.

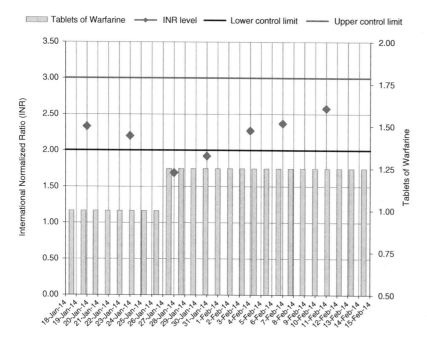

Figure 12.19 Anticoagulant by AVK (antivitamin K).

Data hugs the center line

In Figure 12.20 the sample data hugs the center line and any variation is random. Here the process is in control and the system is conforming to the Taguchi expectations of quality control though perhaps in this case the control limits could be narrower.

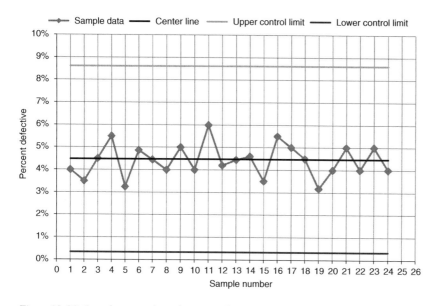

Figure 12.20 Sample means hug the center line: in control.

Wide variations

In Figure 12.21 although the sample data is within the control limits there are wide variations. This is symptomatic of an erratic operation and may be an indicator of assignable variations.

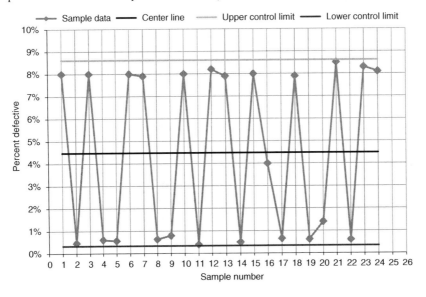

Figure 12.21 Sample means fluctuate widely: in control but erratic operation.

Trend in the data

Figure 12.22 indicates that though the process is in control there is an upward trend that is pointing to a soon to be out of control situation. Again there are probably assignable causes and this should be investigated.

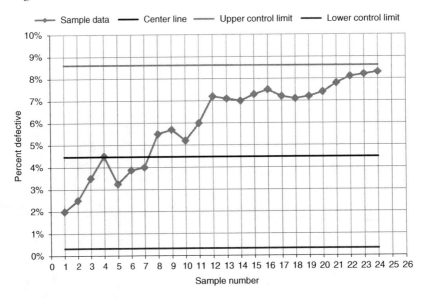

Figure 12.22 Sample means indicate a trend: eventually out of control?

Abrupt changes

Figure 12.23 shows an abrupt change in performance although sample data points are within control limits. At first sample data is between the center line and the upper control limit and then suddenly the performance of the process improved and data points begin to hug the lower limit, indicating assignable reasons.

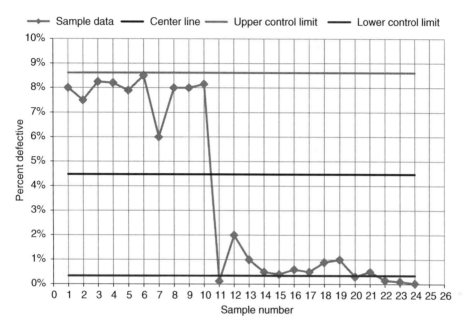

Figure 12.23 Sudden change in sample means: in control but unusual.

Out of control

Figure 12.24 shows that from the beginning, the process is out of control and so should be investigated.

Online statistical process control

With advancement of technology, many production firms have online statistical process control systems continuously measuring data. In filling of beverage cans, yogurt pots, toothpaste, molding of chocolate bars etc., each unit is automatically weighed, and this weight is incorporated into the average weight of all units measured since the operation started. The average weight, the range, and a graphical distribution of all the units are displayed on a computer screen adjacent to the process line or a control room. The operator sees directly, in real time if the process is operating correctly and if it is not can take necessary action. Prior to this technology an operator had to take samples from the production line, make the appropriate measurement, sometimes in a laboratory away from the production area, and then plot this information on the control chart. If the process was indicated to be out of control then appropriate action was taken. During the time that the analysis was being carried out either the process was still running, and perhaps producing faulty units, or the production line was stopped waiting for feedback regarding the quality test. In this case, valuable production time was lost.

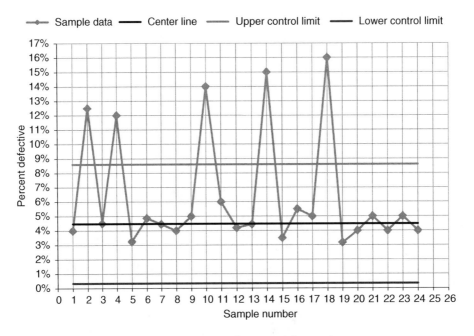

Figure 12.24 Five sample means above upper limit: out of control.

In hotel and food services, online statistical process control in the same manner is not feasible though the information system can give real-time instant information on revenues, rooms sold, etc.

Moving on

This chapter has covered statistical process control, an essential part of operations management. The next and final chapter deals with six-sigma management whose roots are in statistics and which makes use of statistical process control.

Notes

1 "Was luxurious Titanic made on the cheap?" *International Herald Tribune*, April 15, 2008, p. 2.
2 *The Times*, supplement, April 1998 http://www.the-times.co.uk.
3 *Economics Control of Manufactured Products*, Litton Educational Publishing, Van Nostrand Reinhold Co. 1931 (Bell Telephone Laboratories).
4 "Global database on body mass index." *World Health Organization*, 2006.
5 www.dailymail.co.uk/health/article-2920219/How-fat-country-nations-highest-obesity-rates-new-maps-surprise-you.html [accessed November 6, 2015].

13 Six-sigma management

ICEBREAKER: NASA, where was six-sigma?

On September 25, 1992 NASA (National Aeronautical and Space Administration) launched the *Mars Observer Spacecraft*, also known as the *Mars Geoscience/Climatology Orbiter*. It was a 1,018 kg (2,244 lb) robotic space probe designed to study the Martian surface, atmosphere, climate and magnetic field. During the interplanetary cruise phase, communication with the spacecraft was lost on August 21, 1993, three days prior to orbital insertion. Attempts to re-establish communication with the spacecraft were unsuccessful.[1]

After the loss of the *Mars Observer* and with rising costs associated with the future International Space Station, NASA began seeking less expensive, smaller probes for scientific interplanetary missions. In 1994, the Panel on Small Spacecraft Technology was established to set guidelines for future miniature spacecraft. The panel determined that the new line of miniature spacecraft should be under 1,000 kg with highly-focused instrumentation. In 1995, a new Mars Surveyor program began as a set of missions designed with limited objectives, low

Figure 13.0 Mars climate orbiter.

costs, and frequent launches. The primary science objectives of the new mission included: to determine the distribution of water on Mars; monitor the daily weather and atmospheric conditions; record changes on the Martian surface due to wind and other atmospheric effects; determine temperature profiles of the atmosphere; monitor the water vapor and dust content of the atmosphere; and look for evidence of past climate change.

The *Mars Climate Orbiter* space probe measured 2.1 m tall, 1.6 m wide and 2 m deep and weighed 338 kg (750 lb). The internal structure was largely constructed with graphite composite/aluminum honeycomb supports, a design found in many commercial airplanes. With the exception of the scientific instruments, battery and main engine, the spacecraft included dual redundancy on the most important systems. The space probe was powered with a three-panel solar array providing an average of 500 W at Mars. Deployed, the solar array measured 5.5 m in length. Power was stored in 12-cell, 16 Ah nickel-hydrogen batteries. These were intended to be recharged when the solar array received sunlight and thus power the spacecraft as it passed into the shadow of Mars. When entering into orbit around Mars, the solar array was to be utilized in the aero braking to slow the spacecraft until a circular orbit was achieved. The design was largely adapted from guidelines from the small spacecraft technology Initiative outlined in the book *Technology for Small Spacecraft*.

The *Mars Climate Orbiter* was launched by NASA on December 11, 1998 to study the Martian climate atmosphere and surface changes and to transmit information. However, on September 23, 1999, communication with the spacecraft was lost as it was going into orbital insertion. This was due to ground-based computer software that produced output in pound-seconds, instead of the metric units of newton-seconds specified in the contract between NASA and Lockheed. The spacecraft encountered Mars on a trajectory that brought it too close to the planet, causing it to pass through the upper atmosphere and apparently disintegrate or orbiting the sun.[2]

NASA announced that confusion over whether measurements were metric or the British (Imperial) system led to the loss of this $125 million spacecraft as it approached Mars. Lockheed Martin, who built the spacecraft, specified certain measurements in the British system, but NASA scientists thought the measurements were in metric units. The resulting misunderstanding, which went undetected for months during the period that the craft was designed, built, and launched meant that the *Mars Climate Orbiter* was off course by some 97 km (60 miles) as it approached its target destination after a journey of some 670 million km (416 million miles). Besides the enormous costs, the findings of this simple calculation error was a major embarrassment for both Lockheed Martin in Denver Colorado and NASA's Jet Propulsion Laboratory in Pasadena, California, who were in charge of the mission. The issue was not that the design data was wrong but the processes employed did not uncover the discrepancies and make the appropriate corrections. If six-sigma management, with its rigorous checks and verifications, had been appropriately employed, this gaffe would never have occurred.[3]

Chapter subjects

✓ **Elements of six-sigma** • History • Statistical roots • Product and process specifications
✓ **Process capability** • Voice of process and voice of customer • Process capability index • Standard deviations and the z value • Six-sigma performance
✓ **Short- and long-term variation** • Variation • Process shift • Short-term sigma limits and long-term defects per million • Defects per unit
✓ **Opportunities in the product, process, or service** • Operations • Converting defects into long-term sigma capability • First-time yield and the hidden activity • Rolled throughput yield and normalized yield • Cycle time, performance level, and the hidden activity

---

Six-sigma management was conceived in the 1980s, in the same decade as many of the quality management concepts, e.g. lean operations and just-in-time, all of which play a significant role in six-sigma management, which is a rigorous approach to business. As the name suggest, the roots of this technique is statistics.

Elements of six-sigma

The ultimate focus of six-sigma management is on **customer requirements**: what they need; when they need it; and the price they are willing to pay. Thus the firm pays attention to the **voice of the customer** (VOC) as it is this that ultimately drives all the **business processes**. In responding to the customer, the firm also constantly focuses on internal business financial measures including **return on investment (ROI)** and **operating margins.**

History

The "father" of six-sigma was **Bill Smith** (1929–1993) who graduated from the US Naval Academy in 1952 and studied at the then Minnesota University School of Management (now Carlson School of Management). Smith worked for some 35 years in engineering and quality assurance before joining Motorola where he was vice president and senior quality assurance manager for the Land Mobile Products Sector. It was with Motorola that in 1986 Smith formulated the **six-sigma (6σ)** approach that encompasses many of the statistical methods presented in this textbook: check sheet, frequency distributions, Pareto, radar diagram, correlation, sampling, and statistical process control. Six-sigma is a way to systematically improve **processes** by eliminating defects or non-conformity in a process, product, or service. **Mikel Harry** worked closely with Bill Smith at Motorola and demonstrated how to transform the six-sigma theory into practical solutions. The six-sigma approach generated considerable profits for Motorola.[4] Historically six-sigma was a management tool developed for **manufacturing** and first adopted by Motorola and then later by other manufacturers including Honeywell, General Electric, Caterpillar, ABB, Ford, and Stanley Black and Decker. Now the six-sigma management approach is being applied to **hotel and food service management** including Starwood Hotels, Marriott International, Hilton, and Ritz-Carlton. In addition, **financial services** such as American Express, Bank of America, Chase JP Morgan, AIG Insurance also employ six-sigma.

Bill Smith did not invent six-sigma but what he did was to apply the numerous work methods and quality concepts developed during the twentieth century, formerly used in manufacturing shop floor quality assurance, into a combined global tool for management excellence. The backbone of six-sigma is the Deming Cycle with its Plan, Do, Check, and Act procedures. Dr. W. Edwards Deming, who died in 1994, was by training a statistician, a professor at New York University, and a consultant to industry. After World War II he was invited by the Japanese government to Japan to help improve quality of Japanese products. He worked extensively with Japanese industries, and was so successful that the government established the annual Deming Prize for innovation in quality management.[5] Within this Deming Cycle are applications of Total Quality Management (TQM); the 5S Rules—Seiro, remove; Seiton, organize; Seiso, keep clean; Seiketsu, standardize; and Shitsuke, respect the rules; the five and "six" zeros of operations: zero breakdowns, zero defects, zero delays, zero inventory, zero paper, and because of these five, "zero" accidents. Six-sigma also includes Kaizen, the continuous improvement philosophy, lean production, and just-in-time, an Ishikawa study, and even Winslow Taylor's scientific management rigor. Six-sigma is the means to realize the values and philosophy associated with all these theories using **statistics** as the measuring tool. Many of the individual analytical tools, when used together, can show greater overall business improvement. The analogy is equivalent to saying that the music from an orchestra is more pleasurable than individual sounds from a violin, piano, oboe, harp, or drum.

Statistical roots

The roots of six-sigma are in statistical analysis. The sigma (σ) is the Greek letter s and refers to the standard deviation or variation indicating how far a given process, product, or service deviates from the specification, mean, μ, or target level. The higher the value of σ then the greater is the deviation from the target. The sigma deviation is the "defect" in the process, product or service. If you can measure how many defects you have then the logic is that you can systematically determine how to eliminate these defects and thus approach near-perfect quality. To achieve six-sigma quality a process must produce no more than 3.4 non-conforming units per million produced, or 0.00034%. For example, in a restaurant if there are 16 errors in the serving of 240 meals this would only be a 3σ operation. To achieve 6σ the restaurant should only be making 16 errors for every 5,333,000 meals served! An accounting firm that was experiencing 3 errors for every 45 invoices would be only operating at 3σ. To be at 6σ would mean only 3 errors per 1,000,000 invoices! Similarly a packaging firm that makes 5 errors for every 75 orders would only be operating at 3σ. To be at 6σ means to make only 1 error per 300,000 orders! Clearly to arrive at these levels means a new way of thinking; a new way of working; and a new way of measuring. This implies searching for **breakthroughs** or new ideas, which is what six-sigma is all about.

Product and process specifications

A product may be **tangible** such as a meal, an automobile, a pair of shoes, etc. or the "product" might be **intangible** such as a weekend stay in a spa-resort, a flight from London to Los Angeles, an MBA degree program, etc. If its intangible it is referred to as a **service**. Whatever the product it has specifications that may be stated in terms of dimension, weight, volume, a time-frame, price, cost, color, etc. These specifications are often fixed and may be set by the **supplier**, the **customer**, or both—a banquet event; a new hotel; a five-course meal; a hotel room ready for resale; a tailored suit, a revenue target, an automobile etc. The process is the way something is done that produces this "product" using resources such as labor, materials, machines, equipment, time, space, information technology, or money. This process also has a specification: the time and budget to construct an office building, to prepare a hotel banquet, to develop a travel itinerary; the budget for raw materials used in restaurant, development time, and budget for designing a computer program. The process and product are two different entities though closely related. However, the **product specification** is independent of the **process specification**. In six-sigma management the objective is to ensure that the process specifications are at all times at least equal but preferably greater than the product specification such that perfect "products" are always produced. The purpose of **statistical process control** *(Chapter 12)* is to monitor the distribution of a process to verify that it remains stable in terms of the process target value and variation around this target. Data that exceeds the control limits on the process control chart indicate when the mean or variability has changed. However, even if a process is in statistical control it may not necessarily always be producing products or services according to the "product" specifications because the control limits of the process are based on the mean and variation in the sampling distribution of the process and not on the mean and variation of "product" design specifications.

Process capability

The process capability is the ability of the process to consistently meet the design specifications of the "product." Here then are two interrelated elements—the **voice of the process** or the internal operational activity: what you manage and control; and the voice of the customer or the external requirements demanded or expected by the market.

Voice of process and voice of customer

The voice of the process must be at least equal to the voice of the customer. As illustrations:

- An airplane offers a choice of beef and fish on its dinner menu. Before departure the airplane loads up on 100 meals of fish and 120 meals of beef for its 220 passengers. During the flight, 130 passengers ask for fish. The process is not capable.
- A tour group wants to book 50 rooms in the hotel for a certain period. There are only 45 vacant rooms. The process is not capable.
- Your sales team promises delivery of a banquet event in three weeks. The resources available in operations indicate four weeks are needed to make the product. The process is not capable.
- There are 45 students enrolled for a certain business course. The assigned teaching room has 35 places. The process is not capable.
- A travel program has 50 people in its group. The chosen restaurant for lunch as 45 seating spaces. The process is not capable.
- Medical centers or hospitals are sometimes overloaded with patients. Their processes are not capable.
- A teaching room has 30 desks. There are 36 students signed up for the course. The process is not capable.

Where variables are concerned, and a range of values is acceptable, the design product specification is expressed in terms of a mean, nominal, or target value, μ. And, because of random occurrences, there is a tolerance with an acceptable upper and lower specification limits, USL and LSL. For example, as shown in Figure 13.1 the net weight of green coffee in a sack is labeled 10 kg. This is the nominal weight though bag sizes between 9.9 and 10.1 kg may be acceptable. Any weights outside these limits would be non-conforming and unacceptable.

Some process situations may have just a lower specification limit or just an upper limit specification. For example as in Figure 13.2 a firm might set a lower limit specification of, say, $50,000 on daily revenues but there is no upper limit specification. A target or mean value may be set at, say, $100,000 daily revenues though values above this are acceptable. Daily revenues below $50,000 are unacceptable.

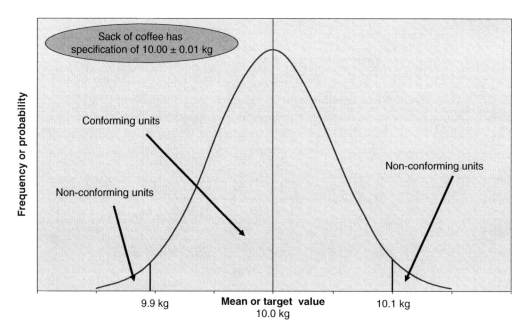

Figure 13.1 Non-conforming units beyond a lower and upper specification.

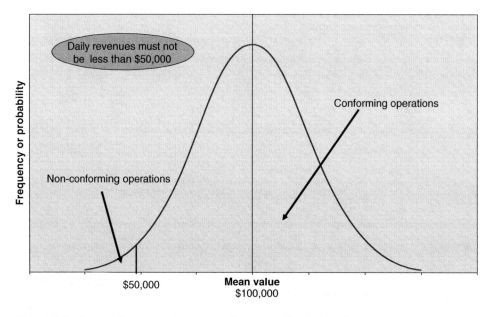

Figure 13.2 Non-conforming units beyond a lower specification level.

Alternatively, as illustrated in Figure 13.3, the firm might set an upper limit specification on annual budget expenditures of, say, $250,000 but there is no lower limit on the budget expenditures. The average, or target value may be, say, $200,000.

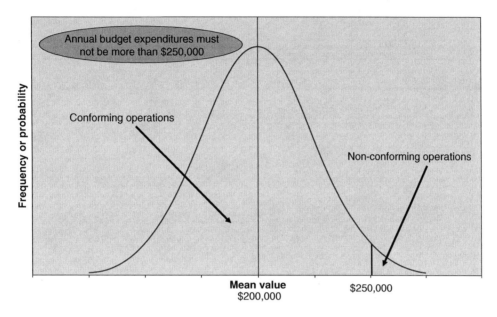

Figure 13.3 Non-conforming units beyond an upper specification level.

In a hotel the specifications may be that a client is checked in and shown to her room in 10 minutes or less (clients do not like to wait). Similarly in a restaurant, where the objective is that a customer receives his main meal within 15 minutes after ordering; a grocery store receives his order from a distribution center within 4 hours of the promised delivery time.

Process capability index

Using the normal distribution as the analytical basis, the process is capable if it has a process distribution whose extreme limits always lie inside the upper and lower specification of the "product" or the service being offered. The range between lower and upper specification limits of the product is the **width**. Consider Figure 13.4 that shows the target or average wait time of 10 min for a hotel customer at the front desk. The upper and lower product specification limits are shown and the distance between the two is the tolerance limit.

Then **centered** on this target value of 10 min is the process average value. This time might be defined by the number of employees at the front desk; their training; the speed of the hotel information system; availability of information; and the customer loading at the front desk. The figure shows three process specifications, all with the same mean value, but with different limits or standard deviations. Note that the product specification and process specifications *are not related*. The product specification is what the customer expects—voice of the customer; an external requirement. The process specification is what the system is able to offer—voice of the process; an internal constraint. Process 1, which has the lowest standard deviation, is capable of meeting customer requirements because its extreme values lie well within the specifications of the front desk service time. Process 2, whose standard deviation is equal to the standard deviation of the "product," is only just capable of meeting requirements because its extreme values lie at the same specification limits of the "product," Process 3, which has the highest standard deviation, is not always capable because its extreme limits lie outside the specifications of the

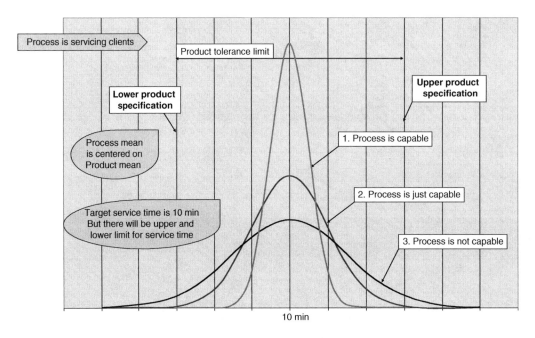

Figure 13.4 Process capability and specifications: front desk service time targeted at 10 mins.

product or service. In this last case, with this process activity at the front desk, some clients will be served or "processed" above and below the specification product tolerance limits. The greater the design margin between the process and the "product" then the lower will be the total "defects" per unit (tdu). The design margin is measured by the **process capability index, C$_p$** defined as:

$$C_p = \frac{Upper\ "product"\ specification - Lower\ "product"\ specification}{Normal\ process\ specification}$$

This can be expressed as:

$$C_p = \frac{"Product"\ specification\ width}{Process\ width}$$

where the difference between the upper and lower "product" specification is the product tolerance range, or product specification width. The process capability index is sometimes referred to as the "process potential" as it describes how capable the process could be when it is centered exactly between the product specifications. Thus in Figure 13.4 visually it is clear that for:

- Process 1: C$_p$ > 1.00 or the process is capable all of the time
- Process 2: C$_p$ = 1.00 or the process is just capable
- Process 3: C$_p$ < 1.00 or the process is not capable all of the time

Going further, assume that in Figure 13.4 that the USL of the customer service is 16 min and the lower specification level is 4 min. This gives a tolerance, or product specification width, of 12 min (16 − 4) that is centered on a target level of 10 min.

- For Process 1 if the standard deviation, $\sigma = 1.50$ min, then $6\sigma = 9.00$ min. This then gives a value of $C_p = 1.33$ (12.00/9.00) or one that is capable most of the time.
- For Process 2 if σ increases now to 2.00 min, then $6\sigma = 12.00$ and $C_p = 1.00$ (12.00/12.00) or only just capable most of the time.
- For Process 3 if σ increases now to 3.00 min, then $6\sigma = 18.00$ and $C_p = 0.67$ (12.00/18.00) or not capable all of the time.

Consider Figure 13.5 that shows a product tolerance limit of $\pm\ 6\sigma$ and a process width of $\pm\ 3\sigma$.

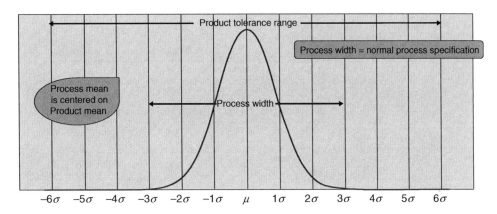

Figure 13.5 Process capability ratio—short term.

For the complete distribution then the process capability ratio is:

$$C_p = \frac{Product\ specification\ width}{Process\ width} = \frac{12\sigma}{6\sigma} = 2.00$$

Alternatively the same value is obtained if we just consider the right (or left) tail of the distribution:

$$C_p = \frac{0.5 * Product\ specification\ width}{0.5 * Process\ width} = \frac{0.5 * 12\sigma}{3\sigma} = 2.00 \qquad 13(i)$$

At this level of C_p the process would be very high-performing.

Standard deviations and the z value

Six-sigma uses the normal distribution as its measurement basis. Figure 13.6 gives again three normal distributions, leptokurtic, mesokurtic, and platykurtic. These represent the sample distribution of a process that produces a product where just the upper specification limit, USL, of the product is indicated.

All these three sample process distributions have the same mean, $\bar{x}$, but a different standard deviation, σ. For any of these distributions, using the normal distribution relationship gives:

$$z = \frac{USL - \bar{x}}{\sigma} \qquad 13(ii)$$

where the value of σ differs according the process. Reorganizing this equation gives:

$$USL - \bar{x} = z\sigma \qquad 13(iii)$$

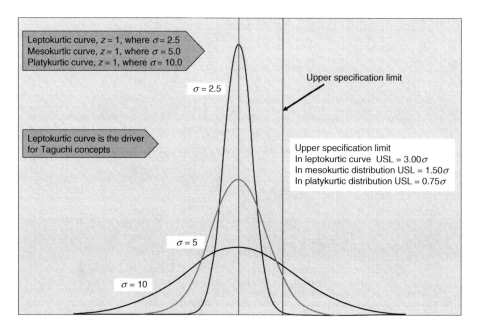

Figure 13.6 Standard deviation and the *z* value.

That is the range of the USL from the mean value of the process is $z\sigma$. Assume that, as shown here, in the leptokurtic distribution the USL for the product is established at 3σ limits from the process. Thus:

- In the leptokurtic distribution, $\sigma = 2.50$ thus when $z = 3$, $(USL - \bar{x}) = z\sigma = 7.50$ $(3*2.50)$.
- For the same range of 7.50 in the mesokurtic distribution, which has a σ of 5.00, then z is 1.50 (7.50/5.00).
- For the same range of 7.50 in the platykurtic distribution, which has a σ of 10.0, then z is 0.75 (7.50/10.0).

In summary, as the standard deviation, σ, increases, in this case from 2.50, to 5.00 and to 10.00 then the z value decreases from 3.00 to 1.50 and to 0.75. Alternatively as the standard deviation, σ, decreases then the z value increases. The z value or the z score, is a measure of the capability of the process. The target in six-sigma is to have a z score numerically equal to 6. For this example, to have the same range, $(USL - \bar{x})$, of 7.50 for a z value of 6.00 means a standard deviation, σ, of 1.25 (7.50/6.00); a low value.

Six-sigma performance

Six-sigma is based on the concept of dispersion where the higher the level of the number of standard deviations, z, then the better is the quality of the process, service, or product. Consider Figure 13.7 that shows a process distribution and the vertical lines the lower and upper specification limits of the product at various numbers of standard deviations, z, from the mean value of z that is 0. That is a normal z-distribution.

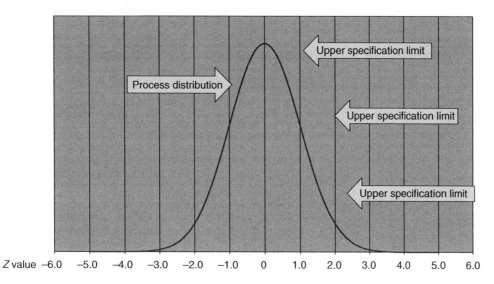

Figure 13.7 Defects per million for short-term capability.

The process distribution is centered on the target value of the product specification. Considering just the right-hand part of the distribution with the product specification width of 12σ and the process width of 6σ then from Equation 13(i):

$$C_p = \frac{0.5 * \textit{Product specification width}}{0.5 * \textit{Process width}} = \frac{0.5 * 12\sigma}{3\sigma} = \frac{6.0}{3.0} = 2.00$$

At this level the percentage of non-conforming units beyond, or to the right of the USL, is 0.00000010% or 0.001 ppm. This would be a high-performing operation. If we consider a product specification width of 10σ, a smaller product tolerance, then:

$$C_p = \frac{0.5 * \textit{Product specification width}}{0.5 * \textit{Process width}} = \frac{0.5 * 10\sigma}{3\sigma} = \frac{5.0}{3.0} = 1.67$$

At this level the percentage of non-conforming units beyond, or to the right of the USL, is 0.00002867% or 0.29 ppm. Again a high-performing operation. At a product specification width of 8σ, a smaller product tolerance, then:

$$C_p = \frac{0.5 * \textit{Product specification width}}{0.5 * \textit{Process width}} = \frac{0.5 * 8\sigma}{3\sigma} = \frac{4.0}{3.0} = 1.33$$

At this level the percentage of non-conforming units beyond, or to the right of the USL, is 0.00316712% or 31.67 ppm. Now there is a measurable quantity of units non-conforming. At a product specification width of 6σ, a much smaller product tolerance, then:

$$C_p = \frac{0.5 * \textit{Product specification width}}{0.5 * \textit{Process width}} = \frac{0.5 * 6\sigma}{3\sigma} = \frac{3.0}{3.0} = 1.00$$

At this level the percentage of non-conforming units beyond, or to the right of the USL, is 0.13498980% or 1,349 ppm, a significant number of non-performing units. Data for other values of z are given in Table 13.1. In conclusion the higher the value of sigma the better performing is the operation.

Table 13.1

z limits (short term)	Percent units non-conforming (above z-limit)	Parts per million non-conforming units	Process capability index C_p
0.00	50.00000000%	500,000.00	0.00
0.50	30.85375387%	308,537.54	0.17
1.00	15.86552539%	158,655.25	0.33
1.50	6.68072013%	66,807.20	0.50
2.00	2.27501319%	22,750.13	0.67
2.50	0.62096653%	6,209.67	0.83
3.00	0.13498980%	1,349.90	1.00
3.50	0.02326291%	232.63	1.17
4.00	0.00316712%	31.67	1.33
4.50	0.00033977%	3.40	1.50
5.00	0.00002867%	0.29	1.67
5.50	0.00000190%	0.02	1.83
6.00	0.00000010%	0.001	2.00

Short- and long-term variation

In most activities there is variation in the short term and this is considered random or common cause variation. There is variation in the time to service a customer; the time to prepare a meal; the time for a delivery vehicle to arrive; the time to complete the phases of a project, etc. This variation may be difficult to reduce though this is an objective within six-sigma.

Variation

Consider the situation in Figure 13.8 that shows the preparation time and delivery for orders from a distribution center.

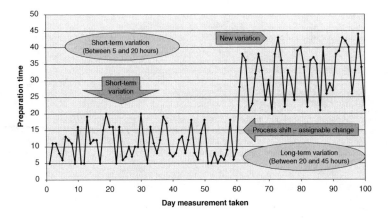

Figure 13.8 Short- and long-term variation—a process shift.

For the first 60 days there is variation that oscillates between 5 and 20 hours—a range of 15 and a midrange of 12.5 hours. This is random short-term variation. Then on the sixty-first day something happened. The variation has shifted upwards and order preparation now varies between some 20 and 45 hours—a range of 25 hours and a midrange of 32.5 hours. Variation is different in the long term than in the short term. Something has happened in the process and this change is **assignable**, meaning that there is a reason that is non-random that this change has occurred. Perhaps the distribution center has suddenly accumulated more orders, meaning resources for the previous orders are now limited; there has been a staff cutback; picking equipment is no longer available; road modifications necessitate a longer route, etc. There has been a process variation or a process shift. In six-sigma management the process must be able to accommodate both short-term and long-term variation, or the **process shift**.

Process shift

Up to this point we have considered only short-term variation in the process. However, long-term changes, or a shift, may occur in the process due, for example, to wear on tools and machines, management changes, labor fatigue, new untrained staff, insufficient staff, delivery delays, new competitors, etc. Six-sigma management is such that long-term changes, or a process shift, can be accommodated. A process shift might occur because the variation has changed as illustrated in Figure 13.9. Here the long-term distribution shows a process variation that is now outside the product specifications limits although the mean value of the process is unchanged. In this case the process is not always capable of producing conforming products all of the time.

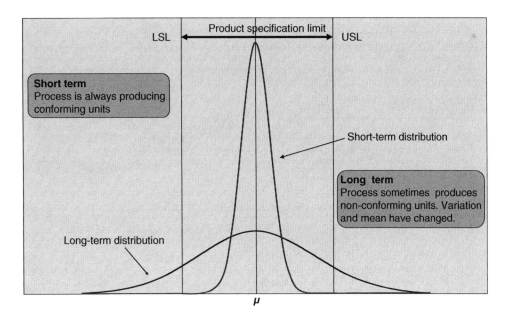

Figure 13.9 Process mean remains fixed but variation has changed.

Consider Figure 13.10 that illustrates a shift to the right for the front desk situation. In the original operation the process mean of 10 min was centered on the product mean of 10 min. The USL limit is still 16 min and the LSL is 4 min giving a tolerance of 12 min. As shown, the process width is equal

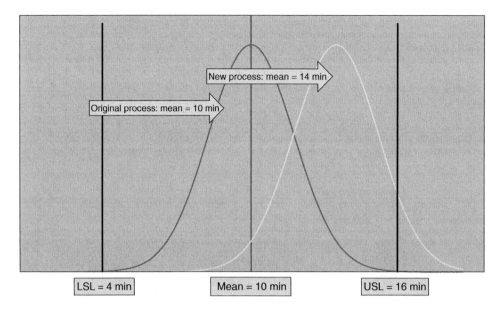

LSL = 4 min   Mean = 10 min   USL = 16 min

Figure 13.10 Process shift to the right.

to the product width and for this situation σ of the process was 2.00 so that $6\sigma = 12.00$. This gives a value of C_p of 1.00 using the following, since we are now considering just half of the distribution:

$$C_p = \frac{0.5 * Product\ specification\ width}{0.5 * Process\ width}$$

With the process shift to the right, part of the process distribution is beyond the USL, indicating that a portion of the process output does not conform. The value of C_p has to be recalculated to measure the true potential of the process defined now as C_{pk}, that is, the minimum calculated from the USL and the LSL. From the left-hand side:

$$C_{pk} = \frac{mean - Lower\ product\ specification}{3\sigma} = \frac{14 - 4}{3 * 2} = \frac{10}{6} = 1.67$$

From the right-hand side:

$$C_{pk} = \frac{Upper\ product\ specification - mean}{0.5 * process\ width} = \frac{16 - 14}{3 * 2.00} = \frac{2.00}{6.00} = 0.33$$

This is the smallest value and is the true process capability index.

Consider the complete situation illustrated by Figure 13.11. The center normal distribution is the short-term process variation with a process width of $\pm 3\sigma$. This width is well within the design specification width of the product so in the short-term conforming units would be produced all of the time. However, in the long term there are process shifts both to the left and right. The mean value of the process has changed but the variation around each mean is the same. However considering the two extreme cases of left and right the total process variation is very close to the product specification width. Thus non-conforming products may be produced some of the time.

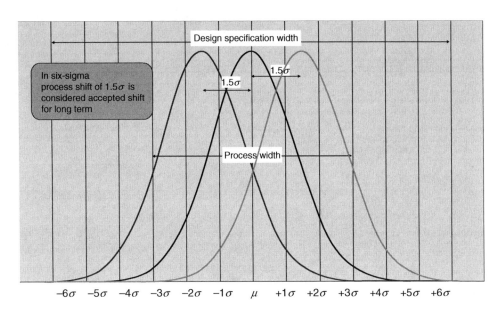

Figure 13.11 six-sigma process capability with a shift in the mean.

In this Figure 13.11 the value of the shift to the right or to the left is equal to 1.5σ with no change in the shape of the process distribution. In six-sigma, a process shift of 1.5σ is considered an accepted shift for long-term process operation. In the long term the short-term process capability ratio C_p as calculated in Equation 13(i)) is adjusted using a factor k to give a process capability index, C_{pk}, where:

$$C_{pk} = C_p(1 - k) \qquad\qquad 13(iv)$$

The value k is the percentage of the product specification width that is consumed by the process shift, or.

$$k = \frac{\text{process shift}}{0.5 * \text{product specificaton width}} \qquad\qquad 13(v)$$

Note that we are just considering a movement, or process shift, of 1.5σ to the right; thus we consider 0.5*production specification width. In this movement to the right the value k is:

$$k = \frac{\text{process shift}}{0.5 * \text{product specificaton width}} = \frac{1.5\sigma}{0.5 * 12\sigma} = \frac{1.5}{6.0} = 0.25$$

Thus:

$$C_{pk} = C_p(1 - k) = 2.00(1 - 0.25) = 1.50$$

where $C_p = 2.00$, the value C_{pk} is that calculated from Equation 13(i).

Another way of determining the value C_{pk} is similar to the equations for the front desk example to consider that $\bar{\bar{x}}$ (mean value) has a value of 0, or equivalent to a standard normal distribution. Thus, in the short term, with no process shift, the value of the process capability, C_p just considering the right side of the distribution, is:

$$C_p = \frac{Upper \text{ product specification} - \bar{\bar{x}}}{0.5 * \text{process width}} = \frac{6\sigma - 0}{3\sigma} = 2.00$$

Now with a process shift the process capability is redefined to give the process capability index and with a process shift to the right of 1.5σ then the value of $\bar{x}$ is now 1.5σ. In this case the value of the process capability index is:

$$C_{pk} = \frac{Upper \text{ product specification} - \bar{\bar{x}}}{0.5 * \text{process width}} = \frac{6\sigma - 1.5\sigma}{3\sigma} = \frac{4.5\sigma}{3\sigma} = 1.50$$

If there was a 1.5σ process shift to the left then the process capability index would be given by:

$$C_{pk} = \frac{\bar{\bar{x}} - \text{Lower product specification}}{3\sigma} = \frac{-1.5\sigma - (-6\sigma)}{3\sigma} = \frac{4.5\sigma}{3\sigma} = 1.50$$

This gives the same value of the process capability index of 1.50. It is this level of process capability that is aimed at in six-sigma quality. The process shift now corresponds to the defect values given by a dispersion of 4.5σ ($6.0\sigma - 1.5\sigma$). Also, since we are considering a shift to the right, we are only considering one side or one tail of the distribution. Thus, the defects per million with the process shift is now according to Table 13.2. With a long-term process shift of 1.5σ to the right (or to the left) there is now a change in the probability of non–conforming products units being produced as the original 6σ limit for the short term, no process shift. This table illustrates why at 6σ limits ($z = 6.00$) with a process shift there are 3.4 parts per million non-conforming units. In order to determine this value we have to use $z = 4.5\sigma$ in order to take into account the considered process shift of 1.5σ. The long-term and short-term values of z are related by:

$$z_{LT} = z_{ST} - 1.5\sigma$$

Note also that the process has a long-term capability index of 1.50 down from a short-term value of 2.00 when there is no process shift.

Table 13.2

z limits (long term)	Percent units non-conforming (above z-limit)	Parts per million non-conforming units	Process capability index C_{pk}
0.00	50.00000000%	500,000.00	0.00
0.50	30.85375387%	308,537.54	0.17
1.00	15.86552539%	158,655.25	0.33
1.50	6.68072013%	66,807.20	0.50
2.00	2.27501319%	22,750.13	0.67
2.50	0.62096653%	6,209.67	0.83
3.00	0.13498980%	1,349.90	1.00
3.50	0.02326291%	232.63	1.17
4.00	0.00316712%	31.67	1.33
4.50	0.00033977%	3.40	1.50

Short-term sigma limits and long-term defects per million

Thus in summary in six-sigma, although we consider the short-term control limits, z_{ST}, the defects per million are determined based on a 1.5σ shift or a long-term control limit z_{LT} and this will give a defect level higher than expected from the short-term situation. This is illustrated in Table 13.3.

Table 13.3

z_{ST}	Percent non-conforming (above z-limit)	ppm units non-conforming	z_{LT} or ($z_{ST} - 1.5\sigma$)	Percent non-conforming (above z-limit)	ppm units non-conforming
0.00	50.00000000%	500,000.00	−1.50	93.31927987%	933,192.80
0.50	30.85375387%	308,537.54	−1.00	84.13447461%	841,344.75
1.00	15.86552539%	158,655.25	−0.50	69.14624613%	691,462.46
1.50	6.68072013%	66,807.20	0.00	50.00000000%	500,000.00
2.00	2.27501319%	22,750.13	0.50	30.85375387%	308,537.54
2.50	0.62096653%	6,209.67	1.00	15.86552539%	158,655.25
3.00	0.13498980%	1,349.90	1.50	6.68072013%	66,807.20
3.50	0.02326291%	232.63	2.00	2.27501319%	22,750.13
4.00	0.00316712%	31.67	2.50	0.62096653%	6,209.67
4.50	0.00033977%	3.40	3.00	0.13498980%	1,349.90
5.00	0.00002867%	0.29	3.50	0.02326291%	232.63
5.50	0.00000190%	0.02	4.00	0.00316712%	31.67
6.00	0.00000010%	0.00	4.50	0.00033977%	3.40

For example, from Table 13.3 if we consider a z value of 3 we would expect 1,350 defects per million (rounded). However, taking into account the 1.5σ shift the long-term defects per million is 66,807. A graph showing the relationship between z_{ST} and long-term defects per million is given in Figure 13.12.

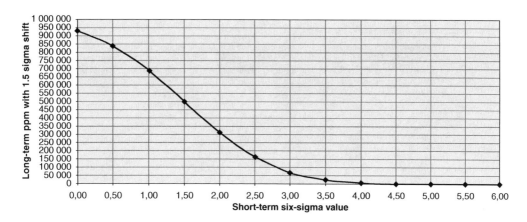

Figure 13.12 Short-term six-sigma and long-term defects per million.

Defects per unit

The yield and defects per unit is a binomial situation either good or bad and the criterion applies to all areas of business: production; sales; purchasing; accounting; administration; healthcare; hospitality, etc. If in a process there is a yield of 95% this implies that there must be 5% defects. When a process or defined characteristic is not within the specification then it is non-conforming or defective. However, just because something is non-conforming it does not necessarily mean that the system or product will not work. For example: A hotel room that is not cleaned according to specifications; a window frame that is cracked; revenues that are not up to target; or a purchasing order with the company name spelled incorrectly. These are defects, or non-compliance with specifications. These may trigger a complaint and in this case represent a "failure" These are not acceptable in six-sigma management and they are costly to rectify. However these units are still operational even though they have a fault.

In six-sigma the item being worked upon and subsequently analyzed is a unit, (u). This unit can be whatever you define. It may be a hotel room; a purchasing order; a month's production of microwave ovens; a hospital patient; daily revenue receipts; a teaching program, etc. An assessment of process capability is then to measure the number of defects (d) in a given number of units to determine the defects per unit, (dpu), which is given by the relationship:

$$\text{Defects per unit (dpu)} = \frac{\text{Number of defects determined (d)}}{\text{Total number of units anlayzed (u)}} \qquad 13(\text{vi})$$

Suppose, for example, that in one month 240 hotel rooms are inspected for order and cleanliness and 131 defects are determined. A defect might be the toilet is not clean; soap is missing; the sheets are crumpled; or the mirror is smeared. These defects may apply to just a few of the 240 rooms and it may be that one room has, say. five defects whereas another has none:

$$\text{Defects per unit} = \frac{131}{240} = 0.5458 \text{ or } 54.58\%$$

This means that on average less than half of the clean rooms are according to specification.

Opportunities in the product, process, or service

In six-sigma we talk about an opportunity, which is a specific characteristic in a product, process, service function, activity, or environment that may turn out as a success or defect. A success is when you "have it right" or the characteristic is within defined specifications, or according to expectations.

Operations

In operations an opportunity might be to stock the correct amount of inventory; to serve the customer within the specified time; to have the correct delivery order from a distribution center to the restaurant; to have the correct mailing address on a billing statement; to have the target daily revenues. The more complex the product, process or unit, then the more opportunities there are "for getting it right," or "for getting it wrong." Putting it another way, the more opportunities then the more complex is the product in question. For example, at a hotel reception the important elements for the customer are: to have baggage carried in by a porter; a quick check-in time; pleasant personnel; and that reception has the correct client booking information. Here there are four opportunities where the client may not be satisfied. The opportunities on a mailing address are limited—the correct spelling of the customer name, the correct street address, or say the correct postal code. However, for a hotel order delivery there might be 60 different products: each product has a different order quantity, each product has a special brand name, and each product has a different volume or weight requirement. For tangible products we might have the number of opportunities on a ball point pen or the number of opportunities on a portable computer. The defects per opportunity are:

$$\text{Defects per opportunity (dpo)} = \frac{\text{Number of defects determined on a unit}}{\text{Number of opportunities on a unit}}$$

Each opportunity must be an individual characteristic that is independent of another characteristic on the unit. For example in the case of a pen, the surface smoothness is an opportunity and a scratch and a dent may not be independent since the dent may occur as the result of the scratch. However, surface smoothness and color are independent opportunities. Assume for example that on a ball point pen the number of opportunities is 37 and the defects, or out of specification characteristics, are 2; for the portable computer there are 3,573 opportunities and 192 defects, or out of specification characteristics, are found:

$$\text{Defects per opportunity} = \frac{2}{37} = 0.0645, \text{ for the ball point pen}$$

$$\text{Defects per opportunity} = \frac{192}{3,573} = 0.0537, \text{ for the portable computer}$$

We have more defects on the computer as there are many more areas where things can be out of specification. Using **defects per opportunity** we can fairly compare the defect rates of systems with different complexities. If we used **defects per unit** then in the above example for individual units we would be comparing 2 defects per unit for the ball point pen compared to 192 defects per unit for the portable computer which at first light gives the impression that the portable computer is very poor. Consider the situation in preparing a meal. In cooking an omelet the number of opportunities is 21 including the temperature, fluffiness, crispness, color, ingredients, taste, etc. There are 2 "defects" determined on the omelet. This gives the defects per opportunity, dpo, of $2/21 = 0.0952$ or 9.52%. In the preparation of a three-course meal the number of opportunities is 79. This applies to starter, main course, and dessert. The number of defects encountered is 8. The defects per opportunity is $8/79 = 0.1013$ or 10.13%. This is not vastly different for the dpo for the omelet. There are more opportunities in a three-course meal. It is more complex and using defects per opportunity, dpo, we fairly compare items of different complexities. It is wrong to compare using defects per unit; here 2 to 8. This would give the impression that the three-course meal is worse than omelet. Also, by corollary you cannot conclude that chef who made the three-course meal is worse than one who made omelet.

When you determine the number of opportunities the number should not be overestimated as this will reduce the defects per opportunity and give the impression that the system is better than reality. For example on a billing form the name may be written as **_DERECK_** instead of Derek. We might say there are five opportunities—the spelling, the font, the case, italic, and underlined. In practice we would probably say that there is only one opportunity—the spelling.

The relationships, defects per unit, and defects per opportunity, only consider one unit or one opportunity. In practice we need to understand values after operating a process, or performing an activity over a long period of time. For example the number of accounting errors after 900 have been processed; the accuracy of orders after 1,200 have been delivered; the cleanliness of hotel rooms after

2,000 have been attended to; the daily revenue targets after six months; the number of truck engines that have been assembled correctly in one month. In this case we refer to the defects per million opportunities. The relationship is:

$$\text{dpmo} = \text{dpo} * 1,000,000$$

Sometimes the defect rate as defined as defective parts per million, or dppm, when material parts are being considered:

$$\text{dppm} = \text{dpo} * 1,000,000$$

Numerically dpmo and dppm are the same. A process that is operating six-sigma has 3.4 defects per million opportunities or 3.4 defective parts per million.

In addition, when determining process capability using opportunities, care must be taken that the calculation is made using the measured opportunities and not actual opportunities. For example, assume that in a certain process we know there are 1,500 opportunities. Of these we only test 850 and find that there are 23 non-conformities. In this case the defects per million is 23*1,000,000/850 or 27,059. It is wrong to say that the defects per million is 23*1,000,000/1,500 or 15,333: this would give the impression that the process capability is better than we are able to describe.

Converting defects into long-term sigma capability

If we know the defects in a certain number of units then we can convert this to the short-term sigma limits by using the following relationship for which an example is given in Table 13.4, where in 175 billing invoices reviewed, 3 errors were discovered.

Table 13.4

Element	Expression	Example
Number of units (invoices) examined	μ	175
Number of units not conforming	d	3
Defects per unit	d/μ	0.017143
Defects per million	$(d/\mu)*1,000,000$	17.143
Percentage of units non-conforming	p	1.7143%
Percentage of units conforming	$1 - p$	98.2857%
Long-term sigma limit, z_{LT}	How calculated (use normal distribution)	2.12
Short-term sigma limit, z_{ST}	$z_{ST} = z_{LT} - 1.5\sigma$ (How expressed)	3.62

First-time yield and the hidden activity

The traditional way of calculating yield rate is output divided by input. However, this can be misleading if not all the process activities are understood. Consider the situation represented by Figure 13.13 where there are 600 meals being prepared for a restaurant banquet.

After the meals are prepared they are inspected to verify that they are according to specifications (portion size, temperature, degree of cooking, presentation, etc.). In the inspection process 120 meals are found not to conform. These go through a rework process and 105 meals are then deemed

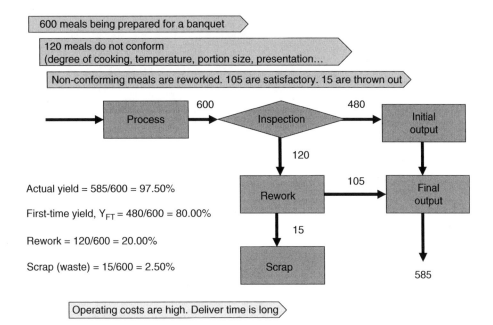

600 meals being prepared for a banquet

120 meals do not conform
(degree of cooking, temperature, portion size, presentation...

Non-conforming meals are reworked. 105 are satisfactory. 15 are thrown out

Actual yield = 585/600 = 97.50%

First-time yield, Y_{FT} = 480/600 = 80.00%

Rework = 120/600 = 20.00%

Scrap (waste) = 15/600 = 2.50%

Operating costs are high. Deliver time is long

Figure 13.13 Yield rate and the hidden activity.

satisfactory and returned to the output lot for a total of 585 good units. Thus the traditional process yield, Y_{TRAD}, is:

$$\text{Traditional process yield is } \frac{585}{600} = 97.50\%.$$

However, this does not tell the whole story regarding the efficiency of the operation. For this process the throughput yield on the first run, or the **First–time yield (Y_{FT})** is:

$$\text{First–time yield is } \frac{480}{600} = 80.00\%.$$

In addition there has been a **rework activity** on the original input units of:

$$\text{Rework activity is } \frac{120}{600} = 20.00\%.$$

On the basis that rework was carried out on all of the 120 non-conforming units before it was clear that 15 had to be scrapped. Further the **scrap rate** or final **defects** is:

$$\text{Scrap rate is } \frac{15}{600} = 2.50\%.$$

This scrap rate level is equivalent to 25,000 defects per million units, which is a long way from the six-sigma criterion of 3.4 defects per million units. The rework activity is the **hidden activity**, meaning the time and money spent by the firm to correct mistakes. This implies that operating costs are higher and that delivery time is longer.

Rolled throughput yield and normalized yield

Processes usually involve many separate activities and this gives rise to the **rolled throughput yield**, Y_{RT}, that takes into account the yield of all of the processes involved in the operation. Consider Figure 13.14 where in a particular hospitality operation there are a total of four processes.

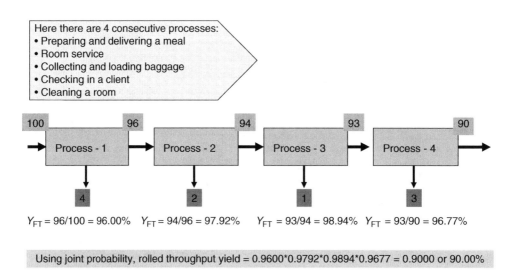

Figure 13.14 Rolled throughput yield.

In the hospitality industry this may involve the cleaning of a hotel room; room service; preparation and delivery of a meal; check-in of a client at the front desk; collecting and loading baggage onto the tour bus; etc. This figure indicates the number of units entering the each process and the number of units that are defective the first time.

- Process 1: units in = 100, units out = 96, Y_{FT} = 96/100 = 96.00%
- Process 2: units in = 96, units out = 94, Y_{FT} = 94/96 = 97.92%
- Process 3: units in = 94, units out = 93, Y_{FT} = 93/94 = 98.94%
- Process 4: units in = 93, units out = 90, Y_{FT} = 90/93 = 96.00%

Then using joint probability, *Chapter 3*, the rolled throughput yield (Y_{RT}) of the four processes is given by:

Rolled throughput yield, Y_{RT} is $0.9600 * 0.9792 * 0.9894 * 0.9600 = 0.9000$ or 90.00%

The First-time yield can also be approximated by the relationship : $Y_{FT} = e^{-dpu}$ 13(vii)

- Process 1: dpu = 4/100 = 0.0400, $Y_{FT} = e^{-0.0400} = 0.9608 = 96.08\%$
- Process 2: dpu = 2/96 = 0.0208, $Y_{FT} = e^{-0.0208} = 0.9794 = 97.94\%$
- Process 3: dpu = 1/94 = 0.0106, $Y_{FT} = e^{-0.0106} = 0.9894 = 98.94\%$
- Process 4: dpu = 3/93 = 0.0323, $Y_{FT} = e^{-0.0323} = 0.9683 = 96.83\%$

The usefulness of this approach is that if you only know the defects per unit you can calculate the First-time yield and subsequently the rolled throughput yield. Using these First-time yield values then the rolled throughput yield is:

Rolled throughput yield, Y_{RT}, is $0.9608 * 0.9794 * 0.9894 * 0.9683 = 0.9015$ or 90.15%

(This is not far from the 90.00% calculated above using the joint probability of First-time yields.)

Note that by taking the logarithmic values on both sides of Equation 13(vii) we can also write:

$$\text{dpu} = -\ln(Y_{FT}) \qquad\qquad 13(\text{viii})$$

Related to the rolled throughput yield is the normalized throughput yield, Y_{NM}, which is that yield when *constant for each process* would give the same rolled throughput yield or:

$$Y_{NM} = \sqrt[n]{(Y_{RT})} = (Y_{RT})^{1/n} \qquad\qquad 13\,(\text{ix})$$

where n is the number of processes. In the four-stage process above the Y_{RT} is 90% and thus substituting this value in Equation 14(vii) gives:

$$Y_{NM} = (0.90000)^{1/4} = 0.9740 \text{ or } 97.40\%$$

This gives a defect level of $1 - 0.9740 = 0.0260$ or 26,000 defects per million. This gives a value of z_{LT} of 1.94. Since $z_{ST} = z_{LT} + 1.5\sigma$ when a process shift is taken into account, then $z_{ST} = 3.44$, which is the σ-level of this particular operation.

In the hospitality industry the system is made up of many sequential processes—a client's hotel stay extends from: arrival, check-in, the room, breakfast, dining, conference amenities, the bar, baggage handling, and final checking out. Here there are many processes and it is important that the First-time yield is close to 100%. Figure 13.15 shows how dramatically the rolled throughput yield can fall according to the number of processes.

For example, from the figure, if we consider that there are 50 processes in a system each with a seemingly high First-time yield of 99%, then:

Rolled throughput yield is $0.9900^{50} = 0.6050$ or only 60.50%

In six-sigma management this situation would be unacceptable and the action required is to look at each process in detail, starting with the one with the lowest First-time yield, with the objective to bring the First-time yield close to 100%. Although this example refers to the hospitality industry it applies to all businesses—manufacturing, distribution process, handling of insurance claims, education, medical services, etc.

Cycle time, performance level, and the hidden activity

The theoretical cycle time is the total time for a unit, either a product or a customer, to pass through all the process steps without delays, waiting, unnecessary transfer, rework or setups. If any of these do occur it takes additional time to inspect, correct, and redo and in this case this additional time adds to the theoretical cycle time. This additional time is the hidden factory. This concept is illustrated in Figure 13.16.

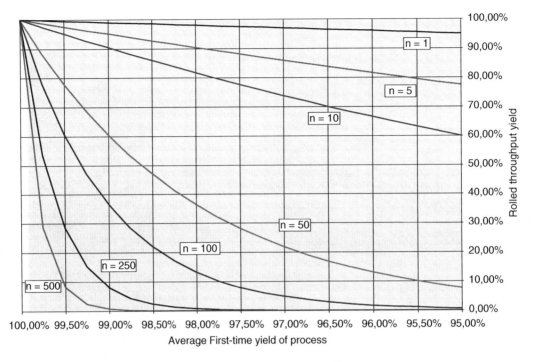

Figure 13.15 Rolled throughput yield according to *n*, the number of processes.

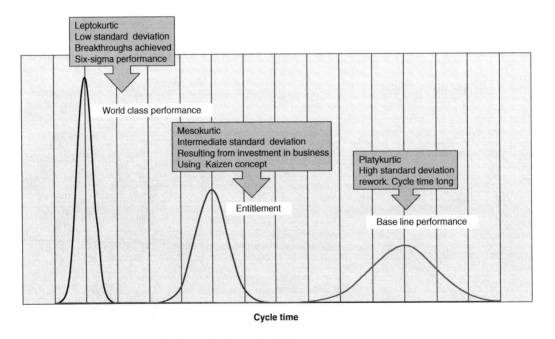

Figure 13.16 Cycle time, performance level, and the hidden activity.

The flat or **platykurtic** curve, with a high standard deviation, illustrates baseline or current level of performance beyond which we need to search for **breakthroughs**. Here there are many delays and unnecessary rework and so the cycle time is long. The intermediate or **mesokurtic** curve illustrates the **entitlement** or the level of performance a firm should be able to attain given the investments already spent on the business. This performance should have been achieved with a **Kaizen** philosophy and it means that you have eliminated all non-random or assignable variables. Going further is difficult at this stage because the random variables are unknown and so entitlement is the performance level with the random level naturally built into the process. The cycle time is intermediate when the entitlement level has been achieved. The slender, or **leptokurtic** distribution represents **world class performance**, a six-sigma level, where you have come up with breakthroughs for the process that include sharply reducing or eliminating common cause variation.

End of story

This is the last chapter of this textbook applying statistics to the business world. The next section gives a compendium of pragmatic case-exercises for all the chapters containing: the Objectives; Situation; Required; and Answers.

Notes

1 https://en.wikipedia.org/wiki/Mars_Observer [accessed May 19, 2016].
2 https://en.wikipedia.org/wiki/Mars_Climate_Orbiter [accessed May 19, 2016].
3 "Two teams, two measures, equaled one lost spacecraft," *The New York Times*, October 1, 1999.
4 Mikel, Harry J, "The Vision of Six Sigma: A roadmap for breakthroughs," Sigma Publishing Co. 5th edn., March 1997.
5 Deming, W.E., "Philosophy Continuous to Flourish," *APICS-The-Performance Advantage*, 1, No. 4, October 1991, p. 20.

Case-exercises

1 Characterizing and defining statistical data

Significance and essentials of numbers

No	Case-exercise	Statistical concept	Application
1	Bernard's restaurant, II	Weighted averages	Hospitality – costing
2	Big Mac hamburgers, I	Characteristics, price units	Hospitality
3	Chemical engineering salaries, I	Characteristics, currencies	Hum. res. manufacturing
4	European food prices	Sorting	Economics, food
5	Innovation rankings, I	Central tendency, dispersion	Innovation, technology
6	Investment options, I	Geometric mean	Finance
7	Madrid hotel, Spain, I	Weighted averages	Hospitality
8	Olympics, London, Summer 2012	Weighted averages, sorting	Sports and leisure
9	Royal Mail package delivery	Weighted averages	Logistics
10	Ski resort restaurant, I	Central tendency, dispersion	Hospitality, sports, project
11	Western USA	Conversion metric/imperial	Travel
12	World wine production, I	Central tendency, dispersion	Hospitality, economics

1. Bernard's restaurant, II

Objective

The objective of this case-exercise is to illustrate the use of weighted averages in costing. Weighted averages takes into consideration the importance, or weighting, of certain information.

Situation

Bernard owns and operates a restaurant in Seattle, Washington, USA. Table BR-1 shows the salaries of his kitchen staff and the time spent preparing the same 100 meals, determined by an audit.

Table BR-1

Category	Master chef	Chef	Assistant chef	Apprentice
Salary, $/hour	40.12	35.20	21.50	12.50
Hours spent per 100 meals	1.50	2.05	2.25	2.65

Required

1   What is the proportion of time that each member of the kitchen staff spends on meal preparation?
2   Using weighted averages, what is the labor rate in $/hour for meal preparation?
3   Using the weighted averages determined in Question 2, what is the labor cost per meal?
4   If there is a banquet of 250 persons, what would be an estimate of total restaurant labor costs for the banquet?
5   From the weighted average calculations, what would have to be the salary of the apprentice such that the labor cost of the banquet is $500.00? All the other salary rates remain the same.
6   If the average cost of the food per 100 meals is $380.00, what would be the variable or operating cost of the banquet (labor + food) using the labor cost of the banquet as in Question 4?

Answers

1   Proportion of time: Master chef, 17.75%; Chef, 24.26%; Assistant chef, 26.63%; Apprentice, 31.36%.
2   Labor rate is $25.31/hour for meal preparation.
3   Labor cost is $2.14/meal.
4   Estimated labor cost for a banquet of 250 persons is $534.60.
5   Salary of apprentice would have to be $7.28/hour to give a labor cost for the banquet of $500.00.
6   Variable or operating cost would be $1,484.60.

2. Big Mac hamburgers, I

Objective

The objective of this case–exercise is to characterize and compare data.

Situation

Table BM-1 shows the price, in local currency, of a Big Mac hamburger in various countries and the conversion rate of the local currency to $US.[1]

Table BM-1

No.	Country	Currency	Price in local currency	Conversion rate to $ US (Jan 11, 2012)
1	Argentina	Peso	20.00	4.31
2	Australia	$Aus	4.80	0.97
3	Brazil	Real	10.25	1.81
4	Britain	£	2.49	0.65
5	Canada	$Can	4.73	1.02
6	Chile	Peso	2,050.00	506.00
7	China (a)	Yuan	15.40	6.32
8	Colombia	Peso	8,400.00	1,851.00
9	Costa Rica	Colones	2,050.00	510.00
10	Czech Republic	Koruna	70.22	20.40
11	Denmark	Danish Kroner	31.50	5.86
12	Egypt	Pound	15.50	6.04
13	Eurozone (b)	Euro	3.49	0.79
14	Hong Kong	$Hong Kong	16.50	7.77
15	Hungary	Forint	645.00	246.00
16	India (c)	Rupee	84.00	51.90
17	Indonesia	Rupiah	22,534.00	9,160.00
18	Israel	Shekel	15.90	3.85
19	Japan	Yen	320.00	76.90
20	Latvia	Lats	1.65	0.55
21	Lithuania	Litas	7.80	2.72
22	Malaysia	Ringgit	7.35	3.14
23	Mexico	Peso	37.00	13.68
24	New Zealand	$New Zealand	5.10	1.26
25	Norway	Norwegian Kroner	41.00	6.04
26	Pakistan	Rupee	260.00	90.10
27	Peru	Sol	10.00	2.69
28	Philippines	Peso	118.00	44.00
29	Poland	Zloty	9.10	3.52
30	Russia	Rouble	81.00	31.80
31	Saudi Arabia	Riyal	10.00	3.75
32	Singapore	$Singapore	4.85	1.29
33	South Africa	Rand	19.95	8.13
34	South Korea	Won	3,700.00	1,159.00
35	Sri Lanka	Rupee	290.00	113.90

(Continued)

Table BM-1 (Continued)

No.	Country	Currency	Price in local currency	Conversion rate to $ US (Jan 11, 2012)
36	Sweden	Swedish Kroner	41.00	6.93
37	Switzerland	Swiss Franc	6.50	0.96
38	Taiwan	$NT	75.00	30.00
39	Thailand	Baht	78.00	31.80
40	Turkey	Lira	6.60	1.86
41	U.A.E.	Dirhams	12.00	3.67
42	Ukraine	Hryvnia	17.00	8.04
43	Uruguay	Peso	90.00	19.45
44	USA	$US	4.20	1.00

(a) Average of five cities; (b) Weighted average price in Eurozone (c) Maharaja Mac (made with chicken instead of beef).

Required

1 Convert the price of the Big Mac hamburger into a common currency of $US using the given exchange rates.
2 Which countries have the minimum and maximum prices of hamburgers in $US? What are these prices in $US? What is the range of price in $US?
3 What is the average price of a hamburger in $US and how many countries have a price below this value?
4 What is the median price of a hamburger in $US? How can you describe the countries of Turkey and the Czech Republic in terms of the median and average price?
5 What is the standard deviation of the hamburger price in $US if it is considered a sample and what is the corresponding coefficient of variation?
6 Determine the quartile boundaries for the price of hamburgers in $US. Where are the first 25% of the listed countries positioned with respect to the quartiles?

Answers

1 The prices of Big Mac hamburgers in $US is given in Table BM-2.
2 The minimum price is India with a price of $1.62 (84 Rupees); maximum price is Norway with an equivalent price of $US 6.79 (41 Norwegian Kroner). The range of prices is $US 5.17.
3 Average price is $US 3.59 and there are 24 countries that have a price below this level.
4 The median price is $US 3.36. The price in the Czech Republic is $US 3.44 and $US 3.55 in Turkey. These prices are below the average price but above the median price.
5 Standard deviation is $US 1.25 and the coefficient of variation is 34.82%.
6 Q_0 $1.62; Q_1 $2.56; Q_2 $3.36; Q_3 $4.26; Q_4 $6.79.
   Argentina, Australia, Brazil, Canada, Colombia, and Denmark are in the fourth quartile; Britain, Chile, Cost Rica, and the Czech Republic are in the third quartile; China is in the first quartile. The quartile positions are shown in the last column of Table BM-2.

Table BM-2

No.	Country	Local currency	Price local currency	Conversion rate to $US (Jan 11, 2012)	Price $US	Quartile position
1	Argentina	Peso	20.00	4.31	4.64	4th quartile
2	Australia	$Aus	4.80	0.97	4.95	4th quartile
3	Brazil	Real	10.25	1.81	5.66	4th quartile
4	Britain	£	2.49	0.65	3.83	3rd quartile
5	Canada	$Can	4.73	1.02	4.64	4th quartile
6	Chile	Peso	2,050.00	506.00	4.05	3rd quartile
7	China (b)	Yuan	15.40	6.32	2.44	1st quartile
8	Colombia	Peso	8,400.00	1,851.00	4.54	4th quartile
9	Costa Rica	Colones	2,050.00	510.00	4.02	3rd quartile
10	Czech Republic	Koruna	70.22	20.40	3.44	3rd quartile
11	Denmark	Danish Kroner	31.50	5.86	5.38	4th quartile
12	Egypt	Pound	15.50	6.04	2.57	2nd quartile
13	Eurozone (c)	Euro	3.49	0.79	4.43	4th quartile
14	Hong Kong	$Hong Kong	16.50	7.77	2.12	1st quartile
15	Hungary	Forint	645.00	246.00	2.62	2nd quartile
16	India (d)	Rupee	84.00	51.90	1.62	1st quartile
17	Indonesia	Rupia	22,534.00	9,160.00	2.46	1st quartile
18	Israel	Shekel	15.90	3.85	4.13	3rd quartile
19	Japan	Yen	320.00	76.90	4.16	3rd quartile
20	Latvia	Lats	1.65	0.55	3.00	2nd quartile
21	Lithuania	Litas	7.80	2.72	2.87	2nd quartile
22	Malaysia	Ringgit	7.35	3.14	2.34	1st quartile
23	Mexico	Peso	37.00	13.68	2.70	2nd quartile
24	New Zealand	$New Zealand	5.10	1.26	4.05	3rd quartile
25	Norway	Norwegian Kroner	41.00	6.04	6.79	4th quartile
26	Pakistan	Rupee	260.00	90.10	2.89	2nd quartile
27	Peru	Sol	10.00	2.69	3.72	3rd quartile
28	Philippines	Peso	118.00	44.00	2.68	2nd quartile
29	Poland	Zloty	9.10	3.52	2.59	2nd quartile
30	Russia	Rouble	81.00	31.80	2.55	1st quartile
31	Saudi Arabia	Riyal	10.00	3.75	2.67	2nd quartile
32	Singapore	$Singapore	4.85	1.29	3.76	3rd quartile
33	South Africa	Rand	19.95	8.13	2.45	1st quartile
34	South Korea	Won	3,700.00	1,159.00	3.19	2nd quartile
35	Sri Lanka	Rupee	290.00	113.90	2.55	1st quartile
36	Sweden	Swedish Kroner	41.00	6.93	5.92	4th quartile
37	Switzerland	Swiss Franc	6.50	0.96	6.77	4th quartile
38	Taiwan	$NT	75.00	30.00	2.50	1st quartile
39	Thailand	Baht	78.00	31.80	2.45	1st quartile
40	Turkey	Lira	6.60	1.86	3.55	3rd quartile
41	UAE	Dirhams	12.00	3.67	3.27	2nd quartile
42	Ukraine	Hryvnia	17.00	8.04	2.11	1st quartile
43	Uruguay	Peso	90.00	19.45	4.63	4th quartile
44	USA	$US	4.20	1.00	4.20	3rd quartile

3. Chemical engineering salaries, I

Objective

The purpose of this case-exercise is to develop an understanding of the characteristics of data.

Situation

Chemical engineering is a profession that covers chemical plants, oil refining, the food industry, nuclear power, waste disposal, the environment, and a host of other industrial and educational activities. For over 30 years, the Institution of Chemical Engineers has been conducting salary surveys worldwide. In 2015 nearly 4,000 members disclosed their earnings and employee benefits. Table CE-1 shows the median salary by age in the United Kingdom; Table CE-2 shows the median salary by age in Australia; Table CE-3 shows the median salary in the United Kingdom by industry sector; and Table CE-4 shows the median salary by country.[2] Table CE-5 shows the exchange rates for various currencies.[3]

Table CE-1

Age range	Median salary £UK
25–29	38,300
30–34	50,000
35–39	61,500
40–44	70,000
45–49	75,000
50–54	80,000
55–59	83,000

Table CE-2

Age range	Median salary $Aus
25–29	95,500
30–34	119,000
35–39	148,000
40–44	175,500
45–49	199,000
50–54	227,284
55–59	249,000

Table CE-3

Industry sector	Median salary £UK
Biochemical engineering	52,000
Chemical and allied products	58,350
Consultancy	51,000
Consumer products	60,400
Contracting	70,000
Education/university	50,000
Finance, insurance, risk	142,500
Food and drink	46,000
Fossil fuel power generation	48,453
Health, safety, and environment	62,172
Industrial gases	64,308
Iron, steel, other metals	54,550
Mining and minerals	62,323
Nuclear decommissioning	48,591
Nuclear power generation	52,328

Table CE-3 (Continued)

Industry sector	Median salary £UK
Oil refining	79,605
Oil/gas exploration and production	87,057
Paper and packaging	47,933
Petrochemicals	72,263
Pharmaceuticals/health care	51,000
Plastics	54,600
Process plant and equipment	43,000
Renewable fuels	52,834
Waste management	49,000
Water	46,373
Biochemical engineering	52,000

Table CE-4

Country	Currency	Median salary
Australia	$Aus	130,000
Canada	$Can	190,000
India	Rupee	1,975,000
Ireland	Euro	72,500
Malaysia	Ringgit	89,900
New Zealand	$New Zealand	110,000
Singapore	$Singapore	120,000
South Africa	Rand	780,000
U.A.E.	Dirham	447,500

Table CE-5

Americas	Currency	Per $US
Argentina	Peso	9.0725
Brazil	Real	3.1129
Canada	$Can	1.2359
Chile	Peso	631.3000
Colombia	Peso	2,555.2400
Ecuador	$US	1.0000
Mexico	Peso	15.5140
Peru	Sol	3.1679
Uruguay	Peso	26.9100
USA	Dollar	1.0000
Venezuela	Bolívar	6.3000

Europe	Currency	Per $US
Bulgaria	Lev	1.7470
Croatia	Kuna	6.7780
Czech Republic	Koruna	24.3070

Europe	Currency	Per $US
Denmark	Krone	6.6634
Eurozone	Euro	0.8931
Hungary	Forint	279.2200
Iceland	Króna	131.9100
Norway	Krone	7.8177
Poland	Zloty	3.7263
Russia	Ruble	54.6530
Sweden	Krona	8.2465
Switzerland	Franc	0.9364
Turkey	Lira	2.6559
Ukraine	Hryvnia	21.1500
United Kingdom	£UK	0.6358

Asia-Pacific	Currency	Per $US
Australia	$Aus	1.2900
China	Yuan	6.2094
Hong Kong	$Hong Kong	7.7522
India	Rupee	63.5137
Indonesia	Rupiah	13,305.0000
Japan	Yen	123.6400
Kazakhstan	Tenge	186.2100
Macau	Pataca	7.9842
Malaysia	Ringgit	3.7540
New Zealand	$New Zealand	1.4476
Pakistan	Rupee	101.7980
Philippines	Peso	45.0750
Singapore	$Singapore	1.3433
South Korea	Won	1,111.9800
Sri Lanka	Rupee	133.8100
Taiwan	$NT	30.9480
Thailand	Baht	33.7600

Middle East and Africa	Currency	Per $US
Bahrain	Dinar	0.3770
Egypt	Pound	7.6246
Israel	Shekel	3.7756
Kuwait	$Kuwait	0.3024
Oman	Rial	0.3850
Qatar	Rial	3.6410
Saudi Arabia	Riyal	3.7502
South Africa	Rand	12.1125
U.A.E.	Dirham	3.6728

Required

1   Determine the annual percentage change in median salaries for the United Kingdom for each four-year period from Table CE-1. What are your conclusions?
2   Determine the annual percentage change in median salaries for Australia for each four-year period from Table CE-2. What are your conclusions?
3   Using the geometric mean as your basis how would you compare the salary increases from the age of 24 to 59 for the United Kingdom and Australia? In which country are you better off salary-wise?
4   For the median salaries according to industry sector for the UK, Table CE-3, sort the data from highest to lowest. Which are the three industry sectors that have the highest median salaries? Which are the three industry sectors that have the lowest median salaries? What is the range of the median salaries?
5   Determine the modal value according to industry sector, Table CE-3. How many times does the modal value occur?
6   Develop the quartile boundary limits for the data in Table CE-3. From these determine: The inter-quartile range, or mid-spread; the quartile deviation (half of the inter-quartile range); mid-hinge (average of first and third quartile).
7   Convert the country median salary data, Table CE-4, into $US. Include Australia and the United Kingdom by using the median values of the data in Tables CE-1 and CE-2 as these countries' median salary. Sort the $US data from highest to lowest and highlight the three highest and three lowest.

Answers

1   From age bracket 25–29 to 30–34 median salary increase is 30.55%. From age bracket 30–34 to 35–39 median salary increase is 23.00%. From age bracket 35–39 to 40–44 median salary increase is 13.82%. From age bracket 40–44 to 45–49 median salary increase is 7.14%. From age bracket 45–49 to 50–54 median salary increase is 6.67%. From age bracket 50–54 to

55–59 median salary increase is 3.75%. Conclusion is that as you get older your rate of salary increase declines.

2 From age bracket 25–29 to 30–34 median salary increase is 24.61%. From age bracket 30–34 to 35–39 median salary increase is 24.37%. From age bracket 35–39 to 40–44 median salary increase is 18.58%. From age bracket 40–44 to 45–49 median salary increase is 13.39%. From age bracket 45–49 to 50–54 median salary increase is 14.21%. From age bracket 50–54 to 55–59 median salary increase is 9.55%. Conclusion is that as you get older your rate of salary increase declines (except from the age bracket 45–49 to 50–54).

3 The geometric mean growth rate of the median salary in the UK from age 24 to 59 is 13.76%; in Australia it is 17.32%. From the salary progression you are better off in Australia.

4 The sorted data is in Table CE-6. The three highest and three lowest industry sectors are shown in italics. The range of the median salaries is £99,500.

5 The modal value is £51,000; it occurs twice; the sectors, Consultancy and Pharmaceuticals/health care.

6 The quartile boundary limits are: $Q_0 = 43,000$; $Q_1 = 49,000$; $Q_2 = 52,834$; $Q_3 = 62,323$; $Q_4 = 142,500$. The inter-quartile range is £13,323; the quartile deviation is £6,661.50; the mid-hinge = £55,661.50.

7 The sorted median salaries by country in $ US are in Table CE-7. The top three and lowest three are given in italics.

Table CE-6

Industry sector	Median salary, £UK
Finance, insurance, risk	*142,500*
Oil/gas exploration and production	*87,057*
Oil refining	*79,605*
Petrochemicals	72,263
Contracting	70,000
Industrial gases	64,308
Mining and minerals	62,323
Health, safety, and environment	62,172
Consumer products	60,400
Chemical and allied products	58,350
Plastics	54,600
Iron, steel, other metals	54,550
Renewable fuels	52,834
Nuclear power generation	52,328
Biochemical engineering	52,000
Consultancy	51,000
Pharmaceuticals/health care	51,000
Education/university	50,000
Waste Management	49,000
Nuclear decommissioning	48,591
Fossil fuel power generation	48,453
Paper and packaging	47,933
Water	*46,373*
Food and drink	*46,000*
Process plant and equipment	*43,000*

Table CE-7

Country	Median salary $US
USA	170,692
Canada	153,734
Australia	136,047
U.A.E.	121,842
UK	110,098
Singapore	89,332
Ireland	81,178
New Zealand	75,988
South Africa	64,396
India	31,096
Malaysia	23,948

4. European food prices

Objective

The objective of this case-exercise is to compare prices that are presented in an indexed form.

Situation

The statistical data in Table EF-1 is a comparable basket of an index of prices in the European Union in 2012. It includes food and non-alcoholic beverages, where the food components are bread and cereals, meat; milk, cheese, and eggs. These individual components are shown in separate columns. Also shown are the index values for alcoholic beverages and tobacco. These index values are relative to a comparable average of 100 for the 27 countries of the European Union. For example if we consider Germany: food and non-alcoholic beverages are 6% more than the European average; bread and cereals are 4% more; meat is 28% more; milk, cheese, and eggs are 92% of the European average or 8% less; alcoholic beverages are 82% of the European average or 18% less; and tobacco is 2% more than the European average. As a further comparison Table EF-2 gives similar index data for countries geographically in Europe but not in the European Union. Again the values shown are relative to 100 or the index value for EU-27.[4]

Required

1   Using the criteria, "Food and non-alcoholic beverages" for the European Union, which three countries in order are the most expensive; the least expensive? Indicate the index value.
2   How would your answer to Question 1 change if you included the ten other countries in Europe that are not part of the European Union? Indicate the index value.
3   Using the criteria, "Milk, cheese, and eggs" for the European Union, which three countries in order are the most expensive; the least expensive? Indicate the index value.
4   How would your answer to Question 3 change if you included the ten other countries in Europe that are not part of the European Union? Indicate the index value.

Table EF-1

No.	Country	Food and non-alcoholic beverages	Bread and cereals	Meat	Milk, cheese, and eggs	Alcoholic beverages	Tobacco
	EU 27	100	100	100	100	100	100
1	Belgium	110	108	118	111	97	95
2	Bulgaria	68	57	59	92	67	57
3	Czech Republic	84	74	73	91	96	69
4	Denmark	143	159	132	117	140	115
5	Germany	106	104	128	92	82	102
6	Estonia	87	84	79	88	102	62
7	Ireland	118	110	110	119	162	199
8	Greece	104	115	91	132	131	74
9	Spain	93	111	83	95	87	83
10	France	109	106	123	100	88	129
11	Italy	111	114	115	126	98	99
12	Cyprus	109	121	89	141	110	82
13	Latvia	87	80	75	96	111	64
14	Lithuania	77	75	63	90	94	55
15	Luxembourg	116	117	129	119	90	81
16	Hungary	81	74	72	88	79	52
17	Malta	98	94	80	113	109	88
18	Netherlands	96	90	117	93	96	108
19	Austria	120	134	132	101	96	86
20	Poland	61	58	55	63	93	58
21	Portugal	90	98	75	105	89	84
22	Romania	67	63	57	93	75	68
23	Slovenia	97	101	93	101	101	67
24	Slovakia	87	82	71	97	91	73
25	Finland	119	130	119	114	175	102
26	Sweden	124	135	126	112	161	132
27	United Kingdom	104	89	100	107	143	194

Table EF-2

Country	Food and non-alcoholic beverages	Bread and cereals	Meat	Milk, cheese, eggs	Alcoholic beverages	Tobacco
Iceland	118	130	119	112	212	131
Norway	186	183	179	214	288	270
Switzerland	155	148	221	133	120	131
Croatia	92	94	75	97	103	57
Montenegro	76	72	67	77	95	34
Former Yugoslav Rep. of Macedonia	58	54	55	67	65	25
Serbia	71	64	64	81	78	30
Turkey	88	70	77	122	205	64
Albania	68	66	52	84	72	27
Bosnia-Herzegovina	76	73	69	78	84	34

5   Using the criteria, "Alcoholic beverages" for the European Union, which three countries in order are the most expensive; the least expensive? Indicate the index value.

6   How would your answer to Question 5 change if you included the ten other countries in Europe that are not part of the European Union? Indicate the index value.

7   If you lived in the United Kingdom how would you compare the prices of the items in the six categories with those of Switzerland?

8   If you lived in the United Kingdom how would you compare the prices of the items in the six categories with those of Turkey?

Answers

1   Most expensive: Denmark (143); Sweden (124); Austria (120). Least expensive: Poland (61); Romania (67); Bulgaria (68).

2   Most expensive: Norway (186); Switzerland (155); Denmark (143). Least expensive: Former Yugoslav Rep. of Macedonia (58); Poland (61); Romania (67).

3   Most expensive: Cyprus (141); Greece (132); Italy (126). Least expensive: Poland (63); Hungary (88) and also Estonia (88).

4   Most expensive: Norway (214); Cyprus (141) Switzerland (133). Least expensive: Poland (63); Former Yugoslav Rep. of Macedonia (67); Montenegro (77).

5   Most expensive: Finland (175); Ireland (162); Sweden (161). Least expensive: Bulgaria (67); Romania (75); Hungary (79).

6   Most expensive: Norway (288); Iceland (212) Turkey (205). Least expensive: Former Yugoslav Rep. of Macedonia (65); Bulgaria (67); Albania (72).

7   In Switzerland compared to the United Kingdom, food and non-alcoholic beverages are 49.04% more expensive; bread and cereals are 66.29% more expensive; meat is 121.00% more expensive; milk, cheese, and eggs are 24.30% more expensive; alcoholic beverages are 16.08% less expensive; and tobacco is 32.47% less expensive.

8   In Turkey compared to the United Kingdom, food and non-alcoholic beverages are 15.38% less expensive; bread and cereals are 21.35% less expensive; meat is 23.00% less expensive; milk, cheese, and eggs are 14.02% more expensive; alcoholic beverages are 43.36% more expensive; and tobacco is 67.01% more expensive.

5. Innovation rankings, I

Objective

This case-exercise illustrates the development of data characteristics in a given database.

Situation

In January 2015 the World Economic Forum was again held in Davos, Switzerland. Innovation has been a major theme at the event since the 1970s when an Austrian professor launched the idea. Countries were awarded an index based on qualitative judgments. In 2015, Bloomberg BusinessWeek developed its own ranking based on six quantifiable factors: intensity of research and development; research personnel; the amount of high-value manufacturing; college graduation rates; patent activity; and the number of high technology companies. Their rankings by country are in Table IR–1.

Table IR-1

Country	Ranking	Country	Ranking
Argentina	50.42	Malaysia	71.54
Australia	81.33	Malta	55.61
Austria	79.21	Morocco	50.07
Belgium	76.01	Netherlands	75.79
Brazil	54.06	New Zealand	77.28
Bulgaria	61.26	Norway	79.86
Canada	83.43	Poland	73.51
China	74.87	Portugal	69.02
Croatia	61.55	Romania	61.89
Czech Republic	68.26	Russia	80.96
Denmark	83.82	Serbia	60.81
Finland	88.38	Slovakia	58.97
France	84.66	Slovenia	73.51
Germany	88.41	Singapore	84.92
Greece	69.64	South Africa	50.16
Hong Kong	64.11	South Korea	96.30
Hungary	65.37	Spain	74.01
Iceland	63.84	Sweden	86.52
Ireland	75.55	Switzerland	79.48
Israel	86.97	Thailand	54.15
Italy	73.66	Tunisia	57.93
Japan	90.58	Turkey	64.04
Latvia	60.28	Ukraine	64.68
Lithuania	57.94	United Kingdom	83.90
Luxembourg	70.00	United States	86.92

Required

1 How many countries are in the database? What is the minimum and maximum ranking?

2 Determine the following characteristics that define central tendency:

   a   average value
   b   median value
   c   modal value and how many
   d   midrange.

3 Determine the following characteristics that illustrate the spread of data:

   a   range
   b   sample variance
   c   population variance
   d   standard deviation considering the data as a sample
   e   standard deviation considering the data as a population
   f   coefficient of variation using the data as a sample
   g   coefficient of variation using the data as a population.

4   Develop the quartile boundary limits. From these boundary limits determine:

   a   the inter-quartile range, or mid-spread
   b   the quartile deviation (half of the inter-quartile range)
   c   mid-hinge (average of first and third quartile).

5   Determine the percentiles for the data.

Answers

1   There are 50 countries in the database. The minimum value of the ranking is 50.07. The maximum is 96.30.
2   (a) Average of the rankings is 71.71; (b) median of the rankings is 73.51; (c) modal value of the rankings is 73.51 and there are two modal values; (d) midrange of the rankings is 73.19.
3   (a) Range, 46.23; (b) sample variance, 144.96; (c) population variance, 142.06; (d) sample standard deviation, 12.04; (e) population standard deviation, 11.92; (f) sample coefficient of variance 16.79%; (g) population coefficient of variance, 16.62%.
4    $Q_0 = 50.07$; $Q_1 = 61.55$; $Q_2 = 73.51$; $Q_3 = 80.96$; $Q_4 = 96.30$.
   (a) Mid-spread, 19.41; (b) quartile deviation, 9.71; (c) mid-hinge, 71.26
5   Table IR–2 gives the percentile values.

Table IR-2

k	Q	k	Q	k	Q	k	Q	k	Q	k	Q
0%	50.07	20%	60.70	40%	68.72	60%	75.65	80%	83.84		
1%	50.11	21%	60.94	41%	69.08	61%	75.76	81%	83.88		
2%	50.16	22%	61.16	42%	69.38	62%	75.87	82%	84.04		
3%	50.28	23%	61.34	43%	69.67	63%	75.98	83%	84.41		
4%	50.41	24%	61.48	44%	69.84	64%	76.47	84%	84.70		
5%	52.06	25%	61.64	45%	70.08	65%	77.09	85%	84.83		
6%	53.84	26%	61.80	46%	70.83	66%	77.94	86%	85.14		
7%	54.10	27%	62.34	47%	71.60	67%	78.88	87%	85.93		
8%	54.14	28%	63.29	48%	72.56	68%	79.30	88%	86.57		
9%	54.75	29%	63.88	49%	73.51	69%	79.43	89%	86.76		
10%	55.46	30%	63.98	50%	73.51	70%	79.59	90%	86.93		
11%	56.51	31%	64.05	51%	73.51	71%	79.78	91%	86.95		
12%	57.65	32%	64.09	52%	73.58	72%	80.17	92%	87.08		
13%	57.93	33%	64.21	53%	73.66	73%	80.71	93%	87.77		
14%	57.94	34%	64.49	54%	73.82	74%	81.06	94%	88.38		
15%	58.30	35%	64.78	55%	73.99	75%	81.24	95%	88.40		
16%	58.81	36%	65.12	56%	74.39	76%	81.83	96%	88.50		
17%	59.40	37%	65.75	57%	74.81	77%	82.86	97%	89.56		
18%	60.04	38%	67.16	58%	75.16	78%	83.52	98%	90.69		
19%	60.44	39%	68.34	59%	75.49	79%	83.71	99%	93.50		
								100%	96.30		

6. Investment options, I

Objective

This case–exercise illustrates the use of the geometric mean in financial investment.

Situation

Susan in California has inherited money from a family member and she intends to invest $50,000 of this inheritance for future needs. Her financial advisor proposes two options. The first option is an investment, principally in overseas bonds and deposits, which has relatively high interest returns in the early years and lower rates in the later years. The second option is an investment, mainly in US bonds and certificates of deposits, that has relatively low interest rates in the early years and higher rates in later years. The rates are given in Table IO-1 for Option 1 and Table IO-2 for the Option 2.

Table IO-1: Option 1

Year	Interest rate
1	6.00%
2	8.50%
3	6.20%
4	5.50%
5	4.90%
6	4.70%
7	4.00%
8	3.70%
9	3.10%
10	2.50%

Table IO-2: Option 2

Year	Interest rate
1	2.50%
2	3.90%
3	3.20%
4	4.20%
5	4.50%
6	5.30%
7	5.70%
8	6.20%
9	6.50%
10	7.70%

Required

1 Determine the value of Susan's investment each year over the ten–year period for Option 1. What is the geometric growth in the mean over the ten years? Verify that by using this value the investment at the end of the tenth year corresponds to the value calculated on an incremental basis.
2 Determine the value of Susan's investment each year over the ten–year period for Option 2. What is the geometric growth in the mean over the ten years? Verify that by using this value the investment at the end of the tenth year corresponds to the value calculated on an incremental basis.
3 Which is the preferred option for Susan based on the value of the investment at the end of ten years? Determine the percentage difference in the gain of the investment between the two options.
4 What would the interest rate have to be in Year 10 for Option 1 such that the value is equal to the value at the end of Year 10 for Option 2? In this case verify that the geometric mean for the two options are the same.

5   If Susan decided to withdraw her investment at the end of Year 5, which option should she select? What is the value of the geometric growth mean for this Option?

Answers

1   Year 1, $53,000.00; Year 2, $57,505.00; Year 3, $61,070.31; Year 4, $64,429.18; Year 5, $67,586.21; Year 6, $70,762.76; Year 7, $73,593.27; Year 8, $76,316.22; Year 9, $78,682.02; Year 10, $80,649.07. Geometric mean is 4.90%. Using this the value of the investment at the end of Year 10 is $80,649.07.
2   Year 1, $51,250.00; Year 2, $53,248.75; Year 3, $54,952.71; Year 4, $57,260.72; Year 5, $59,837.46; Year 6, $63,008.84; Year 7, $66,600.35; Year 8, $70,729.57; Year 9, $75,326.99; Year 10, $81,127.17. Geometric mean is 4.96%. Using this the value of the investment at the end of Year 10 is $81,127.17.
3   Option 2 is the preferred investment. The gain in Option 1 is $30,649.07 and the gain in Option 2 is $31,127.17. Option 2 is an increase of 1.56% over Option 1.
4   The interest rate has to be 3.11% for the tenth year for Option 1. In this case the geometric mean for the two options is the same.
5   Option 1 is the preferred investment with an accumulated value at the end of the fifth year of $67,586.21 compared to a $59,837.46 for Option 2. The geometric mean for Option 1 after five years is 6.21%.

7. Madrid hotel, Spain, I

Objective

The objective of this case-exercise is to show how to evaluate survey results from customer evaluations using a weighting approach according to importance.

Situation

A four-star hotel in Madrid, Spain has a project to improve the quality of its service by incorporating the six-sigma management philosophy. As a first approach, it uses a questionnaire to establish the client's perception of its current facilities. The results of a random sample for 40 clients using this questionnaire are given in Table MH-1. The abbreviations (HR) Human Resources; (F&B) Food and Beverages; (M) Maintenance after each criterion indicate the department to which the personnel are attached. Note, not everybody responded to all of the questions, either because they did not use the facility or because they had no wish to answer.

Required

1   Using weighted averages, which of the criterion listed shows the highest satisfaction response? Indicate the weighted average value.
2   Using weighted averages, which of the criterion listed shows the lowest satisfaction response? Indicate the weighted average value.

Table MH-1

No.	Criteria Score	Very poor 1	Poor 2	Adequate 3	Good 4	Very good 5
1	Cleanliness of hotel room (HR)	5	3	10	11	7
2	Friendliness of hotel staff (HR)	3	7	12	14	2
3	Buffet breakfast selection (F&B)	0	2	7	12	15
4	Restaurant service (HR)	5	3	12	7	7
5	Room service (F&B)	0	4	13	9	6
6	Dinner menu choices (F&B)	0	6	12	10	10
7	Reception and checkout (HR)	0	4	6	14	9
8	Lunch menu choices (F&B)	0	3	8	7	9
9	Bar service (HR)	1	2	12	8	9
10	Wine selections available (F&B)	1	3	13	10	8
11	Fitness center (M)	0	0	5	13	10
12	Swimming pool and area (M)	5	2	12	6	2

3   Using weighted averages, how many of the criterion listed has a score greater than an initial target value of 4.00? Indicate the criterion and the weighted average.

4   Using weighted averages, which criterion that is the responsibility of the Food and Beverage Department, shows the highest satisfaction response? What is this weighted average?

5   Using weighted averages, which criterion that is the responsibility of the Human Resource Department, shows the highest satisfaction response?

6   Assuming that the survey results represent the hotel population, what is a reasonable estimate of the number of clients that use the fitness center?

Answers

1   Fitness Center with a weighted average of 4.18.
2   Swimming pool and area with a weighted average of 2.93.
3   Two criterion: Fitness Center (4.18); Buffet breakfast selection (4.11).
4   Buffet breakfast selection (4.11).
5   Reception and checkout (3.85).
6   About 70%.

8. Olympics, London, Summer 2012

Objective

This case-exercise illustrates that results of statistical data can change according to measurement criteria using sorting and weighted averages depending on the given criteria.

Situation

Table OL-1 gives the final medal count in alphabetical order by country for the Summer Olympics, 2012 held in London, UK.[5]

Table OL-1

Country	Gold	Silver	Bronze	Country	Gold	Silver	Bronze
Afghanistan	0	0	1	Kazakhstan	7	1	5
Algeria	1	0	0	Kenya	2	4	5
Argentina	1	1	2	Kuwait	0	0	1
Armenia	0	1	2	Latvia	1	0	1
Australia	7	16	12	Lithuania	2	1	2
Azerbaijan	2	2	6	Malaysia	0	1	1
Bahamas	1	0	0	Mexico	1	3	3
Bahrain	0	0	1	Moldova	0	0	2
Belarus	2	5	5	Mongolia	0	2	3
Belgium	0	1	2	Montenegro	0	1	0
Botswana	0	1	0	Morocco	0	0	1
Brazil	3	5	9	Netherlands	6	6	8
Britain	29	17	19	New Zealand	5	3	5
Bulgaria	0	1	1	North Korea	4	0	2
Canada	1	5	12	Norway	2	1	1
China	38	27	23	Poland	2	2	6
Colombia	1	3	4	Portugal	0	1	0
Croatia	3	1	2	Puerto Rico	0	1	1
Cuba	5	3	6	Qatar	0	0	2
Cyprus	0	1	0	Romania	2	5	2
Czech Republic	4	3	3	Russia	24	25	33
Denmark	2	4	3	Saudi Arabia	0	0	1
Dominican Republic	1	1	0	Serbia	1	1	2
Egypt	0	2	0	Singapore	0	0	2
Estonia	0	1	1	Slovakia	0	1	3
Ethiopia	3	1	3	Slovenia	1	1	2
Finland	0	1	2	South Africa	3	2	1
France	11	11	12	South Korea	13	8	7
Gabon	0	1	0	Spain	3	10	4
Georgia	1	3	3	Sweden	1	4	3
Germany	11	19	14	Switzerland	2	2	0
Greece	0	0	2	Taiwan	0	1	1
Grenada	1	0	0	Tajikistan	0	0	1
Guatemala	0	1	0	Thailand	0	2	1
Hong Kong	0	0	1	Trinidad and Tobago	1	0	3
Hungary	8	4	5	Tunisia	1	1	1
India	0	2	4	Turkey	2	2	1
Indonesia	0	1	1	Uganda	1	0	0
Iran	4	5	3	Ukraine	6	5	9
Ireland	1	1	3	USA	46	29	29
Italy	8	9	11	Uzbekistan	1	0	3
Jamaica	4	4	4	Venezuela	1	0	0
Japan	7	14	17				

Required

1   If the total number of medals won is the criterion for rating countries, which countries in order are in the first ten? If there is a tie then precedence is given to the country with the most gold.

2   If the number of gold medals won is the criterion for rating countries, which countries in order are in the first ten? If there is a tie then precedence is given to the country with the most medals. What are the changes from Question 1?

3   If there are three points for a gold medal, two points for silver and one point for bronze which countries in order are in the first ten? Indicate the weighted average for these ten countries. How do these results change compared to the results for Question 1?

4   If countries of the European Union were represented as the "United States of Europe", i.e. equivalent to the United States of America, what would be the tally according to total medals obtained? How do these results compare to the results of Question 1?

Answers

1   This is in Table OL-2.

2   This is in Table OL-3. Britain moves up to from fourth to third place; Russia drops from third down to fourth place; Germany moves down from fifth to sixth place; Japan moves down from sixth to tenth place; Australia is not in the first ten places; France climbs from eighth to seventh place; South Korea goes from ninth to fifth place; Italy drops from the list of first ten; Hungary comes into the list now at the ninth place.

3   This is in Table OL-4. Japan drops from sixth to seventh place; Australia drops from seventh to eighth place; France moves up from eighth to sixth place.

4   There are 25 countries from the 28-country European Union, that obtained medals. Austria, Luxemburg, and Malta are not in the list. The results are in Table OL-5. Europe is now well ahead. USA and China each drop a place Russia moves to fourth; Japan to fifth; Australia sixth; South Korea to seventh; and now Ukraine, Canada, and Brazil appear in Table OL-5.

Table OL-2

No.	Country	Gold	Silver	Bronze	Total
1	USA	46	29	29	104
2	China	38	27	23	88
3	Russia	24	25	33	82
4	Britain	29	17	19	65
5	Germany	11	19	14	44
6	Japan	7	14	17	38
7	Australia	7	16	12	35
8	France	11	11	12	34
9	South Korea	13	8	7	28
10	Italy	8	9	11	28

Table OL-3

No.	Country	Gold	Silver	Bronze	Total
1	USA	46	29	29	104
2	China	38	27	23	88
3	Britain	29	17	19	65
4	Russia	24	25	33	82
5	South Korea	13	8	7	28
6	Germany	11	19	14	44
7	France	11	11	12	34
8	Italy	8	9	11	28
9	Hungary	8	4	5	17
10	Japan	7	14	17	38

Table OL-4

No.	Country	Gold 3	Silver 2	Bronze 1	Total	Weighted average
1	USA	46	29	29	104	225
2	China	38	27	23	88	191
3	Russia	24	25	33	82	155
4	Britain	29	17	19	65	140
5	Germany	11	19	14	44	85
6	France	11	11	12	34	67
7	Japan	7	14	17	38	66
8	Australia	7	16	12	35	65
9	South Korea	13	8	7	28	62
10	Italy	8	9	11	28	53

Table OL-5

No.	Country	Gold	Silver	Bronze	Total
1	Europe	95	105	111	311
2	USA	46	29	29	104
3	China	38	27	23	88
4	Russia	24	25	33	82
5	Japan	7	14	17	38
6	Australia	7	16	12	35
7	South Korea	13	8	7	28
8	Ukraine	6	5	9	20
9	Canada	1	5	12	18
10	Brazil	3	5	9	17

9. Royal Mail package delivery

Objective

To illustrate the use of weighted averages.

Situation

The United Kingdom's Royal Mail, whose roots date back to the court of Henry VIII in the 16th century, was privatized in October 2013.[6] The sale, which will provide additional capital, will allow Royal Mail to expand its package delivery business. Like its larger rivals, FedEx of the United States, and DHL of Germany, which is part of Deutsche Post, Royal Mail has benefited for parcel services as more people buy goods online, which are delivered to them by post. As an illustration of the Royal Mail prices, Table RM-1 gives the price of a package according to weight range for delivery from the United Kingdom to anywhere in the United States for guaranteed next day delivery.[7] In addition the table shows for one particular year the number of packages sent to the United States according to weight range.

Table RM-1

Weight, kg	0.10	1.00	2.00	5.00	10.00	20.00	25.00	30.00	40.00	50.00
Price, £	49.99	54.69	64.09	87.54	127.04	186.44	215.94	245.44	304.44	363.44
No. of packages	18,525	13,850	15,245	16,750	14,260	7,250	8,260	4,260	5,250	3,240

Required

1 From the data provided, what are the total revenues received by The Royal Mail?
2 What are the average revenues per the number of packages shipped? Determine your answer using total values and weighted averages according to the proportion by weight.
3 Which package size is the greatest contributor to total revenues received by The Royal Mail? (This can be determined either using total revenues or the proportion of revenues.)
4 What would have to be the number of 50 kg packages shipped at the price level given such that the total annual revenues are £13,000,000? All the other data remains the same.
5 What would have to be the price for a 50 kg package such that the total revenues received by the Royal Mail are £13,000,000? All other data remains the same.

Answers

1 Total revenues are £12,895.24.
2 Average revenues/package are £120.64/package either using total values or sum of proportion by number multiplied by price.
3 Package size that is the biggest contributor to total revenues is the 10 kg package.
4 Number of 50 kg packages such that total revenues are £13,000,000 is 3,528.
5 Price of the 50 kg package such that total revenues are £13,000,000 is £395.77.

10. Ski resort restaurant, I

Objective

The objective of this case-exercise is to illustrate the development of data characteristics in a given database.

Situation

The ski area operators at Val d'Isère, France, are considering building a new restaurant high in the mountains. This restaurant will be a combination full and self-service and will provide breakfast and lunch. There will also be a late bar for those skiers who "close the slopes." Before a decision is made on the construction project, they want to know something about the snow levels at the particular site where the proposed restaurant is to be built. They are concerned that if the snow levels are too deep for any length of time foundation and construction costs would be very high and make the project unfeasible. On the other hand, if the snow level is too low for any length of time, it is felt that this would diminish the number of skiers coming to the area, making the capacity utilization of the restaurant, and hence revenues, low. The data in Table SR–1 is the snow depth in meters taken near the proposed site over the last 15 years. There are 60 items of data as each year snow levels were recorded for the first day in December, January, February, and March. High snow level readings may be as a result of drifting, or minor avalanches.

Table SR-1

3.24	4.72	3.48	4.24	3.80	3.72	4.60	1.91	4.25	4.64
3.12	5.92	3.04	7.18	3.56	1.32	3.44	2.71	4.08	8.19
3.96	3.92	8.20	1.25	7.20	6.04	9.27	0.40	2.60	3.00
8.75	5.70	3.68	4.26	6.80	9.80	7.80	5.75	0.75	6.78
5.93	4.27	6.14	5.78	5.78	1.91	8.37	2.68	7.25	6.10
0.95	6.90	4.12	5.48	5.12	7.20	2.92	1.91	2.12	5.00

Required

1 What is the minimum and maximum of the snow levels from this data?
2 Determine the following characteristics that define central tendency:

    a   average value
    b   median value
    c   modal value and how many
    d   midrange.

3 Determine the following characteristics that illustrate the spread of data:

    a   range
    b   sample variance
    c   population variance
    d   standard deviation considering the data as a sample
    e   standard deviation if the data were considered a population
    f   coefficient of variation using the data as a sample
    g   coefficient of variation using the data as a population.

4 Develop the quartile boundary limits. From these boundary limits determine:

    a   the inter-quartile range, or mid-spread
    b   the quartile deviation (half of the inter-quartile range)
    c   mid-hinge (average of first and third quartile).

5 Determine the percentiles for the data.

Answers

1 Minimum, 0.40 meters; maximum, 9.80 meters.
2 (a) Average, 4.72 m; (b) median, 4.27 m; (c) mode, 1.91 m; 3 modal values; (d) midrange, 5.10 m.
3 (a) Range, 9.40 m; (b) sample variance, 5.1002 m^2; (c) population variance, 5.0152 m^2; (d) sample standard deviation, 2.2584 m; (e) population standard deviation, 2.2395 m; (f) sample coefficient of variance 47.88%; (g) population coefficient of variance, 47.48%.
4 $Q_0 = 0.40$; $Q_1 = 3.10$; $Q_2 = 4.27$; $Q_3 = 6.11$; $Q_4 = 9.80$.
   (a) Mid-spread, 3.01 m; (b) quartile deviation, 1.505 m; (c) mid-hinge, 4.605 m.
5 Table SR-2 gives the percentile values.

Table SR-2

k	Q	k	Q	k	Q	k	Q	k	Q
0%	0.40	20%	2.88	40%	3.94	60%	5.26	80%	6.82
1%	0.61	21%	2.95	41%	3.98	61%	5.48	81%	6.88
2%	0.79	22%	3.00	42%	4.05	62%	5.61	82%	7.01
3%	0.90	23%	3.02	43%	4.09	63%	5.71	83%	7.17
4%	1.06	24%	3.05	44%	4.12	64%	5.74	84%	7.19
5%	1.24	25%	3.10	45%	4.19	65%	5.76	85%	7.20
6%	1.29	26%	3.16	46%	4.24	66%	5.78	86%	7.20
7%	1.40	27%	3.23	47%	4.25	67%	5.78	87%	7.22
8%	1.74	28%	3.34	48%	4.25	68%	5.80	88%	7.25
9%	1.91	29%	3.44	49%	4.26	69%	5.88	89%	7.53
10%	1.91	30%	3.47	50%	4.27	70%	5.92	90%	7.84
11%	1.91	31%	3.50	51%	4.30	71%	5.93	91%	8.07
12%	1.93	32%	3.55	52%	4.49	72%	5.98	92%	8.19
13%	2.05	33%	3.62	53%	4.61	73%	6.04	93%	8.20
14%	2.24	34%	3.68	54%	4.63	74%	6.08	94%	8.28
15%	2.53	35%	3.71	55%	4.68	75%	6.11	95%	8.39
16%	2.64	36%	3.74	56%	4.73	76%	6.13	96%	8.61
17%	2.68	37%	3.79	57%	4.90	77%	6.42	97%	8.87
18%	2.70	38%	3.85	58%	5.03	78%	6.78	98%	9.18
19%	2.75	39%	3.92	59%	5.10	79%	6.79	99%	9.49
								100%	9.80

11. Western USA

Objective

The objective of this case-exercise is compare measurements in the US Imperial measuring system with the metric system and prices in $US and Euros.

Situation

Hélène and Guillaume, students at the University Lyon III in France, are on an exchange program at the University of California, San Diego. In the summer they decide to take a trip by car from San Diego visiting the western part of the USA. Together on the trip were two US students: Mike, a big fellow at six feet five inches and 275 pounds, and Susan, smaller at five feet seven inches and 132 pounds. The itinerary of their trip according to the towns is given in Table WU–1 together with the temperatures at various times of the day in these towns. Table WU–2 gives the price of some food items that they purchased during their trip. They consumed an enormous amount of gasoline on the trip but, at an average price of $3.60 per gallon, they spent much less than if they had covered the same distance in Europe. When they stopped for gas, Guillaume always made sure that the front tire pressure of their rented car was always at 29 psi and 36 psi for the back tires.

Table WU-1

Town	Distance Miles	Temperature °F
San Diego	0	85
Los Angeles	125	92
San Luis Obispo	183	95
San Francisco	231	62
Eureka	280	75
Grants Pass	175	85
Klamath Falls	103	42
Lakeview	96	78
Winnemucca	213	82
Reno	167	62
Ely	321	91
Salt Lake City	241	95
Spanish Forks	50	85
Salina	88	92
Mt. Carmel Junction	148	105
Page	93	104
Grand Canyon	139	110
Williams	59	104
Needles	190	103
Las Vegas	108	100
Barstow	153	105
San Diego	178	97
TOTAL	3,341	

Table WU-2

Food item	Price $US	Quantity
Regular milk	3.77	1 gallon
Fresh white bread	2.18	1 loaf (1 lb)
White rice	1.33	1 lb
Eggs	2.26	Dozen
Local cheese	4.29	1 lb
Chicken breasts	3.49	1 lb
Apples	1.69	1 lb
Oranges	1.65	1 lb
Tomatoes	1.70	1 lb
Potatoes	1.08	1 lb
Lettuce	1.57	1 head

Required

1   Hélène and Guillaume are required to write an account of their trip for their professor at Lyon III. Convert all the numerical data to the metric system and financial data to Euros using the normal basis of presenting price information. Use an exchange rate of €1.00 = $1.36. Appendix A gives other appropriate conversion factors.

Answers

1   Mike is 1,96 meters and weighs 125 kilograms. Susan is 1,70 meters and weighs 60 kilograms. The distances for the itinerary and the temperatures are in Table WU–3

and Table WU-4 gives the food prices. The average price of gasoline was €0,70/liter and the front tire pressure was 2,0 bars and the back tire pressure 2,5 bars.

Note that the numerical data shown here uses a comma for the decimal point, which is the correct way in the French system. However, since this case-exercise is written in English, then the full stop (period) should really be used. (If that isn't confusing?!)

Table WU-3

Town	Distance Kilometers	Temperature °C
San Diego	0,00	29
Los Angeles	201,16	33
San Luis Obispo	294,50	35
San Francisco	371,75	17
Eureka	450,60	24
Grants Pass	281,63	29
Klamath Falls	165,76	6
Lakeview	154,49	26
Winnemucca	342,78	28
Reno	268,75	17
Ely	516,59	33
Salt Lake City	387,84	35
Spanish Forks	80,47	29
Salina	141,62	33
Mt. Carmel Junction	238,18	41
Page	149,66	40
Grand Canyon	223,69	43
Williams	94,95	40
Needles	305,77	39
Las Vegas	173,80	38
Barstow	246,22	41
San Diego	286,46	36
TOTAL	5 376,67	

Table WU-4

Food item	Price €	Quantity
Regular milk	0,73	liter
Fresh white bread	0,88	250g = baguette
White rice	2,16	kg
Eggs	1,66	dozen
Local cheese	6,95	kg
Chicken breasts	5,66	kg
Apples	2,74	kg
Oranges	2,67	kg
Tomatoes	2,76	kg
Potatoes	1,75	kg
Lettuce	1,15	One head

12. World wine production, I

Objective

This case-exercise illustrates the development of data characteristics in a given database.

Situation

The Table WW-1 gives the worldwide wine production, by country, for 2007 and 2010 in thousand liters.[8]

Table WW-1

Country	2007	2010	Country	2007	2010
Albania	17,000	17,000	Latvia	6,200	6,050
Algeria	80,000	70,000	Lebanon	15,000	15,000
Argentina	1,504,600	1,625,000	Lithuania	7,000	7,000
Armenia	5,000	5,000	Luxembourg	13,000	13,500
Australia	955,000	1,073,000	Macedonia	90,000	85,000
Austria	221,340	231,370	Madagascar	9,000	9,000
Azerbaijan	5,000	5,000	Malta	4,790	3,710
Belarus	12,500	10,000	Mexico	100,000	98,000
Bolivia	7,000	7,000	Moldavia	374,400	410,000
Bosnia	5,000	5,000	Morocco	35,000	30,000
Brazil	240,000	245,000	New Zealand	147,600	190,000
Bulgaria	175,700	142,600	Paraguay	6,000	6,000
Canada	50,000	56,000	Peru	60,000	52,000
Chile	828,000	884,000	Portugal	754,200	587,200
China	390,000	425,000	Rumania	501,500	495,740
Croatia	136,537	127,800	Russia	600,000	540,000
Cyprus	18,300	14,652	Slovakia	32,810	34,633
Czech Republic	43,400	54,500	Slovenia	73,840	74,000
Egypt	2,600	3,000	South Africa	851,600	922,000
EU, Other	10,000	9,000	Spain	3,829,000	3,609,700
France	5,212,700	4,626,900	Switzerland	103,900	127,000
Georgia	90,000	80,000	Tajikistan	6,000	6,000
Germany	900,000	932,000	Tunisia	30,000	26,000
Greece	387,430	336,560	Turkey	14,000	14,000
Hungary	314,430	334,370	Turkmenistan	20,000	18,000
Israel	6,500	6,000	Ukraine	210,000	200,000
Italy	4,963,100	4,580,000	United Kingdom	2,500	2,400
Japan	90,000	85,000	United States	2,510,844	2,653,187
Kazakhstan	20,000	20,000	Uruguay	100,000	112,000
Kyrgyzstan	2,000	2,000	Uzbekistan	25,000	25,000

Required

1   What are the minimum and maximum values of wine production in 2007 and 2010? To which countries are these values attributed?
2   Determine the following characteristics that define central tendency both for 2007 and 2010:

   a   average value
   b   median value
   c   modal value and how many
   d   midrange.

3   Determine the following characteristics that illustrate the spread of data for both 2007 and 2010:

   a   range
   b   standard deviation on the basis that the data is all inclusive of world wine production.

4 Develop the quartile boundary limits for both 2007 and 2010. From these boundary limits determine:

    a   inter-quartile range, or mid-spread
    b   mid-hinge (average of first and third quartile).

5 Which ten countries are the leading wine producers? What are your comments about some of these countries when comparing production in 2007 and 2010?
6 What are your comments about global wine production in comparing 2007 and 2010?
7 How would you describe the wine production countries of China, Greece, New Zealand, Switzerland, Cyprus, and Malta?

Answers

1 Year 2007: minimum, 2,000,000 liters (Kyrgystan); maximum, 5,212,700,000 liters (France). Year 2010: minimum, 2,000,000 liters (Kyrgystan); maximum, 4,626,900,000 liters (France).
2 Central tendency:

    a   average 2007 is 453,772,000 liters; average 2010 is 439,748,000 liters
    b   median 2007 is 66,920,000 liters; median 2010 is 63,000,000 liters
    c   mode 2007 is 90,000,000 liters (3 modal values); mode 2010 is 6,000,000 liters (3 modal values)
    d   midrange 2007 is 2,607,350,000 liters; midrange 2010 is 2,314,450,000 liters.

3 Spread:

    a   range 2007 is 5,210,700,000 liters; range 2010 is 4,624,900,000 liters
    b   standard deviation 2007 is 1,063,307,000 liters; Standard deviation 2010 is 990,873,000 liters.

4 The quartile boundary limits in thousand liters are in Table WW-2.
5 The ten leading wine producers are France, Italy, Spain, United States, Argentina, Australia, Germany, South Africa, Chile, and Portugal. The wine production of the "old world" wine producers has declined between 2007 and 2010: France by 11.24%; Italy by 7.72%; Spain by 5.73%; and Portugal by 22.14%. But the wine production of the "new world" producers has increased: Argentina by 8.00%; Australia by 12.36%; South Africa 8.27%; and Chile by 6.76%. This is a threat to the "old world" producers. The production in the USA has also increased by 5.67%, which is mainly in California and also production in Germany by 3.56%.

Table WW-2

Quartile boundary limits	2007	2010
Minimum, Q_0	2,000	2,000
First quartile, Q_1	11,875	9,750
Second quartile, Q_2	66,920	63,000
Third quartile, Q_3	329,423	334,918
Maximum, Q_4	5,212,700	4,626,900

6 Globally the world wine production has declined by an average of 3.09% and the median amount has declined by 5.86%. Since the production of the highest producer France has

fallen the midrange has declined, by 11.23%; the range has fallen by 11.24%; and the standard deviation by 6.81%.

7   Of the six countries identified:

- Wine production of China has increased by 8.97% and it is in the fourth quartile of producers.
- Production in Greece has declined by 13.13%. It is in the fourth quartile.
- Production in New Zealand has increased by 28.73%. It is in the third quartile.
- Production in Switzerland has increased by 22.23%. It's positioned in the third quartile.
- Production in Cyprus has declined by 19.93%. It is positioned in the second quartile.
- Malta's production has declined by 22.55%. It is positioned in the first quartile.

In summary, from the responses to this Question 7 and also Question 5, wine production in Europe, "The Old World" has declined whereas it has increased in many of the "The New World Countries."

Notes

1 Big Mac Index, *The Economist*, January 14, 2012

2 TCE Salary Survey, *The Chemical Engineer*, 888, June 2015, p. 54.

3 Currency rates, London close, June 25, 2015, *The Wall Street Journal, European Edition*, June 26–28, p. 23.

4 "Significant differences in price levels for food, beverages, and tobacco across Europe in 2012", http://epp.eurostat.ec.europa.eu/statistics [accessed June 21, 2013].

5 *International Herald Tribune*, August 13, 2012, p. 13.

6 "Investors excited by Royal Mail", *International Herald Tribune*, October 11, 2013, p. 15.

7 www.parcelforce.com/price-finder [accessed October 14, 2013].

8 http://www.wineinstitute.org/resources/statistics: Source: Trade Data and Analysis [accessed May 1, 2013].

2 Presenting and organizing data
Ways to illustrate information and convince

No	Case-exercise	Statistical concept	Application
1	Big Mac hamburgers, II	Histograms, charts, box plot	Hospitality
2	Charlie's restaurant	Pareto	Hospitality, quality
3	Chemical engineering salaries, II	Box plot, graphs, bar chart	Human resources
4	City Hotel	Spider Web/Radar	Hospitality
5	Fruit distribution	Pareto	Quality, logistics, food
6	Hotel functions, I	Bar charts	Hospitality, operations
7	Innovation rankings, II	Sorting, histogram, box plot	Economics
8	Investment options, II	Line graph, geometric mean	Finance
9	Madrid hotel, Spain, II	Gap bar chart, radar diagram	Hospitality
10	Normandy fishing	Ogives, stem & leaf	Services, food
11	Ski resort restaurant, II	Frequency distributions	Hospitality, sports, project
12	World wine production, II	Histograms	Hospitality, economics

1. Big Mac hamburgers, II

Objective

The objective of this case-exercise is to illustrate the development of histograms, bar charts, and a box and whisker plot. (It uses the same database as the case-exercise Big Mac hamburgers in Chapter 1.)

Situation

Table BM-1 gives the price in local currency of a Big Mac hamburger in various countries and the conversion rate of the local currency to US Dollars.[1]

Required

1 Show the data for the price of a Big Mac hamburger in $US as a bar chart, in alphabetical order.
2 Show the data for the price of a Big Mac hamburger in $US as a histogram, sorted from lowest to highest.
3 Develop the box and whisker plot.
4 Develop a stem-and-leaf display for the data.
5 What are some of the conclusions that you might draw from these visual displays?

Table BM-1

No.	Country	Currency	Price in local currency	Conversion rate to $US (Jan 11, 2012)
1	Argentina	Peso	20.00	4.31
2	Australia	$Aus	4.80	0.97
3	Brazil	Real	10.25	1.81
4	Britain	£	2.49	0.65
5	Canada	$Can	4.73	1.02
6	Chile	Peso	2,050.00	506.00
7	China (a)	Yuan	15.40	6.32
8	Colombia	Peso	8,400.00	1,851.00
9	Costa Rica	Colones	2,050.00	510.00
10	Czech Republic	Koruna	70.22	20.40
11	Denmark	Danish Kroner	31.50	5.86
12	Egypt	Pound	15.50	6.04
13	Eurozone (b)	Euro	3.49	0.79
14	Hong Kong	$Hong Kong	16.50	7.77
15	Hungary	Forint	645.00	246.00
16	India (c)	Rupee	84.00	51.90
17	Indonesia	Rupia	22,534.00	9,160.00
18	Israel	Shekel	15.90	3.85
19	Japan	Yen	320.00	76.90
20	Latvia	Lats	1.65	0.55

Table BM-1 (Continued)

No.	Country	Currency	Price in local currency	Conversion rate to $US (Jan 11, 2012)
21	Lithuania	Litas	7.80	2.72
22	Malaysia	Ringgit	7.35	3.14
23	Mexico	Peso	37.00	13.68
24	New Zealand	$New Zealand	5.10	1.26
25	Norway	Norwegian Kroner	41.00	6.04
26	Pakistan	Rupee	260.00	90.10
27	Peru	Sol	10.00	2.69
28	Philippines	Peso	118.00	44.00
29	Poland	Zloty	9.10	3.52
30	Russia	Rouble	81.00	31.80
31	Saudi Arabia	Riyal	10.00	3.75
32	Singapore	$Singapore	4.85	1.29
33	South Africa	Rand	19.95	8.13
34	South Korea	Won	3,700.00	1,159.00
35	Sri Lanka	Rupee	290.00	113.90
36	Sweden	Swedish Kroner	41.00	6.93
37	Switzerland	Swiss Franc	6.50	0.96
38	Taiwan	$NT	75.00	30.00
39	Thailand	Baht	78.00	31.80
40	Turkey	Lira	6.60	1.86
41	U.A.E.	Dirhams	12.00	3.67
42	Ukraine	Hryvnia	17.00	8.04
43	Uruguay	Peso	90.00	19.45
44	USA	$US	4.20	1.00

(a) Average of five cities; (b) Weighted average price in Eurozone and (c) Maharaja Mac (made with chicken instead of beef).

Answers

1 The bar chart is in Figure BM-1. Note only alternate countries are shown—22 rather than 44—as there is insufficient space. The alphabetical order of the countries starts from bottom to top.
2 The sorted histogram is in Figure BM-2.
3 The box and whisker plot is in Figure BM-3.
4 The stem-and-leaf display is in Figure BM-4.
5 Of the 44 countries surveyed, hamburger prices are generally above the median price in the OECD (rich) and/or English speaking countries. Prices are generally below the median price $3.36 in developing countries (Figure BM-2 and Figure BM-3). If we use the price of the Big Mac hamburger as a benchmark, the information can be translated into saying that the cost of living is lower in developing countries. From the box and whisker plot of Figure BM-4 the data is right-skewed. The average hamburger price ($US 3.59) is greater than the median price ($US 3.36). Most hamburger prices (41%) are between two and three $US and this is confirmed by the stem-and-leaf display of Figure BM-5. (Also, check with the calculations of Big Mac Hamburgers, Part I.)

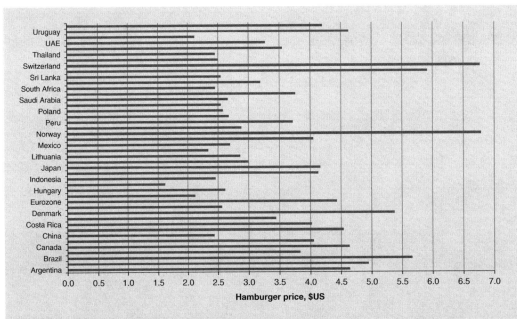

Figure BM-1 Big Mac hamburgers: bar chart of prices.

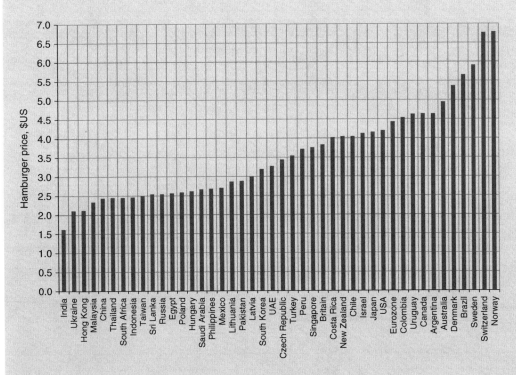

Figure BM-2 Big Mac hamburgers: histogram of sorted data.

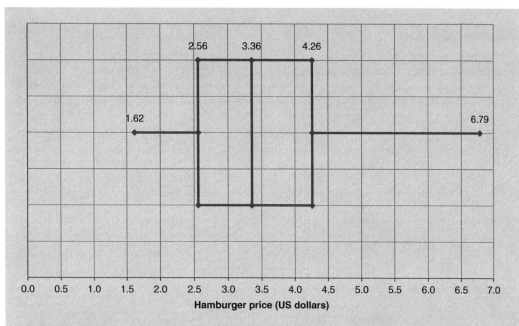

Figure BM-3 Big Mac hamburgers: box and whisker plot.

STEM	LEAVES																
1	62																
2	11	12	34	44	45	45	46	50	55	55	57	59	62	67	68	70	87	89
3	0	19	27	44	55	72	76	83									
4	2	5	5	13	16	20	43	54	63	64	64	95					
5	38	66	92														
6	77	79															

Figure BM-4 Big Mac hamburgers: stem–and–leaf display.

2. Charlie's restaurant

Objectives

This case-exercise illustrates the development of a Pareto diagram, an analytical tool useful in auditing operations.

Situation

Charlie's restaurant has heard many customer complaints about its service. Further, the facility is often below seating capacity, even though in the same geographical area other restaurants are always full. The owner identifies problem areas for the restaurant and records for lunch and dinner, over a three-month period, how often these problems occur, recording them on a check sheet as shown in Table CR–1.

Table CR-1

Area of dissatisfaction	Frequency of occurrence in period
Prices too high	11
Waiting too long to be seated	24
Menu choice limited	19
Food sometimes cold	8
Wine selection limited	5
Serving staff unpleasant	31
Billing errors	7
Salad not fresh	14
Cramped seating	21
Tableware not clean	2

Required

1 Construct a Pareto diagram for this information.
2 What is the problem that happens most often and what is the percentage occurrence? This is the problem area that you would probably tackle first. What are possibly underlying reasons?
3 Which are the four problem areas that constitute a little under 80 percent of the quality problems in the restaurant?

Answers

1 The sorted data from which the Pareto diagram is constructed is in Table CR-2.
2 From this the leading area of dissatisfaction is 'serving staff unpleasant' at 21.83%. This may be a result of lack of motivation, poor work conditions, low pay, or autocratic management relationship.
3 The problem areas are: serving staff unpleasant, 21.83%; waiting too long to be seated, 16.90%; cramped seating, 14.79%; and menu choice limited, 13.38%.

Table CR-2

Area of dissatisfaction	Frequency of occurrence	Percent of total	Cumulative percentage
Serving staff unpleasant	31	21.83	21.83
Waiting too long to be seated	24	16.90	38.73
Cramped seating	21	14.79	53.52
Menu choice limited	19	13.38	66.90
Salad not fresh	14	9.86	76.76
Prices too high	11	7.75	84.51
Food sometimes cold	8	5.63	90.14
Billing errors	7	4.93	95.07
Wine selection limited	5	3.52	98.59
Tableware not clean	2	1.41	100.00
Total	142	100.00	

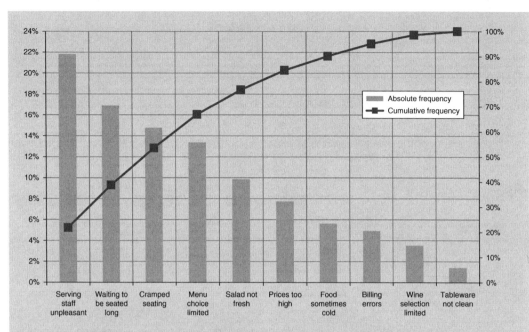

Figure CR-1 Charlies Restaurant: Pareto diagram.

3. Chemical engineering salaries, II

Objective

The purpose of this case-exercise is to illustrate ways to visually present data (it uses the same data files as Chemical engineering salaries in *Chapter 1.*)

Situation

Chemical engineering is a profession that covers chemical plants, oil refining, the food industry, nuclear power, waste disposal, the environment, and a host of other industrial and educational activities. For over 30 years, the Institution of Chemical Engineers has been conducting salary surveys worldwide. In 2015 nearly 4,000 members disclosed their earnings and employee benefits. Table CE-1 shows the median salary by age in the United Kingdom; Table CE-2 shows the median salary by age in Australia; Table CE-3 shows the median salary in the United Kingdom by industry sector; and Table CE-4 shows the median salary by country.[2] Table CE-5 shows exchange rates for the various currencies.[3]

Required

1 Draw line graphs of median salaries by age group for the UK (Table CE-1) and Australia (Table CE-2) in £UK. What are your conclusions?
2 For the median salaries, according to industry sector for the UK, Table CE-3, draw a histogram for salaries from highest to lowest. Which are the three industry sectors that have the highest median salaries? Which are the three industry sectors that have the lowest median salaries?

Table CE-1

Age range	Median salary £UK
25–29	38,300
30–34	50,000
35–39	61,500
40–44	70,000
45–49	75,000
50–54	80,000
55–59	83,000

Table CE-2

Age range	Median salary $Aus
25–29	95,500
30–34	119,000
35–39	148,000
40–44	175,500
45–49	199,000
50–54	227,284
55–59	249,000

Table CE-3

Industry sector	Median salary £UK
Biochemical engineering	52,000
Chemical and allied products	58,350
Consultancy	51,000
Consumer products	60,400
Contracting	70,000
Education/university	50,000
Finance, insurance, risk	142,500
Food and drink	46,000
Fossil fuel power generation	48,453
Health, safety, and environment	62,172
Industrial gases	64,308
Iron, steel, other metals	54,550
Mining and minerals	62,323
Nuclear decommissioning	48,591
Nuclear power generation	52,328
Oil refining	79,605
Oil/gas exploration and production	87,057
Paper and packaging	47,933
Petrochemicals	72,263
Pharmaceuticals/health care	51,000
Plastics	54,600
Process plant and equipment	43,000
Renewable fuels	52,834
Waste management	49,000
Water	46,373
Biochemical engineering	52,000

3 Develop the quartile boundary limits and draw the box and whisker plot. What are your conclusions from this visual display? (You may have determined the quartile boundary limits from Part I of this case exercise.)

4 Draw a bar chart of the median salaries by country (Table CE-4) in $US, from lowest to highest. Include Australia and the UK by using the median values of the data in Tables CE-1 and CE-2 as these countries' median salaries. Which three countries have the lowest median salaries? Which three countries have the highest median salaries? (You may have performed the currency conversion from Part I of this case-exercise.)

Table CE-4

Country	Currency	Median salary
Australia	$Aus	130,000
Canada	$Can	190,000
India	Rupee	1,975,000
Ireland	Euro	72,500
Malaysia	Ringgit	89,900
New Zealand	$New Zealand	110,000
Singapore	$Singapore	120,000
South Africa	Rand	780,000
U.A.E.	Dirham	447,500

Table CE-5

Americas	Currency	Per $US
Argentina	Peso	9.0725
Brazil	Real	3.1129
Canada	$Can	1.2359
Chile	Peso	631.3000
Colombia	Peso	2,555.2400
Ecuador	$US	1.0000
Mexico	Peso	15.5140
Peru	Sol	3.1679
Uruguay	Peso	26.9100
USA	Dollar	1.0000
Venezuela	Bolivar	6.3000

Europe	Currency	Per $US
Bulgaria	Lev	1.7470
Croatia	Kuna	6.7780
Czech Republic	Koruna	24.3070
Denmark	Krone	6.6634
Eurozone	Euro	0.8931
Hungary	Forint	279.2200
Iceland	Króna	131.9100
Norway	Krone	7.8177
Poland	Zloty	3.7263
Russia	Ruble	54.6530
Sweden	Krona	8.2465
Switzerland	Franc	0.9364
Turkey	Lira	2.6559
Ukraine	Hryvnia	21.1500
United Kingdom	£UK	0.6358

Asia-Pacific	Currency	Per $US
Australia	$Aus	1.2900
China	Yuan	6.2094
Hong Kong	$Hong Kong	7.7522
India	Rupee	63.5137
Indonesia	Rupiah	13,305.0000
Japan	Yen	123.6400
Kazakhstan	Tenge	186.2100
Macau	Pataca	7.9842
Malaysia	Ringgit	3.7540
New Zealand	$New Zealand	1.4476
Pakistan	Rupee	101.7980
Philippines	Peso	45.0750
Singapore	$Singapore	1.3433
South Korea	Won	1,111.9800
Sri Lanka	Rupee	133.8100
Taiwan	$NT	30.9480
Thailand	Baht	33.7600

Middle East and Africa	Currency	Per $US
Bahrain	Dinar	0.3770
Egypt	Pound	7.6246
Israel	Shekel	3.7756
Kuwait	$Kuwait	0.3024
Oman	Rial	0.3850
Qatar	Rial	3.6410
Saudi Arabia	Riyal	3.7502
South Africa	Rand	12.1125
U.A.E.	Dirham	3.6728

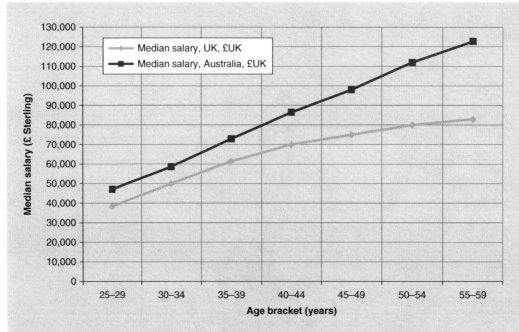

Figure CE-1 Median salaries: UK and Australia.

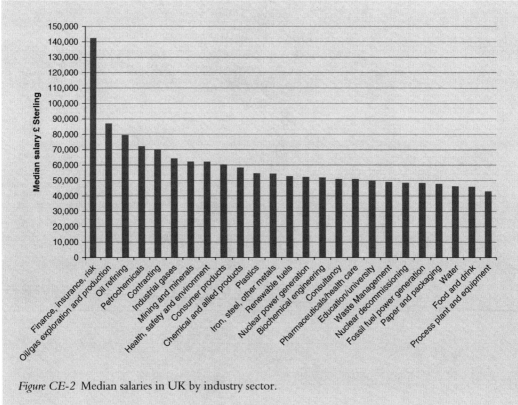

Figure CE-2 Median salaries in UK by industry sector.

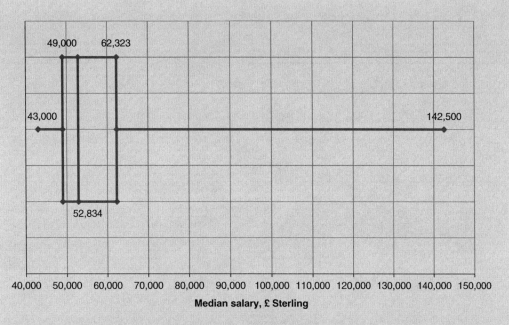

Figure CE-3 Box and whisker plot of median salaries, by industry sector, in the UK.

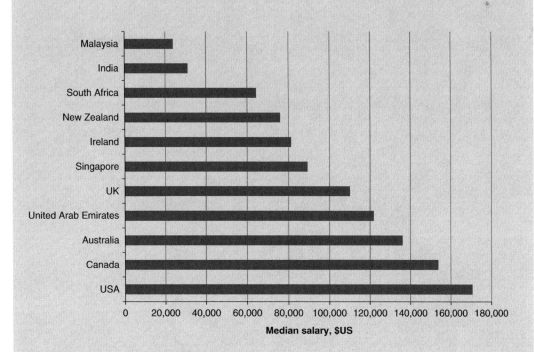

Figure CE-4 Median salary by country, $US.

Answers

1   The line graph is in Figure CE-1. On the same currency basis, salaries are higher in Australia than in the United Kingdom. For both countries the rate of salary increase declines with employee age but less so in Australia. This confirms the conclusion in the same-name case-exercise of Chapter 1.

2   The histogram is in Figure CE-2. The highest salaries are for finance, insurance, risk; oil/gas exploration; and oil refining. The lowest are for water; food and drink; and process plant and equipment.

3   The box and whisker plot is in Figure CE-3. The plot is right-skewed, meaning that there are higher data values above the median of the "median values." The median value is £52,834 (renewable fuels sector). The average of the median values is £60,266, or higher than the median value, which is an indicator of a right-skewed distribution.

4   The bar chart for the median salaries by country is in Figure CE-4. Malaysia, India, and South Africa have the lowest median salaries; USA, Canada, and Australia have the highest median salaries.

4. City Hotel

Objective

The objective of this case-exercise is to illustrate the development of a spider web (radar) diagram for quality management.

Situation

The City Hotel wishes to benchmark its facility with other hotels in the city. The criteria for analyses are: experienced staff, cleanliness of facilities, variety of food, speed of service, fitness facilities, internet/WiFi/fax, location, and relative price. It carries out an audit by using a customer questionnaire over a three-month period. This questionnaire is based on a points score from 1 to 5, with 5 being very good and 1 being poor. The weighted average results of this audit are given in Table CH-1.

Table CH-1

Criteria	Weighted average score
Experienced staff	4.72
Cleanliness of facilities	3.14
Variety of food	2.32
Speed of service	3.91
Fitness facilities	0.97
Internet/WiFi/fax	2.92
Location	3.84
Relative price	2.25

Required

1   Show this information as a spider web diagram.
2   Develop a gap analysis histogram based on the maximum possible score of 5.00. This compares your results with the "world's best"—a six-sigma target. A gap analysis is a tool in ISI-2000.
3   If you felt that at present the world's best was not feasible but you believed a target of at least 4.00 was realistic, what needs to be done and what might be some of the constraints? Show the actual results, and the desired results on the same radar (spider web diagram) of Question 1.

Answers

1   The radar diagram is developed directly using the data in Table CH-1.
2   The data for a gap analysis histogram based on the maximum possible score of 5.00 is in Table CH-2.
3   Weighted unit average scores of criteria need to be increased according to the following except for "Experienced Staff". The radar diagram is developed directly using the data in Table CH-1 plus the add-ons.

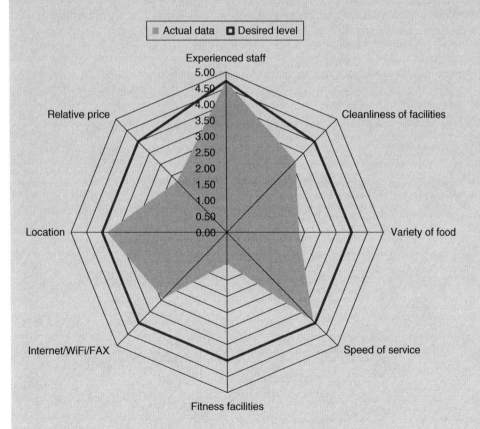

Figure CH-1 City Hotel: spider web diagram.

Table CH-2

Criteria	Weighted average score	Gap
Experienced staff	4.72	0.28
Cleanliness of facilities	3.14	1.86
Variety of food	2.32	2.68
Speed of service	3.91	1.09
Fitness facilities	0.97	4.03
Internet/WiFi/fax	2.92	2.08
Location	3.84	1.16
Relative price	2.25	2.75

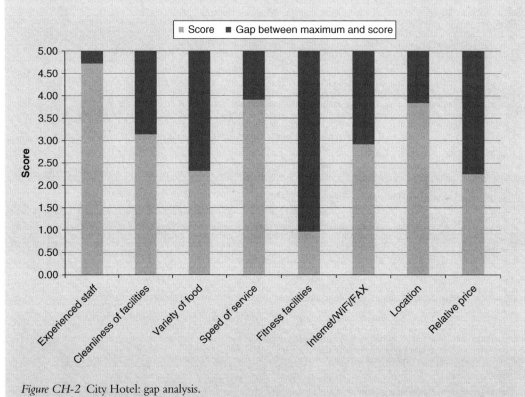

Figure CH-2 City Hotel: gap analysis.

- Cleanliness of facilities—critical aspect part of HAACP. Staff must pay attention. 0.86 needs to be added to the score.
- Variety of food—expand the menu and rotate. 1.68 needs to be added to score.
- Speed of service—improve planning and organization and perhaps take on more staff. 0.09 needs to be added to the score.
- Fitness facilities—constraint is perhaps there is no room available, increased capital cost. 3.03 needs to be added to the score.
- Internet/WiFi/fax—more cabling, increased cost. 1.08 needs to be added to the score.

- Location—not much can be done but maybe spruce up current location. 0.16 needs to be added to score.
- Relative price—if quality is improved perhaps prices can be maintained, if not lower price. Will impact net income. 1.75 needs to be added to score.

5. Fruit distribution

Objective

The objective of this case-exercise is to develop a Pareto diagram for quality and auditing purposes.

Situation

A fruit wholesaler has been receiving complaints from retail clients on the quality of fresh fruit that have been delivered. In order to monitor the situation the wholesaler employed a student to rigorously take note of the problem areas and to record the number of times these problems occurred over a three-month period. Table FD-1 gives the recorded information over the three-month period. The column "reasons" in the table is considered exhaustive, i.e. all of the possible problem areas have been covered.

Table FD-1

Reason	No. of occurrences in three months
Bacteria on some fruit	9
Boxes badly loaded	62
Boxes damaged	17
Client documentation incorrect	23
Fruit not clean	25
Fruit squashed	74
Fruit too ripe	14
Labeling wrong	11
Orders not conforming	6
Route directions poor	30

Required

1 Construct a Pareto curve for this information.
2 What is the problem that happens most often and what is the percentage occurrence?
3 Is the answer to Question 2 the problem area that you would tackle first? Explain.
4 What are the problem areas that cumulatively constitute about 80% of the quality problems in delivery of the fresh fruit?

Answers

1 The Pareto diagram is in Figure FD-1.
2 Fruit squashed; 27.31%.

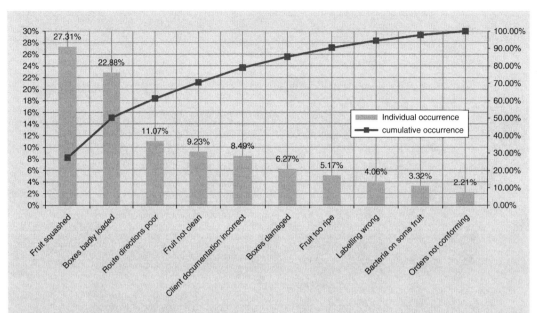

Figure FD-1 Fruit distribution: Pareto diagram.

3   No, bacteria on the fruit that occurs 3.32% of the time is a serious quality problem—part of the HAACP quality criteria—even though it is not the highest value. This problem should be investigated first.
4   Fruit squashed (27.31%) Boxes badly loaded (22.88%); Route directions poor (11.07%); Fruit not clean (9.23%); Client documentation incorrect (8.49%). This equals 78.98% frequency of occurrence.

6. Hotel functions, I

Objective

The objective of this case-exercise is to illustrate the development of stacked histograms.

Situation

A random survey of four- and five-star hotel employees in Austria, France, Germany, Italy, Spain, and the United Kingdom was made to determine in what departments the people were working. This information is given in the Table HF-1.

Required

1   Develop a stacked histogram by hotel functions for this information.
2   Develop a stacked histogram by country for this information.
3   What are your immediate observations from the graphs that you have developed?

Table HF-1

	Austria	France	Germany	Italy	Spain	United Kingdom
Accounting	27	25	34	65	36	28
Conferences & events	20	26	56	82	32	29
Engineering & maintenance	25	26	23	24	32	38
Food and beverage	108	91	152	162	165	109
Front office/reception	35	36	87	45	78	125
Human resources	17	25	32	65	45	56
Marketing/reservation	91	36	78	58	57	69
Purchasing	65	73	45	89	57	89
Recreational facilities	22	32	45	12	32	56
Retail outlets	5	8	19	21	23	5
Room service	58	65	98	45	23	72
Security	18	19	21	32	45	10
Uniformed services	25	26	25	27	31	42

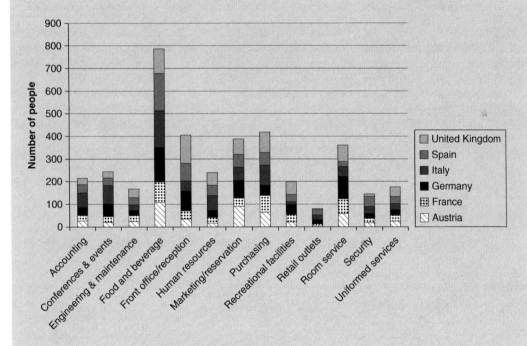

Figure HF-1 Hotel functions: stacked histogram by function.

Answers

1   Figure HF-1 is a stacked histogram by hotel functions.
2   Figure HF-2 is a stacked histogram by country.
3   From the histogram in Figure HF-1, the majority of people surveyed work in food and beverages and this is verified on a country basis in the histogram of Figure HF-2. From the

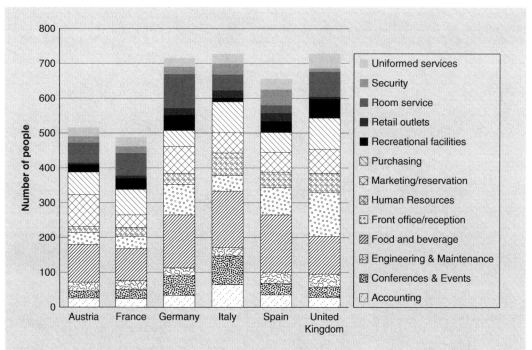

Figure HF-2 Hotel functions: stacked histogram by country.

histogram in Figure HF-1, fewer people surveyed work in retail outlets associated with the hotel and this is verified on a country basis in the histogram of Figure HF-2. This is not surprising since not all hotels have retail stores on their premises.

7. Innovation rankings, II

Objective

The objective of this case-exercise is to demonstrate ways to present quantitative data.

Situation

The World Economic Forum (WEF), held each year in Davos, Switzerland brings together key industrialists, bankers, economists, politicians, and others. to discuss the world situation. In 2015, starting on January 21, it had for its major theme, *Innovation*. Although not a new theme, innovation is considered a key to a country's competitiveness. In its 2014–15 report the Forum ranked Finland first for innovation followed by Switzerland, Israel, Japan, USA Germany, Sweden, the Netherlands, Singapore, and Taiwan. The WEF's ranking is rather qualitative as it relies heavily on an opinion poll of 15,000 executives who are asked to grade nations on capacity for innovation and the degree of cooperation between industry and university researchers. Bloomberg Business has a different ranking procedure, which zeroes in on six quantifiable inputs to the innovation process for the country. These are: intensity of research and development, number of research personnel, number of high technology enterprises,

Table IR-1

Country	Ranking	Country	Ranking
Argentina	50.42	Malaysia	71.54
Australia	81.33	Malta	55.61
Austria	79.21	Morocco	50.07
Belgium	76.01	Netherlands	75.79
Brazil	54.06	New Zealand	77.28
Bulgaria	61.26	Norway	79.86
Canada	83.43	Poland	73.51
China	74.87	Portugal	69.02
Croatia	61.55	Romania	61.89
Czech Republic	68.26	Russia	80.96
Denmark	83.82	Serbia	60.81
Finland	88.38	Singapore	84.92
France	84.66	Slovakia	58.97
Germany	88.41	Slovenia	73.51
Greece	69.64	South Africa	50.16
Hong Kong	64.11	South Korea	96.30
Hungary	65.37	Spain	74.01
Iceland	63.84	Sweden	86.52
Ireland	75.55	Switzerland	79.48
Israel	86.97	Thailand	54.15
Italy	73.66	Tunisia	57.93
Japan	90.58	Turkey	64.04
Latvia	60.28	Ukraine	64.68
Lithuania	57.94	United Kingdom	83.90
Luxembourg	70.00	United States	86.92

extent of high-value manufacturing, patent activity, and postsecondary graduates. Using these criteria they assigned an index for each of the 50 countries in their survey and this information, given alphabetically, for 2014 is in Table IR-1.[4]

Required

1 Sort the data from the Bloomberg survey from the highest to the lowest. Plot this sorted data as a histogram. Which countries are in the top ten? Compare this information to the World Economic Forum results.

2 Develop the boundary limits of the quartiles for the Bloomberg data and draw the box and whisker plot. In the listing there are 30 countries geographically in Europe. Position them according to their quartile value. Which countries have the median value of the index?

Answers

1 The comparative rankings are in Table IR-2. The histogram of the sorted data is in Figure IR-1.

2 The quartile boundary limits are: $Q_0 = 50.07$; $Q_1 = 61.64$; $Q_2 = 73.51$; $Q_3 = 81.24$; and $Q_4 = 96.30$. The box and whisker plot is in Figure IR-2. European countries within the

Table IR-2

No.	Country Bloomberg	Ranking Bloomberg	No.	Country WEF
1	South Korea	96,30	1	Finland
2	Japan	90,58	2	Switzerland
3	Germany	88,41	3	Israel
4	Finland	88,38	4	Japan
5	Israel	86,97	5	United States
6	United States	86,92	6	Germany
7	Sweden	86,52	7	Sweden
8	Singapore	84,92	8	Netherlands
9	France	84,66	9	Singapore
10	United Kingdom	83,90	10	Taiwan

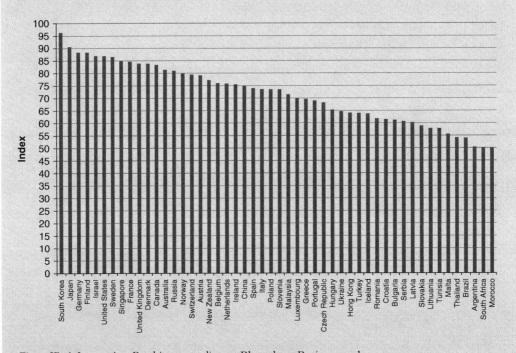

Figure IR-1 Innovation Ranking according to Bloomberg Businessweek.

fourth quartile are Germany, Finland, Sweden, France, United Kingdom, and Denmark. Within the third quartile are Norway, Switzerland, Austria, Belgium, Netherlands, Ireland, Spain, Italy, Poland, and Slovenia. Within the second quartile are Luxembourg, Greece, Portugal, Czech Republic, Hungary, Iceland, and Romania. Within the first quartile are Croatia, Bulgaria, Serbia, Latvia, Slovakia, Lithuania, and Malta.

Poland and Slovenia both have the median value of the index of 73.51.

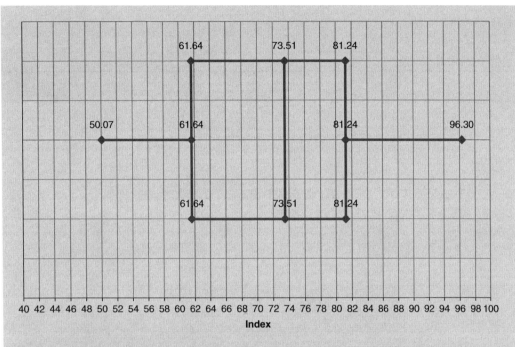

Figure IR-2 Bloomberg Businessweek Innovation Index: box and whisker plot.

8. Investment options, II

Objective

To show graphically the central tendency of the geometric mean compared to individual interest rates.

Situation

Susan in California has inherited money from a family member and she intends to invest $50,000 of this inheritance for future needs. Her financial advisor proposes two options. The first option is an investment principally in overseas bonds and deposits and has relatively high interest returns in the early years and lower rates in the later years. The second option is an investment mainly in US and bonds and certificates of deposits that has relatively low interest rates in the early years and higher rates in later years. The rates are given in Table IO-1. (This is the same information as the case-exercise Investment options in *Chapter 1*.)

Table IO-1

Year	1	2	3	4	5	6	7	8	9	10
Option 1, rate	6.00%	8.50%	6.20%	5.50%	4.90%	4.70%	4.00%	3.70%	3.10%	2.50%
Option 2, rate	2.50%	3.90%	3.20%	4.20%	4.50%	5.30%	5.70%	6.20%	6.50%	7.70%

Required

1 Determine the geometric mean rate for Option 1 and Option 2 (you may have done this in Part I of the case-exercise in *Chapter 1*).
2 Plot the actual interest rates for each option and plot this on a line graph with both of the geometric mean growth rates for the two options. This gives a visual presentation of the central tendency in a financial situation.

Answers

1 The geometric mean growth rate for Option 1 is 4.90%; for Option 2 it is 4.96%.
2 The graph of the percentage rates of the two investment options together with the geometric mean growth rates is given in Figure IO-1.

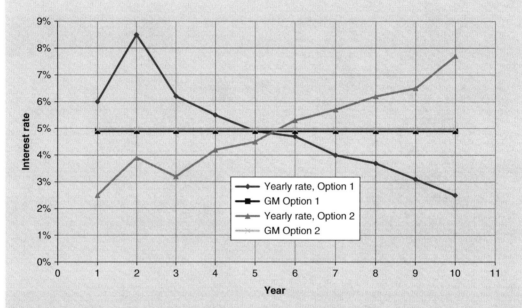

Figure IO-1 Investment options: interest rates.

9. Madrid hotel, Spain, II

Objective

The objective of this case-exercise is to show how to visually present survey results obtained from customer evaluations. (This is the second part of the case-exercise Madrid Hotel, Spain, given in *Chapter 1*.)

Situation

A four-star hotel in Madrid, Spain has a project to improve the quality of its service by incorporating the six-sigma management philosophy. As a first approach, it uses a questionnaire to establish the client's perception of its current facilities. The results of a random sample for 40 clients using this questionnaire are given in Table MH-1. The abbreviations (HR) Human Resources; (F&B) Food and Beverages; (M) Maintenance after each criterion indicate the department to which the personnel are attached. Note that not everybody responded to all of the questions, either because they did not use the facility, or they had no wish to answer.

Table MH-1

No.	Criteria	Very poor 1	Poor 2	Adequate 3	Good 4	Very good 5
1	Cleanliness of hotel room (HR)	5	3	10	11	7
2	Friendliness of hotel staff (HR)	3	7	12	14	2
3	Buffet breakfast selection (F&B)	0	2	7	12	15
4	Restaurant service (HR)	5	3	12	7	7
5	Room service (F&B)	0	4	13	9	6
6	Dinner menu choices (F&B)	0	6	12	10	10
7	Reception and checkout (HR)	0	4	6	14	9
8	Lunch menu choices (F&B)	0	3	8	7	9
9	Bar service (HR)	1	2	12	8	9
10	Wine selections available (F&B)	1	3	13	10	8
11	Fitness center (M)	0	0	5	13	10
12	Swimming pool and area (M)	5	2	12	6	2

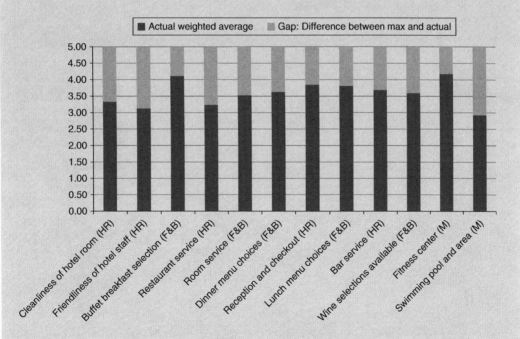

Figure MH-1 Weighted averages vertical histogram: Madrid Hotel, Spain.

Required

1   Present the weighted averages of this data on a vertical histogram that also shows the "gap" between the actual result and the maximum possible score.
2   Present the weighted averages of this data on a radar or spider web diagram.
3   What are your conclusions from the two visual aids developed in Questions 1 and 2?

Answers

1   The vertical histogram is in Figure MH–1.
2   The radar or spider web diagram is in Figure MH–2.
3   The best performing areas are the fitness center and the buffet breakfast, with scores over 4.00. The worst performing is the swimming pool area, with a score below 3.00.

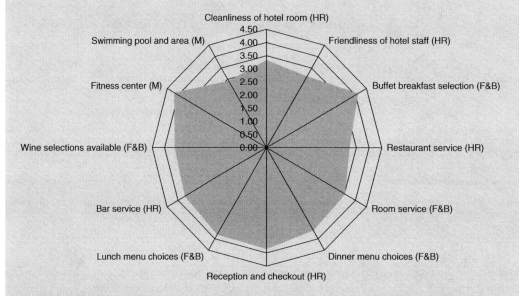

Figure MH-2 Weighted averages spider web diagram: Madrid Hotel, Spain.

10. Normandy fishing

Objective

The purpose of this case-exercise is to demonstrate the presentation and organization of data for easier interpretation.

Situation

Alexander Kowlow fishes off the Normandy coast between England and France. He believes that to financially break even he must catch at least 4,500 kg of fish per trip. Table NF-1 gives a sample of the kilograms of fish caught in the last 30 fishing trips.

Table NF-1

6,175	6,365	3,230	3,420	1,900	2,189
6,650	5,320	4,275	7,600	4,750	4,864
4,370	7,695	6,175	8,550	3,990	5,472
4,560	6,650	7,125	5,700	5,130	3,648
2,736	6,080	6,840	4,560	5,240	3,583

Required

1   Draw the ogives for this data using logical maximum and minimum values for the class limits to the nearest 1,000 kg, increasing by 1,000 kg. From the appropriate ogive, approximately what proportion of the trips break even for Alexander?
2   What is the approximate amount of catch in kg that Alexander exceeds 75% of the time?
3   About 5% of Alexander's catch is less than how many kg?
4   What is the median value of Alexander's catch in kg?
5   Develop a stem-and-leaf display for this data using thousands for the stem and hundreds for the leaves. In what frequency range does the fish catch occur most often?

Answers

1   The ogives are given in Figure NF-1. From the "greater than" ogive the proportion of trips where Alexander breaks even, or catches at least 4,500 kg of fish, is about 64%. (The value is 63.33% by linear interpretation.)
2   From the "greater than ogive" at least 75% of the time Alexander catches 3,800 kg of fish or more. (The value is 3,900 kg by linear interpretation.)

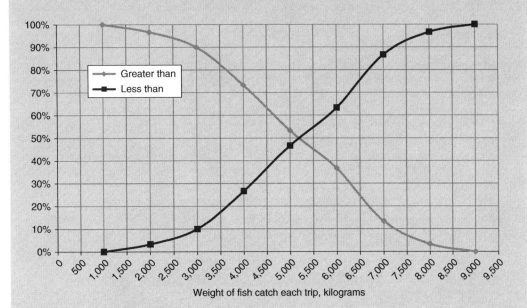

Figure NF-1 Normandy fishing: ogives.

3   From the "less than ogive" 5% of the time or less, Alexander catches less than 2,300 kg of fish. (The value is 2,250 kg by linear interpretation.)
4   From the intersection of the ogives, the median value of Alexandra's catch is about 5,200 kg of fish. (The median value direct from the data is 5,185 kg.)
5   The stem-and-leaf display is shown in Figure NF-2. It is in the frequency range ≥ 6,000 kg to < 7,000 kg that Alexandra's fish catch occurs most often.

Stem	Leaves						
1,000	900						
2,000	189	736					
3,000	230	420	583	648	990		
4,000	275	370	560	560	750	864	
5,000	130	240	320	472	700		
6,000	80	175	175	365	650	650	840
7,000	125	600	695				
8,000	550						

Figure NF-2 Normandy fishing: stem-and-leaf diagram.

11. Ski resort restaurant, II

Objective

The objective of this case-exercise is to illustrate various ways to visually present a dataset. (This uses the same data file as contained in the case-exercise Ski resort restaurant, Part I, given in *Chapter 1*).

Situation

The ski area operators at Val d'Isère, France, are considering building a new restaurant high in the mountains. This restaurant will provide breakfast and lunch, and will be a combination full and self-service. There will also be a late bar for those skiers who "close the slopes." Before a decision is made on the construction project, they want to know something about the snow levels at the particular site where the proposed restaurant is to be built. They are concerned that if the snow levels are too deep for any length of time, foundation and construction costs would be very high and make the project unfeasible. On the other hand, if the snow level is too low for any length of time, it is felt that this would diminish the number of skiers coming to the area, making the capacity utilization of the restaurant, and hence revenues, low. The data in Table SR-1 is the snow depth in meters taken near the proposed site over the last 15 years. There are 60 items of data as each year snow levels were recorded for the first day in December, January, February, and March. High snow level readings may be as a result of drifting, or minor avalanches.

Required

1   Develop the frequency distribution table for this dataset.
2   From the frequency distribution table construct the absolute frequency distribution.

Table SR-1

3.24	4.72	3.48	4.24	3.80	3.72	4.60	1.91	4.25	4.64
3.12	5.92	3.04	7.18	3.56	1.32	3.44	2.71	4.08	8.19
3.96	3.92	8.20	1.25	7.20	6.04	9.27	0.40	2.60	3.00
8.75	5.70	3.68	4.26	6.80	9.80	7.80	5.75	0.75	6.78
5.93	4.27	6.14	5.78	5.78	1.91	8.37	2.68	7.25	6.10
0.95	6.90	4.12	5.48	5.12	7.20	2.92	1.91	2.12	5.00

3   From the frequency distribution table display the relative frequency distribution.
4   From the frequency distribution table construct the absolute frequency polygon using the mid-points of the class ranges as the x-axis.
5   From the frequency distribution table construct the relative frequency polygon using the mid-points of the class ranges as the x-axis.
6   From the frequency distribution draw the greater than and less than ogives on a single display. What can you say about snow level depths of 7 and 8 m?
7   From the quartile values that you have developed according to *Chapter 1*, draw the box and whisker plot. What does this plot demonstrate?
8   From the raw data develop a stem-and-leaf display. What can you conclude from this display?
9   From the percentile values that you have developed according to *Chapter 1*, plot these as a histogram.

Answers

1   The frequency distribution is in Table SR-2. This is interpreted as follows:

- There are 3 or 5.00% of the values in the boundary limits > 0 and ≤ 1.00 m
- There are 5 or 8.33% of the values in the boundary limits > 1 and ≤ 2.00 m
- There are 6 or 10.00% of the values in the boundary limits > 2 and ≤ 3.00 m
- Etc.

The frequency distribution table is the basis for the frequency distributions and the ogives.

Table SR-2

Class number	Boundary limits, meters	Number of values	Percentage of values
0	0	0	0.00%
1	1.00	3	5.00%
2	2.00	5	8.33%
3	3.00	6	10.00%
4	4.00	11	18.33%
5	5.00	10	16.67%
6	6.00	8	13.33%
7	7.00	6	10.00%
8	8.00	5	8.33%
9	9.00	4	6.67%
10	10.00	2	3.33%

2 The absolute frequency distribution is in Figure SR-1.
3 The relative frequency distribution is in Figure SR-2.

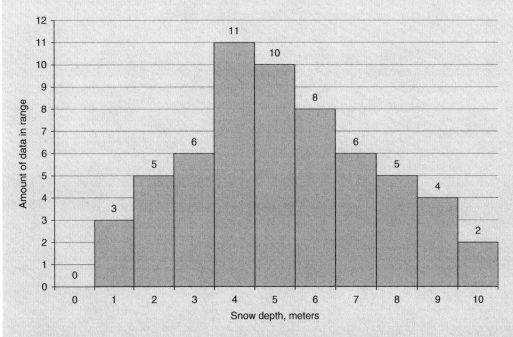

Figure SR-1 Ski resort restaurant: absolute frequency distribution.

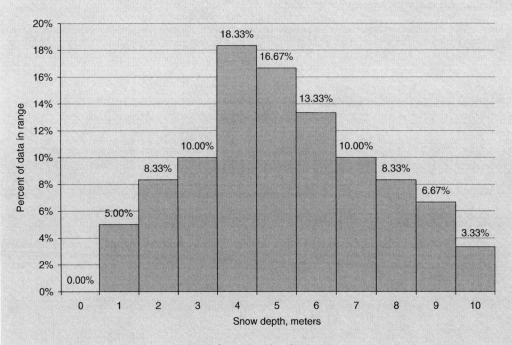

Figure SR-2 Ski resort restaurant: relative frequency distribution.

4   The absolute frequency polygon is in Figure SR–3.
5   The relative frequency polygon is in Figure SR–4.

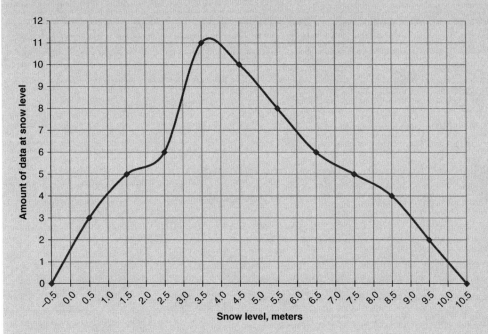

Figure SR-3 Ski resort restaurant: absolute frequency polygon.

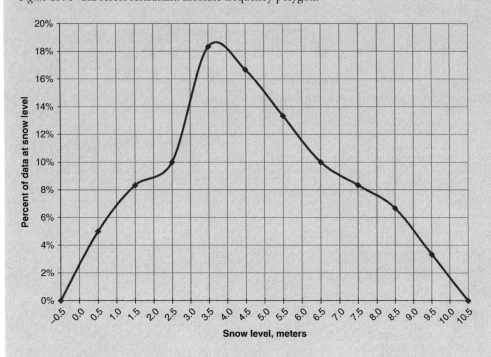

Figure SR-4 Ski resort restaurant: relative frequency polygon.

6   The ogives are in Figure SR–5. From these ogives about 80% of the snow level is less than 7 m. About 10% of the snow level is greater than 8 m.

7   The box and whisker plot is in Figure SR–6. It shows the snow level depths in meters for the minimum (Q_0), first quartile (Q_1), median (Q_2), third quartile (Q_3), and maximum (Q_4). It is right-skewed: the distance from Q_2 to Q_4 is greater than the distance from Q_0 to Q_2;

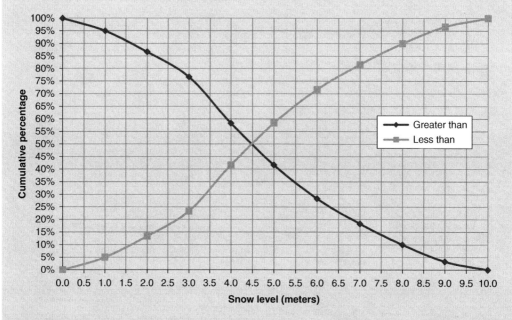

Figure SR-5 Ski resort restaurant: ogives.

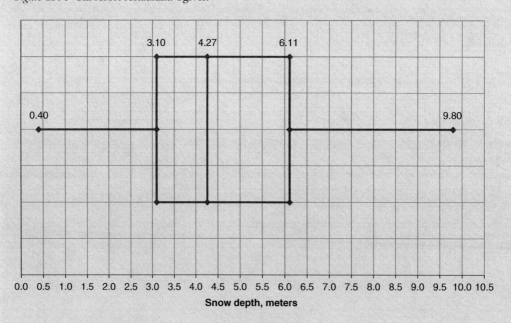

Figure SR-6 Ski resort restaurant: box and whisker plot.

and the average (4.27 m) is greater than the median (4.27 m). The data values greater than 4.27 m are more dispersed than those less than 4.27 m. This is verified by the frequency distribution curves of Figures SR–1 through SR–4.

8 The stem-and-leaf display is in Figure SR–7. This confirms what frequency distribution curves tell us: that most of the snow level measurements are at a depth of between 3 and 4 m. Also it shows that the modal value of the data is 1.91 m.

9 The percentiles are given in Figure SR–8.

STEM   **LEAVES**

0	40	75	95								
1	25	32	91	91	91						
2	12	60	68	71	92						
3	0	4	12	24	44	48	56	68	72	80	92	96
4	8	12	24	25	26	27	60	64	72		
5	0	12	48	70	75	78	78	92	93		
6	4	10	14	78	80	90					
7	18	20	20	25	80						
8	19	20	37	75							
9	27	80									

Figure SR-7 Ski resort restaurant: stem–and–leaf diagram.

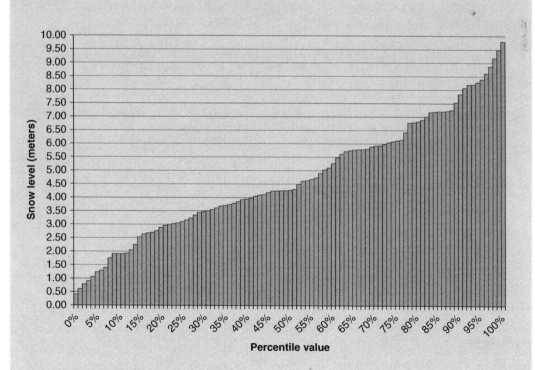

Figure SR-8 Ski resort restaurant: percentiles.

12. World wine production, II

Objective

This case-exercise illustrates the development of side-by-side histograms and bar charts. (This is the second part of the case-exercise, World wine production, Part I given in *Chapter 1*.)

Situation

The Table WW-1 gives the worldwide wine production, by country for 2007 and 2010 in thousand liters.[5]

Table WW-1

Country	2007	2010	Country	2007	2010
Albania	17,000	17,000	Latvia	6,200	6,050
Algeria	80,000	70,000	Lebanon	15,000	15,000
Argentina	1,504,600	1,625,000	Lithuania	7,000	7,000
Armenia	5,000	5,000	Luxembourg	13,000	13,500
Australia	955,000	1,073,000	Macedonia	90,000	85,000
Austria	221,340	231,370	Madagascar	9,000	9,000
Azerbaijan	5,000	5,000	Malta	4,790	3,710
Belarus	12,500	10,000	Mexico	100,000	98,000
Bolivia	7,000	7,000	Moldavia	374,400	410,000
Bosnia	5,000	5,000	Morocco	35,000	30,000
Brazil	240,000	245,000	New Zealand	147,600	190,000
Bulgaria	175,700	142,600	Paraguay	6,000	6,000
Canada	50,000	56,000	Peru	60,000	52,000
Chile	828,000	884,000	Portugal	754,200	587,200
China	390,000	425,000	Rumania	501,500	495,740
Croatia	136,537	127,800	Russia	600,000	540,000
Cyprus	18,300	14,652	Slovakia	32,810	34,633
Czech Republic	43,400	54,500	Slovenia	73,840	74,000
Egypt	2,600	3,000	South Africa	851,600	922,000
EU, Other	10,000	9,000	Spain	3,829,000	3,609,700
France	5,212,700	4,626,900	Switzerland	103,900	127,000
Georgia	90,000	80,000	Tajikistan	6,000	6,000
Germany	900,000	932,000	Tunisia	30,000	26,000
Greece	387,430	336,560	Turkey	14,000	14,000
Hungary	314,430	334,370	Turkmenistan	20,000	18,000
Israel	6,500	6,000	Ukraine	210,000	200,000
Italy	4,963,100	4,580,000	United Kingdom	2,500	2,400

Table WW-1 (Continued)

Country	2007	2010	Country	2007	2010
Japan	90,000	85,000	United States	2,510,844	2,653,187
Kazakhstan	20,000	20,000	Uruguay	100,000	112,000
Kyrgyzstan	2,000	2,000	Uzbekistan	25,000	25,000

Required

1 Sort the entire data from highest to lowest in terms of wine production for 2010. For the first 20 countries in the list, plot a side-by-side-histogram for 2010 and 2007.
2 Sort the entire data from highest to lowest in terms of wine production for 2010. For the first 20 countries in the list, plot a side-by-side-bar chart for 2010 and 2007.
3 What are your overall conclusions both from Questions 1 and 2 covering the years 2007 and 2010?

Answers

1 The side-by-side histogram is given in Figure WW-1.
2 The side-by-side bar chart is given in Figure WW-2.
3 Wine production in the "old wine countries" France, Italy, Spain, Portugal, Greece has declined in the period 2007 to 2010 whereas it has increased in the "New world countries," United States, Argentina, Australia, South Africa, Chile, and New Zealand.

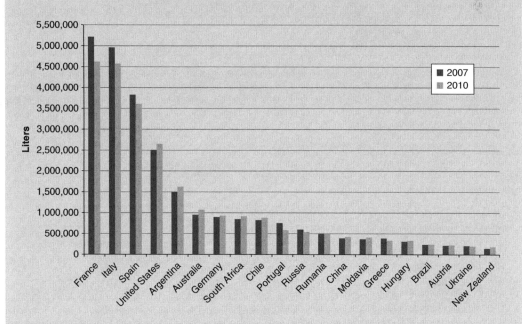

Figure WW-1 Side-by-side histogram: world wine production in 2010 and 2007.

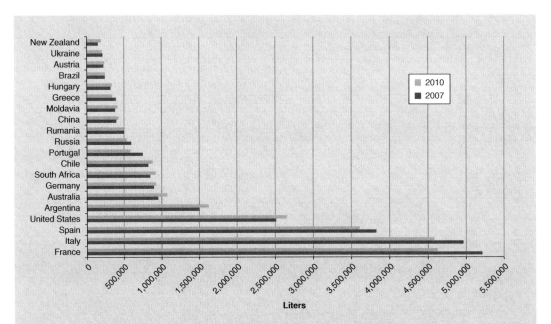

Figure WW-2 Side-by-side bar chart: world wine production in 2010 and 2007

Notes

1 Big Mac Index: *The Economist*, January 14, 2012.
2 TCE Salary Survey, *The Chemical Engineer*, Issue 888, June 2015, p. 54.
3 Currency rates, London close, June 25, 2015, *The Wall Street Journal, European Edition*, June 26–8, p. 23.
4 "Focus on Davos," *Bloomburg Businessweek*, January 19–25, 2015, p. 49.
5 http://www.wineinstitute.org/resources/statistics (source: Trade Data and Analysis, May 1, 2013).

3 Essentials of probability and counting rules

No	Case-exercise	Statistical concept	Application
1	Bottle filling	Basic probability	Operations, food
2	Cutlery	Basic probability	Hospitality
3	David's pizzas	Counting rules	Hospitality
4	Drinking and driving	Basic probability	Human resources, food
5	Electronic circuits	Series/parallel arrangement	Manufacturing
6	Hospitality management school	Venn diagram	Education, hospitality
7	Jardins de la Beffe	Counting rules	Environment, food
8	Menu choices	Counting rules	Hospitality
9	Television quiz game	Basic probability	Entertainment
10	Wedding outfit	Counting rules	Retail
11	Weekly business publication	Basic probability	Publishing
12	Wine choices	Counting rules	Hospitality

1. Bottle filling

Objective

This case-exercise demonstrates the basic probability rules in statistics.

Situation

On an automatic combined beer bottling and capping machine there are two major problems that occur; overfilling of the bottles, and caps not fitting correctly on the bottle top. From collected past data it is known that 2% of the bottles are overfilled. Further, if a bottle is overfilled, then 25% of the bottles are not capped correctly as the pressure differential between the bottle and the capping machine is too low. In addition, even if a bottle is filled correctly, then 1% of the bottles are still not properly capped.

Required

1 Enumerate the four simple events in this bottling situation.
2 Enumerate the four joint events for this situation.
3 What is the percentage of bottles that will be faulty capped and thus have to be rejected before final packing? A faulty cap means that air will get into the product and may be a health risk.
4 Show this information on a cross-classification table using 10,000 bottles as the sample size.

Answers

1 The four simple events are:

   - an overfilled bottle
   - a normally filled bottle
   - an incorrectly capped bottle
   - a correctly capped bottle.

2 The four joint events are:

   - an overfilled bottle and correctly capped
   - an overfilled bottle and incorrectly capped
   - a normally filled bottle and correctly capped
   - a normally filled bottle and incorrectly capped.

3 1.48%.
4 This is in Table BF-1.

Table BF-1

Filling condition	Bottles capped correctly	Bottles capped incorrectly	Total
Bottles correctly filled	9,702	98	9,800
Bottles overfilled	150	50	200
Total	9,852	148	10,000

2. Cutlery

Objective

This case-exercise illustrates some basic probability concepts.

Situation

A small village restaurant keeps its cutlery of knives and forks in one box: there are 22 knives and 29 forks. One day, in preparing for the lunch period, the server reaches into the box without looking and selects the cutlery at random. Each item is withdrawn one at a time.

Required

1   If two utensils are selected at random, one at a time from the box, without replacement, what is the probability that both items selected will be a fork?
2   If two utensils are selected at random from the box, without replacement, what is the probability that a correct place setting, a knife and a fork, will be withdrawn?
3   If three items are selected at random from the box, without replacement, what is the probability that all three will be knives?
4   If three items are selected at random from the box, with replacement, what is the probability that all three will be knives?
5   If three items are selected at random from the box, without replacement, what is the probability that all three will be forks?
6   If three items are selected at random from the box, with replacement, what is the probability that all three will be forks?
7   If two items are selected at random from the box, with replacement, what is the probability that the items will be a correct place setting of a knife and a fork?

Answers

1   If two utensils are selected at random, one at a time from the box, without replacement, the probability that both items selected will be a fork is 31.84%.
2   If two utensils are selected at random from the box, without replacement, the probability that a correct place setting, a knife and a fork, will be withdrawn is 50.04%.
3   If three items are selected at random from the box, without replacement, the probability that all three will be knives is 7.39%.
4   If three items are selected at random from the box, with replacement, the probability that all three will be knives is 8.03%.
5   If three items are selected at random from the box, without replacement, the probability that all three will be forks is 17.55%.
6   If three items are selected at random from the box, with replacement, the probability that all three will be forks is 18.39%.
7   If two items are selected at random from the box, with replacement, the probability that the items will be a correct place setting of a knife and a fork is 49.06%.

3. David's pizzas

Objective

The objective of this case-exercise is to show the use of the counting rules to determine the number of choices available.

Situation

David sells pizzas in the village of Dardilly in the Rhône Department, France on Tuesdays and Fridays. The sizes of the pizzas are 15, 30, or 40 cm in diameter. The toppings that David offers with any of the diameters are as in Table DP-1.

Table DP-1

Goats cheese	Ham		Tomatoes	Salmon	Olives
Mushrooms	Sausage		Red pepper	Lentils	Walnuts
Anchovies	Mozzarella cheese		Gherkin		

Required

1   If a client asks for a 30 cm pizza with three toppings, how many possible different combinations of this size of pizza can be offered to the client?
2   If a client asks for a 30 cm pizza with five toppings, how many possible different combinations of this size of pizza can be offered to the client?
3   If a client asks for a 30 cm pizza, what is the maximum number of different combinations of this size of pizza can be offered to the client?
4   David plans to make a display of six of his bestselling 30 cm size pizzas. This display will be on his front counter in a linear (straight line) array. How many different ways can these six pizzas be organized on his counter?
5   How many different combinations of pizza are possible for the three sizes, 15, 30, and 40 cm if five toppings are selected from all of those possible?

Answers

1   From Table If a client asks for a 30 cm pizza with three toppings the possible different combinations of this size of pizza that can be offered to the client is 286.
2   If a client asks for a 30 cm pizza with five toppings, the possible different combinations of this size of pizza that can be offered to the client is 1,287.
3   If a client asks for a 30 m pizza, the maximum number of different combinations of this size of pizza that can be offered to the client is 1,716.
4   If David's display is on his front counter in a linear (straight line) array then the different ways these six pizzas be organized on his counter is 720.
5   The number of different combinations of pizza possible for the three sizes, 15, 30, and 40 cm if five toppings are selected from all of those possible is 3,861.

4. Drinking and driving

Objective

This case-exercise illustrates basic probability concepts.

Situation

Statistics indicate that, on average, 68% of people drive after drinking more than the legal amount of alcohol. Further, past data indicates that if a person drinks more than the legal amount of alcohol and drives, there is an 80% chance that a person will be involved in an injurious or lethal accident. Even if a person does not drink and drive, there is still a 10% chance that a person will be involved in an accident resulting in injury or fatality.

Required

1   What are the four simple events in this situation?
2   What are the joint events for this situation?
3   What is the probability of an individual being in an automobile accident?
4   If the analysis were made looking at a sample of 5,000 people, how would this information appear in a cross-classification table?

Answers

1   Do not drink and drive; do drink and then drive; involved in an accident; not involved in an accident.
2   Drink before driving and then involved in an accident; drink before driving and not involved in an accident; do not drink before driving and then involved in an accident; drink before driving and not involved in an accident.
3   The probability of an individual being in an automobile accident is 57.60%.
4   The cross classification is in Table DD-1.

Table DD-1

Event	Drink before driving	Do not drink before driving	Total
In an accident	2,720	160	2,880
Not in an accident	680	1,440	2,120
Total	3,400	1,600	5,000

5. Electronic circuits

Objective

This case-exercise demonstrates the application of series and parallel arrangements of basic probability.

Situation

A manufacturer assembles printed electronic circuits that are used in the automobile industry for a variety of applications. For four different applications there are four components A, B, C, and D that have individual reliabilities of 99%, 96%, 90%, and 92% respectively. In these four applications the four components are assembled according to the following flow schemes.

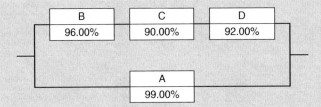

Figure EC-1 Electronic circuits: application 1

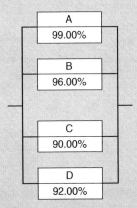

Figure EC-2 Electronic circuits: application 2

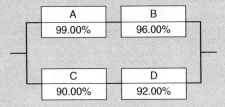

Figure EC-3 Electronic circuits: application 3

Figure EC-4 Electronic circuits: application 4

Required

1   Determine the system reliability of each of the four applications.
2   What is the probability of system failure for each of the four applications?

Answers

1   Application 1: 78.6931%; Application 2: 99.7949%; Application 3: 99.9997%; Application 4: 99.1469%.
2   Application 1: 21.3069%; Application 2: 0.2051%; Application 3: 0.0003%; Application 4: 0.8531%.

6. Hospitality management school

Objective

The objective of this case-exercise is to show the application of the Venn diagram to illustrate probability. A Venn diagram gives a visual presentation of a given probability situation.

Situation

A hospitality management school offers three-year degree programs specializing in hotel management, culinary art, tourism, casino operation, and health spa management. There is a combined option that specializes in joint hotel management and tourism for which the registered students spend additional class time. In one particular year there are 120 students enrolled in the hospitality degree programs. Of these 120 students, 30 are taking classes in tourism, 38 are taking classes in hotel management. Included in these figure are 16 who have elected to specialize jointly in hotel and tourist management. This information is representative of the general profile of the school's hospitality degree management program.

Required

1   Illustrate the enrolment on a Venn diagram.
2   From the Venn diagram, what is the probability that a randomly selected student specializes exclusively in tourism?
3   From the Venn diagram, what is the probability that a randomly selected student specializes exclusively in hotel management?
4   From the Venn diagram, what is the probability that a randomly selected student specializes in hotel management and tourism?
5   From the Venn diagram, what is the probability that a randomly selected student will be taking a speciality in tourism?
6   From the Venn diagram, what is the probability that a randomly selected student will be taking a speciality in hotel management?
7   From the Venn diagram, what is the probability that a randomly selected student specializes in hotel management or tourism?
8   From the Venn diagram, given that a randomly selected student is in hotel management, what is the probability that the student also specializes in tourist management? Verify your answer using Bayes' theorem.
9   From the Venn diagram, given that a student is in tourism, what is the probability that a randomly selected student is also specializing in hotel management? Verify your answer using Bayes' theorem.

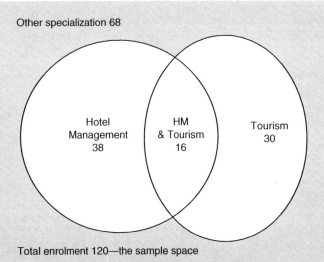

Other specialization 68

Total enrolment 120—the sample space

Figure HM-1 Hospitality management: Venn diagram.

Answers

1　The Venn diagram is in Figure HM-1.
2　The probability that a randomly selected student specializes *exclusively* in tourism is 11.67%.
3　The probability that a randomly selected student specializes *exclusively* in hotel management is 18.33%.
4　The probability that a randomly selected student specializes in hotel management *and* tourism is 13.33%.
5　The probability that a randomly selected student will be taking a speciality in tourism is 25.00%.
6　The probability that a randomly selected student will be taking a speciality in hotel management is 31.67%.
7　The probability that a randomly selected student specializes in hotel management *or* tourism is 43.33%.
8　Given that a randomly selected student is studying hotel management, the probability that the selected student specializes in tourism is 42.11%.
9　Given that a randomly selected student is studying tourism, the probability that the selected student specializes in hotel management is 53.33%.

7. Jardins de la Beffe

Objective

The objective of this case-exercise is to illustrate the use of the combination and permutation rules for determining number of available choices.

Situation

The village of Dardilly, in the Rhone region of France, acquired an agricultural terrain of approximately 11,000 m^2 in 2007, mostly occupied by an orchard of 600 apple trees that had been abandoned for 12 years. The remainder of the land was covered with brambles and weeds; it was eventually cleared and developed into small plots each of 50 m² for allotments. The orchard has four varieties of apple—Melrose, Granny Smith, Canadian, and Golden Delicious. This whole area was the basis for a voluntary association called *Les Jardins de la Beffe* (The Gardens of the Beffe), where the current President is Joël Varlet. The objective of this association is an allotment where residents of Dardilly can grow vegetables for their personal consumption providing they also take care of a certain number of apple trees. A portion of the fruit from these trees goes to the local retirement home. The rest is for personal consumption. It must not be sold. All growing activity in the Garden has to be biological and natural; no pesticides are allowed. In order to identify the fruit trees Joel uses colored beads in various combinations attached on a cord like a necklace to each tree. Table JB-1 gives the bead colors that are available.

Table JB-1

Red	Yellow	Grey	Turquoise	Navy blue
Sky Blue	Purple	Black	Indigo	White
Green	Rose	Orange	Violet	Brown

Required

1 If all beads listed in Table JB-1 are used in a combination arrangement, what is the minimum number that must be selected in order to uniquely identify each tree? How many trees could in fact be identified by this combination arrangement?

2 Using a combination arrangement, what is the minimum number of beads from Table JB-1 needed to uniquely identify each tree? In this case, how many beads from this selection have to be chosen? How many trees could in fact be identified by this combination arrangement?

3 In a combination arrangement, what is the minimum number of beads needed to uniquely identify each tree if only two are selected from the beads? Again, how many trees in fact could be uniquely identified?

4 If all beads in Table JB-1 are used, what is the minimum number in a permutation arrangement that must be chosen to uniquely identify each tree? Again, how many trees in fact could be uniquely identified?

5 If beads were organized in a permutation combination, what is the minimum number needed to uniquely identify each tree and how many must be chosen from this number. Again, how many trees in fact could be uniquely identified?

6 There are 30 members in the garden association and each member is assigned an equal number of trees. In this case, in a combination arrangement, what is the minimum number of beads, and how many in a combination arrangement are needed to uniquely identify trees? Each member has the same color code.

7 There are 30 members in the garden association and each member is assigned an equal number of trees. In this case, in a permutation arrangement, what is the minimum number of beads, and how many in a permutation arrangement are needed to uniquely identify trees? Each member has the same color code.

Answers

1 Minimum number in the selection out of 15 beads is 4 to give 1,365.
2 Minimum number of colors is 12 with 5 each time from this selection to give 792.
3 There are 36 colors needed if only 2 beads are chosen and this gives a value of 630.
4 In a permutation arrangement using all the 15 beads 3 must be chosen from the selection to give a value of 2,730.
5 If beads were organized in a permutation combination the minimum number needed to uniquely identify each tree is 6 with 5 being chosen from this number. In fact this could uniquely identify 720 trees.
6 The minimum number of beads for 30 members in a combination arrangement is 6 with 3 beads being used each time in the selection to give 20.
7 The minimum number of beads for 30 members in a permutation arrangement is 5 with 2 beads being used each time in the selection to give 20.

8. Menu choices

Objective

This case-exercise illustrates the application of the counting rules in statistics.

Situation

A hotel in France offers a three-course dinner in its restaurant. In Table MC-1 are the meal choices available. For €16 a customer can select one option each from the *entrée*, *plat garni*, and *dessert*. For €14 a client can choose a *plat garni*, and *dessert* but not an entrée. Alternatively for €12 a client can chose only a *plat garni* taking neither an *entrée* nor *dessert*.

Table MC-1

Entrées	Plat Garni	Dessert
Potage	Escalope milanaise	Salade de fruits
Oeuf mimosa	Poulet basquais	Oeufs à la neige
Feuilleté de champignons	Steak grillé	Coupe de glace
Salade composée	Sauté de veau Maringote	Crème caramel
Jambon de pays	Fricassé de volaille	Mousse au chocolat
Assiette de crudité	Escalope aux champignons	Corbeille de fruits
Pâté de campagne	Longe de porc aux olives	Tarte aux pommes
Jambon de Paris	Escalope savoyarde	Gâteau au chocolat
Salade savoyarde	Filet de Dorade au citron vert	Tranche napolitaine
	Truite à l'Estragon	
	Darne de Saumon meunière	

Required

1 How many different choices are available if a client pays €16?
2 How many different choices are available if a client pays €14?
3 How many different choices are available if a client pays €12?

Answers

1 If a client pays €16 there are 891 choices.
2 If a client pays €14 there are 99 choices.
3 If a client pays €12 there are 11 choices.

9. Television quiz game

Objective

This case-exercise demonstrates the application of basic probability.

Situation

In a television quiz game there are three contestants. These contestants have to make a selection, in turn, choosing one box from ten closed boxes. One box contains a check for €100,000; three boxes each contain a check for €10; and the remaining boxes are empty.

Required

1 Before the game starts, what is the probability that the third contestant selects the box containing the check for €100,000?
2 The game starts, and neither the first nor the second contestant select the box with the €100,000 check. Now, what is the probability that the third contestant selects the box containing the check for €100,000?
3 The game starts. What is the probability that each contestant wins a check for €10.00?
4 The game starts. What is the probability that no contestant wins the check for €100,000?
5 The game starts. What is the probability that all contestants open an empty box?
6 The game starts. What is the probability that only one contestant wins a check for €10.00? The others open another box.

Answers

1 Before the game starts, the probability that the third contestant selects the box containing the check for €100,000 is 10%.
2 After the game starts, the probability that the third contestant selects the box containing the check for €100,000 is 12.50%.
3 After the game starts, the probability that each contestant wins a check for €10.00 is 0.83%.
4 After the game starts, the probability that no contestant wins a the check for €100,000 is 70.00%.
5 After the game starts, the probability that all contestants open an empty box is 16.67%.
6 After the game starts, the probability that just one contestant wins a check for €10 is 52.50%.

10. Wedding outfit

Objective

The objective of this case-exercise is to demonstrate the application of counting rules.

Situation

Christine is choosing a wedding outfit for her son Guillaum's wedding. She will buy a blouse, jacket, skirt, shoes, and a hat. In the store where she intends to buy the outfit there is a choice of nine blouses, ten jackets, fifteen skirts, thirteen pairs of shoes, and fifteen hats. Guillaume is not too thrilled about his mum wearing a hat and so Christine is thinking possibly of not including a hat.

Required

1   If Christine buys a hat with her outfit, how many combinations are possible?
2   With the option of having or not having a hat with her outfit, how many combinations are possible?
3   If the hat is definitely not in the selection, how many combinations are possible?

Answers

1   If Christine buys a hat with her outfit, there are 263,250 possible combinations.
2   With the option of having or not having a hat with her outfit, there are 280,800 possible combinations.
3   If the hat is definitely not in the selection, there are 17,550 possible combinations.

11. Weekly business publication

Objective

The objective of this case-exercise is to illustrate to calculation of basic probability.

Situation

A weekly business publication in Europe performs a survey or its readers and classifies the survey responses according to the person's country of origin and type of work. This information according to the number or respondents is given in Table WBP-1.

Table WBP-1

Country	Consultancy	Engineering	Investment banking	Product marketing	Architecture
Denmark	852	232	541	452	385
France	254	365	842	865	974
Spain	865	751	695	358	845
Italy	458	759	654	587	698
Germany	598	768	258	698	568

Required

1   What is the probability that a survey response taken at random comes from a reader in Italy?
2   What is the probability that a survey response taken at random comes from a reader in Italy and who is working in engineering?
3   What is the probability that a survey response taken at random comes from a reader who works in consultancy?
4   What is the probability that a survey response taken at random comes from a reader who works in consultancy and is from Germany?
5   What is the probability that a survey response taken at random from those who work in investment banking comes from a reader who lives in France?
6   Given that a reader lives in France, what is the probability that in a survey response taken at random is working in investment banking?
7   Given that a reader lives in France, what is the probability that in a survey response taken at random is working in engineering or architecture?

Answers

1   The probability that a survey response taken at random comes from a reader in Italy is 20.60%?
2   The probability that a survey response taken at random comes from a reader in Italy and who is working in engineering is 4.95%.
3   The probability that a survey response taken at random comes from a reader who works in consultancy is 19.76%
4   The probability that a survey response taken at random comes from a reader who works in consultancy and is from Germany is 3.90%.
5   The probability that a survey response taken at random from those who work in investment banking comes from a reader who lives in France is 28.16%.
6   Given that a reader lives in France, the probability that in a survey response taken at random is working in investment banking is 25.52%.
7   Given that a reader lives in France, the probability that in a survey response taken at random is working in engineering or architecture is 40.58%.

12. Wine choices

Objective

The objective of this case-exercise is to illustrate the application of the counting rules in statistics.

Situation

A restaurant has in its wine cellar six different types of white wine, seven different types of rosé wine, and ten different types of red wine.

Required

1   If a client asks for a white, rosé, and red wine, how many combinations can the sommelier offer?
2   If a client asks for only a white and red wine, how many combinations can the sommelier offer?
3   What is the minimum number of different red wines that the restaurant must keep in the wine cellar in order that the sommelier can offer a minimum of 1,000 combinations of white, rosé, and red wine? The choices of white and rosé wine remain unchanged as in the original situation.
4   What is the minimum number of different white wines that the restaurant must keep in the wine cellar in order that the sommelier can offer a minimum of 1,000 combinations of white, rosé, and red wine? The choices of rosé and red wine remain unchanged as in the original situation.
5   What is the minimum number of different rosé wines that the restaurant must keep in the wine cellar in order that the sommelier can offer a minimum of 1,000 combinations of white, rosé, and red wine? The choices of white and red wine remain unchanged as in the original situation.
6   Besides just the choices to the customer, what are considerations that the restaurant must consider in its operations?

Answers

1   If a client asks for a white, rosé, and red wine, 420 combinations can be offered.
2   If a client asks for only a white and red wine, 60 combinations can be offered.
3   The minimum number of different red wines that the restaurant must keep in the wine cellar is 24 and this would give 1,008 choices.
4   The minimum number of different white wines that the restaurant must keep in the wine cellar is 15 and this would give 1,050 choices.
5   The minimum number of different rosé wines that the restaurant must keep in the wine cellar is 17 and this would give 1,020 choices.
6   The more wine that is kept in storage, the greater is the stocking cost or the working capital required. Also, more storage space is needed.

4 Discrete data and probability

Chance and risk applied to integer values

No	Case-exercise	Statistical concept	Application
1	Bicycle rental	Expected values	Maintenance
2	Coffee shop	Poisson distribution	Services, food
3	Diploma	Binomial	Education
4	Fenwicks	Poisson-Binomial	Logistics
5	Hair driers	Binomial	Operations
6	Newspaper delivery	Binomial/Poisson	Delivery service
7	Petrol pump	Poisson	Services
8	Registered and dining	Binomial	Hospitality
9	Rental cars	Expected, random variables	Services
10	Restaurant customers	Expected values	Hospitality
11	Restaurant food sales	Expected values	Hospitality
12	Restaurant staff	Binomial, Poisson	Hospitality

1. Bicycle rental

Objective

The objective of this case-exercise is to demonstrate the application of randomness and expected values for decision making.

Situation

A company in Switzerland rents out all-terrain bicycles during the summer month for use in the Alps. At one location the firm has just one employee who devotes his time to the rental arrangement. Since he is alone, he has not time to perform preventive maintenance but only repairs the bikes when a client brings them back after they have broken. If this happen, the firm estimates it loses SF 30.00 each time because of customer irritation and also it is obliged to let the client have a longer rental time than that for which they have actually paid. Based on data for the last summer period, Table BR-1 gives the breakdown frequency for the bikes.

   The firm is thinking of employing another person on a part-time basis whose sole function will be to perform preventive maintenance on the fleet of bikes. This would cost the firm SF 280.00 per month. Even with this preventive maintenance there is still expected to be, on average, one bike broken down per week. Base your analysis on a four-week month.

Table BR-1

No of bikes broken down	1	2	3	4	5	6	7
Number of weeks this level occurred	2	3	5	4	6	2	2

Required

1   What is the expected number per week of bikes that are broken down?
2   What is the estimated cost per week due to bicycles that are broken down?
3   Using expected values make a decision as to whether the firm should hire a second person part-time to perform preventive maintenance or stay with its present policy. Justify your response.

Answers

1   Expected value per week of the number of bikes that are broken down is 3.96.
2   Estimated cost per week due to bicycles that are broken down is SF 118.80.
3   The rental agency should hire part-time labor. Estimated weekly cost with the existing arrangement is SF 118.80. If part-time labor is employed estimated cost is SF 100.00 per week, or less costly.

2. Coffee shop

Objective

This case-exercise illustrates the application of the Poisson distribution.

Situation

The owner of a small coffee shop on a certain stretch of highway knows that on average nine people per hour come in for service. Sometimes the only waitress in the shop is very busy, and sometimes there are only a few customers.

Required

1   Develop the complete Poisson distribution for this situation.
2   The owner has decided that if there is greater than a 10 percent chance that there will be at least 13 clients coming into the coffee shop in a given hour, the manager will hire another waitress. Develop the information to help the manager make a decision.

Answers

1   The Poisson distribution is in Table CS-1.

Table CS-1

Random variable x	Exact probability of x	Cumulative probability of x
0	0.01%	0.01%
1	0.11%	0.12%
2	0.50%	0.62%
3	1.50%	2.12%
4	3.37%	5.50%
5	6.07%	11.57%
6	9.11%	20.68%
7	11.71%	32.39%
8	13.18%	45.57%
9	13.18%	58.74%
10	11.86%	70.60%
11	9.70%	80.30%
12	7.28%	87.58%
13	5.04%	92.61%
14	3.24%	95.85%
15	1.94%	97.80%
16	1.09%	98.89%
17	0.58%	99.47%
18	0.29%	99.76%
19	0.14%	99.89%
20	0.06%	99.96%
21	0.03%	99.98%
22	0.01%	99.99%
23	0.00%	100.00%
Total	100.00%	

2   The manager should hire another waitress as the probability of having at least 13 customers coming in the shop is 12.42 percent (100% − 87.58%). That is greater than the benchmark value of 10 percent.

3. Diploma

Objective

This case-exercise demonstrates the application of the binomial distribution.

Situation

A certain business school has a double diploma program for its students. In this program, participants are able to spend a year studying at the following universities: INSEAD, France; Bocconi, Italy; Lancaster, England; Texas, USA. Connecticut, USA; and Toronto, Canada. The competition for this program is high. Candidates are selected on their language ability, motivation, and GMAT (Graduate Management Admission Test) score. Past data indicates that 60 percent of the candidates are accepted for this program. Acceptance or rejection follows a Bernoulli process.

Required

1   Develop the probability distribution of the exact and cumulative probabilities of acceptance if 15 candidates apply for this program. Show the exact data on a histogram.
2   If 15 candidates apply, what is the probability that exactly 5 candidates will be accepted?
3   If 15 candidates apply, what is the probability that exactly 10 candidates will be accepted?
4   If 15 candidates apply, what is the probability that at least 10 candidates will be accepted?
5   If 15 candidates apply, what is the probability that no more than 10 candidates will be accepted?
6   If 15 candidates apply, what is the probability that fewer than 10 candidates will be accepted?

Answers

1   The probability distribution is in Table D-1.
2   If 15 candidates apply, the probability that exactly 5 candidates will be accepted is 2.45%.
3   If 15 candidates apply, the probability that exactly 10 candidates will be accepted is 18.59%.
4   If 15 candidates apply, the probability that at least 10 candidates will be accepted is 40.32%.
5   If 15 candidates apply, the probability that no more than 10 candidates will be accepted is 78.27%.
6   If 15 candidates apply, the probability that fewer than 10 candidates will be accepted is 59.68%.

Table D-1

Random variable, x or exact number of successes	P(x) or probability of exactly x	P(x) cumulative
0	0.00%	0.00%
1	0.00%	0.00%
2	0.03%	0.03%
3	0.16%	0.19%
4	0.74%	0.93%
5	2.45%	3.38%
6	6.12%	9.50%
7	11.81%	21.31%
8	17.71%	39.02%
9	20.66%	59.68%
10	18.59%	78.27%
11	12.68%	90.95%
12	6.34%	97.29%
13	2.19%	99.48%
14	0.47%	99.95%
15	0.05%	100.00%
Total	100.00%	

4. Fenwicks

Objective

The purpose of this case-exercise is to demonstrate the Poisson-binomial approximation.

Situation

A distribution centre has a fleet of 25 Fenwick trolleys that it uses every day for unloading and putting into storage products it receives on pallets from its suppliers. The same Fenwicks are used as needed to take products out of storage and transfer them to the loading area. These 25 Fenwicks are battery driven and at the end of the day they are plugged into the electric supply for recharging. From past data it is known that, on a daily, basis on average one Fenwick will not be properly recharged and thus not available for use.

Required

1   Develop the Poisson distribution for this situation.
2   Develop the binomial distribution for this situation.
3   Using both the Poisson and binomial distribution what is the probability that on any given day, three of the Fenwicks are out of service? Compare the answers.

Answers

1   The Poisson distribution is in Columns 2 and 3 of Table F-1 according to the value of *x*.
2   The binomial distribution is in Columns 4 and 5 of Table F-1 according to the value of *x*.
3   From the Poisson distribution the probability is 6.1313%. For the binomial distribution the probability is 5.9962%. The values are not vastly different. Rounded to the nearest whole number they are both 6%.

Table F-1

Random variable, x	Poisson exact	Poisson cumulative	Binomial exact	Binomial cumulative
0	36.7879%	36.7879%	36.0397%	36.0397%
1	36.7879%	73.5759%	37.5413%	73.5810%
2	18.3940%	91.9699%	18.7707%	92.3517%
3	6.1313%	98.1012%	5.9962%	98.3478%
4	1.5328%	99.6340%	1.3741%	99.7220%
5	0.3066%	99.9406%	0.2405%	99.9624%
6	0.0511%	99.9917%	0.0334%	99.9958%
7	0.0073%	99.9990%	0.0038%	99.9996%
8	0.0009%	99.9999%	0.0004%	100.0000%
9	0.0001%	100.0000%	0.0000%	100.0000%
10	0.0000%	100.0000%	0.0000%	100.0000%
11	0.0000%	100.0000%	0.0000%	100.0000%
12	0.0000%	100.0000%	0.0000%	100.0000%
13	0.0000%	100.0000%	0.0000%	100.0000%
14	0.0000%	100.0000%	0.0000%	100.0000%
15	0.0000%	100.0000%	0.0000%	100.0000%
16	0.0000%	100.0000%	0.0000%	100.0000%
17	0.0000%	100.0000%	0.0000%	100.0000%
18	0.0000%	100.0000%	0.0000%	100.0000%
19	0.0000%	100.0000%	0.0000%	100.0000%
20	0.0000%	100.0000%	0.0000%	100.0000%
21	0.0000%	100.0000%	0.0000%	100.0000%
22	0.0000%	100.0000%	0.0000%	100.0000%
23	0.0000%	100.0000%	0.0000%	100.0000%
24	0.0000%	100.0000%	0.0000%	100.0000%
25	0.0000%	100.0000%	0.0000%	100.0000%
Total	100.00%		100.00%	

5. Hair driers

Objective

This case-exercise applies the concept of the binomial distribution.

Situation

Historical data indicates that by simply replacing the heating coil, 65 percent of defective hair driers assembled by an electrical company can be repaired. Since the units can either be repaired, or not, the probability of success is considered to follow a binomial distribution.

Required

1   Develop a distribution of the probability of repairing a hair drier if there is a quantity of 10 units that are defective.
2   What is the probability that six of the defective units can be repaired?
3   What is the probability that at least seven can be repaired?
4   What is the probability that no more than six can be repaired?

Answers

1 The exact and the cumulative probabilities are given in Table HD-1.
2 The probability that six of the defective units can be repaired is 23.77%.
3 The probability that at least seven can be repaired is 51.38%.
4 The probability that no more than six can be repaired is 48.62%.

Table HD-1

No. that can be repaired x	Probability this number x can be repaired	Cumulative probability of x that can be repaired
0	0.00%	0.00%
1	0.05%	0.05%
2	0.43%	0.48%
3	2.12%	2.60%
4	6.89%	9.49%
5	15.36%	24.85%
6	23.77%	48.62%
7	25.22%	73.84%
8	17.57%	91.40%
9	7.25%	98.65%
10	1.35%	100.00%
Total	100.00%	

6. Newspaper delivery

Situation

This case-exercise illustrates the application of the binomial and the Poisson distribution.

Situation

A business professor in France has a subscription to the *International New York Times*, which is delivered to his house in the morning. For three months, August, September, and October of 2015, he noted whether or not the paper was delivered. This information is given in Table ND-1. A "yes" means a delivery is made and "no" means it is not made. There is no delivery on Sunday. This data is to be considered representative of any three-month or one-quarter delivery period.

Required

1 What is the characteristic probability for the newspaper not being delivered?
2 From a binomial distribution, in any one-quarter period what is the probability that there will be exactly four days when the newspaper is not delivered?
3 From a binomial distribution, in any one-quarter period what is the probability that there will be more than four days when the newspaper is not delivered?
4 From a binomial distribution, in any one-quarter period what is the probability that there will be fewer than four days when the newspaper is not delivered?

Table ND-1

August	Mon	Tue	Wed	Thur	Fri	Sat	Sun
Week 1						Yes	No delivery
Week 2	Yes	No	Yes	Yes	Yes	Yes	No delivery
Week 3	Yes	Yes	Yes	Yes	Yes	Yes	No delivery
Week 4	Yes	Yes	Yes	Yes	Yes	Yes	No delivery
Week 5	Yes	Yes	Yes	Yes	Yes	Yes	No delivery
Week 6	Yes						

September	Mon	Tue	Wed	Thur	Fri	Sat	Sun
Week 1	Yes	Yes	Yes	Yes	Yes	Yes	No delivery
Week 2	Yes	Yes	Yes	Yes	Yes	Yes	No delivery
Week 3	Yes	Yes	Yes	Yes	Yes	No	No delivery
Week 4	Yes	Yes	Yes	Yes	Yes	Yes	No delivery
Week 5	Yes	Yes	Yes				

October	Mon	Tue	Wed	Thur	Fri	Sat	Sun
Week 1				Yes	Yes	Yes	No delivery
Week 2	Yes	No	Yes	Yes	Yes	Yes	No delivery
Week 3	Yes	Yes	Yes	Yes	Yes	No	No delivery
Week 4	Yes	Yes	Yes	Yes	Yes	Yes	No delivery
Week 5	Yes	Yes	Yes	Yes	Yes	Yes	No delivery

5   From a binomial distribution, in any one-quarter period what is the probability that there will be at least four days when the newspaper is not delivered?

6   From a Poisson distribution, what is the average number of days in a quarter when the newspaper is not delivered?

7   What are the probabilities for Questions 2 through 5 using a Poisson distribution? Show the results in a summary table comparing with the binomial distribution results. What are your comments?

Answers

1   The characteristic probability is 5.06%.

2   Probability that there will be exactly four days in a given quarter when the newspaper is not delivered is 20.05%.

3   Probability there will be more than four days in a given quarter when the newspaper is not delivered is 37.11%.

4   Probability there will be fewer than four days in a given quarter when the newspaper is not delivered is 42.84%.

5   Probability there will be at least four days in a given quarter when the newspaper is not delivered is 57.16%.

6   From a Poisson distribution, the average number of days in a given quarter when the newspaper is not delivered is 4.

7   The probabilities for Questions 2 through 5 using a Poisson distribution are in Table ND-2. The probabilities are very close. If the probabilities are rounded, they are the same.

Table ND-2

Question	Binomial (Excel function)	Poisson (Excel function)	Binomial (rounded)	Poisson (rounded)
Exactly 4	20.05%	19.54%	20%	20%
More than 4	37.11%	37.12%	37%	37%
Less than 4	42.84%	43.35%	43%	43%
At least 4	57.16%	56.65%	57%	57%

7. Petrol pump

Objective

The objective of this case–exercise is to demonstrate the application of the Poisson distribution.

Situation

A petrol service station is interested to learn about the utilization of its single automatic petrol pump, which is operated by the insertion of a credit card. In order to know if he is recuperating his investment, the franchise owner of this service station wants some assurance that there is a probability of greater than 50 percent that 10 or more customers in any hour use the automatic pump. Past data indicates that on average eight customers per hour use the automatic pump.

Table PP-1

Customers x	P(x) *exactly*	P(x) *cumulative*
0	0.03%	0.03%
1	0.27%	0.30%
2	1.07%	1.38%
3	2.86%	4.24%
4	5.73%	9.96%
5	9.16%	19.12%
6	12.21%	31.34%
7	13.96%	45.30%
8	13.96%	59.25%
9	12.41%	71.66%
10	9.93%	81.59%
11	7.22%	88.81%
12	4.81%	93.62%
13	2.96%	96.58%
14	1.69%	98.27%
15	0.90%	99.18%
16	0.45%	99.63%
17	0.21%	99.84%
18	0.09%	99.93%
19	0.04%	99.97%
20	0.02%	99.99%
21	0.01%	100.00%
Total	100.00%	

Required

1   Develop the Poisson distribution table for the utilization of this petrol pump.
2   Plot the Poisson distribution table as a histogram.
3   Should the franchise owner be satisfied with the utilization, based on the data given?

Answers

1   The Poisson distribution data is in Table PP-1.
2   The Poisson distribution as a histogram is in Figure PP-1.
3   No, the franchise owner should not be satisfied, as the percentage of ten or more persons using the pump is only 28.34%.

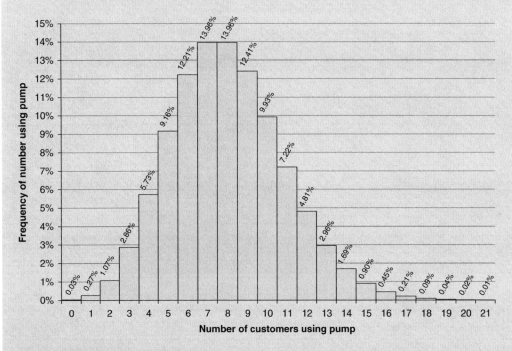

Figure PP-1 Petrol pump: Poisson distribution—lambda = 8.

8. Registered and dining

Objective

The objective of this case-exercise is to illustrate the development and analysis of a binomial distribution.

Situation

A small hotel has a restaurant for its guests where it serves dinner and breakfast. Most of the guests have breakfast in the hotel but not all dine at the restaurant, preferring to eat at restaurants in the town. Past records show that 70 percent of the guests that are registered at the hotel eat an evening meal in the hotel restaurant. The registration and dining at the hotel follows a Bernoulli process. Knowing how many hotel guests eat at the restaurant is important. It may be an indicator of the quality of the meals and of course registration and dining adds to revenues.

Required

1   Develop a table that shows all the possible exact probabilities of dining at the hotel if 40 guests are registered and the corresponding data for cumulative probabilities.
2   Show as a histogram the table developed in Question 1.
3   If 40 guests are registered, what is the probability that exactly 25 will have dinner at the hotel?
4   If 40 guests are registered, what is the probability that at least 25 will have dinner at the hotel?
5   If 40 guests are registered, what is the probability that no more than 25 will have dinner at the hotel?
6   If 40 guests are registered, what is the probability that fewer than 25 will have dinner at the hotel?
7   By increasing the menu choices, lowering the price, and generally sprucing up the restaurant area, the hotel is able to increase the probability of guests dining at the hotel from 70 percent to 80 percent. In this case what is the probability that at least 25 will have dinner at the hotel?

Table RD-1

Column 1 Random variable x	Column 2 Probability of exactly x or P(x)	Column 3 Probability of cumulative or ΣP(x)	Column 1 Random variable x	Column 2 Probability of exactly x or P(x)	Column 3 Probability of cumulative or ΣP(x)
0	0.00%	0.00%	21	0.85%	1.48%
1	0.00%	0.00%	22	1.72%	3.20%
2	0.00%	0.00%	23	3.14%	6.33%
3	0.00%	0.00%	24	5.18%	11.51%
4	0.00%	0.00%	25	7.74%	19.26%
5	0.00%	0.00%	26	10.42%	29.68%
6	0.00%	0.00%	27	12.61%	42.28%
7	0.00%	0.00%	28	13.66%	55.94%
8	0.00%	0.00%	29	13.19%	69.13%
9	0.00%	0.00%	30	11.28%	80.41%
10	0.00%	0.00%	31	8.49%	88.90%
11	0.00%	0.00%	32	5.57%	94.47%
12	0.00%	0.00%	33	3.15%	97.62%
13	0.00%	0.00%	34	1.51%	99.14%
14	0.00%	0.00%	35	0.61%	99.74%
15	0.00%	0.00%	36	0.20%	99.94%
16	0.01%	0.01%	37	0.05%	99.99%
17	0.02%	0.03%	38	0.01%	100.00%
18	0.06%	0.09%	39	0.00%	100.00%
19	0.16%	0.24%	40	0.00%	100.00%
20	0.38%	0.63%	Total	100.00%	

Answers

1   Exact probabilities are shown in Column 2 of Table RD–1 and the cumulative probabilities in Column 3.
2   The histogram of the probabilities is in Figure RD–1.
3   If 40 guests are registered, the probability that exactly 25 will have dinner at the hotel is 7.74%.
4   If 40 guests are registered, the probability that at least 25 will have dinner at the hotel is 88.49%.
5   If 40 guests are registered, the probability that no more than 25 will have dinner at the hotel is 19.26%.
6   If 40 guests are registered, the probability that fewer than 25 will have dinner at the hotel is 11.51%.
7   By revamping conditions, the probability that at least 25 will have dinner at the hotel is 99.71%.

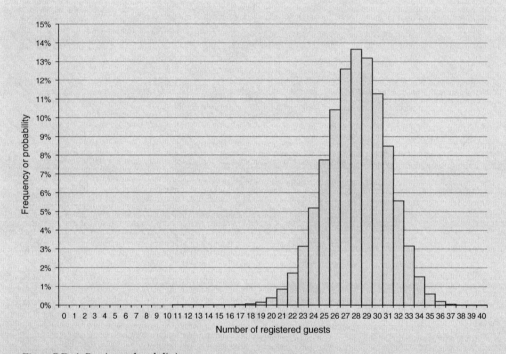

Figure RD-1 Registered and dining.

9. Rental cars

Objective

The purpose of this case-exercise is to demonstrate the application of random numbers.

Situation

Roland Ryan operates a car leasing business in Wyoming, USA and he has ten outlets in this state. He is developing his budgets for the following year and is proposing to use historical data for estimating his profits for the coming year. For the previous year he has accumulated data from two of his agencies: one in Cheyenne, given in Table RC-1; and the other in Laramie, given in Table RC-2. This data gives the number of cars leased, and the number of days at which this level of cars were leased during the 250 days per year when the leasing agencies were open. Each car that was leased generated $22.00 in profit.

Table RC-1 Cheyenne.

No. of cars leased	Days this number of cars leased
20	2
21	9
22	12
23	14
24	14
25	18
26	24
27	26
28	29
29	27
30	25
31	20
32	15
33	8
34	6
35	1

Table RC-2 Laramie.

No. of cars leased	Days this number of cars leased
20	1
21	1
22	2
23	2
24	12
25	20
26	38
27	49
28	50
29	37
30	19
31	13
32	2
33	2
34	1
35	1

Required

1 Using the data from the Cheyenne agency, what is a reasonable estimate of the average number of cars leased per day during the year the analysis was made? What is the variance and the standard deviation for this data?

2 From the average daily number of cars leased determined in Question 1, what is a reasonable estimate of the daily and also annual profit for the Cheyenne agency?

3 Using the data from the Laramie agency, what is a reasonable estimate of the average number of cars leased per day during the year the analysis was made? What is the variance and the standard deviation for this data?

4 From the average daily number of cars leased determined in Question 3, what is a reasonable estimate of the daily and also annual profit for the Laramie agency?

5 What are your comments about the calculations that you have made?

6 For estimating future activity for the leasing agency, which of the data from Cheyenne or Laramie would be the most reliable? Justify your response visually by plotting an absolute frequency distribution for each of the two agencies. Corroborate your conclusions quantitatively using the coefficient of variation.

Answers

1   Using the data from the Cheyenne agency, a reasonable estimate of the average number of cars leased per day during the year is 22.50. The variance is 11.2340 (cars/day)^2 and the standard deviation is 3.3517 cars/day.

2   A reasonable estimate of the daily profit for the Cheyenne agency is $605.00. An estimate of the annual profit is $151,250.00.

3   Using the data from the Laramie agency, a reasonable estimate of the average number of cars leased per day during the year is 22.50. The variance is 4.5700 (cars/day)^2 and the standard deviation is 2.1378 cars/day.

4   A reasonable estimate of the daily profit for the Laramie agency is $605.00. An estimate of the annual profit is $151,250.00.

5   The average value of the number of cars rented, and the estimated daily and annual profit are the same for both agencies. However, the variation and the standard deviations are different.

6   The absolute frequency distributions for the two agencies are given in Figure RC-1 and Figure RC-2. Visually it can be seen that the data for the Laramie agency is less dispersed; more data is clustered around the mean value of 27.50. This makes this data more reliable for estimating purposes. This conclusion is corroborated by a coefficient of variation of 7.77 percent for Laramie compared to a coefficient of variation of 12.19 percent for Cheyenne. A lower coefficient of variation for comparing datasets indicates less dispersion.

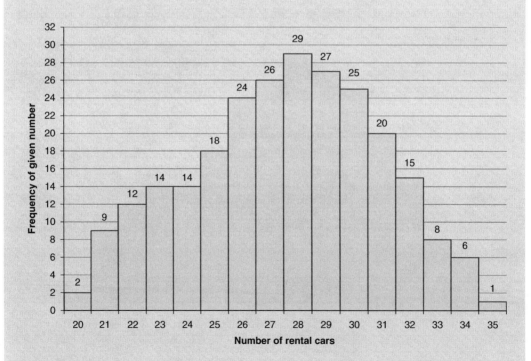

Figure RC-1 Rental cars: Cheyenne agency.

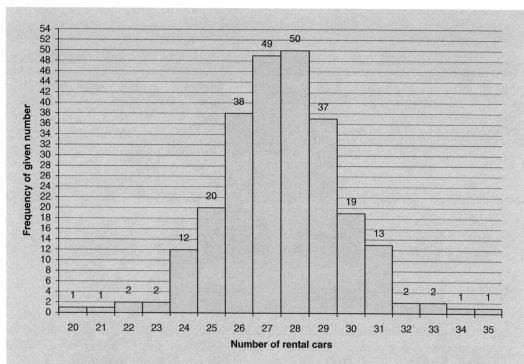

Figure RC-2 Rental cars: Laramie agency.

10. Restaurant customers

Objective

To illustrate the use of random variables in probability.

Situation

A small restaurant with a maximum capacity of 30 clients has recorded the information shown in Table RE-1 regarding the number of customers who have come in for lunch over the last 12-month period. The arrival of clients at the restaurant is considered random. The restaurant is open 340 days per year.

Required

1 Develop a table for the random variable calculation in order to compute the average number of clients and the standard deviation. From this table construct the histogram of the number of clients and the number of days that this number of clients comes into the restaurant.

2 Using the concept of expected values and probability and to the nearest two decimal places, what is the average number of customers per day in the restaurant over the last 12-month period?

3 What is the standard deviation of the average number of customers per day in the restaurant?

4 If the average meal price is €27.00, what is an estimate of the average receipts per day? What is an estimate for a 12-month period based on the opening days of the restaurant?

5 Using the average meal price of €27.00, what are the receipts per day when the restaurant is full?

Table RE-1

Number of clients x	Days with this number of clients	Number of clients x	Days with this number of clients
9	7	20	28
10	8	21	26
11	9	22	25
12	10	23	23
13	12	24	20
14	14	25	19
15	15	26	16
16	17	27	10
17	19	28	9
18	22	29	4
19	24	30	3

Table RE-2

Customers x	Days occurred	Relative frequency, P(x)	$x*P(x)$	$(x - \mu)$	$(x - \mu)^2$	$P(x)*(x - \mu)^2$
9	7	2.06%	0.19	−10.62	112.73	2.3210
10	8	2.35%	0.24	−9.62	92.50	2.1765
11	9	2.65%	0.29	−8.62	74.26	1.9658
12	10	2.94%	0.35	−7.62	58.03	1.7067
13	12	3.53%	0.46	−6.62	43.79	1.5456
14	14	4.12%	0.58	−5.62	31.56	1.2994
15	15	4.41%	0.66	−4.62	21.32	0.9407
16	17	5.00%	0.80	−3.62	13.09	0.6544
17	19	5.59%	0.95	−2.62	6.85	0.3829
18	22	6.47%	1.16	−1.62	2.62	0.1693
19	24	7.06%	1.34	−0.62	0.38	0.0269
20	28	8.24%	1.65	0.38	0.15	0.0120
21	26	7.65%	1.61	1.38	1.91	0.1461
22	25	7.35%	1.62	2.38	5.68	0.4173
23	23	6.76%	1.56	3.38	11.44	0.7739
24	20	5.88%	1.41	4.38	19.21	1.1297
25	19	5.59%	1.40	5.38	28.97	1.6189
26	16	4.71%	1.22	6.38	40.73	1.9169
27	10	2.94%	0.79	7.38	54.50	1.6029
28	9	2.65%	0.74	8.38	70.26	1.8599
29	4	1.18%	0.34	9.38	88.03	1.0356
30	3	0.88%	0.26	10.38	107.79	0.9511
TOTAL	340	100.00%	19.62			24.6538

6 What would the average meal price have to be if the restaurant wants receipts to be an estimated €600.00 per day, using the average value calculated from the random variable calculation?

7 If for the next 12-month period the restaurant proposes to increase meal prices such that the average price goes up by 5%, what would be an estimate of annual receipts using the expected value determined in Question 2?

Answers

1 The random variable calculation is given in Table RE-2. The histogram in Figure RE-1 is developed from Column 1 (*x*-axis) and Column 2 (*y*-axis) of the relative frequency probability *P(x)*.

2 Average number customers per day in the restaurant over the last 12-month period is 19.62.

3 Standard deviation of the average number of customers per day in the restaurant is 4.9653.

4 Estimate of the average receipts per day is €529.68. Estimate for a 12-month period is €180,090.00.

5 Estimate of receipts per day when the restaurant is full is €810.00.

6 Average meal price needs to be €30.58 if the restaurant wants an estimated €600.00 receipts per day.

7 An estimate of annual receipts at a 5% increase is €189,094.50.

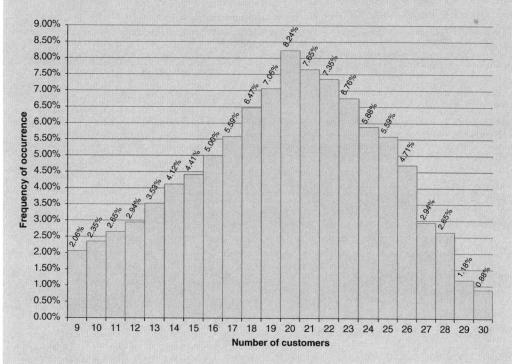

Figure RE-1 Restaurant customers: relative frequency distribution.

11. Restaurant food sales

Objective

This case-exercise demonstrates the application of random variables and how they can be used in estimating future outcomes.

Situation

A restaurant in Lyon, France has an average markup on its meals, not including beverages, of 30 percent. The average food costs per meal are €6.75. The owner carried out an audit over the last 300 days to determine the number of meals sold per day, and the number of days for which this amount was sold. To simplify the audit the number sold was rounded to the nearest 5 number of meals. The results are summarized in Table RF-1. For example, where it is shown 100 meals sold per day for 12 days, this would include all meals sold between 100 and 102.

Table RF-1

Meals sold/day x	Days this amount sold
90	18
95	22
100	12
105	13
110	17
115	16
120	12
125	18
130	17
135	25
140	45
145	35
150	50

Required

1 Using the random variable relationship, what is the average number of meals sold per day?
2 Determine the standard deviation of the number of meals sold per day.
3 Using this information for forecasting, what is an estimate of future operating income on the sales of restaurant meals?

Answers

1 Average number of meals sold per day is 127.12.
2 Standard deviation of the number of meals sold per day is 19.77.
3 An estimate of future operating income is €2,000 (2,002.09).

12. Restaurant staff

Objective

To illustrate the use of a Poisson distribution and its relationship to the binomial distribution.

Situation

A restaurant in London has 25 full time staff not including the manager and the principal chef. From past data, the absenteeism rate of these staff is 4.5%.

Required

1 Develop the exact and cumulative Poisson distribution for the data. Plot the exact distribution data as a histogram.
2 Using the Poisson distribution, what is the probability that on any given day exactly three staff members do not show up for work?
3 Using the Poisson distribution, what is the probability that fewer than three staff members do not show up for work?
4 Using the Poisson distribution, what is the probability that more than three staff members do not show up for work?
5 Develop an exact and cumulative binomial distribution for the data. Plot the exact probability distribution as a histogram.
6 Using the binomial distribution, what is the probability that on any given day exactly three staff members do not show up for work?
7 Using the binomial distribution, what is the probability that fewer than three staff members do not show up for work?
8 Using the binomial distribution, what is the probability that more than three staff members do not show up for work?
9 What are your comments about the two frequency distributions that you have developed, and the probability values that you have determined?

Answers

1 The Poisson distributions are given in Columns 2 and Column 3 of Table RS–1. It is from the data in Column 2 that the histogram is developed as shown in Figure RS–1.
2 Probability that on any given day exactly 3 staff members do not show up for work is 7.70%.
3 Probability that fewer than three staff members do not show up for work is 89.53%.
4 Probability that more than three staff members do not show up for work is 2.76%.
5 The binomial distributions are given in Columns 4 and Column 5 of Table RS–1. It is from the data in Column 4 that the histogram is developed as shown in Figure RS–2.
6 Probability that on any given day exactly 3 staff members do not show up for work is 7.61%.
7 Probability that fewer than three staff members do not show up for work is 89.96%.
8 Probability that more than three staff members do not show up for work is 2.43%.
9 They both tail off rapidly to the right. The values are close. Since $P < 5\%$ and sample size is greater than 20 it is reasonable to use the Poisson–binomial approximation.

Table RS-1

Column 1 Random variable x	Column 2 Exact Poisson probability P(x)	Column 3 Poisson probability cumulative	Column 4 Exact binomial probability P(x)	Column 5 Binomial probability cumulative
0	32.4652%	32.4652%	31.6289%	31.6289%
1	36.5234%	68.9886%	37.2592%	68.8881%
2	20.5444%	89.5331%	21.0680%	89.9561%
3	7.7042%	97.2372%	7.6110%	97.5671%
4	2.1668%	99.4040%	1.9725%	99.5396%
5	0.4875%	99.8915%	0.3904%	99.9299%
6	0.0914%	99.9830%	0.0613%	99.9912%
7	0.0147%	99.9976%	0.0078%	99.9991%
8	0.0021%	99.9997%	0.0008%	99.9999%
9	0.0003%	100.0000%	0.0001%	100.0000%
10	0.0000%	100.0000%	0.0000%	100.0000%
11	0.0000%	100.0000%	0.0000%	100.0000%
12	0.0000%	100.0000%	0.0000%	100.0000%
13	0.0000%	100.0000%	0.0000%	100.0000%
14	0.0000%	100.0000%	0.0000%	100.0000%
15	0.0000%	100.0000%	0.0000%	100.0000%
16	0.0000%	100.0000%	0.0000%	100.0000%
17	0.0000%	100.0000%	0.0000%	100.0000%
18	0.0000%	100.0000%	0.0000%	100.0000%
19	0.0000%	100.0000%	0.0000%	100.0000%
20	0.0000%	100.0000%	0.0000%	100.0000%
21	0.0000%	100.0000%	0.0000%	100.0000%
22	0.0000%	100.0000%	0.0000%	100.0000%
23	0.0000%	100.0000%	0.0000%	100.0000%
24	0.0000%	100.0000%	0.0000%	100.0000%
25	0.0000%	100.0000%	0.0000%	100.0000%
Total	100.00%		100.00%	

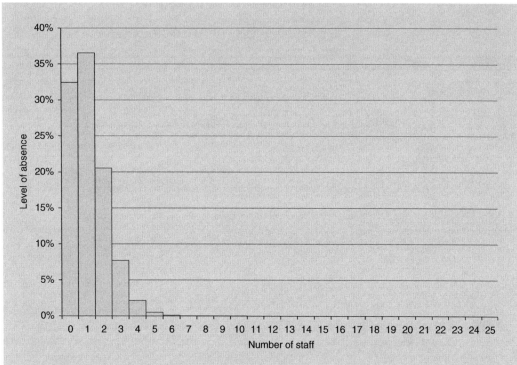

Figure RS-1 Restaurant staff: Poisson distribution.

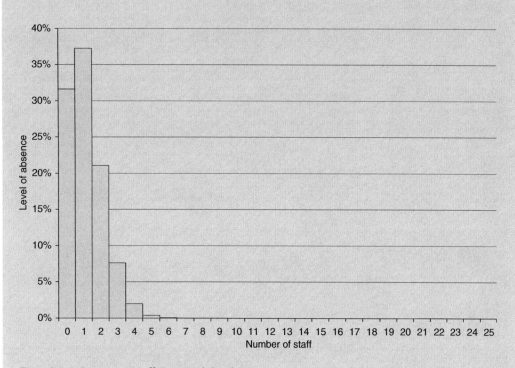

Figure RS-2 Restaurant staff: Binomial distribution.

5 Continuous distributions and probability

Chance and risk in uninterrupted information

No	Case-exercise	Statistical concept	Application
1	Apricot jam	Normal distribution	Food
2	Automobile repair time	Normal distribution	Services
3	Butter	Normal distribution	Restaurant, food
4	Chateau renovation	Exponential distributions	Hospitality, operations
5	Delivery vehicles	Normal distribution	Logistics
6	Electrical resistors	Normal distribution	Electrical engineering
7	Grass seeds	Normal distribution	Operations
8	Ham and food catering	Normal distribution	Hospitality, restaurant
9	Peanuts	Normal distribution	Operations, food
10	Restaurant service time	Normal distribution	Hospitality
11	Tires	Normal	Production
12	Training program	Normal distribution	Human resources

1. Apricot jam

Objective

This case-exercise illustrates the application of the normal distribution.

Situation

A family-owned company near Lyon in southeastern France makes jam. One of its products during the summer period is apricot jam, and the production time of this variety of jam is 90 days. Each day it produces a production lot of 50,000 jars of apricot jam that is labeled *Net weight 500 grams*. It buys the apricot fruit from local producers at a price of €250.00 per metric ton. On a particular day an operator takes a large sample of filled jars from the production line and determines that the average weight of this sample is 502.75 grams, with a standard deviation of 3.22 grams. These two statistics represent the filling operation for the total amount of apricot jam produced during the 90-day production period. The filling operation is considered to follow a normal distribution.

Required

1   What proportion of jars of apricot jam will be below the stated net weight of 500 g?
2   The jars of apricot jam are packed in boxes of twelve jars per box. In this case what proportion of boxes will be below 6 kg in net weight?
3   On a daily basis what is the expected cost of the apricot fruit to the jam producer, using the labeled net weight?
4   On a daily basis what is the actual cost of the apricot fruit to the jam producer, using the statistical sampling information?
5   For the total production period, what is the additional cost of the apricot fruit to the jam producer between what is produced according to the labeling requirements, and the actual cost?
6   How would the situation change if, with more rigorous quality control, the company was able to have an average sample weight of filled jars for the operation of 501.15 g and a standard deviation of 1.58 g?

Answers

1   The proportion of jars of apricot jam that will be below the stated net weight of 500 g is 19.65%.
2   The proportion of boxes that will be below 6 kg in net weight is 0.15%.
3   On a daily basis the expected cost of the apricot fruit to the jam producer using the labeled net weight is €6,250.00.
4   On a daily basis the actual cost of the apricot fruit to the jam producer using the statistical sampling information is €6,284.38.
5   For the total production period, the additional cost of the apricot fruit to the jam producer between what is produced according to the labeling requirements, and the actual cost is €3,093.75.

6   The proportion of jars of jam below the stated net weight is 23.34%; proportion of boxes below 6 kg in net weight is 0.58%; additional cost for the total production period over and above the expected value from the net weight drops to €1,293.75.

2. Automobile repair time

Objective

This case-exercise applies the concept of the normal distribution.

Situation

Christian Auclair owns a garage in Belleville, north of Lyon, France. His son, Pascal, works for him and does all the automobile service and repair work. Christian performs the body repair of automobiles. Many of the cars that come into the Auclair garage are Renault models. The length of time taken by Pascal to fully service a Renault automobile including oil change, verification of the electronic systems, brake inspection, checking engine performance, synchronization of gears, wheel alignment, positioning of lights, etc. takes 3 hours and 45 minutes. The standard deviation of this activity is 25 minutes. This service time is considered to follow a normal distribution. Pascal works an eight-hour day.

Required

1   A Renault automobile is brought into the garage in the morning. What is the probability that the service time for this car will be more than 4.0 hours?
2   On a certain day, at 08:00 Pascal has two Renault automobiles to service. What is the probability that Pascal will have to work overtime in order to finish both cars?

Answers

1   Probability that the service time for this car will be more than 4.0 hours is 27.43%.
2   Probability that Pascal will have to work overtime in order to finish both cars is 19.81%.

3. Butter

Objective

This case-exercise demonstrates the application of the normal distribution in operations management.

Situation

Butter made from cows' milk is prepared in molds on a production line. The nominal net weight of a certain format of a block of unsalted butter, as indicated on the packaging, is 500 grams. The filling machines for this butter are set to the nominal weight and the standard deviation of the filling operation is 4.85 grams. The supplier sells this format of butter to a restaurant for €1.50 per block.

Required

1   What percentage of blocks of butter can be expected to have a net weight between 502 and 500 g?
2   What percentage of blocks of butter can be expected to have a net weight between 502 and 515 g?
3   What is the combined percentage of blocks of butter that can be expected to have a net weight that is either no more than 494 g or at least 510 g?
4   If there are 40,000 blocks of butter in a production lot how many of them would have a net weight between 507 and 518 g?
5   What is the net weight of blocks of butter above which 85% are greater than this net weight?
6   What are the symmetrical limits of the net weight between which 99% of the blocks of butter lie?
7   The blocks of butter are packed in cases of one dozen blocks per case. What proportion of cases will be above 6.02 kg in net weight?

Answers

1   Percentage of blocks of butter expected to have a net weight between 502 g and 500 g is 16.00%.
2   Percentage of blocks of butter expected to have a net weight between 502 g and 515 g is 33.90%.
3   Combined percentage of blocks of butter expected to have a net weight that is either no more than 494 g grams or at least 510 g is 12.76%.
4   If there are 40,000 blocks of butter in a production lot 2,975 (2,974.60) would have a net weight between 507 and 518 g.
5   Net weight of blocks of butter above which 85% are greater than this net weight is 495 g (494.97).
6   Symmetrical limits between which 99% of blocks of butter lie are 487.51 g and 512.49 g.
7   Proportion of cases that will be above 6.02 kg in net weight is 11.69%.

4. Chateau renovation

Objective

The objective of this case-exercise is to illustrate the development of the learning curve and its application to **time** and **cost** in a business activity, here related to the hospitality industry.

Situation

A group of five doctors conceived a project to renovate a sixteenth century chateau in Clermont-Ferrand, France that they planned to rent out for banquets, weddings, and as a conference center, using a food catering service in Clermont. The group wanted to refurbish the chateau exactly how it was when it was first constructed. One part of the project was to hire artisans to manufacture the furniture. This was not easy as artisans in the immediate area were limited as many of the techniques had been lost. They found one artisan who could make straight-backed wooden chairs with woven straw and rope seats; an artisan who could make upholstered armchairs; and an artisan who could make armoires that would be used for storing kitchen supplies, dry goods, table linen, cutlery, and the like. Labor costs for all of the artisans was €15.20/hour. Since making this type of furniture was a new concept to the artisans, it was felt that it would be appropriate to apply the learning curve concept for evaluating time and cost.

Required

Straight-backed chairs

1   The artisan for the straight-backed chairs took 20 hours for the first chair and 18 hours for the second. Using these figures, determine the percentage learning rate.
2   Develop an exponential learning curve for the production of these straight-backed chairs using 256 chairs, as the upper limit. Convert this exponential curve into a logarithmic graph.
3   Calculate how long it takes to make the twentieth chair and verify your answer from the graph that you have developed in Question 2.
4   Using the learning curve production times and the artisan labor costs, determine the total labor costs for 120 chairs.
5   Assume that the artisan wanted to make a profit margin of 10% on the price of each chair that he sells. What would be the total price of the total number of 120 chairs to the owners?
6   What is the average unit price of the chair to the owners?
7   What would be your answers to Questions 5 and 6 if the time to produce the first chair was used in your calculation? That is, the learning curve criteria was not employed.

Upholstered chairs

8   The artisan for the upholstered chairs took 35 hours to make the first unit. His learning rate was 80%. Develop an exponential learning curve for the production of 50 chairs. Convert this exponential curve into a logarithmic graph.
9   Calculate how long it takes to make the twentieth chair and verify your answer from the graph that you have developed in the previous question.
10   Using the learning curve production times and the artisan labor costs, determine the total labor cost for 50 chairs.
11   Assume that this artisan wanted to make a profit margin of 15% on the price of each chair that he sells. Using the learning curve criteria what would be the price of the total number of 50 chairs to the owners?
12   What is the average unit price of the upholstered chair to the owners?
13   What would be your answers to Questions 11 and 12 if the time to produce the first upholstered chair was used in your calculation? That is, the learning curve reduction times were not used.

Armoires

14   The artisan for the armoires took 40 hours for the second armoire with a learning rate of 85%. How long did it take to make the first armoire?
15   Develop an exponential learning curve for the production of 20 armoires. Convert this exponential curve into a logarithmic graph.

16 Calculate how long it takes to make the twentieth armoire and verify your answer from the graph that you have developed in the previous question.

17 Using the learning curve production times and the artisan labor costs, determine the total labor cost for 20 armoires.

18 Assume that this artisan wanted to make a profit margin of 20% on the price of each armoire that he sells. What would be the total price of the total number of 20 armoires to the owners?

19 What is the average unit price of the armoires to the owners?

20 What would be your answers to Questions 18 and 19 if the time to produce the first armoire was used in your calculation? That is, the learning curve reduction times were not used.

Answers

Straight-backed chairs

1 90.00%.

2 The data for developing the exponential curve is in Table CR–1 based on doubling output each time. The exponential form of the curve is in Figure CR–1 and the logarithmic presentation in Figure CR–2.

Table CR-1

Number of chairs	1	2	4	8	16	32	64	128	256
Hours for nth unit	20.00	18.00	16.20	14.58	13.12	11.81	10.63	9.57	8.61

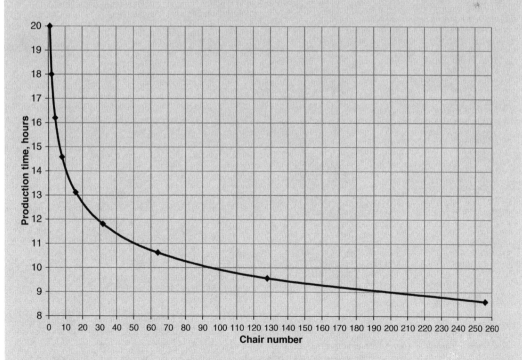

Figure CR-1 Chateau renovation, learning curve: straight-backed chairs.

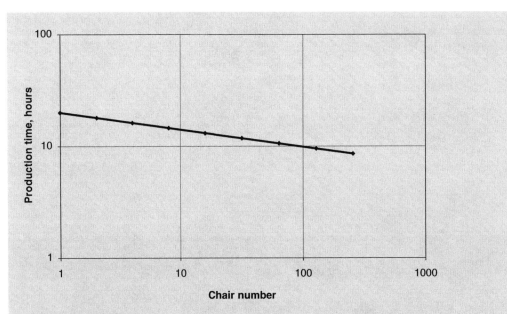

Figure CR-2 Chateau renovation, logarithmic learning curve: straight-backed chairs.

Table CR-2

Number of chairs	1	2	4	8	16	32	64
Hours for nth unit	35.00	28.00	22.40	17.92	14.34	11.47	9.18

Table CR-3

Number of armoires	1	2	4	8	16	32
Hours for nth unit	47.06	40.00	34.00	28.90	24.57	20.88

3   12.68 hours.
4   €20,649.34.
5   €22,943.71.
6   €191.20.
7   €40,533.33; €337.78.

Upholstered chairs

8   The data for developing the exponential curve is in Table CR-2 based on doubling output each time. The exponential form of the curve is in Figure CR-3 and the logarithmic presentation in Figure CR-4.
9   13.34 hours.
10   €10,704.75.
11   €12,593.83.
12   €251.88.
13   €31,294.12; €625.88.

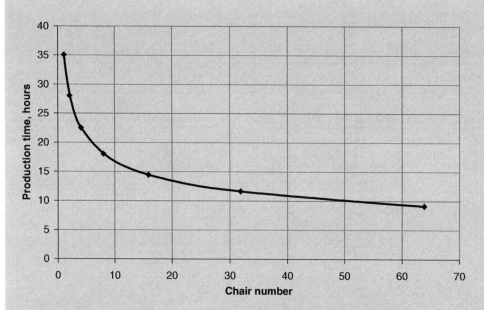

Figure CR-3 Chateau renovation, learning curve: upholstered chairs.

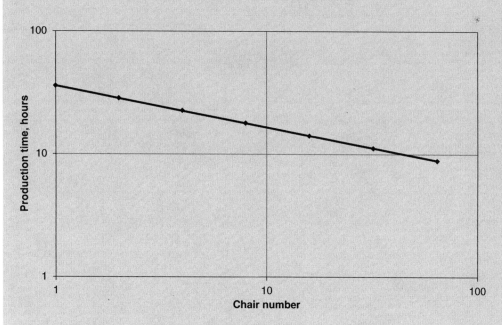

Figure CR-4 Chateau renovation, logarithmic learning curve: upholstered chairs.

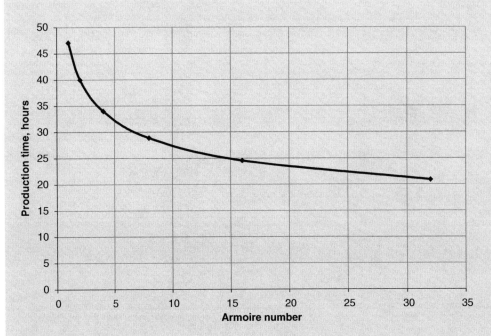

Figure CR-5 Chateau renovation: exponential learning curve: armoires.

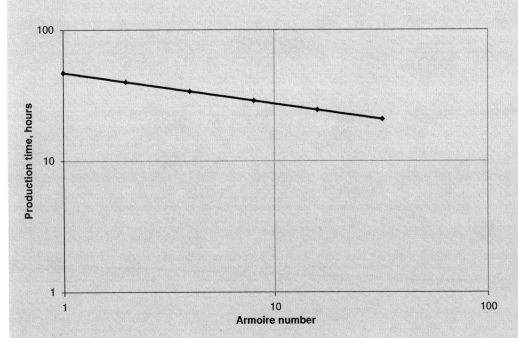

Figure CR-6 Chateau renovation: logarithmic learning curve: armoires.

Armoires

14   47.06 hours.
15   The data for developing the exponential curve is in Table CR-3 based on doubling output each time. The exponential form of the curve is in Figure CR-5 and the logarithmic presentation in Figure CR-6.
16   23.31 hours.
17   €8,871.28.
18   €11,089.10.
19   €554.45.
20   €17,882.35; €941.18.

5. Delivery vehicles

Objective

This case-exercise illustrates the application of the normal distribution.

Situation

A food distribution company has a fleet of 1,500 trucks that it uses for delivering products from its distribution center to retail outlets. Based on historical records, the trucks travel on average 120,000 km per year before their mechanical condition indicates they must be replaced. The standard deviation of this replacement distance is 25,000 km. The distance traveled by the trucks approximates a normal distribution.

Required

1   Estimate what proportion of trucks can be expected to travel between 75,000 and 130,000 km per year.
2   What is the probability that a randomly selected truck travels less than 60,000 km per year?
3   What percentage of trucks can be expected to travel more than 145,000 km per year?
4   How many of the trucks in the fleet are estimated to travel between 40,000 and 90,000 km per year?
5   What is the minimum distance that will be traveled by at least 90% of the trucks?
6   What are the symmetrical limits between which 75% of the trucks will travel?

Answers

1   The proportion of trucks that can be expected to travel between 75,000 and 130,000 km/year is 61.95%.
2   The probability that a randomly selected truck travels less than 60,000 km/year is 0.82%.
3   The percentage of trucks that can be expected to travel more than 145,000 km/year is 15.87%.
4   About 172 trucks in the fleet are estimated to travel between 40,000 and 90,000 km/year.
5   The minimum distance traveled by at least 90% of the trucks is 87,961 kilometers.
6   The symmetrical limits between which 75% of the trucks will travel are 91,241 km (91,241.27) and 148,759 km (148,758.73).

6. Electrical resistors

Objective

This case-exercise illustrates the application of the normal distribution.

Situation

A resistor in an electrical circuit has an average resistance of 100 Ω and a standard deviation of 4 Ω. The tolerance limit for the resistors in this circuit is 100 $\pm$ 6 Ω and their manufacture follows a normal distribution. The firm has 20,000 of these resistors in its inventory.

Required

1   What percentage of the inventory will have a resistance above 110 Ω?
2   What percentage of the inventory will be outside the tolerance limit of 100 $\pm$ 6 Ω?
3   Ten resistors are connected in a series circuit to make a nominal 1,000 Ω resistor. In this case, what is the probability that the resistance in the circuit will be less than 975 Ω?
4   One part of the circuit requires two matched resistors that differ by no more than 1 Ω. If the first resistor of a randomly selected pair has a resistance of 102 Ω, what is the probability that the second one will be a match according to the criteria given?

Answers

1   The percentage of inventory that has a resistance above 110 Ω is 0.62%.
2   The percentage of the inventory that will be outside the tolerance limit is 13.36%.
3   With ten resistors connected in a series circuit to make a nominal 1,000 Ω resistor the probability that the resistance in the circuit will be less than 975 Ω is 2.41%.
4   If one part of the circuit requires two matched resistors that differ by no more than 1 Ω and the first resistor of a randomly selected pair has a resistance of 102 Ω, then the probability that the second one will be a match according to the given criteria is 17.47%.

7. Grass seeds

Objective

This case exercise demonstrates the application of the normal distribution.

Situation

Robert has just planted grass seed in his back garden to make a new lawn. According to the information on the bag containing the seeds the average germination period is ten days with a standard deviation of three days. Seed germination follows a normal distribution. Germination is indicated by green shoots of grass just visible on the surface of the soil.

Required

1 After five days, to the nearest two decimal places, how much of Robert's lawn will be covered in green shoots?
2 After six days, to the nearest two decimal places, how much of Robert's lawn still needs to be covered in green shoots?
3 After twelve days, to the nearest two decimal places, how much of Robert's lawn is estimated to be covered in green shoots?
4 Between what symmetrical limits, in whole number of days, will 80% of the lawn have been covered in green shoots?
5 To the nearest whole number, in how many days will 95% of the lawn be covered?

Answers

1 After five days 4.78% of Robert's lawn will be covered in green shoots.
2 After six days 90.88% of Robert's lawn still needs to be covered in green shoots.
3 After twelve days 74.75% of Robert's lawn is estimated to be covered in green shoots.
4 Between the symmetrical limits of six and fourteen days 80% of the lawn will be covered in green shoots.
5 After about fifteen days 95% of the lawn will be covered.

8. Ham and food catering

Objective

This case-exercise illustrates the use of a normal distribution where there is a target, or average value, and a distribution of the data.

Situation

A food caterer bases its banquet price to its clients on food costs, labor costs, and time for the event. For one project, "Helen and Willies wedding," the client has asked for a lunch where the meat is ham. The caterer bases its proposal on serving two slices of ham per guest where the caterer bases its proposal on the basis that the average weight of each ham slice is 70 g. It knows from past activity that the standard deviation of the cutting of ham in the machine is 2.25 g and the machine cutting process follows a normal distribution. However, in carrying out an audit of the activity by taking samples it is found that the average weight of the ham slices is in fact 74.75 g. The ham costs the caterer €3.25 per kg. The proposal is based on 2,250 guests at the wedding.

Required

1 For this event what is the expected total cost for the ham served to the guests, the actual total cost of the ham served, and the difference in the two values?
2 What is the probability that for the proposed event, a slice of ham will be more than the desired weight of 70 g?
3 How many guests will receive portions of ham weighing less than the 70 g?

4   What is the minimum and maximum weight of ham per slice in the middle 99% of the average weight?
5   How many guests will receive a slice of ham that weighs between 70.00 and 74.75 g?

Answers

1   Expected total cost for the ham served to the guests is €1,023.75. Actual total cost of the ham served is €1,093.22. The difference in the two values is €69.47.
2   Probability that a slice of ham will be more than the desired weight of 70 g is 98.26%.
3   Number of guests that will receive portions of ham weighing less than the 70 g is 39 (39.11).
4   Minimum and maximum weight in the middle 99% of average weighs is 68 g (68.95) and 81 g (80.55).
5   Number of guests that will receive a slice of ham between 70.00 and 74.75 g is 1,086 (1,085.89).

9. Peanuts

Objective

The objective of this case-exercise is to demonstrate the application of probability using the normal distribution.

Situation

Salted peanuts sold in a store have a nominal net weight of 125 g, as indicated on the packaging. Tests at the production site indicate that the average weight in a package is 126.75 g with a standard deviation of 1.25 g.

Required

1   If you buy a packet of these peanuts at a store, what is the probability that your packet will contain more than 127 g?
2   If you buy a packet of these peanuts at a store, what is the probability that your packet will contain less than the nominal indicated weight of 125 g?
3   What is the minimum and maximum weight of a packet of peanuts in the middle 99% of the packet weights?
4   In the packets of peanuts 95% will contain at least how much in weight?

Answers

1   Probability that a packet will contain more than 127 grams is 42.07%.
2   Probability that a packet will contain less than the nominal indicated weight of 125 grams is 8.08%.
3   Minimum and maximum weight in the middle 99% of the peanuts is 123.53 g and 129.97 g.
4   95% of the packets contain at least 124.69 g of peanuts.

10. Restaurant service time

Objective

The objective of this case–exercise is to illustrate the application of the normal distribution in probability analysis.

Situation

The profitability of a restaurant depends on how many customers can be served and the price paid for a meal. Thus, a restaurant should service the customers as quickly as possibly but at the same time provide them quality service in a relaxed atmosphere. A certain restaurant in New York, in a three month study, had the data in Table RS-1 regarding the time taken to service clients. It believed that it was reasonable to assume that the time taken to service a customer, from showing to the restaurant table and seating, to clearing the table after the client had been serviced, could be approximated by a normal distribution.

Table RS-1

Activity	Average time min	Variance
Showing to table, and seating client	4.24	1.1025
Selecting from menu	10.21	5.0625
Waiting for order	14.45	9.7344
Eating meal	82.14	378.3025
Paying bill	7.54	3.4225
Getting coat and leaving	2.86	0.0625
Clearing table	3.56	0.7744

Required

1 What is the average time and standard deviation to serve a customer such that the restaurant can then receive another client?
2 What is the probability that time to service a customer lies between 90 and 125 min?
3 What is the probability that time to service a customer lies between 70 and 140 min?
4 What is the combined probability that time to service a customer lies in 70 min or less or at least 140 min?
5 If in the next month it is estimated that 1,200 customers will come to the restaurant, to the nearest whole number, what is a reasonable estimate of the number of customers that can be serviced between 70 and 140 min?
6 Again, on the basis that 1,200 customers will come to the restaurant in the next month, 85% of the customers will be serviced in a minimum of how many minutes?

Answers

1 Average service time is 125.00 minutes; standard deviation is 19.96 min.
2 Probability that time to service a customer lies between 90 and 125 min is 46.02%.
3 Probability that time to service a customer lies between 70 and 140 min is 77.09%.
4 Probability that time to service a customer lies in 70 minutes or less or at least 140 min is 22.91%.

5   Estimate of the number of customers that can be serviced between 70 and 140 min is 925.
6   On the basis that 1,200 customers arrive, 85% of the customers will be serviced in a minimum of 104.31 min.

11. Tires

Objective

This case-exercise demonstrates the application of the normal distribution.

Situation

Michelin Co. determined that, for a certain truck tire used in the United States, the average distance for a tire before the treads indicated that it should be replaced was 50,000 miles. The company determined that on an annual basis, the distance traveled per truck was normally distributed, with a standard deviation of 12,000 miles. In the analysis there were 1,000 tires.

Required

1   What proportion of trucks can be expected to travel between 34,000 and 50,000 miles per year before the tires need to be replaced?
2   What is the probability that a randomly selected truck travels between 34,000 and 38,000 miles per year before the tires need to be replaced?
3   What percentage of trucks can be expected to travel below 30,000 and above 60,000 miles per year before the tires need to be replaced?
4   How many of the 1,000 trucks in the fleet, are expected to travel between 30,000, and 60,000 miles in the year before the tires need to be replaced?
5   How many miles will be traveled by at least 80% of the trucks before the tires need to be replaced?

Answers

1   Proportion of trucks that can be expected to travel between 34,000 and 50,000 miles per year before the tires need to be replaced is 40.88%.
2   Probability that a randomly selected truck travels between 34,000 and 38,000 miles per year before the tires need to be replaced is 6.74%.
3   Percentage of trucks can be expected to travel below 30,000 and above 60,000 miles per year before the tires need to be replaced is 25.01%.
4   Of the 1,000 trucks in the fleet, about 750 are expected to travel between 30,000, and 60,000 miles in the year before the tires need to be replaced.
5   About 39,900 miles can be traveled by at least 80% of the trucks before the tires need replacing.

12. Training program

Objective

This case-exercise illustrates the application of the normal distribution.

Situation

An automobile company has installed an enterprise resource planning system (ERP) to better manage the firm's supply chain. The human resource department has been instructed to develop a training program for the employees to fully understand how the new system functions. This training program has a fixed lecture period and at the end of the program there is a self-paced, online, practical exam that the participants have to pass before they are considered competent with the new ERP system. If they fail the exam they are able to retake it as many times as they wish in order to pass. When the employee passes the exam they are considered competent with the ERP system and they immediately receive a 2% salary increase. During the last several months, average completion of the program, which includes passing the exam, has been 56 days, with a standard deviation of 14 days. The time taken to pass the exam is considered to follow a normal distribution.

Required

1   What is the probability that an employee will successfully complete the program between 40 and 51 days?
2   What is the probability that an employee will successfully complete the program in 35 days or less?
3   What is the combined probability that an employee will successfully complete the program in no more than 34 days or more than 84 days?
4   What is the probability that an employee will take at least 75 days to complete the training program?
5   What are the lower and upper limits in days within which 80% of the employees will successfully complete the program?

Answers

1   The probability that an employee will successfully complete the program between 40 and 51 days is 23.39%.
2   The probability that an employee will successfully complete program in 35 days or less is 6.68%.
3   The combined probability that an employee will successfully complete the program in no more than 34 days or more than 84 days is 8.08%.
4   The probability that an employee will take at least 75 days to complete the training program is 8.74%.
5   The lower and upper limits in days within which 80% of the employees will successfully complete the program are about 38 and 74 days.

6 Methods and theory of statistical sampling

Correct testing can improve reliability

No	Case-exercise	Statistical concept	Application
1	Adult males in Northern Europe	Sampling, mean value	Life sciences, demographics
2	Automobile salvage	Sampling, mean value	Services
3	Child mortality	Sampling, proportions	Demographics
4	City of London	Sampling, mean value	Operations, restaurants
5	Croissants	Sampling, mean value	Food, retail services
6	Food bags	Sampling, mean value	Production, food
7	Online grocery shopping	Sampling, mean value	Food, retail services
8	Online hotel reservations	Sampling, mean value	Hospitality, technology
9	Telemarketing	Sampling, mean value	Service operations
10	Tertiary education	Sampling, proportions	Education, demographics
11	Unemployment rate	Sampling, proportions	Economics, demographics
12	Wine dispensing unit	Sampling, mean value	Food, hospitality

1. Adult males in Northern Europe

Objective

The objective of this case–exercise is to demonstrate the use of statistical sampling.

Situation

The average height of adult males in ten of the countries of Northern Europe is shown in Table AM-1.[1] These averages were determined using the same population count for each of the ten countries. The collected data follows a normal distribution with a standard deviation of this height data being 4.48 cm.

Table AM-1

Country	Average height, cm
Denmark	180.6
Estonia	179.1
Finland	178.2
Germany	178.1
Iceland	181.7
Lithuania	176.3
Netherlands	184.8
Norway	179.9
Sweden	180.0
United Kingdom	176.8

Required

1 What is the median height of adult males in Northern Europe according to Table AM-1?
2 If one adult male is selected at random from the Netherlands, what is the probability that he will be over the median height of the adult males in Northern Europe given by Question 1?
3 What are the lower and upper symmetrical limits of heights between which 90% of the population of adult males in the Netherlands will lie?
4 If samples of two adult males are selected at random in the Netherlands, what is the probability that the average will be over the median height of adult males in Northern Europe as given by the answer to Question 1?
5 What are the lower and upper symmetrical limits of heights between which 90% of the averages of the population of adult males in the Netherlands will lie, for samples of size two?
6 If samples of four adult males are selected at random in the Netherlands, what is the probability that the average will be over the median height of adult males in Northern Europe as given by the answer to Question 1?
7 What are the lower and upper symmetrical limits of heights between which 90% of the averages of the population of adult males in the Netherlands will lie, for samples of size four?

8   If samples of eight adult males are selected at random in the Netherlands, what is the probability that the average will be over the median height of adult males in Northern Europe as given by the answer to Question 1?

9   What are the lower and upper symmetrical limits of heights between which 90% of the averages of the population of adult males in the Netherlands will lie, for samples of size eight?

10   Explain the differences in the results.

Answers

1   The median height of adult males in Northern Europe is 179.1 cm.

2   If one adult male is selected at random from the Netherlands, the probability is 89.88% that he will be over the median height of the adult males in Northern Europe as given by the answer to Question 1.

3   The lower and upper symmetrical limits of heights between which 90% of the population of adult males in the Netherlands will lie are 177.44 cm and 192.16 cm or a range of 14.71 cm.

4   If samples of two adult males are selected at random in the Netherlands, the probability is 96.43% that the average will be over the median height of adult males in Northern Europe as given by the answer to Question 1.

5   The lower and upper symmetrical limits of heights between which 90% of the averages of the population of adult males in the Netherlands will lie, for samples of size two, are 179.60 cm and 190.00 cm or a range 10.40 cm.

6   If samples of four adult males are selected at random in the Netherlands, the probability is 99.46% that the average will be over the median height of adult males in Northern Europe as given by the answer to Question 1.

7   The lower and upper symmetrical limits of heights between which 90% of the averages of the population of adult males in the Netherlands will lie, for samples of size four, are 181.12 cm and 188.48 cm or a range of 7.36 cm.

8   If samples of eight adult males are selected at random in the Netherlands, the probability is 99.98% that the average will be over the median height of adult males in Northern Europe as given by the answer to Question 1.

9   The lower and upper symmetrical limits of heights between which 90% of the averages of the population of adult males in the Netherlands will lie, for samples of size eight, are 182.20 cm and 187.40 cm or a range of 5.20 cm.

10   The average height of the adult males in the Netherlands is 184.8 cm. We are comparing this value to 179.1 cm, the median height of adult males in Europe. This value is to the left of the mean value for the Netherlands. As the sample size increases according to Questions 2, 4, 6, and 8 then more data will cluster around the mean as evidenced by the increase in the percentage values. In a similar relationship, as the sample size increases, the range of the symmetrical limits gets smaller. These are a consequence of the central limit theorem.

2. Automobile salvage

Objectives

The purpose of this case-exercise is to illustrate the application of statistical sampling.

Situation

Joe and three colleagues have a created a small automobile salvage company. Their work consists of visiting sites that have automobile wrecks and recovering those parts that can be resold. Often from these wrecks they recoup engine parts, computers from the electrical systems, scrap metal, and batteries. From past work salvaged components, on average, generate €198.00 per car with a standard deviation of €55.00. Joe and his three colleagues pay themselves wages of €15.00/hour, which they consider as their operating costs. They work 40 hours per week in a five-day week and between them they are able to complete the salvage work on four cars per day. One particular period they carry out salvage work at a site near Hamburg, Germany where there are 72 wrecked cars.

Required

1 What is the corrected standard error for this situation?
2 What is the probability that after one week the team will have collected enough parts to generate total revenue of €4,200?
3 On the assumption that the probability outcome in Question 2 is achieved, what would be the profit to each team member at the end of the week?

Answers

1 The correct standard error is €10.5250.
2 The probability that after one week the team will have collected enough parts to generate total revenue of €4,200 is 87.29%.
3 On the assumption that the probability outcome in Question 2 is achieved, the profit to each team member at the end of the week is €450.00.

3. Child mortality

Objective

The purpose of this case exercise is to illustrate the concept of sampling from an infinite population.

Situation

According to the United Nations the deaths of children under the age of five per 1,000 live births of the population of children born in 2013 was as shown in Table CM-1 for selected countries in Africa and Asia.[2]

Required

1 For Mali, to the nearest two decimal places, what is the population proportion of the number of deaths of children under five?
2 If a random sample was taken for Mali, to the nearest two decimal places, what would be the proportion between 5% and 15% of the number of deaths of children under five?
3 If a random sample of 50 was taken for Mali, to the nearest four decimal places, what would be the standard error of the proportion of the number of deaths of children under five?

4   If a random sample of 50 was taken for Mali, to the nearest two decimal places, what would be the proportion between 5% and 15% of the number of deaths of children under five?

5   If a random sample of 100 was taken for Mali, to the nearest two decimal places, what would be the proportion between 5% and 15% of the number of deaths of children under five?

6   How do you explain the differences in the answers for Questions 2 to 5?

Table CM-1

Country	Child mortality under five per 1,000 live births
Uganda	66
Tanzania	52
Ethiopia	64
Kenya	71
Mali	123
Nigeria	117
Congo	119
Chad	148
Afghanistan	97
Pakistan	86
Ghana	78

Answers

1   For Mali the population proportion of the number of deaths of children under five is 12.30%.

2   If a random sample was taken for Mali the proportion between 5% and 15% of the number of deaths of children under five is 12.07%.

3   If a random sample of 50 was taken for Mali the standard error of the proportion of the number of deaths of children under five is 0.0464.

4   If a random sample of 50 was taken for Mali the proportion between 5% and 15% of the number of deaths of children under five is 66.15%.

5   If a random sample of 100 was taken for Mali, the proportion between 5% and 15% of the number of deaths of children under five is 78.14%.

6   The larger the sample size the greater is the proportion of data that lies around the population mean value. Note that the boundaries of 5% and 15% lie on either side of the population mean value.

4. City of London

Objectives

The purpose of this case-exercise is to illustrate the application of statistical sampling.

Situation

The clients of a restaurant in the City of London during the lunch periods from Monday to Friday are almost exclusively those who work in the financial district, and they have a tight lunch break. The restaurant manager has a target of serving the client from seating to being presented the bill of 45 minutes. However, the manager knows there is a variance of 72.25 (minutes)^2 in this service time due to the speed of a client selecting a menu, the number of clients in the restaurant, and the weather. Service time is considered to follow a normal distribution.

Required

1 To the nearest two decimal places, what is the probability that the time taken to service one client selected at random will be between 38 and 52 minutes?
2 If five clients are selected at random, to the nearest two decimal places, what is the probability that the time taken to service them will be between 38 and 52 minutes?
3 If ten clients are selected at random, to the nearest two decimal places, what is the probability that the time taken to service them will be between 38 and 52 minutes?
4 If fifteen clients are selected at random, to the nearest two decimal places, what is the probability that the time taken to service them will be between 38 and 52 minutes?
5 What is the explanation for the change in answers for Questions 1 through 5?
6 How might you characterize the sample data taken?

Answers

1 The probability that the time taken to service one client selected at random will be between 38 and 52 minutes is 58.98%.
2 If five clients are selected at random, the probability that the time taken to service them will be between 38 and 52 minutes is 93.44%.
3 If ten clients are selected at random, the probability that the time taken to service them will be between 38 and 52 minutes is 99.08%.
4 If fifteen clients are selected at random, the probability that the time taken to service them will be between 38 and 52 minutes is 99.86%.
5 As the sample size increases more of the data clusters around the population mean as a consequence of the central limit theorem. Note that the values of 38 and 52 minutes lie either side of the mean value.
6 The sample data can be considered strata as it represents a relatively homogeneous set of information; business people in them City of London, having lunch in a restaurant at a particular time.

5. Croissants

Objective

This purpose of case-exercise is the demonstrate the concept of sampling for the mean value in statistical analysis.

Situation

The croissant is the staple of the French *petit déjeuner*. A particular bakery in Douarnenez, Brittany, who has being operating for some 30 years, sells an estimated 2,000 croissants per day. Butter is one of the main ingredients. The target weight of the croissants is 70 g but there is variation due to the baking process, humidity, and the volume of butter used. From past experience he baker knows that that the standard deviation of the weight of the croissants is 3 g.

Required

1   If a customer buys one croissants from the bakery, what is the probability that it will weigh more than 71 g?
2   If a customer buys four croissants from the bakery, what is the probability that the average weight of the croissants will be more than 71 g?
3   If a customer buys eight croissants from the bakery, what is the probability that the average weight of the croissants will be more than 71 g?
4   If a customer buys sixteen croissants from the bakery, what is the probability that the average weight of the croissants will be more than 71 g?
5   If a customer buys one croissant from the bakery, what is the probability that it will weigh between 68 and 73 g?
6   If a customer buys four croissants from the bakery, what is the probability that the average weight of the croissants will be between 68 and 73 g?
7   If a customer buys eight croissants from the bakery, what is the probability that the average weight of the croissants will be between 68 and 73 g?
8   If a customer buys sixteen croissants from the bakery, what is the probability that the average weight of the croissants will be between 68 and 73 g?
9   Explain the differences between the answers to Questions 1 through 4.
10 Explain the differences between the answers to Questions 5 through 8. Why is the progression the reverse of what you see for the answers to Questions 1 through 4?

Answers

1   Probability that the croissant will weigh more than 71 g is 36.94%.
2   Probability that the average weight of the four croissants is more than 71 g is 25.25%.
3   Probability that the average weight of the eight croissants is more than 71 g is 17.29%.
4   Probability that the average weight of the sixteen croissants is more than 71 g is 9.12%.
5   Probability that the weight of the croissant is between 68 and 73 g is 58.89%.
6   Probability that the average weight of the four croissants is between 68 and 73 g is 88.60%.
7   Probability that the average weight of the eight croissants is between 68 and 73 g is 96.80%.
8   Probability that the average weight of the sixteen croissants is between 68 and 73 g is 99.61%.
9   Here the values decrease with sample size. Since it is asking more than 71 g the larger the sample size the smaller amount of data will be away from the mean or target value.
10 Here the values of 68 and 73 g are either side of the mean value. As the sample size increases a larger amount of data clusters around the mean value, as a consequence of the central limit theorem.

6. Food bags

Objective

The objective of this case-exercise is to demonstrate the concept of sampling in statistical analysis.

Situation

A paper company in Finland manufactures treated double-strength bags used for holding up to 20 kg of dry food such as rice, potatoes, barley, etc. These bags have a nominal breaking strength of 8 kg/cm^2 with a production standard deviation of 0.70 kg/cm^2. The manufacturing process of these food bags follows a normal distribution.

Required

1　What percentage of the bags produced have a breaking strength between 8.0 and 8.5 kg/cm^2?
2　What percentage of the bags produced have a breaking strength between 6.5 and 7.5 kg/cm^2?
3　What proportion of the sample means of samples of size 10 will have breaking strength between 8.0 and 8.5 kg/cm^2?
4　What proportion of the sample means of samples of size 10 will have breaking strength between 6.5 and 7.5 kg/cm^2?
5　Compare the answers of Questions 1 and 3, and 2 and 4.
6　What distribution would the sample means follow for samples of size 10?

Answers

1　Percentage of the bags that have a breaking strength between 8.0 and 8.5 kg/cm^2 is 26.25%.
2　Percentage of the bags that have a breaking strength between 6.5 and 7.5 kg/cm^2 is 22.15%.
3　Proportion of sample means of samples of size 10 with a breaking strength between 8.0 and 8.5 kg/cm^2 is 48.81%.
4　Proportion of sample means of samples of size 10 with a breaking strength between 6.5 and 7.5 kg/cm^2 is 1.19%.
5　Since the sample is of size 10 rather than size one from the population the data clusters around the mean. In Questions 1 and 3 the percentage increases as the limits are around the mean. In Questions 2 and 4 the percentage decreases since the limits are away from the mean.
6　Normal as the population is normal.

7. Online grocery shopping

Objective

This case-exercise demonstrates the concept of sampling for the mean value in statistical analysis.

Situation

A large supermarket chain in the United Kingdom has instituted an online grocery shopping service where customers can purchase orders through the Internet. Orders placed before 12:00

will be delivered to the customer at the latest by 22:00 the same day. The average sale on groceries using online shopping is £125.00 per customer with a standard deviation of £31.25. The distribution of sales though online shopping is considered to follow a normal distribution. All of the information for online sales per customer is stored on a database.

Required

1   What is the probability that if one online customer sale is selected at random the sample average customer sale will lie below £95.00?
2   What is the probability that if four online customer sales are selected at random the sample average customer sale will lie below £95.00?
3   What is the probability that if nine online customer sales are selected at random the sample average customer sale will lie below £95.00?
4   What is the probability that if one online customer sale is selected at random the sample average customer sale will lie between £95.00 and £140.00?
5   What is the probability that if four online customer sales are selected at random the sample average customer sale will lie between £95.00 and £140.00?
6   What is the probability that if nine online customer sales are selected at random the sample average customer sale will lie between £95.00 and £140.00?
7   What is the explanation for the answers to Questions 1 though 3 and Questions 4 through 6.

Answers

1   The probability that if one online customer sale is selected at random the sample average customer sale will lie below £95.00 is 16.85%.
2   The probability that if four online customer sales are selected at random the sample average customer sale will lie below £95.00 is 2.74%.
3   The probability that if nine online customer sales are selected at random the sample average customer sale will lie below £95.00 is 0.20%.
4   The probability that if one online customer sale is selected at random the sample average customer sale will lie between £95.00 and £140.00 is 51.59%.
5   The probability that if four online customer sales are selected at random the sample average customer sale will lie between £95.00 and £140.00 is 80.40%.
6   The probability that if nine online customer sales are selected at random the sample average customer sale will lie between £95.00 and £140.00 is 92.31%.
7   In Questions 1 through 3 we are asking about the probability that online sales will be below £95. As the sample size increases the proportion gathers around the mean value and so there is a lower proportion in the tail below £95. Thus the values decrease with sample size. In Questions 4 through 6 the values of £95 and £140.00 lie on either side of the mean. Thus with an increase in sample size the proportions cluster around the mean or the values increase with sample size. This is a consequence of the central limit theorem.

8. Online hotel reservations

Objective

The purpose of this case-exercise is to demonstrate the theory of statistical sampling for the mean value.

Situation

A large hotel company based in McLean, Virginia, USA is examining ways to improve its online reservation system. As a first step, it carried out an extensive sampling program and determined that the average time a client was connected to its website was 18 min with a standard deviation of 4 min. A large sample of data was taken and the distribution was approximately normal.

Required

1   To the nearest two decimal places, what is the percentage of online reservations that lasted between 17 and 20 min?
2   If a random sample size of 10 online reservations were taken, then to the nearest two decimal places, what is the percentage of connections that last between 17 and 20 min?
3   If a random sample size of 25 online reservations were taken, then to the nearest two decimal places, what is the percentage of connections that last between 17 and 20 min?
4   If a random sample size of 50 online reservations were taken, then to the nearest two decimal places what is the percentage of connections that last between 17 and 20 min?
5   What is the most plausible reason that explains the differences in the answers of the four immediate previous questions?
6   To the nearest whole number, what is the sample size at a 99% confidence such that the error is ±1.00 minutes of the true population value?

Answers

1   The percentage of online reservations that lasted between 17 and 20 min is 29.02%.
2   If a random sample size of 10 online reservations were taken, then the percentage of connections that last between 17 and 20 min is 72.85%.
3   If a random sample size of 25 online reservations were taken, then the percentage of connections that last between 17 and 20 min is 88.81%.
4   If a random sample size of 50 online reservations were taken, then the percentage of connections that last between 17 and 20 min is 96.12%.
5   As the sample size increases more data is grouped around the population mean value. Note that the values of 17 and 20 min are on either side of the population mean value.
6   The sample size at a 99% confidence such that the error is ±1.00 min of the true population value is 107 (106.16).

9. Telemarketing

Objective

This case–exercise is designed to illustrate the concept of sampling for mean values.

Situation

A telemarketing firm wants to maximize the number of calls made to clients in order to improve potential customer contact with the objective of increasing sales. It establishes an average time of the calls made by its operators as 4 minutes (240 seconds) and it knows from practice that the standard deviation of the call-time is 0.50 minutes (30 seconds).

Required

1 What is the probability that a single call taken at random will last between 234 and 246 seconds?
2 What is the probability that a single call taken at random will last between 225 and 240 seconds?
3 If random samples of 25 calls are selected what is the probability that a call taken at random will last between 234 and 246 seconds?
4 If random samples of 25 calls are selected what is the probability that a call taken at random will last between 225 and 240 seconds?
5 If random samples of 100 calls are selected what is the probability that a call taken at random will last between 234 and 246 seconds?
6 If random samples of 100 calls are selected what is the probability that a call taken at random will last between 225 and 240 seconds?
7 Explain the difference in the results between Questions 1, 3, and 5 and Questions 2, 4, and 6. Why is the progression not the same?

Answers

1 Probability that a single call taken at random will last between 234 and 246 seconds is 15.85%.
2 Probability that a call taken at random will last between 225 and 240 seconds is 19.15%.
3 If random samples of 25 calls are selected the probability that a call taken at random will last between 234 and 246 seconds is 68.27%.
4 If random samples of 25 calls are selected the probability that a call taken at random will last between 225 and 240 seconds is 49.38%.
5 If random samples of 100 calls are selected the probability that a call taken at random will last between 234 and 246 seconds is 95.45%.
6 If random samples of 100 calls are selected the probability that a call taken at random will last between 225 and 240 seconds is 50.00%.
7 For Question 1, 3, and 5 the random limits of 234 and 246 seconds are on either side of the mean value of 240 seconds. Increasing from sample sizes of 1, to 25, and then to 100 the percentage of values cluster around the mean and this progression is from 15.85%, to 68.27% and finally 95.45%, or close to 100%. For Questions 2, 4, and 6 the limits of 225 and 240 seconds are on the left side of the mean with the centerline being the right-hand boundary limit. Increasing from sample sizes of 1 to 25, and then to 100 still gives data clustering around the mean progressing from 19,15%, to 49.38% and finally to 50.00%. However, the proportion cannot go beyond 50% since the other 50% is on the right side of the mean between limits of 240 and 255 seconds.

10. Tertiary education

Objective

The objective of this case-exercise is to demonstrate the application of sampling in proportions.

Situation

According to the OECD the percentage of the population in the age range 25–34 with a tertiary education (at least a university degree) in 1970 and 2013 was according to Table TE-1.[3]

Table TE-1

Country	1970	2013
Japan	16.00%	58.00%
Britain	18.00%	48.00%
USA	29.00%	45.00%
France	13.00%	44.00%
Spain	9.00%	42.00%
Greece	8.00%	38.00%
Germany	20.00%	30.00%
Brazil	8.00%	16.00%

Required

1 If random samples of 50 people were taken in Japan in 2013 what proportion between 50% and 64% would have a tertiary education?
2 If random samples of 100 people were taken in Japan in 2013 what proportion between 50% and 64% would have a tertiary education?
3 If random samples of 200 people were taken in Japan in 2013 what proportion between 50% and 64% would have a tertiary education?
4 If random samples of 50 people were taken in Britain in 2013 what proportion between 50% and 64% would have a tertiary education?
5 If random samples of 100 people were taken in Britain in 2013 what proportion between 50% and 64% would have a tertiary education?
6 If random samples of 200 people were taken in Britain in 2013 what proportion between 50% and 64% would have a tertiary education?
7 How do you explain the progression of data in the answers to Questions 1 through 3 and Questions 4 through 6?

Answers

1 If random samples of 50 people were taken in Japan in 2013 the proportion between 50% and 64% that would have a tertiary education is 67.91%.
2 If random samples of 100 people were taken in Japan in 2013 the proportion between 50% and 64% that would have a tertiary education is 83.54%.
3 If random samples of 200 people were taken in Japan in 2013 the proportion between 50% and 64% that would have a tertiary education is 94.63%.
4 If random samples of 50 people were taken in Britain in 2013 the proportion between 50% and 64% that would have a tertiary education is 37.68%.
5 If random samples of 100 people were taken in Britain in 2013 the proportion between 50% and 64% that would have a tertiary education is 34.38%.
6 If random samples of 200 people were taken in Britain in 2013 the proportion between 50% and 64% that would have a tertiary education is 28.56%.
7 In the case of Japan, the recorded population proportion is 58%. The benchmark limits are 50% and 64%. These values lie on either side of the population proportion. Thus as the sample size increases from 50 to 200 the proportion of data that clusters around the

population proportion increases. For Britain, the recorded population proportion is 48%. The benchmark values of 50% and 64% are both on the right side of the population proportion. As the sample size increases from 50 to 200 the proportion of data clusters around the mean and so there is less away from the mean. This explains why the percentage decreases with sample size.

11. Unemployment rate

Objective

The objective of this case-exercise is to demonstrate the application of sampling in proportions.

Situation

According to the published statistics of the World Bank for the period 2010 to 2013, the percentage unemployment rate for selected countries was as shown in Table UR-1. Unemployed means those available for and seeking work.[4]

Table UR-1

Year	2010	2011	2012	2013
Australia	5.2	5.1	5.2	5.7
Canada	8.0	7.4	7.2	7.1
Croatia	11.8	13.4	15.8	17.7
Cyprus	6.3	7.9	11.8	15.8
France	9.3	9.2	9.9	10.4
Germany	7.1	5.9	5.4	5.3
South Africa	24.7	24.7	25.0	24.9
UK	7.9	7.8	8.0	7.5
USA	9.7	9.0	8.2	7.4

Required

1   If random samples of 100 people were taken in France in 2013 what proportion between 4% and 8% would be unemployed?
2   If random samples of 200 people were taken in France in 2013 what proportion between 4% and 8% would be unemployed?
3   If random samples of 100 people were taken in Germany in 2013 what proportion between 4% and 8% would be unemployed?
4   If random samples of 200 people were taken in Germany in 2013 what proportion between 4% and 8% would be unemployed?
5   If random samples of 100 people were taken in the UK in 2013 what proportion between 4% and 8% would be unemployed?
6   If random samples of 200 people were taken in the UK in 2013 what proportion between 4% and 8% would be unemployed?
7   What is an explanation for the paired results of each country?

Answers

1 If random samples of 100 people were taken in France in 2013 the proportion between 4% and 8% who would be unemployed is 19.79%.
2 If random samples of 200 people were taken in France in 2013 the proportion between 4% and 8% who would be unemployed is 13.16%?
3 If random samples of 100 people were taken in Germany in 2013 the proportion between 4% and 8% who would be unemployed is 60.51%.
4 If random samples of 200 people were taken in Germany in 2013 the proportion between 4% and 8% who would be unemployed is 74.99%.
5 If random samples of 100 people were taken in the UK in 2013 the proportion between 4% and 8% who would be unemployed is 48.33%.
6 If random samples of 200 people were taken in the UK in 2013 what proportion between 4% and 8% would be unemployed is 57.57%.
7 By the central limit theorem as the sample size increases more data clusters around the mean. For France, the sample proportion values of 4% and 8% lie to the left of the population proportion value of 10.4%. As the sample size increases more data clusters around the mean thus correspondingly less at the extremities. This explains why in going from a sample size of 100 to 200, the percentage amount *decreases*. However, in the case of Germany and the UK the sample proportion values of 4% and 8% lie either side of the population proportions (5.3% for Germany and 7.5% for the UK). Thus with increase in sample size the proportion around the mean *increases*.

12. Wine dispensing unit

Objective

The objective of this case–exercise is to illustrate sampling in statistical analysis.

Situation

A wine dispensing unit in a restaurant is regulated so that the amount dispensed into the wine pitcher is on average 25 cl. The filling operation is normally distributed and the standard deviation of the dispensing unit is 1.0 cl. no matter the setting of the mean value.

Required

1 What is the volume that is dispensed such that only 5% of the wine pitchers contain this amount or less?
2 If the dispensing unit is regulated such that only 5% of the wine pitchers contained 20 cl. or less by how much could the nominal value of the setting be reduced? In this case, on average, a customer would be receiving what percentage less of beverage?
3 With a nominal dispensing unit setting of 25 cl. if samples of 10 wine pitchers are taken what is the volume that will be exceeded by 95% of sample means?
4 There is a maintenance rule such that if the sample average content of 10 wine pitchers falls below 24.50 cl., a technician will be called out to check the machine settings. In this case, how often would this happen at a nominal machine setting of 25 cl.?

5   What should the nominal machine setting be to ensure that no more than 1% maintenance calls are made? In this case, on average, customers will be receiving how much more wine?

Answers

1   The volume dispensed, such that only 5% of wine pitchers contain this amount or less, is 23.36 cl.
2   If the machine is regulated, such that only 5% of the wine pitchers contain 20 cl. or less, the machine setting be reduced by 3.36 cl. In this case, on average a customer would be receiving 13.42% less wine.
3   With a nominal machine setting of 25 cl., for samples of 10 wine pitchers the volume that will be exceeded by 95% of sample means is 24.48 cl.
4   A technician will be called out to check the machine settings 5.69% of the time.
5   Nominal machine setting to ensure that no more than 1% maintenance calls are made is 25.24 cl. In this case on average customers will be receiving almost 1% more of wine (0.94%).

Notes

1   http://www.disabled-world.com/artman/publish/height-chart.shtml (accessed January 18, 2016).
2   *The Economist*, March 14, 2015, p. 83.
3   *The Economist*, "Special Report; The Young," January 23, 2016, p. 6.
4   www.worldbank.org (accessed July 31, 2015).

7   Estimating population characteristics

Giving confidence to your evaluations

No	Case-exercise	Statistical concept	Application
1	Automobile insurance	Mean, t and z, size	Services
2	California Households	Confidence, mean value	Marketing
3	Eco-halogen light bulbs	Mean, student-t	Production, sales
4	Glass bottles	Confidence, proportions, size	Production
5	Hardware store	Confidence, mean, student-t	Retail store
6	Hiking in the Andes	Confidence, proportions, size	Tourism, education
7	Hotel room prices	Mean, normal-z, size	Hospitality
8	Neck ties	Mean, auditing, student-t	Retail sales
9	Saint Emilion vines	Confidence, mean, auditing	Hotel and food services
10	State taxes	Mean, student-t, size	Government
11	Third work shift	Proportions	Human resources
12	Travel agency, Corfu	Mean, auditing, size	Hospitality, tourism

1. Automobile insurance

Objective

The objective of this case-exercise is to estimate population characteristics for mean values.

Situation

An automobile insurance company is examining its records of insurance payments for automobile repairs to customers to see whether its premiums are in line with its costs. An analyst with the firm took a random sample of insurance payments in the last twelve months of drivers in the 20–30 year age range. This information, in Euros, is in Table AI-1.

Table AI-1

4,012	2,148	13,857	18,253	2,235	12,546
2,001	12,589	25,432	6,587	3,500	1,456
15,234	42,800	4,569	4,589	10,536	5,468

Required

1. To the nearest whole number, what would be the best point estimate of the mean amount of insurance payments to drivers in the 20–30 year age range?
2. To the nearest whole number, determine the 80%, 95%, and 99% confidence limits of the amounts paid out by the insurance company to drivers in the 20–30 year age range?
3. What is an explanation of the differences in the limits as determined in Question 2?
4. If you used (incorrectly) the normal z distribution, rather than the correct Student's t what are the 80%, 95%, and 99% confidence limits? What is the purpose of doing this?
5. Using the Student's t distribution, what would be the best estimate of sample size, to the nearest whole number, in order to be 80% confident to be within $\pm$ €1,000 of the mean value of insurance payments to drivers in the 20 to 30 age group? What is the value if a normal z distribution is used? Why is there a difference?
6. In performing a more detailed analysis of insurance payments of all their clients the company found that the average payment was less than that indicated by the initial sample data. What is the most plausible explanation for this difference?

Answers

1. A point estimate of the mean amount paid out to drivers in the 20–30 year age range is €10,434.
2. The 80% confidence limits are €7,144 and €13,724; 95% confidence limits are €5,229 and €15,639; and 99% confidence limits are €3,284 and €17,584.
3. The higher the confidence limits required then the broader is the range.
4. Using the normal z distribution the 80% confidence limits are €7,232 and €13,596; the 95% confidence limits €5,599 and €15,269; the 99% confidence limits are €4,079 and €16,789. The differences between the values obtained with the Student's t distribution and the normal z distribution are not large. For example, the difference between the Student's t distribution upper confidence level at 80% and the corresponding value for

the normal z distribution is less than 1%. From a business perspective this is probably not significant.

5. Using the Student's t distribution the best estimate of sample size, to the nearest whole number, in order to be 80% confident to be within ± €1,000 of the mean value of insurance payments to drivers in the 20 to 30 age group is 195. Using a normal z distribution it is 180. The difference is due to the difference in the values of the Student's t and the normal z. At a sample size of 195 a normal z distribution would be used.

6. The data given is a stratified sample for people in the 20–30 age group. They are a higher risk group regarding automobile accidents.

2. California households

Objective

The objective of this case exercise is to demonstrate the use of sampling to estimate population mean values.

Situation

A random sample of 175 families of four in Pasadena, California were sampled on the amount they spent per week on grocery shopping. It only included food; no other items such as cleaning fluids, and the like were included. The sample average was $249.50 and the sample standard deviation $27.25.

Required

1. Determine an 85% confidence interval and the range for the average amount spent by all households in Pasadena, California.
2. Determine a 90% confidence interval and the range for the average amount spent by all households in Pasadena, California.
3. Determine a 95% confidence interval and the range for the average amount spent by all households in Pasadena, California.
4. Determine a 99% confidence interval and the range for the average amount spent by all households in Pasadena, California.
5. Explain the differences between the answers to Questions 1 through 4.

Answers

1. For an 85% confidence interval the lower limit is $246.53 and the upper limit is $252.47 or a range of $5.93.
2. For a 90% confidence interval the lower limit is $246.11 and the upper limit is $252.89 or a range of $6.78.
3. For a 95% confidence interval the lower limit is $245.46 and the upper limit is $253.54 or a range of $8.07.
4. For a 99% confidence interval the lower limit is £$244.19 and the upper limit is $254.81 or a range of $10.61.
5. The higher the confidence level, going from 85% to 99%, the further apart are the limits to offset uncertainties.

3. Eco-halogen light bulbs

Objective

The objective of this case-exercise is to demonstrate the application of estimating.

Situation

A Chinese firm manufactures eco-halogen light bulbs that are sold in the European market. The manufacturer claims that one particular 40 watt light has a life of 2,000 hours before they fail. The European distributor of these bulbs takes a random sample of 14 from a production lot and tests them until they fail. The time in hours when these bulbs failed is in Table EH-1.

Table EH-1

2,001	1,954	2,025	2,020	2,015	2,013	1,973
1,999	1,996	1,955	2,032	1,979	1,953	2,029

Required

1. What distribution is the correct one to use? Explain your reasoning.
2. From the sample data what is an estimate of the average life of the light bulbs? What is an estimate of the population standard deviation?
3. How would the distributor describe the 80% confidence intervals for the mean life of the light bulbs?
4. Determine the 90%, confidence intervals for the mean length of the life of the light bulbs.
5. Determine the 99% confidence intervals for the mean length of the life of the light bulbs.
6. What is an explanation for the different ranges of the lives of the light bulbs given for the answers to Questions 1, 2, and 3?
7. What might be some preliminary observations on this experiment?

Answers

1. A Student's *t* distribution should be used since the sample size is less than 30 and there is no information about the population standard deviation of the life of the light bulbs.
2. The estimate of the average life of the light bulbs is 1,996 hours and an estimate of the population standard deviation is 28.55 hours. This number is the same as the sample value. (There is no information about the population.)
3. The distributor might say, "From my random sample experiment I estimate that the mean life of the light bulbs is 1,996 hours and I am 80% confident that the life is between 1,985.70 and 2,006.30 hours."
4. The 90% confidence limits are 1,982.49 and 2,009.51 hours.
5. The 99% confidence limits are 1,973.01 and 2,018.99 hours.
6. The higher the certainty then the broader is the range for the given confidence in order to cover for a high perceived trust in the experiment. The range at 80% confidence is 20.61 hours; a range of 27.03 hours at 90% confidence; and a range of 45.97 hours at a 99% confidence.

7. From the experiment it appears that the manufacturer is a little optimistic about the claim since the average value at 1,996 is just below the 2,000 hour claim (albeit by only 0.2%). It would be more reassuring if the average life from the experiment was above the 2,000 hour claim. The distributor could run another experiment to verify the results. Alternatively, the manufacturer could sell this lot of light bulbs at a discounted price to the distributor.

4. Glass bottles

Objective

The objective of this case-exercise is to estimate population characteristics for proportions.

Situation

A production filling operation for *l'eau gazeuse* has a filling line for one liter glass bottles that are dedicated for restaurants and hotels. Before the bottles are filled they pass through an optical detector to ensure that there are no imperfections such as cracks, air holes in the glass, chips on the bottle top, or other faults. Any glass bottle that is identified as being faulty is automatically ejected from the line. In one particular production run 10,250 glass bottles passed through the optical detector and 81 were ejected.

Required

1. In the production filling operation, what is a point estimate of the proportion of glass bottles that are defective?
2. Determine 90% confidence intervals for the proportion of defective bottles in this production filling operation. What is the margin of error assuming that the sample proportion is an estimate of the population proportion?
3. Determine 98% confidence intervals for the proportion of defective bottles in this production filling operation. What is the margin of error assuming that the sample proportion is an estimate of the population proportion?
4. Explain the differences between the confidence limits determined in Question 2 and 3.
5. If an estimate of the proportion of defective bottles to within a margin of error of $\pm$ 0.0016 ($\pm$ 0.16%) of the population proportion at 90% confidence is required, and using the sample proportion as an estimate of the population proportion, how many bottles, or what sample size, should pass through the optical detector? What are your observations from this answer compared to that for Question 2?
6. If an estimate of the proportion of defectives to within a margin of error of $\pm$ 0.0010 ($\pm$ 0.10%) of the population proportion at 90% confidence is required, and using the sample proportion as an estimate of the population proportion, how many bottles, or what sample size, should pass through the optical detector? What are your observations from this answer compared to that for Question 2?
7. Assume that no sampling information is available. In this case, at a 90% confidence level, how many bottles, or what sample size, should pass through the optical detector for a conservative estimate, to within a margin of error of $\pm$ 0.001 ($\pm$ 0.10%)?
8. What are your comments about the answers obtained in Question 6 and 7 and generally for this sampling process?

Answers

1. A point estimate of the proportion of bottles that are defective in the production filling operation is 0.79%.
2. The 90% confidence intervals for the proportion of defective bottles in the production filling operation are 0.65% and 0.93%. The margin of error is ± 0.14%.
3. The 98% confidence intervals for the proportion of defective bottles in the production filling operation are 0.59% and 0.99%. The margin of error is ± 0.20%.
4. The higher confidence level in Question 3 compared to Question 2 requires a broader confidence range.
5. At a 90% confidence level the sample size is 8,286 glass bottles for a margin of error of ± 0.16%. This is lower than the original sample size of 10,250. This is not surprising since the margin of error of ± 0.16% is numerically greater than the margin of error of ± 0.14% observed from the original sample information.
6. At a 90% confidence level the sample size is 21,212 glass bottles for a margin of error of ± 0.10%. This is greater than the original sample size of 10,250 but is not surprising since the margin of error of ± 0.10% is numerically less than the margin of error of ± 0.14% observed from the sample information.
7. The sample size, or number of bottles that should pass through the optical detector, for a conservative estimate at a 90% confidence level is 676,386.
8. The conservative values require a very high inspection quantity. However, since the inspection process is automatic perhaps this is not a problem. Note that this is not a random inspection as the device samples every bottle.

5. Hardware store

Objective

The objective of this case-exercise is to estimate population characteristics for the mean value.

Situation

A hardware store in San Francisco purchases a truckload of ceramic floor tiles from a supplier, knowing that the tiles are imperfect. An imperfect tile means that the color may not be uniform, there may be surface hairline cracks, or there may be air pockets on the surface finish. The hardware store intends to sell the tiles in its stores at a marked-down price. An assistant in the store takes a random sample of tiles from the truck and counts the number of imperfections on the tiles. This information is given in Table HS-1.

Table HS-1

17	9	3	1	7
10	9	8	6	18
12	11	6	8	5
9	4	14	15	13
17	5	9	5	4

Required

1. What is the best estimate of the mean number of imperfections on the truckload of tiles? This would be the point estimate.
2. What is the best estimate of the population standard deviation? On what are you basing your estimate?
3. How would you describe the 80% confidence limits?
4. How would you describe the 99% confidence limits? Why are these results different from those for Question 3?
5. What have you taken into account in developing your analysis?
6. What are your comments to the results of this sampling experiment?

Answers

1. The best estimate is that there are 9 imperfections per tile.
2. The estimate of the population standard deviation, σ, is 4.6637 imperfections. This is numerically the same as the sample standard deviation s. The sample data is all that we have available.
3. The mean number of imperfections is 9 per tile and I am 80% confident that the number of imperfections lies between 8 and 10 imperfections per tile (7.77 and 10.23).
4. The mean number of imperfections is 9 per tile and I am 99% confident that the number of imperfections lies between 6 and 12 imperfections per tile (6.39 and 11.61). These limits are broader because we are giving a higher confidence level.
5. The sample size is less than 30 so we use a Student's t distribution rather than the normal distribution.
6. The number of imperfections is quite high. The tiles perhaps would be more appropriate for decorating a laundry room or bathroom rather than a living room as the imperfect tiles would be less visible.

6. Hiking in the Andes

Objective

The objective of this case-exercise is show the application of estimating for proportions.

Situation

The student bureau of a certain university in California, where there are 20,000 students, plans to organize a hiking trip to the Andes Mountains. Susan Reading, the head of the student's bureau, wishes to have an estimate of the number of students who will be coming on the trip. Susan selects at random 120 students and of these 24 indicate they will be coming.

Required

1. What is the sample proportion of students who say that they will not be coming?
2. Is the finite population multiplier needed in this problem?
3. Determine the standard error of the proportion for this sampling analysis.

4. Determine the margin of error in this experiment.
5. How would Susan express the sampling experiment at 85% confidence level for the proportion of students coming on the trip?
6. How would Susan express the sampling experiment at 99% confidence level for the proportion of students coming on the trip? Explain why there is a difference in the confidence limits for this question and Question 4.
7. Assume that Susan requires an estimate of the number of students coming on the trip to within a margin of error of ± 3% and at an 85% confidence level. In which case what is the sample size Susan should take if the sample proportion is considered an estimate of the population proportion?
8. Assume that Susan requires an estimate of the number of students coming on the trip to within a margin of error of ± 3% and at an 85% confidence level. What is the conservative sample size Susan should take?

Answers

1. The sample proportion who says that they will not be coming is 80%.
2. The finite population multiplier is not needed as the ratio of n/N is less than the benchmark value of 5%.
3. The standard error of the proportion is 0.0365.
4. The margin of error is ± 5.26%.
5. Susan would say that of the student population at the university she estimates that 20% would be coming on the trip and that she is 85% confident that the proportion lies between 14.74% and 25.26% of the student population.
6. Susan would say that of the student population at the university she estimates that 20% would be coming on the trip and that she is 99% confident that the proportion lies between 10.59% and 29.41% of the student population. The limits are much broader because Susan is using a higher confidence level.
7. The sample size Susan should take is 369 students (368.40).
8. The sample size Susan should take is 576 students (575.63).

7. Hotel room prices

Objective

The objective of this case-exercise is to develop a range of estimates and appropriate sample sizes according to certain criteria.

Situation

John wishes to develop a website to describe the hotels in the Malaga area of Spain. The purpose of this website is to improve tourism in this area particularly for those from England and the Netherlands. One piece of information he needs is an estimate of hotel room prices in the region. John takes a random sample of hotel room prices per person for a double room and this information, in Euros, is shown in Table HR-1.

Table HR-1

111	81	115	90	102	93	119	139	95	88	123	144	81
149	152	79	142	84	80	108	88	113	92	61	107	98
148	136	105	103	150	118	150	152	77	111	85	152	98
124	132	131	133	80	103	126	144	85	87	134	82	124
81	89	96	136	108	114	118	83	107	139	131	91	113

Required

1. Based on this information, what is the best estimate of the average hotel room price in the Malaga region?
2. What would be John's 80%, 90%, and 95% confidence intervals for the average price estimate that you have developed in Question 1? Also indicate the price range.
3. How do you explain the differences in the range levels determined in Question 2?
4. What sample size would you need to take in order to be within ± €2 at a confidence level of 90%?
5. If the first value in the list was €351 instead of €111 the sample size, n, would be close to 1,000. What is the explanation for this?

Answers

1. An estimate of the average price is €110.92.
2. The 80% intervals are €107.06 and €114.78 giving a range of €7.72; the 90% intervals are €105.97 and €115.88 giving a range of €9.91; the 95% intervals are €105.02 and €116.83 giving a range of €11.81.
3. The higher the certainty of your probability, the greater is the range level.
4. A sample size of about 400 prices (399.24).
5. The value of €351 is high relative to the other values in the dataset. This increases the estimated standard deviation, meaning that there is a higher variation in the data. Thus, a higher sample size would be required to nullify this uncertainty.

8. Neck ties

Objective

The purpose of this case-exercise is to apply to concept of estimating for population characteristics used in auditing.

Situation

Tie-rack, a store at the King's Cross Station in London, sells ties of different colors and designs to suit the business person as well as the well-to-do dandy. It is currently making an end-of-year inventory evaluation. It takes a random sample from all of the ties in its store and records the price of ties in this sample. This is given in Table NT-1 in pounds sterling (£). The inventory records file indicates that there are a total of 1,250 ties in the store.

Table NT-1

22.75	25.00	25.50	42.00	37.00
26.95	20.00	21.00	28.25	24.50
31.28	15.50	32.50	18.00	
35.90	16.00	21.00	44.00	
23.00	29.50	22.50	32.00	

Required

1. From this sample, to the nearest two decimal places, what is an estimate of the average price of the ties in the store? What is an estimate of the total value of neck ties in the store?
2. From this sample, to the nearest four decimal places, what is an estimate of the standard error of the sample mean?
3. For statistical analysis, should you use the normal distribution criteria or the Student's t approach? Explain why and in which case, what is the value of the appropriate statistic for a 95% confidence level?
4. What are the 95% confidence limits for the average value of ties in the store?
5. What are the 95% confidence limits for the total value of ties in the store?
6. If you performed the statistical analysis for the owner of the store, how would you express your findings to the owner for the total value of neck ties in the store?

Answers

1. An estimate of the average price of neck ties in the store is £27.01. An estimate of the total value of neck ties is £33,757.39.
2. An estimate of the standard error of the mean for the neck ties is £1.6753.
3. A Student's t distribution should be used as the sample size is 22 or less than 30 and the population standard deviation is unknown. The value of the Student's t is ± 2.0796 for a 95% confidence level.
4. The lower 95% confidence level for the average value of neck ties is £23.52; the upper 95% confidence level is £30.49.
5. The lower 95% confidence level for the total value of neck ties is £29,402.43; the upper 95% confidence level is £38,112.34.
6. I estimate that the total value of neck ties in the store is £33,757 and based on my sampling experiment, I am 95% confident that the total value lies between £29,402 and £38,112.

9. Saint Emilion vines

Objective

The objective of this case-exercise is to illustrate the application of estimating population characteristics using sampling data and extending it to an auditing situation.

Situation

Saint-Emilion, the name of a well-known wine, is a commune in southwestern France in the Gironde Department in the region of Aquitaine-Limousin-Poitou-Charentes. A particular grape owner in this locality wants to estimate the yield from his 6,000 grape vines so that he can make some forecast of revenues from the sale of his wine. He samples 80 of the grape vines at random and counts the number of grape bunches on these vines. This information is in Table SE-1.

Table SE-1

24	37	26	30	34	30	14	20	22	34	18	17	38	22	31	30
14	22	26	32	37	38	21	32	12	29	26	35	10	14	11	21
31	21	22	14	17	18	36	33	28	12	38	11	31	18	20	21
28	19	25	27	28	24	38	26	23	13	20	25	11	19	24	16
36	20	32	24	34	24	16	16	31	21	20	10	29	20	25	22

Required

1. What is an estimate of the average number of bunches of grapes on all of the vines?
2. Develop a 95% confidence limit for the average bunch of grapes for the total of the 6,000 grape vines.
3. Using the information developed in Questions 1 and 2 give an estimate of the total bunches of grapes and the corresponding 95% confidence limits. That is the complete vineyard is being audited.
4. Assume now that a smaller random sample was taken, represented by the first row of Table SE-1. What now is an estimate of the average number of bunches of grapes on all of the vines?
5. Using the information developed in Question 4 give an estimate of the average bunches of grapes and the corresponding 95% confidence limits.
6. Using the information developed in Question 4, give an estimate of the total bunches of grapes and the corresponding 95% confidence limits. That is the complete vineyard is being audited
7. What are your comments?

Answers

1. An estimate of the average number of bunches of grapes on each of the 6,000 vines is 24.05.
2. The estimated limits of the average number of grapes on the vines, at a 95% confidence level, is at the lower level 22.33 and the upper level 25.77.
3. An estimate of the total bunches of grapes in the vineyard is 144,300 with the 95% confidence limits at 133,996 and 154,604 bunches of grapes.
4. An estimate of the average number of bunches of grapes on each of the 5,200 vines is 26.69.
5. The estimated limits of the average number of grapes on the vines, at a 95% confidence level, is at the lower level 22.75 and the upper level of 30.63.
6. An estimate of the total bunches of grapes in the vineyard is 160,125 with the 95% confidence limits at 136,494 and 183,756 bunches of grapes.
7. Regarding the estimate of the total number of grape bunches in the vineyard, the sample size of 16 gives a value that is almost 11% (10.97%) higher compared to the experiment with a sample size of 80. For an estimate perhaps a 11% difference is not large. In practice the larger sample size should give the most reliable result, though the smaller sample size is quicker to perform and less costly in terms of labor charges. Note that for a sample size of 80 we use a normal z distribution; for a sample size of 16 we use a Student's t distribution.

10. State taxes

Objective

The objective of this case-exercise is to estimate population characteristics for the mean value.

Situation

In order to estimate the total annual revenues to be collected from its residents for budget purposes, the Tax Commissioner in a certain state in the USA collects a random sample of 20 tax returns. The taxes to be paid in $US according to these returns are shown in Table ST-1.

Table ST-1

14,500	1,750	10,250	2,250	17,500
5,250	9,000	72,000	23,000	15,000
15	12,520	5,260	6,980	13,520
5,230	620	12,580	1,950	6,280

Required

1. What is the value of the correct test statistic to be used for analysis?
2. To the nearest whole number what is a point estimate of the average state tax returns?
3. To the nearest whole number determine the 80%, 95%, and 99% confidence intervals for the mean state tax returns.
4. How do you explain the differences in these intervals and what does it say about confidence in decision making?
5. Using as an example the 95% confidence interval how would present your analysis to your superior?
6. What sample size needs to be taken at an 80% confidence level to be within ± $1,000 of the total tax returns for the state?

Answers

1. To be statistically correct, a Student's t distribution should be used as the sample size is less than 30 and the population standard deviation is unknown. The value of the Student's t is ± 1.3277.
2. The point estimate of the average state tax returns is $11,773.
3. The 80% confidence limits are $7,179 and $16,367. The 95% limits are $4,531 and $19,015. The 99% limits are $1,874 and $21,672.
4. The higher the confidence limits, going from 80% to 99%, then the broader is the interval range in $US.
5. I estimate that the mean value of the tax returns for the year in question is $11,773 and I am 95% confident that they will lie between $4,531 and $19,015.
6. A random sample size of 423 tax returns is needed in order to be 80% confident that the level is within ± $1,000 of the true population mean. In this case a normal z distribution would need to be used.

11. Third work shift

Objective

The objective of this case-exercise is the application of estimation for proportions.

Situation

The management of an automobile factory in Germany with 15,000 employee is considering the introduction of a third work shift that would be from 22:00 to 06:00 to accommodate the increased product demand. The advantage of this third shift would be to provide a 20% salary premium to those workers involved and would nullify the alternative of outsourcing some production to China. The human resource department took a random sample of 900 employees and found that there were 225 who were not in favor of a night shift.

Required

1.   What is the proportion of employees who are in favor of a night shift?
2.   What are the 90% confidence limits for the population who are not in favor?
3.   What are the 90% confidence limits for the proportion who are in favor of a night shift?
4.   What are the 99% confidence limits for the population who are not in favor?
5.   What are the 99% confidence limits for the proportion who are in favor of a night shift?
6.   What is your explanation of the difference between the answers to Questions 2 and 4 and Questions 3 and 5?
7.   What sample size should be taken in order that at a 99% confidence level the company would be within ± 2% of the true population mean if the sample data was used for the calculation?
8.   What conservative sample size should be taken in order that at a 99% confidence level the company would be within ± 2% of the true population mean?

Answers

1.   The proportion of employees who are in favor of a night shift is 0.75 or 75.00%.
2.   The 90% confidence limits for the population who are not in favor are 22.63% and 27.37% or a range of 4.75%.
3.   The 90% confidence limits for the population who are in favor are 72.63% and 77.37%.
4.   The 99% confidence limits for the population who are not in favor are 26.23% and 33.77% or a range of 7.44%.
5.   The 99% confidence limits for the population who are in favor are 66.23% and 73.77%.
6.   The higher the confidence limits, the broader is the range.
7.   A sample size of 3,110 (3,110.11 calculated).
8.   A sample size of 4,150 (4,146.81 calculated).

12. Travel agency, Corfu

Objective

The objective of this case-exercise is to use estimating to examine population characteristics.

Situation

John runs a travel agency that deals exclusively with vacations in Greece where one popular destination is the island of Corfu. In order that he can respond to his client's requests on hotel prices he asks his partner on the island to perform a sampling analysis on the daily room prices, in the period June through September, for a standard three-star hotel. The agent provides, in Euros/day, the information shown in Table TA-1.

Table TA-1

141	125	158	70	114	141	208	93	92	117	201	121	110
152	79	108	112	207	119	193	179	101	150	102	174	132
209	135	149	197	171	114	144	154	106	100	132	199	144
150	177	147	162	132	109	161	92	133	133	94	75	90
158	100	76	129	127	205	135	200	188	143	165	86	192
109	195	99	124	82	74	81	153	206	123	65	122	125

Required

1. Based on the sample data, and to the nearest whole number, what is the best estimate of the average daily room price?
2. To the nearest two decimal places, what are John's 80% confidence limits of the average daily room price?
3. To the nearest two decimal places, what are John's 99% confidence limits of the average daily room price?
4. What is the most plausible explanation for the difference in answers for the immediate previous two questions?
5. To the nearest whole number, what sample size of room prices, for this hotel standard, should be taken in order to be within ± €2.00 of all these hotels on the island with 95% confidence?

Answers

1. Based on the sample data, the best estimate of the average daily room price is €136.
2. To the nearest two decimal places, John's 80% confidence limits of the average daily room price are €129.78 and €141.25.
3. To the nearest two decimal places, John's 99% confidence limits of the average daily room price are €123.99 and €147.05.
4. The more certain you are in a situation, the broader the confidence limits that you will give.
5. The sample size of room prices, for this hotel standard, that should be taken in order to be within ± €2.00 of all these hotels on the island with a 95% confidence is 1,500.

8 Hypothesis testing for a single population

Giving assurance to assumptions that are made

No	Case-exercise	Statistical concept	Application
1	Buildings in Instabul	Proportions	Construction, safety
2	Chocolate bars	Mean value	Production, food
3	Christmas tree lights	Mean value	Production, retail
4	Emergency services	Mean value	Social services
5	Gender wage gap	Proportions	Demographics, human resources
6	Obesity in children	Mean value	Health, children, food
7	Sandwich sales	Mean value	Retail, hospitality
8	Spotlights	Mean value	Project, production
9	Tomato ketchup	Mean value	Food, operations
10	Trucks in Romania	Proportions	Logistics, safety
11	UK students	Proportions	Education, demographics
12	Young people and employment	Proportions	Demographics, human resources

1. Buildings in Istanbul

Objective

The objective of this case–exercise is to demonstrate hypothesis testing for proportions from a single population.

Situation

The mayor of Istanbul, Turkey, says that 90% of its buildings are built according to the building codes. A consulting firm in the European Union takes a random sample of 275 buildings and of these determines that 238 are built according to code.

Required

1 Express the hypothesis situation for the question: "Is there evidence of a difference?"
2 To the nearest two decimal places, what proportion of the buildings sampled is considered not built according to code? What is the proportion considered constructed to code?
3 To the nearest four decimal places what is the standard error of the proportion of the sampling process?
4 To the nearest four decimal places in this sampling process what is the value of the test statistic for the number of buildings considered safe?
5 What are your conclusions at a 5% significance level to test if there is a difference in the proportion of buildings in Istanbul constructed according to code?
6 Justify your response to Question 5 using the *p*-value approach.
7 Justify your response to Question 5 by determining the confidence levels at 5% significance.
8 Express the hypothesis situation for the question: "Is there evidence that the proportion of buildings built to code is less than the hypothesized value?"
9 What are your conclusions at a 5% significance level testing for evidence that the proportion of buildings built according to code is less than the hypothesized value? Justify your response using the normal *z* statistic, the *p*-value and the confidence limit.

Answers

1 Null hypothesis of proportion built to code: $H_0:p = 90\%$: Alternative hypothesis: $H_1:p \neq 90\%$.
2 The proportion of buildings not built according to code is 13.45%. Proportion constructed according to code is 86.55%.
3 The standard error of the proportion of the sampling process is 0.01809.
4 The value of the test statistic for the number of buildings constructed according to code is −1.9096.
5 At a 5% significance level testing for a difference, there is no evidence to suggest that we should not accept the null hypothesis. The critical limits of *z* are ± 1.9600. The sample value of −1.9096 falls inside these limits.
6 At a 5% significance level testing for a difference there is an area of 2.5% in each tail. The *p*-value is 2.81% and this is greater than 2.50% and this corroborates our answer to Question 5.

7 The confidence levels at 5% significance for the testing of a difference are 86.45% and 93.55%. Since our sample value of 86.55% falls between these limits this corroborates our answer to Question 5.

8 Null hypothesis of proportion built to code: $H_0: p \geq 90\%$: Alternative hypothesis: H_1: $p < 90\%$.

9 At a 5% significance level testing for less than, there is evidence to suggest that we should reject the null hypothesis. The critical limit of z is -1.6449. The sample z value of -1.9096 falls outside this limit. The p-value is still 2.81% and this is less than the significance level of 5%. The confidence limit is 87.02% and the sample value of 86.55% is lower than this value. All of these confirm the conclusion.

2. Chocolate bars

Objective

The objective of this case-exercise is to illustrate the concept of hypothesis testing from a single population.

Situation

A confectionery company near Birmingham, England indicates on the label of one format of its chocolate bars that the net weight is 100 g. A government weights and measures inspector weighs a random sample of chocolate bars from the production line to see if the labeling information conforms to specifications. The sampling information of the weights of the chocolate bars in grams is given in Table CB-1.

Table CB-1

106.5	104.8	99.8	104.0	94.8
104.3	79.3	88.1	104.3	88.9
99.1	99.3	100.8	87.8	99.9
92.3	106.3	88.2	99.5	101.3
101.3	102.1	82.3	85.7	100.2

Required

1 If you are asking the question, "Is there evidence of a significant difference?" how would the null and alternative hypothesis be expressed?

2 What is the sample mean value of the weight of chocolate bars?

3 At a 5% significance, using the test statistic, is there evidence that the weight of chocolate bars is different than 100 g?

4 Confirm the answer to Question 2 using the p-value approach.

5 Do the limits of the weight of the chocolate bars also confirm your conclusions from Question 2?

6 If you now are asking the question, "Is here evidence of the weight being less than the labeled weight?" how would the null and alternative hypothesis be expressed?

7   At a 5% significance, using the test statistic, is there evidence that the weight of chocolate bars is less than 100 g?

8   Confirm the answer to Question 7 using the *p*-value approach.

9   Do the limits of the sample weight of the chocolate bars also confirm your conclusions to the answer of Question 7?

10   Why is this type of test important?

Answers

1   Null hypothesis: $H_0 = 100$ g; Alternative hypothesis: $H_1 \neq 100$ g.

2   Sample mean weight is 96.8360 grams.

3   There is no evidence of a difference. The critical Student's *t* value is ± 2.0639. The sample test Student's *t* value is -2.0094. This lies on the left side of the distribution and lies between the critical value and the mean value of the distribution.

4   At a 5% significance for the test of a difference there is 2.50% in each of the tails. The *p*-value is 2.85%, which is greater than 2.50%. This result confirms the conclusion of Question 2.

5   The lower and upper limits are 96.73 grams and 103.27 g respectively. The sample weight of 96.8360 grams lies between these limits. This again confirms the answer to Question 2.

6   Null hypothesis: $H_0 \geq 100$ g; Alternative hypothesis: $H_1 < 100$ g.

7   The critical value of the Student's *t* is now -1.7109. The sample test value of the Student's *t* is -1.998. This value is numerically greater than the critical value so for a 5% significance there is evidence that the weight of the chocolate bars is significantly less than the labeled amount of 100 grams.

8   At a 5% significance for the test of "less than" there is 5.00% in the left-hand tail. The *p*-value of the sample test value is is 2.85%, which is less than 5.00%. This result confirms the conclusion of Question 7.

9   At a 5% significance for the test of "less than" the lower limit of the weight is 97.29 grams. The sample weight of 96.8360 is less than this amount. This again confirms the answer to Question 7.

10   If producer is producing significantly less than the labeled 100 grams according to specifications then this is erroneous labeling. If production was significantly more than specifications, it would be costing the producer more than necessary.

3. Christmas tree lights

Objective

The objective of this case-exercise is to demonstrate the application of hypothesis testing for small samples from a single population.

Situation

In March a European retailer is sourcing suppliers for Christmas products for the following holiday season. It contacts a manufacturing firm of Christmas tree lights in order to see if its product would satisfy its retail outlets. The Chinese manufacturer says that the mean life of its lights is 120 hours before they fail. The retailer, before it negotiates a contract for the purchase of

500,000 of these lights, wants to verify the manufacturer's claim. It takes a random sample of 25 lights and tests them until they burn out. The time in hours of these lights before they failed is given in Table CL-1.

Table CL-1

127.8	125.7	119.7	124.8	113.7
125.1	95.2	105.7	125.2	106.7
118.9	119.2	120.9	105.3	119.9
110.8	127.5	105.8	119.4	121.6
121.5	122.5	98.7	102.8	120.2

Required

1   Write an expression for the hypothesis test if you are analyzing for a difference.
2   To the nearest two decimal places, what is the sample mean of the time in hours before the Christmas tree lights fail?
3   At a 5% significance level, testing for a difference using the test statistics, is there evidence to suggest that the manufacture's claim is not correct? Quantify your response.
4   Validate your conclusions to Question 3 using the *p*-value.
5   Corroborate your conclusions from Questions 3 and 4 by determining the acceptance levels and positioning the sample mean value within these levels.
6   Write an expression for the hypothesis test if you are analyzing to see if the mean life of the Christmas lights is less than the manufacturer's claim.
7   At a 5% significance to test if the life of the bulbs is less than the manufacturer's claim what are your conclusions? Use the Student's *t* statistics, *p*-value, and the acceptance limits to justify your answer.

Answers

1   The null hypothesis is: $H_0{:}\mu = 120$ hours. The alternative hypothesis is: $H_1{:}\mu \neq 120$ hours.
2   The sample mean of the time before the Christmas tree lights fail is 116.18 hours.
3   No, there is no evidence at 5% testing for a difference to suggest that the mean life of the light is significantly different than 120 hours. The critical Student's *t* limits are ± 2.0639 and the sample test Student's *t* of -2.0094 lies within the boundary of these critical limits. Accept the null hypothesis.
4   The *p*-value of the sample test experiment is 2.79% and this is greater than 2.50% since in testing for a difference there is 2.50% in each tail. This confirms the conclusions to Question 3.
5   The lower acceptance level is 116.08 hours and the upper is 123.92 hours. The sample mean value of 116.18 hours lies between these boundary levels.
6   The null hypothesis is: $H_0{:}\mu \geq 120$ hours. The alternative hypothesis is: $H_1{:}\mu < 120$ hours.
7   At a 5% significance there is evidence indicating that the life of light is less than 120 hours.

- The critical Student's *t* value is -1.7109. The sample Student's *t* value is -2.0094. This is below the critical limit.
- The *p*-value for the sample is 2.79% and this is less than 5%.
- The lower acceptance limit is 116.75 hours. The sample value of 116.18 hours is less.

4. Emergency services

Objective

The objective of this case–exercise is to examine hypothesis testing from a single population.

Situation

A district in Texas is testing the response time of its emergency services to respond to crisis calls such as a fire, automobile accident, or personal illness, but particularly heart attacks. The longer the response time the higher is the probability that lives of individuals involved may be seriously jeopardized. The district has a target of 8 minutes response time but this depends on traffic circulation when the call is made, and the number of emergency staff available. A random sample of the response time in minutes taken on one particular day is given in Table ES-1.

Table ES-1

7	11	13	12	6	9
10	6	5	11	8	7
10	15	18	7	10	6

Required

1 Determine the sample size and the mean value of the sample response time.
2 At the 5% significance level, using the concept of critical value testing, does this sample data indicate that the response time is different from 8 minutes?
3 At the 5% significance level, using the p-value concept, does your answer corroborate the conclusion of Question 2? Give your reasoning.
4 At the 5% significance level what are the confidence intervals when the test is asking for a difference? How do these intervals confirm your answers to Questions 2 and 3?
5 At the 5% significance level, using the concept of critical value testing, does this data indicate that the response time for an emergency call is greater than 8 minutes?
6 At the 5% significance level, using the p-value concept, does your answer corroborate the conclusion of Question 5? Give your reasoning.
7 At the 5% significance level what is the confidence limit when the test is asking "greater than"? How does this interval confirm your answers to Questions 5 and 6?
8 Which of these two tests is the most important? What are your comments?

Answers

1 Sample size is 18; mean value of response time is 9.50 minutes.
2 No, there is no evidence to suggest there is a difference; critical value of t is ± 2.1098 and the test value of t is $+1.8349$. This is a two-tailed test with 2.5% in each tail.
3 Yes, the p-value confirms the answer to Question 2. The p-value is 8.41%, which is greater than 5.0%.
4 The limits are 6.28 and 9.72 minutes. The sample value of 9.50 minutes is contained within this interval.

5   Yes, there are indications that response time is greater than 8 min. Critical value of t is $+1.7396$ and the test value of t is $+1.8349$. This is a one-tailed test with 5.0% in the right-hand tail.
6   Yes, the p-value is 4.20% which is less than 5.0%.
7   The upper confidence limit is 9.42 min. The sample value of 9.50 min is outside or above this limit.
8   The test for being greater than 8 min. If this is the case, the district is not meeting its objectives for emergency services. Other random samples could be taken to validate, or otherwise, the conclusions.

5. Gender wage gap

Objective

The objectivee of this case-exercise is to illustrate hypothesis testing of proportions.

Situation

Across racial and ethnic groups the gender wage gap in the United States has not seen significant improvements in recent years. In 2014 women's median weekly earnings for full time work were \$719 compared with \$871 for men. In fact, controlling for inflation, neither women's nor men's earnings significantly increased between 2013 and 2014.[1] A random sample of 105 women is taken and of these 28 had salaries less than their male counterparts for the same type of work.

Required

1   According to the data what is the percentage difference of women's earnings compared to men's? Express this value as the null hypothesis and indicate the alternative hypothesis.
2   Using the critical value approach at a 1% significance level, is there evidence to suggest that the salaries of women are different than the population percentage difference determined in Question 1?
3   Using the p-value approach are you able to corroborate your conclusions from Question 2? Explain your reasoning.
4   What are the confidence limits at the 1% level? How do they agree with your conclusions of Questions 2 and 3?
5   Repeat the procedures for Questions 2 through 4 but now using a significance level of 5% in testing for a difference. What are your conclusions?
6   At a 1% level, test if there is evidence to suggest that percentage difference of salaries of men to women is significantly more than reported. Write the appropriate hypothesis relationships and use the criteria of Questions 2 though 4 to justify your response.

Answers

1   According to the published information, women's salaries are 17.45% less than men's. Written as a hypothesis for a difference this is expressed as $H_0 = 17.45\%$; $H_1 \neq 17.45\%$.
2   No, at a 1% significance level, there is no evidence to suggest that the salaries of women are different than men's. The benchmark critical limits, z, are ± 2.5758. The sample test statistic, z, is 2.4880. This value is within the boundaries of the critical limits.

3   The *p*-value of the sample test is 0.64% or larger than the 0.5% in the upper tail. This corroborates the conclusions to Question 2.
4   The confidence limits at a 1% significance for a difference are 7.91% and 26.99%. The sample value of 26.67% lies between these limits and corroborates the conclusion to Question 2.
5   Here we have increased the significance or importance of the test for a difference from 1% to 5% and the statistical evidence indicates that salaries of women are different than men's.

- The benchmark critical limits, z, are ± 1.9600. The sample test statistic, z, is 2.4880 and this value is outside the boundaries of the critical limits.
- The *p*-value of the sample test is 0.64% or smaller than the 2.5% in the upper tail.
- The confidence limits are 10.19% and 24.71%. The sample value, 26.67%, lies outside these limits.

6   The hypothesis relationships are $H_0 \leq 17.45\%$: $H_1 > 17.45\%$. This is a right-tailed test since we are asking "Is there evidence that the percentage is more than reported?" There is 1% in the upper tail and in this case there is evidence that the salary disparity is more than reported.

- The benchmark critical limit z, is 2.3263. The sample test statistic, z, is 2.4880. This sample value is outside or above the boundary of the critical limit.
- The *p*-value of the sample test is 0.64% or smaller than the 1.00% in the upper tail.
- The confidence limit is 26.07%. The sample value, 26.67%, lies outside or above this limit.

6. Obesity in children

Objective

The objective of this case-exercise is to examine hypothesis testing from a single population.

Situation

There is a big problem worldwide in many countries—the USA, Britain, Australia, countries of Africa, and the Middle East—where people are obese, particularly children. In a particular school in England the administrators have asked the owner of the two vending machines in the establishment to replace the chocolate bars with oranges. Unlike chocolate bars that are processed and thus the average weight is easy to control, oranges vary quite markedly in weight. The vending firm asks its supplier of oranges to sort them before they are delivered as it wants the average weight to be 300 g. The criterion for this is that the vending firm wants to be reasonably sure that each child who purchases an orange is going to get one of equivalent weight and level of vitamin C. A truckload of oranges arrives at the vendor's depot and an inspector takes a random sample of 27 oranges. Table OC-1 gives the weight of each orange in the sample.

Required

1   Express a hypothesis test for this situation when you are analyzing for a difference.
2   At the 5% significance level, using statistical values measuring for a difference, does it appear that the vendor is able to provide the oranges according to the established criteria?

Table OC-1

286	294	292	301	281	289
312	272	328	278	308	314
282	286	315	288	288	
282	281	287	322	278	
275	284	331	302	294	

3   Using the *p*-value concept, can you corroborate your answer to Question 2? Give your reasoning.
4   At the 5% significance level what are the confidence intervals when the test is asking for a difference in the weight? How do these intervals confirm your answers to Questions 1 and 2?
5   Express a hypothesis test for this situation when you are analyzing to see if the weight is less than the required specifications.
6   At the 5% significance level, using the concept of critical value testing, does this sample data indicate that the weight of the oranges in this truckload is less than the desired 300 g?
7   At the 5% significance level, using the *p*-value concept, does your answer corroborate the conclusion of Question 6? Give your reasoning.
8   At the 5% significance level what are the confidence intervals when the test is asking for "less than" in the weight. How do these intervals confirm your answers to Questions 6 and 7?
9   What are your comments concerning the criteria for this hypothesis test?

Answers

1   $H_0:\mu = 300$ g; $H_1:\mu \neq 300$ g.
2   No, there is no evidence that there is a significant difference. Critical value of *t* is ± 2.0555; sample test value of *t* is -1.9192. This is a two-tailed test with 2.5% in each tail.
3   Yes. The *p*-value of the test statistic is 6.60% which is greater than 5.0%. Alternatively, half of this value or 3.30% is greater than 2.50%.
4   The confidence intervals are 293.45 g and 306.55 g. The sample average value of the weight of the oranges, 293.88 g, is contained within these intervals.
5   $H_o: \mu \geq 300$ g; $H_1:\mu < 300$ g. We are only interested if the weight is significantly less than the specifications.
6   Yes. Critical value of *t* is -1.7056 and the test value of *z* is -1.9192. This is a one-tailed test with 5.0% in the left-hand tail. The sample value is outside the benchmark critical value.
7   Yes. The *p*-value is 3.30% which is less than 5.0% benchmark value.
8   The lower level is 294.56 g and there is no defined upper level. The sample test value of 293.88 g is outside this lower level boundary.
9   The test "less than" has more meaning than a difference. The school is more interested in understanding if the children are getting less than the given requirements. If they are getting an orange that weighs more than 300 g this is a nutritional advantage.

7. Sandwich sales

Objective

The objective of this case-exercise is to apply hypothesis testing from a single population.

Situation

A baker's store close to a business school in Lyon, France, reports to its owner that it sells 1,600 sandwiches per day to a mostly student and professor population. It makes a study over a one-month (30 day) period and records that on average it sells 1,550 sandwiches per day. It knows from past sales that the standard deviation of sandwich sales is 115. The sales of sandwiches is considered to follow a normal distribution.

Required

1   At a significance level α of 0.01 or 1.00% for analysis, using the critical value method, is there evidence that the sale of sandwiches is different from the claimed value of 1,600?
2   If you use the *p*-value for testing are you able to verify your conclusions in Question 1? Explain your reasoning.
3   What are the confidence limits corresponding to a significance level of 1%. How do these values corroborate your conclusions for Questions 1 and 2?
4   At a significance level α of 1.00% for analysis, using the critical value method, is there evidence that the sale of sandwiches is less than 1,600?
5   If you use the *p*-value for testing are you able to verify your conclusions in Question 4? Explain your reasoning.
6   What are the confidence limits corresponding to a significance level of 1.00%. How do these values corroborate your conclusions for Questions 4 and 5?

Answers

1   No, there is no evidence that the number of sandwiches sold is significantly different from the claim of 1,600 at the 1% level; critical value of *z* is ± 2.5758 and the test value of *z* is -2.3814, which is within the critical boundaries. This is a two-tailed test with 0.5% in each tail.
2   Yes, the *p*-value in each tail is 0.86% which is more than 0.50% Alternatively, the total *p*-value is 1.72% which is more than the α value of 1.00%.
3   Lower value is 1,545.92 and upper value is 1,654.08. The test value of 1,550 sandwiches is contained within these limits.
4   Yes, there is evidence that the sales is significantly less at the 1% level; critical value of *z* is -2.3263 and the test value of *z* is -2.3814, which is outside the critical boundary. This is a one-tailed test with 1.00% in the left tail.
5   Yes, the *p*-value is 0.86% which is less than 1.00%.
6   Lower value is 1,551.6 and the test value of 1,550 sandwiches is outside or lower than this limit.

8. Spotlights

Objective

The objective of this case-exercise is to apply the concept of hypothesis testing for a single population.

Situation

The Bumbo Company is building a new office facility west of Paris. The project manager in charge of the construction has contracted with the Zippy Company to supply a large quantity of low energy spotlights for the offices. Zippy indicates to the project manager that the average life of the spotlights made by its company is 600 hours with a standard deviation of 100 hours. The project manager wishes to know if the life of these lights is different from the 600 hours indicated by Zippy. For this purpose, he takes a sample of 75 of these spotlights and tests them until they go out. The average life of the sample of these lights is 580 hours.

Required

1 Define the null and alternative hypothesis for this situation.
2 What is the sample value of the test statistic?
3 What are your conclusions from this hypothesis test in terms of the test statistic?
4 Confirm your results to Question 3 using the *p*-value approach.
5 Determine the confidence limits in hours. How do these results confirm your answer to Question 3?
6 How would your answers change for Questions 2 through 5 if the significance level was 10% in a test for the difference?

Answers

1 $H_0:\mu = 600$ hours and $H_1:\mu \neq 600$ hours.
2 The sample value of the test statistic is $z = -1.7321$.
3 At a 5% significance level for a difference, meaning that there is an area of 2.50% in each tail, the critical boundary limits of z are ± 1.9600. Since the sample value falls within these boundary limits we accept the null hypothesis and conclude that the claim of Zippy Company is correct.
4 The area indicated by the sample statistics is 4.16%. This value is greater than the critical area of 2.50% in the lower tail and so this confirms the answer to Question 3.
5 The confidence limits of this hypothesis test are 577.37 and 622.63 hours. As the average sample test value of 580 hours falls within these limits this again confirms the results of Question 3.
6 At a 10% significance:

- The sample value of the test statistic remains unchanged at $z = -1.7321$.
- There is now an area of 5% in each tail that gives a critical value of z of ± 1.6449.
- The sample test statistic falls outside the boundary limit and so we reject the null hypothesis.
- The *p*-value of the sample statistic is still 4.16% and this is less than 5%. This confirms the rejection of the null hypothesis.
- The confidence limits at 10% significance are now 581.01 and 618.99 hours. Thus the sample value of 580 hours falls outside these confidence limits (580 is less than 581.01) and this again confirms the rejection of the null hypothesis.

9. Tomato ketchup

Objective

The objective of this case-exercise is to illustrate the concept of hypothesis testing from a single population.

Situation

A company makes a bio type of tomato ketchup from a rice-based syrup. A particular format of one of the bottles of this tomato ketchup indicates a net weight of 560 g. As this filling operation has been made many times the standard deviation is known at 4.0 g. For quality purposes, to verify that the quantity of ketchup in the bottles is according to specifications, an inspector takes a random sample of bottles from the filling line and weighs their contents. This information is in Table TC-1.

Table TC-1

567	553	554	557	559	569	572
568	567	552	565	565	556	556
557	565	565	552	555	553	566
566	558	550	560	552	567	552
569	567	562	571	564	572	558

Required

1   Responding to the question "Is there evidence of a significant difference?", how would you express the null and alternative hypothesis?
2   What is the sample mean value of the weight of the tomato ketchup?
3   At a 5% significance, using the test statistic, is there evidence that the weight of tomato ketchup is different than 560 g?
4   Confirm the answer to Question 2 using the *p*-value approach.
5   Do the limits of the weight of the tomato ketchup also confirm your conclusions of Question 2?
6   Repeat Questions 2 through 4 using a significance level of 10%.
7   Redo the hypothesis test for significance levels, using the test statistic but employing the sample standard deviation rather than the population standard deviation in your calculation.
8   What are some of your comments about this sampling experiment? Why is this type of test important?

Answers

1   Null hypothesis: $H_0 = 560$ g: Alternative hypothesis: $H_1 \neq 560$ g.
2   Sample mean weight is 561.17 g.
3   At a 5% significance level there is no evidence of a difference. The critical values of z are ± 1.9600. The sample test z value is 1.7326. This lies between the critical values of z.
4   At a 5% significance for a difference, there is an area of 2.5% in each of the distribution tails. The *p*-value of the test sample is 4.16%, which is greater than 2.50%. This result confirms the conclusion of Question 2.
5   The lower and upper limits are 558.67 g and 561.33 g respectively. The sample weight of 561.17 g lies between these limits. This again confirms the answer to Question 3.
6   At a 10% significance testing for a difference there is evidence that the weight of the ketchup is greater than the net value indicated on the label.

- The critical z values at 10% are $\pm$ 1.6449. The sample test value of 1.7326 is above the upper limit.
- The area in the tails is 5%. The sample p-value of 4.16% is less than 5%.
- The critical limits are 558.89 g and 561.11 g. The sample weight of 561.17 g is outside these limits.

7 The sample standard deviation is 6.75 g. Using this value gives a sample test statistic of 1.0268 that is within the critical limits at both a 5% and 10% significance.

8 This type of test is important to justify that the average weight of ketchup in the bottles is not significantly less than the net weight, thus cheating customers. Alternatively, it is not significantly more, and thus an avoidable additional cost to the producer. For example, in this case the difference is about 1 gram more per bottle. Assume a producer fills 500,000 bottles. It costs €2/kilogram to make the ketchup or €0.0020/gram. The extra cost to the producer for this excess amount is €1,000 (500,000*1*0.002).

There is a difference between the sample standard deviation and the recorded population. The sample value is 69% more than the population value. This should be examined. Is the population value from old data? Perhaps the sample was not completely random. The distribution nozzles of the filling machine are gumming; etc.

10. Trucks in Romania

Objective

The objective of this case-exercise is to demonstrate hypothesis testing for proportions from a single population.

Situation

The government of Romania reports that 95% of its trucks on the roads of Romania and Europe are safe and conform to European security regulations. A consulting firm in the European Union takes a random sample of 180 trucks and of these determines that only 166 are considered safe.

Required

1 Express the hypothesis situation for the question: "Is there evidence of a difference?"
2 What is the proportion of trucks that are considered unsafe? What is the proportion of trucks considered safe? What is the criterion for your analysis?
3 To the nearest four decimal places, what is the standard error of the proportion of the sampling process?
4 At a 5% significance using the z values testing for a difference, is there evidence that the proportion of trucks not conforming to specifications is different from 95%? In which case what is your decision?
5 Confirm you answer to Question 4 using the p-value approach.
6 Do the sample limits confirm your answer to Question 4?
7 Express the hypothesis situation for the question: "Is there evidence of that the proportion safe is less than the hypothesized value?"

8   At a 5% significance, using the z values testing for "less than", is there evidence that the proportion of trucks not conforming to specifications is less than 95%? In which case what is your decision?

9   Confirm you answer to Question 4 using the p-value approach.

10   Do the sample limits confirm your answer to Question 4?

Answers

1 [*] Null hypothesis of proportion being safe $H_0:p = 95\%$: Alternative hypothesis $H_1:p \neq 95\%$.

2   The proportion of trucks unsafe is 7.78%. The proportion of trucks considered safe is 92.22%. The situation is binomial. Either the trucks are safe or they are not.

3   The sampling error of the proportion is 0.0162.

4   No, there is no evidence to suggest that the safety level of the trucks is different from the hypothesized value. Sample test statistic z is -1.7100; critical value of statistics z is -1.9600. Thus we should accept the null hypothesis.

5   At a 5% significance, there is 2.50% in each tail when testing for a difference. The sample p-value is 4.36% and this is greater than 2.50%.

6   Yes, the lower limit of trucks considered safe is 91.82% and upper limit is 98.18%. The sample value of 92.22% lies between these values.

7   Null hypothesis of proportion being safe $H_0:p \geq 95\%$: Alternative hypothesis $H_1:p < 95\%$.

8   Yes. The test statistic is still $z = -1.7100$ but the critical value is now $z = -1.6449$. Reject the null hypothesis and accept the alternative hypothesis.

9   At a 5% significance for "less than" there is 5.00% in the left tail. The sample p-value is 4.36% and this is less than 5.00%.

10   Yes. The lower critical limit is 92.33%. The sample value of 92.22% is less than this amount.

11. UK students

Objective

The objective of this case-exercise is to examine hypothesis testing for proportions.

Situation

According to the Higher Education Statistics Agency (HESA), the number of international students at universities in the United Kingdom, in the academic year 2013–14, was according to Table UK-1.[2]

Table UK-1

Country	International students	Total student body
England	355,580	1,871,474
Scotland	48,360	230,286
Wales	25,610	134,789
Northern Ireland	5,950	54,091

Random samples of 155 students were selected from each of the four countries. Of these, the foreign students were: 39 from England, 25 from Scotland, 41 from Wales, and 19 from Northern Ireland.

Required

1 In the given period, what is the total student body in the United Kingdom?
2 From the data, what are the recorded population proportions of the foreign students in each of the four countries? These are the hypothesized proportions of the foreign students in each country.
3 From the random sampling data what are the sample proportions of foreign students in each country?
4 From the sampling information, determine the standard error of the proportion for each of the four countries.
5 At a 5% significance level, using the test statistic, does the sample information indicate any evidence to suggest that the proportions of foreign students in countries of the UK are different from the reported data? Justify your response. What are your comments?
6 Justify your response to Question 5 using the *p*-value approach.
7 Determine the confidence limits for the sampling data. How do these values corroborate your response to Question 5?

Answers

1 In the given period the total student body in the United Kingdom is 2,290,640.
2 The recorded population proportions of the foreign students, and thus the null hypothesis values, are: England, 19.00%; Scotland, 21.00%; Wales, 19.00%; and Northern Ireland, 11.00%.
3 The random sample proportions of the foreign students are: England, 25.16%; Scotland, 16.13%; Wales, 26.45%; and Northern Ireland, 12.26%.
4 From the sampling data, the standard errors of the proportions are: England, 0.03151; Scotland, 0.03272; Wales, 0.03151; and Northern Ireland, 0.02513.
5 At a 5% significance level testing for the difference there is an area of 2.50% in each tail.

- This gives a critical test statistic z of ± 1.9600.
- The sample z value statistics for the four countries are: England, 1.9553; Scotland, -1.4889; Wales, 2.3648; and Northern Ireland, 0.5006.
- For England, Scotland, and Northern Ireland there is no sample evidence to suggest that the proportion of foreign students in these countries is different from the reported data since the sample z value falls within the critical boundaries. However, this is not the case for Wales as the sample z value of 2.3648 falls outside is critical boundary limits of ± 1.9600. This either indicates that the reported data for Wales is not correct or that there was some bias in the collected sampling data for Wales.

6 The sample *p*-values are: England, 2.53%; Scotland, 6.83%; Wales, 0.90%; and Northern Ireland, 30.83%. Except for Wales all these values are greater than 2.50%, the critical significance area. For Wales, 0.90% is less than 2.50% thus again indicating an inconsistency. These percentage values corroborate the answers to Question 5.

7   The confidence levels are: England, 12.82% and 25.18%; Scotland, 14.59% and 27.41%; Wales, 12.82% and 25.18%; and Northern Ireland, 6.07% and 15.93%. Except for Wales the sample percentages all fall within these confidence limits. For Wales, the sample proportion of 26.45% falls outside the confidence limits. This again corroborates the answers to Question 5.

12. Young people and employment

Objective

The objective of this case-exercise is to examine hypothesis testing for proportions.

Situation

Roughly one quarter of the world's people, some 1.8 billion, have turned 15 but not yet reached the age of 30. In many ways are they are the most fortunate young adults to have lived; they are richer than previous generations; live in a world with no world wars (though there are many other awful sectarian hostilities); no Hitler; Stalin; or Mao Zedong; better educated; better nutrition; more intelligent than their elders. If they are female or gay they enjoy greater freedom in more countries that their elders ever imagined. With vast improvements in technology many can expect to live beyond 100. However, they suffer a big barrier in moving forward. Even with a good education they cannot find a job. In many countries government policies favor the old over the young. The young have difficulty finding a place to live. According to statistics for 2014[3] young people aged 15–29 not in employment, education, or training (NEETs) as a percentage of population is according to Table YP-1.

Random samples of 95 adults in the age range 15–29 students were selected from each of Britain and Nigeria. Of these in Britain 17 said that they were unemployed; in Nigeria 36 said that they were unemployed.

Table YP-1

Country	Percentage unemployed, age 15–29
Brazil	19
Britain	13
China	11
France	14
Germany	9
India	29
Japan	4
Nigeria	28
Russia	12
USA	16

Required

1 How would you present the null and alternative hypotheses for the employment situation among the young in Britain and Nigeria?
2 From the random sample data what are the sample proportions of young people that are unemployed in Britain and Nigeria?
3 From the sampling information, determine the standard error of the proportion for Britain and Nigeria.
4 At a 5% significance level, using the test statistic, does the sample information indicate any evidence suggesting that the proportion of young people in Britain is different from the reported information?
5 Justify your response to Question 4 using the *p*-value approach for a 5% significance level.
6 Determine the critical limits for the sampling data for Britain at a 5% significance level. How do these values corroborate your response to Question 4?
7 At a 10% significance level, using the test statistic, does the sample information indicate any evidence to suggest that the proportion of young people in Britain is different from the reported information?
8 Justify your response to Question 7 using the *p*-value approach for a 10% significance level.
9 Determine the critical limits for the sampling data for Britain at a 10% significance level. How do these values corroborate your response to Question 4?
10 Repeat Questions 4, 5, and 6 at a 5% significance level for Nigeria.
11 Repeat Questions 7, 8, and 9 at a 10% significance level for Nigeria.
12 What are your comments on the results?

Answers

1 Britain $H_0 = 13\%$ and $H_1 \neq 13\%$: Nigeria $H_0 = 28\%$ and $H_1 \neq 28\%$.
2 The sample proportion of young people that are unemployed in Britain is 17.89%; for Nigeria it is 37.89%.
3 The standard error of the proportion for Britain is 3.45%; for Nigeria it is 4.61%.
4 No, there is no evidence to suggest from the sample that there is any difference from the published information. The critical limits of z at a 5% significance level are ± 1.9600. The value z of the sample test statistic is 1.4186 and this value lies within the boundaries of the critical limits.
5 At a 5% significance level, there is an area of 2.50% in each tail. The *p*-value of the sample test is 7.80%. This is larger than the 2.50% critical value and so this corroborates the answer to Question 5.
6 The critical limits are 6.24% and 19.76%. The sample value of 17.89% lies within these limits. Again this validates the conclusion of Question 5.
7 No there is no evidence to suggest from the sample that there is any difference from the published information. The critical limits of z at a 10% significance level are ± 1.6449. The value z of the sample test statistic is 1.4186 and this value lies within the boundaries of the critical limits.
8 At a 10% significance level, there is an area of 5.00% in each tail. The *p*-value of the sample test is 7.80%. This is larger than the 5.00% critical value and so this corroborates the answer to Question 7.

9   The critical limits are 7.32% and 18.68%. The sample value of 17.89% lies within these limits. Again this validates the conclusion of Question 7.

10 For Nigeria, at a 5% significance level there is evidence to suggest a difference in the data.

   • The critical limits for z at 5% significance level are $\pm$ 1.9600. The sample test statistic of z is 2.1479. This value is outside the critical boundary limits.
   • The area in the tails for the critical limits are 2.50%. The p-value for Nigeria is 1.59%. This value is less than the critical area of 2.50%.
   • The critical limits are 18.97% and 37.03%. The sample value of 37.89% lies outside these limits.

11 For Nigeria, at a 10% significance level there is evidence to suggest a more pronounced difference with the published data.

   • The critical limits for z at 10% significance level are $\pm$ 1.6449. The sample test statistic of z is 2.1479. This value is outside the critical boundary limits.
   • The area in the tails for the critical limits are 5.00%. The p-value for Nigeria is 1.59% This value is less than the critical area of 5.00%.
   • The critical limits are 20.42% and 35.58%. The sample value of 37.89% lies outside these limits.

12 The sample experiment taken in Britain seems to corroborate the published data but not in the case of Nigeria. This could be because it is more difficult to obtain reliable information in emerging economies than in OECD countries that have a more rigorous control on their statistical information.

Notes

1 *Institute for Women's Policy Research*, "The gender wage gap, 2014," www.iwpr.org [accessed May 15, 2016]
2 www.ukcisa.org.uk/Info-for-universities-colleges–schools/Policy-research–statistics/Research–statistics/International-students-in-UK-HE/ [accessed January 16, 2016].
3 Special Report "The Young," *The Economist*, January 23, 2016, p. 5.

9   Hypothesis testing for different populations

Validating further your assumptions

No	Case-exercise	Statistical concept	Application
1	Beer drinking	Chi-squared	Retail sales
2	Depression and work environment	Proportions	Health, employment
3	Diet pills	Mean, paired difference, student–t	Health, life style
4	Driving time in Paris and Rome	Proportions	Logistics
5	Engineers' salaries in Germany/Italy	Difference of means, large samples	Demographics, employment
6	Giant slalom	Mean, paired difference, student–t	Sports, leisure
7	Marathon running	Mean, paired, small samples	Sports, leisure
8	Music preference	Chi-squared	Social sciences, demographics
9	Pharmaceutical industry	Mean, student–t	Retail, health
10	Productivity levels	Mean, student–t	Manufacturing, productivity
11	Ski sweaters	Two populations, mean	Retail sales
12	Smartphone consultation	Chi-squared	Social sciences, demographics

1. Beer drinking

Objective

The objective of this case-exercise is to demonstrate the application of the chi-squared method for hypothesis testing.

Situation

A survey was performed in England, Germany, Italy, Portugal, Norway, and Spain on weekly drinking beer habits among adults. The consumption patterns were categorized according to: Never drink; From 0 to 250 cl, From 250 to 500 cl, From 500 to 1,000 cl, Greater than 1,000 cl. The sampling information is given in Table BD-1. It is to be analyzed, using the chi-squared test, to see if there is a relationship between drinking habits and the country of origin. The null hypothesis is that there is no relation between country of residence and beer drinking patterns.

Table BD-1

Country	England	Germany	Italy	Portugal	Norway	Spain
Never drink	32	29	35	25	25	21
From 0 to 250 cl	34	32	70	55	68	56
From 250 to 500 cl	40	45	42	42	48	37
From 500 to 1,000 cl	45	52	60	38	75	62
Greater than 1,000 cl	62	72	77	62	80	46

Required

1 Define the null and alternative hypothesis for this situation.
2 From the total sample data, and to the nearest two decimal places, what is the proportion of adults who say they consume from 500 to 1,000 cl. of beer per week?
3 What is the value of the sample chi-squared value? Determine from the chi-squared equation and also using the Excel functions.
4 If you were to make a test at a 1% significance level, what would be the value of the critical, or benchmark, chi-squared value? What is your interpretation of this value when comparing it to the sample chi-squared value?
5 If you were to make a test at a 5% significance level, what would be the value of the critical, or benchmark, chi-squared value? What is your interpretation of this value when comparing it to the sample chi-squared value?
6 To the nearest two decimal places, what is the area under the chi-squared distribution, as dictated by the sample data, at which you would be unable to neither statistically accept nor reject the null hypothesis?

Answers

1 Null hypothesis: $H_0: p_E = p_G = p_I = p_P = p_N = p_S$. Alternative: $H_1: p_E \neq p_G \neq p_I \neq p_P \neq p_N \neq p_S$

Here the indices refer to the first letter of the country. The null hypothesis is that beer drinking habits are not dependent on the country of residence. The alternative hypothesis is that beer drinking patterns are dependent on the country of residency.

2   The proportion of adults who say they consume from 500 to 1,000 cl of beer per week is 22.63%.

3   The value of the sample chi-squared value is 34.2787.

4   The value of the critical, or benchmark, chi-squared value at a 1% significance level is 37.5662. Since the sample chi-squared value is less than the critical value we accept the null hypothesis and say that there is no statistical evidence indicating that the beer drinking patterns are dependent on the country.

5   The value of the critical, or benchmark, chi-squared value at a 5% significance level is 31.4104. Since the sample chi-squared value is greater than the critical value we reject the null hypothesis and say that there is statistical evidence indicating beer drinking patterns are dependent on the country.

6   The area under the chi-squared distribution, as dictated by the sample data, at which you would be unable to neither statistically accept nor reject the null hypothesis, is 2.43%.

2. Depression and work environment

Objective

The objective of this case-exercise is to illustrate hypothesis testing for proportions for different populations.

Situation

There is a lot of talk in the media about depression being associated with the work environment. Depression is when people are unable to cope with the stress of what is considered normal daily life and may involve behavior like being unable to work, drinking, fear of surroundings, and other abnormal behavior. A study was made in Europe to see if depression might be associated with the type of firm where a person is employed; a large monopolistic company with a bureaucratic rigid structure, or a smaller firm where there are fewer established rules. The criterion for the experiment was that a person was depressed if they had been off work for at least one week and was under the care of a doctor. An anonymous random sample of 282 people was selected from a large company where 129 people indicated that they had suffered depression. An anonymous random sample of 204 people was selected from smaller companies and 76 indicated depression problems.

Required

1   Develop the relationships for this hypothesis test to see if there is a difference in the proportions of people at each type of firm suffering depression.

2   What is the value for the estimated standard error of the difference in proportions?

3   Using a z value approach, at a 5% significance level, is there evidence to suggest that the proportion of depression in employees from large companies is different from employees in smaller companies?

4   Confirm your conclusion to Question 3 using the *p*-value approach.
5   Validate your conclusions to Questions 3 and 4 by testing results with the acceptance levels.
6   Indicate the hypothesis relationship to test to see if the proportion of depression is more predominant at larger firms rather than at smaller ones.
7   Using a *z* value approach, at a 5% significance level, is there evidence to suggest that the proportion of depression is higher at large firms than at smaller ones? Validate your results using the *p*-value, and acceptance level.
8   What are your comments on these results and this study?

Answers

1   Sample proportion for large firms is p_{large} and sample proportion for small firms is p_{small}. Null hypothesis is that there is no difference or $H_0 : p_{large} = p_{small}$. The alternative hypothesis is that there is a difference, or $H_1 : p_{large} \neq p_{small}$.
2   Estimated standard error of the difference is 4.50%.
3   Critical value of *z* at a 5% signification is $\pm$ 1.9600. Sample value of the test statistic *z* is +1.8862. The value 1.9600 > 1.8862. There is no statistical evidence to suggest that there is a difference of depression levels between the two types of firms. Accept the null hypothesis.
4   For the sample value, the area in the upper tail is 2.96% and this value is > 2.50%, the critical value. This is a two-tailed test with an area of 2.5% in each tail. This confirms acceptance of the null hypothesis.
5   The acceptance limits are −8.90% (in practice 0) and 8.90%. The difference in the sample proportions of 8.49% lies within this range.
6   Null hypothesis is that depression at large firms is not greater than at smaller ones, or $H_0 : p_{large} \leq p_{small}$. The alternative hypothesis is that depression is greater at larger firms, or $H_1 : p_{large} > p_{small}$.
7   At a 10% significance level there is evidence to suggest that depression is more predominant at large firms compared to smaller ones:

   •   Sample test value of *z* remains unchanged at +1.8862. Critical *z* value at 5% is 1.6449. The value 1.8862 > 1.6449.
   •   The *p*-value of sample test is 2.96% and this area is less than 5% the upper tail area.
   •   The acceptance level is 7.40%. This is lower than the difference in the sample proportions of 8.49%.

8   Reports indicate that people who work for large monopolistic bureaucratic companies have less freedom of thought and expression. They have to abide by the rules whether or not they agree with them. However, with smaller firms there is more freedom so that people can use their own ideas. (They can move up according to Maslow's hierarchy of needs.) The stress in these companies is that there is more market competition and many of them are "start ups."

3. Diet pills

Objective

The objective of this case-exercise is to illustrate hypothesis testing from different populations in a before and after situation.

Situation

There is an enormous amount of publicity for over-the-counter diet pills. Manufacturers claim that by taking diet pills a person will considerably reduce their body weight. For one particular product, a manufacturer claimed in its program that by taking one of its pills each day the average weight loss after three months would be at least 10 kg. An independent researcher obtained permission to follow 12 patients who followed this program. This information is in Table DP-1. It gives the weight of the patient before they started taking the pills; their weight after three months during which time they took one of the pills each day; then their weight six months later after stopping the program. In this six-month period they did not take any diet-related pills.

Table DP-1

Patient number	1	2	3	4	5	6	7	8	9	10	11	12
Weight before, kg	99	99	106	104	98	99	107	96	107	99	107	101
Weight after 3 months, kg	85	94	101	78	87	72	102	70	95	86	90	78
Weight after 6 months, kg	92	97	99	90	96	83	105	91	96	87	92	88

Required

1 Indicate the null and alternative hypotheses for this situation to show if the manufacturer's claim of losing on average at least 10 kg is valid.
2 Using the critical value approach at a 5% significance level, does it appear that the manufacturer's claim is valid? Quantify your decision.
3 Validate your conclusion in Question 2 by using the *p*-value criteria.
4 Confirm your conclusions to Questions 2 and 3 by determining the upper limit of acceptance.
5 Redo your analysis using the weight levels six months after finishing the program, comparing it to the patients' weights before they started taking the diet pills. Quantify your results. What are your comments?

Answers

1 Null hypothesis is that the weight loss is not more than 10 kg: $H_0:\mu \leq 10$ kg: Alternative hypothesis is that the weight loss is more than 10 kg: $H_1:\mu > 10$ kg.
2 At a 5% significance level the critical Student's *t* value is 1.7959. The sample value of the Student's *t* for testing a difference in weight is 2.1903. Since this value is greater than the critical value there is evidence that taking the diet pills does reduce weight loss by more than 10 kg.
3 The *p*-value of the sample statistic is 2.55%, which is less than 5% and validates the manufacturer's claim.
4 The upper limit of acceptance of a difference in weight loss is 14.37 kg. The average weight loss difference in the researcher's experiment is 15.33 kg. This value is higher than 14.37 kg and so again confirms our conclusion.

5   In this case the sample statistic is now −0.7617 or less than the critical value of 1.7959. The average value of the weight loss difference from the original weights is now 8.83 kg, which is less than the claim of more than 10 kg, and also less than the upper value of acceptance of 12.75 kg. It seems that the program is valid if patients are taking the diet pills. However, once they stop they put the weight back on again.

4. Driving time in Paris and Rome

Objective

The objective of this case-exercise is to illustrate hypothesis testing for proportions for different populations.

Situation

A study was made in Europe to see if there was a significant difference between the driving time of people working in the center of Paris, France and the driving time of people working in the center of Rome, Italy. The benchmark for commuting time was at least two hours per day. A random sample of 375 people was selected from Paris and 205 said that they had a daily commute of at least two hours per day. A random sample of 286 people was selected in Rome and 137 replied that they had a commute of at least two hours.

Required

1   Develop the relationships for this hypothesis test.
2   What is the value for the estimated standard error of the difference?
3   Using a z value approach, at a 5% significance level, is there evidence to suggest that the driving time of people in Paris is different from that of people in Rome?
4   Confirm your conclusion to Question 2 using the p-value approach.
5   Using a z value approach, at a 5% significance level, is there evidence to suggest that the driving time for workers in Paris is greater than the driving time for those in Rome?
6   Confirm your conclusion to Question 4 using the p-value approach.
7   What are your comments on this case-exercise?

Answers

1   Sample proportion for Paris is p_1 and sample proportion for Rome is p_2. Null hypothesis is that there is no difference, or $H_0: p_1 = p_2$. The alternative hypothesis is that there is a difference, or $H_1: p_1 \neq p_2$.
2   Estimated standard error of the difference is 3.92%.
3   Critical value of z at a 5% signification is ± 1.9600. Sample value of the test statistic z is +1.7275. The value $1.9600 > 1.7275$. There is no statistical evidence to suggest that there is a difference between the driving times in the two cities. Accept the null hypothesis.
4   For the sample value, the area in the upper tail is 4.20%. This area of 4.20% > 2.50%, the critical value, and so again we accept the null hypothesis.

5   The sample test value of z remains unchanged at $+1.7275$. Critical z value at 5% is 1.6449. The value $1.7275 > 1.6449$. There is statistical evidence to suggest that driving time is greater in Paris than in Rome. Reject the null hypothesis and accept the alternative.

6   Using the p-value approach, the area in the upper tail corresponding to a sample test value of 1.7275 is still 4.20%. Now this value is less than the 5.00% significant value and so the conclusion is the same: there is evidence to suggest that the commuting time for those in Paris is greater than for those in Rome.

7   The conclusions drawn from hypothesis testing are sensitive to the significant levels used and the criteria of the hypothesis test.

5. Engineers' salaries in Germany and Italy

Objective

The objective of this case-exercise is to examine the difference between two independent populations

Situation

A study was made by the European Union on the salaries of engineers, in the 30–40 age range, in Germany and Italy to see if there was a significant difference between them. This is bearing in mind that Germany is often considered the leader in the engineering field. Random samples were taken in both countries for salaries in Euros and the data for Germany is in Table ES-1, and for Italy is in Table ES-2.

Required

1   The first test was to see if, at a 10% significance level, there is a difference between the engineers' salaries in Germany and Italy. In this case express the appropriate hypothesis relationships.

2   Determine the critical z values at a 10% significance level for a difference.

Table ES-1 Germany

44,579	54,939	50,175	43,460	49,829	55,896	59,499
44,384	48,628	46,457	46,758	39,307	54,142	38,292
54,231	37,866	54,185	55,665	56,064	44,822	44,171
53,507	59,012	50,732	55,462	48,613	53,051	50,263
59,265	53,115	35,559	46,020	56,428	40,669	48,856
55,979	40,323	44,335	48,050	43,809	44,530	43,128
57,012	46,278	53,793	59,152	51,440	38,672	42,694
40,240	53,799	55,687	52,586	55,018	49,266	47,533
43,928	46,184	49,056	33,926	43,980	54,322	54,735
57,380	41,262	52,546	44,861	47,184	46,621	50,893
44,044	47,342	58,420	41,751	60,146	43,323	48,278

Table ES-2 Italy

47,487	52,566	54,156	41,841	55,836	52,131	49,683
49,812	47,628	59,586	50,799	54,048	51,198	45,270
52,704	50,787	45,684	45,807	43,578	44,694	52,467
50,379	45,795	45,852	46,767	36,978	41,370	60,240
46,161	46,371	55,125	40,920	40,329	49,728	54,870
38,703	44,583	51,681	53,946	34,923	44,862	44,658
52,278	45,555	46,242	40,164	42,975	50,937	43,461
43,896	56,847	49,086	51,123	44,922	51,615	48,684
32,349	39,465	47,754	53,847	41,094	42,438	53,676
48,276	52,182	48,147	45,066	47,415	54,423	37,263
41,334	59,829	47,202	49,953	56,970	57,261	53,466
53,757	44,787	36,093	42,909	42,018	51,663	52,527
50,775	43,002	47,805	38,358	39,864	43,137	48,870

3   Using the sample test statistic as your measure, is there evidence to suggest that there is a difference in the salaries between Germany and Italy? Justify your response.

4   Verify your response to Question 3 using the p-value approach.

5   What are the acceptance limits for the salary differences? How does the sample difference compare to the acceptance limits and thus corroborate your answers to Questions 3 and 4?

6   Define the null and alternative hypothesis for a test to see if the salaries in Germany are significantly greater than those in Italy.

7   Perform the hypothesis test at a 10% significance level to establish whether salaries are greater in Germany than Italy. Use the test statistics, p-value, and acceptance limit to justify your answers.

Answers

1   The null hypothesis is that there is no significant difference in the salaries of the two countries:

- $H_0:\mu_{GERMANY} = \mu_{ITALY}$.

The alternative hypothesis is that there is a significant difference in the salaries of the two countries:

- $H_1:\mu_{GERMANY} \neq \mu_{ITALY}$.

2   The critical values of z are ± 1.6449. There is 5% of the area in each tail.

3   There is no evidence to suggest that there is a difference in the salaries between the two countries. The sample test statistics z is 1.3568 and this is within the boundaries of the critical limits.

4   The sample p-value is 8.74%. This is greater than the tail area of 5%. This verifies the conclusion to Question 3.

5   The acceptance limits are $\pm$ €1,567.05. The sample difference of the salaries is €1,292.62 and this amount lies within the acceptance limits.

6   The null hypothesis is that salaries in Germany are not significantly greater than those in Italy:

- $H_0:\mu_{GERMANY} \leq \mu_{ITALY}$.

The alternative hypothesis is that salaries in Germany are significantly greater than those in Italy:

- $H_1 : \mu_{GERMANY} > \mu_{ITALY}$.

7 There is statistical evidence to indicate that salaries in Germany are greater than those in Italy.

- Critical value of z is 1.2816. There is 10% of the area in the tail. The sample test value of z is 1.3568 and this is greater than the critical value.
- The sample p-value is 8.74% and this is less than the critical area of 10%.
- The acceptance level for being greater is €1,220.93. The sample value of the difference is €1,292.62 and this is greater than the acceptance level.

6. Giant slalom

Objective

This case-exercise demonstrates paired differences in hypothesis testing.

Situation

A ski team has eight members who compete in the giant slalom downhill ski race. In order to improve their performance, the team undergoes a rigorous training program. In Table GS-1 are the completion times in seconds for a giant slalom race for the eight team members before and after the rigorous training program. In this training program, the ski instructor believes the time for finishing the giant slalom can be reduced by at least five seconds.

Table GS-1

Team member	1	2	3	4	5	6	7	8
Before, time (sec)	85.12	78.45	79.02	91.25	95.21	98.21	87.00	92.00
After, time (sec)	79.04	73.06	81.02	81.65	89.12	81.12	74.28	78.46

Required

1 Define the null and alternative hypothesis for this situation.
2 To the nearest two decimal places what is average difference of the racing time for the eight ski members?
3 To the nearest four decimal places what is the estimated standard error of the mean for the time differences?
4 Comparing the appropriate test statistics, at a 5% significance level is there evidence that the instructors claim is correct? Verify your answer using the probability values. How is your answer also corroborated by the acceptance level?
5 Comparing the appropriate test statistics, at a 10% significance level is there evidence that the instructors claim is correct? Verify your answer using the probability values. How is your answer also corroborated by the acceptance level?

Answers

1   Null hypothesis: H_0: $\mu_1 - \mu_2 \leq 5.00$ seconds.
     Alternative hypothesis: H_1: $\mu_1 - \mu_2 > 5.00$ seconds.
2   The average difference of the racing time for the eight ski members is 8.5638 seconds.
3   The estimated standard error of the mean for the time differences is 2.1119 seconds.
4   The sample Student's t value is 1.6875. The critical Student's t value at a 5% significance level is 1.8946. Since the sample Student's t value is less than the critical value, we accept the null hypothesis, indicating that at this level the instructors claim is not correct.
     The p-value of the sample value is 6.77% and this is greater than the 5.00% significance. The acceptance level is 9.0011 seconds. The average difference of the racing time for the team members is 8.5638 seconds and this is less than the acceptance level, thus confirming the conclusions.
5   The sample Student's t value is still 1.6875. The critical Student's t value at a 10% significance level is 1.4149. Since the sample Student's t value is outside the boundary of the critical value, we reject the null hypothesis and accept the alternative hypothesis indicating that at this significance level the instructors claim is correct. The p-value of the sample value is 6.77% and this is less than the 10.00% significance. The acceptance level is 7.9882 seconds. The average difference of the racing time for the team members is 8.5638 seconds and above the acceptance level, thus confirming the conclusions.

7. Marathon running

Objective

The objective of this case-exercise is to illustrate hypothesis testing from different populations in a before and after situation using small sample sizes.

Situation

A group of seven friends who work in the Los Angeles basin, California run marathons together (a marathon is 26.219 miles or 42.195 kilometers). They first ran together the Los Angeles Marathon. The names of the runners and the finishing time for this competition in hours and minutes is given Table MR–1. After this they consulted with a veteran marathon runner who indicated that in his experience, with considerable training, one should be able to reduce the time for running a marathon by at least 10 minutes for each marathon race entered. The two next marathons that the friends entered was one in Santa Monica and the other at the Pasadena Rose Bowl both in Southern California. After these two competitions they then entered in the Boston Marathon. Their finish times for this race are in the last line of Table MR–1.

Table MR-1

Runner	Bob	Ed	Derek	Phil	Mike	John	Bill
Los Angeles Marathon	4h 15m	4h 06m	3h 30m	5h 18m	4h 12m	3h 48m	4h 12m
Boston Marathon	3h 24m	3h 30m	3h 00m	3h 51m	3h 36m	3h 09m	3h 30m

Required

1   Express the null and alternative hypotheses for this situation to indicate if the veteran marathon runner's experience is valid.
2   Using the critical value approach, at a 1% significance level, does it appear that the veteran marathon runner's experience is valid? Quantify your decision.
3   Validate your conclusion of Question 2 by using the *p*-value criteria.
4   Confirm your conclusions to Questions 2 and 3 by determining the upper limit of acceptance and position the sample value within this limit.
5   Do your conclusions change if you use a 5% significance level? Quantify your results using the test statistic, *p*-value, and the acceptance limit.
6   What are some of your comments on an experiment of this nature?

Answers

1   Null hypothesis after three marathons is that the time reduction is not greater than 30 minutes:

- H_0: $\mu_{\text{Los Angeles}} - \mu_{\text{Boston}} \leq 0.50$ hours

Alternative hypothesis is that the time reduction is at least 30 minutes:

- H_0: $\mu_{\text{Los Angeles}} - \mu_{\text{Boston}} > 0.50$ hours

2   At a 1% significance level the critical Student's *t* value is 3.1427. The sample value of the Student's *t* is 2.1777. Since this value is less than the critical value there is no evidence that the veteran's claim is valid.
3   The *p*-value of the sample statistic is 3.61%, which is greater than 1% and validates the conclusions of Question 2.
4   The upper limit of acceptance is for a time difference is 0.8814 hours. The sample difference in completion time is 0.7643 hours and this is less than the upper limit.
5   Yes, the conclusions change:

- At a 5% significance level the critical Student's *t* value is now 1.9432. The sample value of the Student's *t* is 2.1777, which is greater than the critical value.
- The *p*-value of the sample statistic is 3.61%, which is less than the 5% critical area of significance.
- The upper limit of acceptance is for a time difference is 0.7558 hours. The sample difference in completion time is 0.7643 hours and this is more than the upper limit.

6   In marathon running there are many variables. Even though the distance is the same, the course terrain varies; elevation changes are different; weather conditions—hot or cold; dry or wet. And there is of course the physical and mental condition at the time of the runner! This is similar for most sporting events so that a statistical analysis although a good indicator can be clouded by other variables.

8. Music preference

Objective

The objective of this case-exercise is to demonstrate the application of the chi-squared method for hypothesis testing.

Situation

A survey was performed in New York State, USA on the music preference of people according to age ranges. The music classifications were: Classical, Jazz, Country, Hard Rock, and Pop. The sampling information is given in Table MP-1. It is to be analyzed, using the chi-squared test, to see if there is a relationship between music preference and age range. The null hypothesis is that there is no relation between age range and music preference.

Table MP-1

Age Range	*15–25*	*25–35*	*35–45*	*45–65*	*Over 65*
Classical	22	24	32	19	38
Jazz	17	32	29	17	32
Country	15	10	18	14	16
Hard rock	22	17	21	25	22
Pop	25	15	30	31	12

Required

1 From the total sample data, and to the nearest two decimal places, what is the proportion of people who prefer jazz?
2 If you were to make a test at a 1% significance level, what would be the critical, or benchmark, chi-squared value?
3 What is the value of the sample or test chi-squared value? Determine the value using the equation for chi-squared and also the Excel functions.
4 At the 1% significance level what are your conclusions about this statistical analysis?
5 At the 3% significance level what are your conclusions about this statistical analysis?

Answers

1 The proportion of people who prefer jazz is 22.88%.
2 At a 1% significance, the critical, or benchmark, chi-squared value is 31.9999.
3 The sample test chi-squared value is 29.6283.
4 The sample chi-squared value is less than the critical, or benchmark chi-squared value, and thus the null hypothesis should be accepted. There does not appear to be any correlation between age and music preference.
5 The sample chi-squared value is greater than the critical, or benchmark chi-squared value, and thus the null hypothesis should be rejected. There appears that there is a relationship between age and music preference.

9. Pharmaceutical industry

Objective

The objective of this case-exercise is to apply the concept of hypothesis testing when there are two populations.

Situation

Pharmacies keep limited stocks of prescription medicines in their stores because, with internet connections, they can order directly from their distributor. In urban locations an order made before 09:00 is delivered to the store by 14:00 the same day, or a time delay of about a half day. However, in more remote country locations this is not the case because the distributor is some distance from the pharmacy. In these cases delivery is not until the following day, a 24-hour delay. As a result, for good customer service, country pharmacies usually hold a higher inventory of prescription medicines. A study was made to see if this was in fact the case. Random sampling was done from 13 urban stores and 12 country stores to see, on one particular day, the number of boxes of heart-related medicines, for example, for blood pressure, anti-coagulants, and for cholesterol the pharmacy had in stock. This data is given in Table PI-1.

Table PI-1

Store N°	1	2	3	4	5	6	7	8	9	10	11	12	13
Pharmacy (country)	28	30	50	54	50	38	30	34	50	62	44	52	
Pharmacy (urban)	24	26	38	24	42	32	50	24	47	44	38	22	20

Required

1 Indicate appropriate null and alternative hypotheses if we wanted to show if those pharmacies in the country have a higher inventory of heart-related medicines than urban pharmacies.
2 Using the critical value method, at a 1% significance level, does this data indicate that pharmacies in the country have a higher level of inventory than those in urban areas?
3 Confirm your conclusions to Question 2 using the *p*-value approach.
4 Validate your conclusions to Questions 2 and 3 positioning the difference in inventory levels with the acceptance value.
5 What are your conclusions at a 5% significance level testing for "greater than"? Justify your response using the critical value approach, *p*-value and acceptance level.
6 What are the implications of these results for inventory management for the pharmacies?

Answers

1 Null hypothesis is $H_0: \mu_{country} \leq \mu_{urban}$. Alternative hypothesis is $H_1: \mu_{country} > \mu_{urban}$. This is a one-tailed test.
2 No, at a 1% significance, the critical value of the Student' *t* is 2.4999 and the sample test statistic *t* is 2.3868 which is less than the critical value.
3 The *p*-value of the sample test is 1.28% and this is greater than the critical 1% area.
4 The upper acceptance level of the difference in inventory is 10.8362 units. The inventory difference from the sample is 10.3462 units and this value is less than the acceptance limit.
5 At a 5% significance there is statistical evidence that inventory levels at country pharmacies are greater than at urban pharmacies:

- The critical value of the Student's *t* is 1.7139 and the sample test statistic *t* is 2.3868 which is greater than the critical value.

- The *p*-value of the sample test is 1.28% and this is greater than the critical 1% area.
- The acceptance level of the difference in inventory is 7.4291 units. The inventory difference from the sample is 10.3462 units and this value is greater than the acceptance limit.

6 Keeping higher inventory means higher inventory holding costs—inventory is cash. It also requires more storage space.

10. Productivity levels

Objective

The objective of this case exercise is to demonstrate hypothesis testing of different populations with small sample sizes.

Situation

A manufacturing firm of engine parts has a factory in England and a similar factory in Poland. It is considering restructuring its organization, but before it embarks on a program it wishes to examine the productivity at each site to see if there is any significant difference. For England, data for 16 weeks, as shown in Table PL-1, was obtained according to the number of units produced each week, the number of operators in that period, plus the hours these operators worked in that week. Differences are due to absenteeism, overtime, planning, and productivity of the operators. Similar data for Poland is shown in Table PL-2 except the data was only obtained for a 13-week period.

Table PL-1: England

Week	1	2	3	4	5	6	7	8	9	10	11	12	13	14	15	16
Units	200	193	194	203	208	207	203	204	207	199	206	208	207	195	192	203
Operators	9	7	8	6	8	6	9	6	7	7	8	7	7	7	9	8
Hours/week	41	41	42	44	44	42	42	44	44	42	44	43	44	41	44	41

Table PL-2: Poland

Week	1	2	3	4	5	6	7	8	9	10	11	12	13
Units	197	199	200	204	204	218	214	193	201	219	204	217	208
Operators	8	8	9	8	10	10	9	8	11	8	9	8	10
Hours/week	40	39	43	39	43	44	41	41	45	43	41	40	44

Required

1 Determine the average productivity for England and Poland in units produced per labor hour.
2 Write an expression for a hypothesis test to see if there is a difference in productivity between the two countries.

3   Using the test statistic, at a 1% level, is there evidence that the productivity is significantly different between the two countries? Quantify your conclusions.
4   Confirm your conclusions to Question 3 using the *p*-value approach.
5   What are the acceptance limits for a 1% significance testing for a difference? How do these validate your conclusions to Questions 3 and 4?
6   Redo the hypothesis testing to see if, at a 1% significance, there is any evidence that the productivity in England is significantly greater than the productivity in Poland. First develop the hypothesis expression and then quantify your conclusions using the test statistic, the *p*-value, and the acceptance limits. What are your conclusions?
7   In a restructuring program for the firm, how might this statistical data be used?

Answers

1   The average productivity for England is 0.6477 units per labor hour. For Poland the productivity is 0.5622 units per labor hour.
2   The null hypothesis is that there is no significant difference in the productivity between the two countries or $H_0: \mu_{ENGLAND} = \mu_{POLAND}$. The alternative hypothesis is that there is a significant difference in the productivity of the two countries or $H_1: \mu_{ENGLAND} \neq \mu_{POLAND}$.
3   There is no statistical evidence indicating that there is a significant difference in productivity levels between the two countries. The critical Student's *t* values are ± 2.7707. The sample test Student's *t* is 2.5809. This value lies within the boundary of the critical Student's *t* values.
4   At 1% significance testing for a difference there is an area of 0.5% in each tail. The sample *p*-value is 0.78% and this is greater than 0.5% and so confirms the conclusions of Question 3.
5   The calculated lower acceptance limit is -0.0918 units per labor hour (in practice 0). The upper acceptance limit is 0.0918 units per labor hours. The sample difference in productivity between the two countries is 0.0855 units per labor hour and this value lies between the acceptance levels. This confirms the conclusions of Questions 3 and 4.
6   The null hypothesis is that the productivity in England is not significantly greater than the productivity in Poland or $H_0: \mu_{ENGLAND} \leq \mu_{POLAND}$. The alternative hypothesis is that the productivity in England is significantly greater than the productivity in Poland, or $H_1: \mu_{ENGLAND} > \mu_{POLAND}$.

- The upper critical Student's *t* is 2.4727. The sample Student's *t* is 2.5809. This value is beyond the critical limit.
- There is now 1% area in the tail. The sample *p*-value is 0.78% and this area is less than 1%.
- The upper acceptance level is 0.0819 units per labor hour. The sample difference in productivity between the two countries is 0.0855 units per labor hour. This value is outside the acceptance level.

In conclusion there is statistical evidence indicating that the productivity in England is greater than that in Poland.
7   If, in the restructuring program of the firm, there are plans to reduce overall production it would suggest that the activity in England be retained in preference to Poland. However, there are other factors to consider such as labor costs, quality of the work, and logistics regarding transporting end products to client locations.

11. Ski sweaters

Objective

The objective of this case-exercise is to demonstrate the application of hypothesis testing for two different populations.

Situation

A Norwegian clothing manufacturer wants to know if there is a difference in the price of a particular model of its ski sweater sold in stores in Norway and the same model sold in stores in Switzerland. The manufacturer took random samples from 37 stores in Norway and found that the average price of the sweater was 875.85 Kroner with a variance of 359.0494 (Kroner)^2. In Switzerland a correspondent took a sample of the same model from 35 stores in Switzerland and found that the average price was 92.75 Euros with a variance of 4.8400 (Euros)^2. Perform the analysis in Euros using an exchange rate of 1 Euro = 9.57 Kroner.

Required

1   Indicate appropriate null and alternative hypotheses for this situation if we wanted to know if there is a significant difference in the price of ski sweaters in the two countries.
2   Using the critical value method, at a 1% significance level, does the data indicate that there is a significant difference in the price of ski sweaters in the two countries?
3   Confirm your conclusions to Question 2 using the *p*-value approach.
4   Using the critical value method would your conclusions change at a 5% significance level?
5   Confirm your conclusions to Question 4 using the *p*-value approach.
6   Indicate appropriate null and alternative hypotheses for this situation if we wanted to test if the price of ski sweaters is significantly greater in Switzerland than in Norway.
7   Using the critical value method, at a 1% significance level, does the data indicate that the price of ski sweaters is significantly lower in Norway than in Switzerland?
8   Confirm your conclusions to Question 7 using the *p*-value criterion.

Answers

1   Null hypothesis is H_0:$\mu_N = \mu_S$. Alternative hypothesis is H_1:$\mu_N \neq \mu_S$, where μ_N is the mean value for Norway and μ_S is the mean value for Switzerland. This is a two-tailed test.
2   No, at 1% significance, the critical values of z are ± 2.5758 and the sample test statistic for a difference is -2.4881. This sample value is within the critical boundaries and so we accept the null hypothesis.
3   Significance level is 1% with 0.50% in each tail. The sample *p*-value is 0.64% and this area is greater than 0.5% and this confirms the acceptance of the null hypothesis.
4   Yes, the conclusions change. At 5% significance, the critical values of z are now ± 1.9600. The sample test statistic is still -2.4881 and this value is outside the critical boundaries. In conclusion, reject the null hypothesis as there is evidence of a significant price difference.
5   Significance level is 5% with 2.50% in each tail. The *p*-value is still 0.64% and this value is less than 2.50%.

6   Null hypothesis is written $H_0 : \mu_N \geq \mu_S$. Alternative hypothesis is $H_1 : \mu_N < \mu_S$, where μ_N is the mean value for Norway and μ_S is the mean value for Switzerland. This is a one-tailed test.

7   Yes, at 1% significance, the critical value of z is −2.3263 and the sample test statistic is −2.4881. This value is less than the critical value and so we reject the null hypothesis. There is evidence that the price of the sweaters are lower in Norway than Switzerland for the given criteria.

8   Significance level is 1% with 1.00% in the tail. The p-value is still 0.64%, which is less than 1.00%.

12. Smartphone consultation

Objective

This case-exercise demonstrates the application of the chi-squared method for hypothesis testing.

Situation

A random sample study was carried out in France to determine the number of times per day that individuals, according to their age range, consulted their Smartphone. This information is given in Table SC-1.

Table SC-1

Age range	15–25	25–35	35–45	45–65	Over 65
More than 30 times/day	34	24	22	19	26
Between 20 and 30 times/day	25	26	24	15	20
Between 10 and 20 times/day	20	22	18	14	18
Between 5 and 10 times/day	9	17	21	25	10
Less than 5 times per day	4	5	12	18	6

Required

1   What are expressions for the null and alternative hypotheses?

2   What is the sample or test chi-squared value? Determine the number using the equation for chi-squared and also the Excel functions. What is the p-value corresponding to the sample test?

3   Using the chi-squared test, at a 5% significance level, is there evidence that the age of the user has some bearing on the number of times that the Smartphone is consulted? Quantify your conclusions.

4   Using the chi-squared test, at a 1% significance level, is there evidence that the age of the user has some bearing on the number of times that the Smartphone is consulted? Quantify your conclusions.

5   Using the chi-squared test, at a 0.1% significance level, is there evidence that the age of the user has some bearing on the number of times that the Smartphone is consulted? Quantify your conclusions.
6   At what level of significance is the sample or test chi-squared value equal to the critical value and is thus a breakeven situation?
7   In the cross-classification data of Table SC-1 assign a score of 5 to a consultation of "more than 30 times/day"; 4 for "between 20 and 30 times/day"; 3 for "between 10 and 20 times/day"; 2 for "between 5 and 10 times/day"; and 1 for "less than 5 times/day". Then determine the weighted average score according to age range and plot as a histogram.
8   What are your comments on your results?

Answers

1   The null hypothesis H_0 is that the daily utilization of the Smartphone has no bearing on the age of the user. The alternative hypothesis H_1 is that the age of the user has a bearing on the utilization.
2   The sample or test chi-squared value is 36.0362. The corresponding p-value is 0.29%.
3   Yes, at a 5% significance level, there is evidence that the age of the user has some bearing on the number of times that the Smartphone is consulted. The critical chi-squared value is 26.2962. The sample value of 36.0362 is greater than the critical value. And the 0.29% tail area of the sample value is less than 5%.
4   Yes, at a 1% significance level, there is evidence that the age of the user has some bearing on the number of times that the Smartphone is consulted. The critical chi-squared value is 31.9999. The sample value of 36.0362 is greater than the critical value. And the 0.29% tail area of the sample value is less than 1%.
5   No, at a 0.1% significance level, there is no evidence that the user's age has any bearing on the number of times that the Smartphone is consulted. The critical chi-squared value is 39.2524. The sample value of 36.0362 is less than the critical value. And the 0.29% tail area of the sample value is more than 0.1%.
6   This is when the sample p-value of 0.29% is the same as the critical value. At this point both the sample and critical chi-squared values are 36.0362.
7   The weighted average scores are in Table SC-2. The histogram is in Figure SC-1.
8   The fact that it is only at a very low significance level of 0.1% that the null hypothesis can be accepted indicates that age does seems to have a bearing on the daily utilization of the Smartphone. Figure SC-1 underscores this conclusion. People in the lower age bracket (teenagers to young adults) are heavy users (a weighted average score of 3.83 out of 5.00). This falls gradually in subsequent age brackets probably because people are working and have less time for consultation of their Smartphone. Over 65, however, people are retired and have lots of free time on their hands. Thus they use their Smartphone more frequently, as evidenced by the weighted average score of 3.63.

Table SC-2

Age range	15–25	25–35	35–45	45–65	Over 65
Weighted average score	3.83	3.50	3.24	2.91	3.63

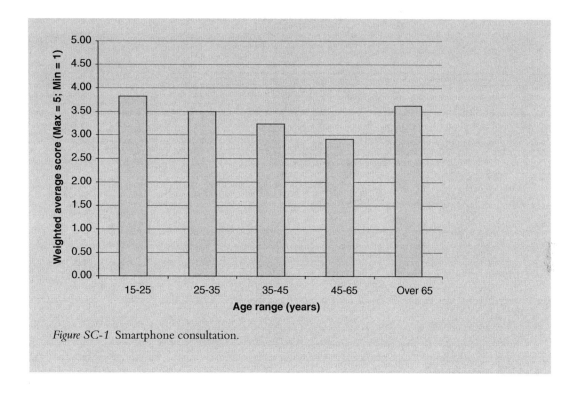

Figure SC-1 Smartphone consultation.

10 Forecasting from correlated data

Making objective predictions to plan your operations

No	Case-exercise	Statistical concept	Application
1	Catalog sales	Multiple regression	Marketing, retail
2	Consulting	Multiple regression	Marketing, consulting
3	Grade scores and class attendance	Causal linear regression	Education
4	Green Thumb	Seasonal regression	Leisure, retail
5	Hotel Franchise, I	Linear regression	Hospitality
6	Hotel Franchise, II	Polynomial regression	Hospitality
7	Hotel-spa complex	Multiple regression	Hospitality
8	Los Angeles restaurant	Linear regression	Hospitality, food
9	Pub lunches	Seasonal regression	Hospitality, food
10	Public transport	Causal linear regression	Community management
11	Westwood Village restaurant	Linear and polynomial regression	Food, hospitality
12	Wine sales	Linear and causal regression	Hospitality, food and beverage

1. Catalog sales

Objective

The objective of this case–exercise is to illustrate multiple regression in forecasting.

Situation

The Chelsea Company, with stores in the San Francisco area, is a distributor of consumer products including televisions, kitchen utensils, audiovisual, and computer equipment. The company produces a catalog of its products that is a key to its sales and is available either on the Internet or as a bound copy. Customer contact is either direct in its stores, telephone, or on the Internet. The Chelsea Company wants to see if there is any relationship between sales revenues, advertising budget, sales staff, and client contact made either directly in its stores, by telephone, or by the Internet. Table CS-1 gives monthly historical data.

Table CS-1

Sales $US	Advertising Budget, $US	Sales Persons	No. of sales Contacts
72,120	7,200	42	27,500
47,000	4,712	21	18,412
57,000	5,512	28	22,478
51,000	4,985	22	20,554
31,540	3,000	22	15,487
58,750	6,245	32	18,724
61,580	6,352	35	22,845
59,450	5,847	35	23,448
57,450	4,897	28	22,045
26,500	3,000	16	9,998

Required

1 Develop a two-independent variable multiple regression model for the sales volume as a function of the advertising budget, and the number of sales person. Does the relationship appear strong? Quantify.
2 From the answer of Question 1, assume for a particular month the budget allocation for advertising is $4,000 and 30 sales persons will be use. What would be an estimate of the sales for that month?
3 What are the 90% confidence intervals for Question 2?
4 Develop a three-independent variable multiple regression model for the sales volume as a function of the advertising budget, the number of sales persons, and the number of sales contacts made. Does the relationship appear strong? Quantify.
5 From the answer developed in Question 4, assume for a particular month it is proposed to allocate a budget of $US 4,000 to use 30 sales persons, with a target to make 21,000 sales contacts, then what would be an estimate of the sales for that month?
6 What are the 90% confidence intervals for Question 5?

Answers

1   The equation is $\hat{y} = 1,198.09 + 8.76 *$ advertising budget $+ 204.02 *$ sales persons. Coefficient of multiple determination r^2 is 0.9413 that indicates a strong relationship.
2   The estimate of sales for the month is $42,339.
3   Lower limit is $35,091 and upper limit is $49,587.
4   $\hat{y} = -3,909.69 + 6.53 *$ advertising budget $- 99.88 *$ sales persons $+ 1.25 *$ sales contacts. Coefficient of multiple determination r^2 is 0.9824 that indicates a strong relationship.
5   The estimate of sales for the month is $45,442.
6   Lower limit is $41,042 and upper limit is $49,843.

2. Consulting

Objective

The objective of this case-exercise is to illustrate the use of multiple regression for forecasting.

Situation

A consulting company, which has a major activity related to the installation and integration of systems software for enterprise resource planning, wishes to see if there is any correlation between the profits generated in this sector and various variables related to its practice. The variables for this particular sector of the work are the average number of consultants on projects, their average experience, and the percent of time that the consultant spends away from his office, which means the consultant is either traveling, or working at the client's facility. Table C-1 gives the historical data for profits on a quarterly basis

Table C-1

Profits, $US	Years of experience, x_1	Average number of consultants, x_2	% Time away from office, x_3
4,501,000	15	27	55
3,241,000	6	45	27
8,745,100	18	112	75
6,984,210	23	80	62
3,421,800	8	35	55
5,845,600	10	98	57
6,854,200	8	80	75
3,589,500	5	30	49
9,542,300	21	120	82
5,689,250	10	54	62

Required

1   Develop a two-independent variable multiple regression model for profits as a function of years of experience and the number of consultants on the projects.

2   Does the relationship in Question 1 appear strong? Quantify.
3   From the model developed in Question 1, estimate profit for a particular quarter, if the average years of experience of consultants on projects were 15, and the average number 50?
4   What are the 90% confidence intervals for Question 3?
5   Develop a three-independent variable multiple regression model for profits as a function of years of experience, number of consultants on the projects, and percent of time spent away from the office.
6   Does the relationship for Question 5 appear strong? Quantify.
7   Using the model developed in Question 5, what would be an estimate of the profit for a particular quarter, if the average years of experience of consultants on projects was 15, the average number 50, and the percent of time away from the office 60%?
8   What are the 90% confidence intervals for Question 7?

Answers

1   The equation is $\hat{y} = 1,322,338.43 + 112,534.93x_1 + 45,868.20x_2$.
2   Yes, $r^2 = 0.9029$. This value is greater than 0.8 so the relationship is strong.
3   The estimated profit is $5,303,772.
4   The lower interval is $3,823,885; the upper interval is $6,783,660.
5   The equation is $\hat{y} = -537,191.06 + 78,970.70x_1 + 32,686.11x_2 + 52,978.74x_3$.
6   Yes, $r^2 = 0.9679$. This value is greater than 0.8 so the relationship is strong and there is better relationship than for a two-independent variable.
7   The estimate of the profit is $5,460,399.
8   The lower interval is $4,517,521; the upper interval is $6,403,277.

3. Grade scores and class attendance

Objective

The objective of this case-exercise is to demonstrate the application of forecasting from correlated statistical data.

Situation

A business professor at a university in California gives a course in applied statistics for business using Microsoft Excel. The class format is 26 sessions, twice a week with 1½ hours per class. In a particular semester where there were 60 students, some participants complained that they did not have a good grade and asked for an explanation. The professor looked at the grade scores, out of 100%, and noted the number of classes that each student attended. This data is given in Table GS-1.

Required

1   Plot a scatter diagram of the grade scores versus the percentage absenteeism. Show on this scatter diagram the linear regression line and equation.
2   Using the scatter diagram and the regression line how might the professor explain the low grades to students?

Table GS-1

Student	1	2	3	4	5	6	7	8	9	10
Classes attended	25	26	21	23	22	26	24	25	21	22
Grade (%)	91.40	99.20	79.80	76.50	68.10	92.25	87.35	94.00	61.29	87.15
Student	11	12	13	14	15	16	17	18	19	20
Classes attended	20	11	24	22	20	22	22	24	23	16
Grade (%)	65.46	47.67	67.00	86.46	84.32	81.19	81.12	98.25	86.27	28.18
Student	21	22	23	24	25	26	27	28	29	30
Classes attended	24	20	19	21	2	20	18	22	16	25
Grade (%)	83.17	73.22	82.45	99.00	5.80	73.19	45.45	84.27	56.21	82.00
Student	31	32	33	34	35	36	37	38	39	40
Classes attended	26	18	10	11	14	25	26	25	15	26
Grade (%)	94.32	66.00	38.78	45.20	29.50	95.00	95.80	88.65	49.32	100.00
Student	41	42	43	44	45	46	47	48	49	50
Classes attended	6	16	24	23	22	23	20	23	23	20
Grade (%)	21.83	45.72	84.45	87.29	86.18	74.00	66.32	73.23	64.27	54.56
Student	51	52	53	54	55	56	57	58	59	60
Classes attended	23	19	25	25	21	18	25	26	21	24
Grade (%)	83.00	69.54	93.27	98.20	72.56	56.28	83.25	98.28	72.00	89.37

3   Develop the linear regression equation that describes this situation and also indicate the coefficient of determination and the coefficient of correlation. Explain the meaning of these terms.
4   If a student is absent for 20% of the classes, what is an estimate of the student's grades?
5   If a student is absent for 50% of the classes, what is an estimate of the student's grades?
6   What are your comments on using this type of correlation for grade estimation?

Answers

1   The scatter diagram with the regression line is given in Figure GS-1.
2   There is a reasonable correlation between the grade score and absenteeism. As the percentage absenteeism increases the grade drops.
3   The regression equation is $\hat{y} = 0.9356 - 0.9948x$. The coefficient of determination r^2 is 0.8068 and the coefficient of correlation r is -0.8982. Here $\hat{y}$ is the forecast grade according to an absentee rate x. It means that when a student is present at all the 26 classes that student is forecast to have a grade of 0.9356 or 93.56%. The grade declines by almost one percentage point (0.9948) for every 1% of time that a student is absent.
4   If a student is absent for 20% of the classes, then an estimate of the student's grade is 73.67%.
5   If a student is absent for 50% of the classes, then an estimate of the student's grade is 43.83%.
6   Students have a different capacity for grasping quantitative subjects. A student who understands logic can get a good grade even though she/he misses several classes. Alternatively, a student who is present for all the classes may not have a good grade as that person has difficulty understanding the concepts.

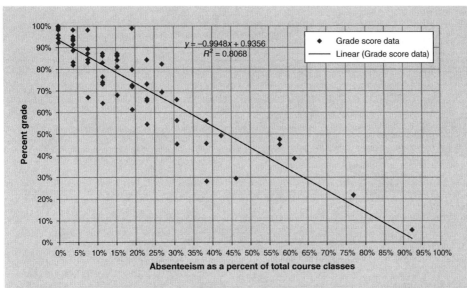

Figure GS-1 Grade score and absenteeism.

4. Green Thumb

Objective

This case–exercise illustrates the use of seasonal regression as a forecasting tool.

Situation

The Green Thumb franchise garden store sells everything for the amateur gardener including all types of plants, tools, clothing, fertilizer, and the like. Table GT-1 gives the sales by quarter since 2008. All data is in €000s.

Table GT-1

Year	Quarter	Sales	Year	Quarter	Sales
2008	Winter	20,344	2012	Winter	23,731
	Spring	21,919		Spring	25,463
	Summer	23,792		Summer	26,938
	Fall	21,507		Fall	24,597
2009	Winter	21,395	2013	Winter	24,805
	Spring	21,956		Spring	26,345
	Summer	23,929		Summer	27,256
	Fall	21,213		Fall	24,147
2010	Winter	21,375	2014	Winter	25,789
	Spring	22,651		Spring	27,452
	Summer	23,998		Summer	27,147
	Fall	22,237		Fall	25,204
2011	Winter	22,785	2015	Winter	26,751
	Spring	24,030		Spring	27,853
	Summer	25,844		Summer	27,585
	Fall	23,897		Fall	25,965

Required

1 Plot the data using coded values for the seasons. What are your observations?
2 Using centered moving averages, what are the seasonal indices that describe this data after the seasonal effect has been removed? How would you interpret these indices?
3 Plot the data after the seasonal effect has been removed. What is the linear regression equation that describes the sales of lunches after the seasonal effect has been removed?
4 Is the regression line a good representation of sales from which the seasonal effect has been removed? Explain your reasoning.
5 Develop a forecast by quarter for 2016.

Answers

1 There is a seasonal variation with highest sales in the spring and summer months and the lowest in the fall and winter months. In addition, the sale of garden products is increasing over the years.
2 The season indices are: winter 0.9709; spring 1.0176; summer 1.0612; fall 0.9567. This implies that winter sales are 2.91% below the yearly average; spring sales are 1.76% above the yearly average; summer sales are 6.12% above the yearly average; fall sales are 4.33% below the yearly average
3 The linear regression equation is $\hat{y} = 201.45x + 21,010.70$ where x is the code value of the quarter.
4 Yes, the coefficient of determination, r^2, is 0.9197. This is greater than a benchmark value of 0.8.
5 In €000s: winter €26,855; spring 28,351; summer 29,779; fall 27,039.

5. Hotel franchise, I

Objective

The objective of this exercise is to develop a linear regression forecast model.

Situation

A hotel franchise in the United States has collected the sales revenue data in Table HF-1 for the several hotels in its franchise.

Table HF-1

Year	Revenues $Millions
2006	35
2007	37
2008	44
2009	51
2010	50
2011	58
2012	59
2013	82
2014	91
2015	104

Required

1 From the given information develop a linear regression model of the time period against revenues.
2 What is the coefficient of determination for the relationship developed in Question 1?
3 What is the annual revenue growth rate based on the given information?
4 From the relationship in Question 1, forecast the revenues in 2016 and give the 90% confidence limits.
5 From the relationship in Question 1, forecast the revenues in 2024 and give the 90% confidence limits.
6 What are your comments related to making forecasts for 2016 and 2024?

Answers

Table HF-2 gives the linear regression statistics from which the answers are developed.

1 The equation for the linear regression model is: $\hat{y} = 7.400x - 14,801.00$.
2 The coefficient of determination, $r^2 = 0.9099$.
3 The estimated annual revenue growth is $7.4 million per year.
4 The forecast is $101.80 million. The lower limit is $87.90 million and upper $115.70 million.
5 The forecast is $161.00 million. The lower limit $147.10 million and upper $174.90 million.
6 Earlier year is closer to the period when the data was measured and so is more reliable. The further away in the time horizon that forecasts are made, the more unreliable is the forecast.

Table HF-2

b	7.4000	−14,801.0000	a
se_b	0.8231	1,653.2983	se_a
r^2	0.9099	7.4766	s_e
F	80.8175	8.0000	df
SS_{reg}	4,517.7000	447.2000	SS_{resid}

6. Hotel franchise, II

Objective

The objective of this exercise is to develop a polynomial regression forecast model.

Situation

A hotel franchise in the United States has collected the sales revenue data in Table HF-1.1 for the several hotels in its franchise. (This is the same data as Hotel Franchise Part I).

Required

1 From the information develop a two-degree polynomial regression model of time against revenues. Show the information graphically.

2   What is the coefficient of determination for the relationship developed in Question 1?
3   From the relationship in Question 1, forecast the revenues in 2016.
4   From the relationship in Question 1, forecast the revenues in 2024.
5   Compare these results with Part I of this case-exercise
6   What are your comments related to making forecast for 2016 and 2024 and in particular using a polynomial model?

Table HF-1.1

Year	Revenues $millions
2006	35
2007	37
2008	44
2009	51
2010	50
2011	58
2012	59
2013	82
2014	91
2015	104

Answers

1   Forecast model is $\hat{y} = 0.7652x^2 - 3.066.2136x + 3,071,868.3818$.
2   Coefficient of determination is 0.9722.
3   Forecast revenues in 2016 are \$118.63 million.
4   Forecast revenues in 2024 are \$294.14.
5   Table HF-2.1 compares the results with Hotel Franchise Part I.
6   Earlier year (2016) is closer to the period when the data was measured and so is more reliable. The further away in the time horizon that forecasts are made (2024) the more unreliable is the forecast. The coefficient of determination for the polynomial model is closer to unity so it suggests a better model. However, using a polynomial model assumes that growth is increasing rapidly. This may not be sustainable. Forecasts for equivalent years for the polynomial model are higher. The polynomial model is *extremely* sensitive to the number of decimal points assigned to the coefficients and that explains why different results may be obtained.

Table HF-2.1

Year	Linear	Polynomial
Forecast 2016	\$101.80 million	\$118.63 million
Forecast 2024	\$161.00 million	\$294.14 million
Coefficient of determination	0.9099	0.9722

7. Hotel–spa complex

Objective

The objective of this case-exercise is to demonstrate the application of multiple regression.

Situation

A hotel complex in France has a health spa with many treatments including massage, aqua gym, envelopment in mud or algae, etc. Clients may stay at the hotel for one week or just for a long weekend. The health spa is open to residents of the hotel who may take a package deal of hotel and health spa treatments, and to outside people who use just the health spa but do not stay at the hotel. In the hotel there is a restaurant that serves both lunch and dinner. Hotel residents may have lunch and dinner if they have full-pension or just dinner if they are booked half-pension. In addition, people who are not staying in the hotel can also take lunch or dinner with no reservation Table HS-1 gives the revenues in Euros (€) generated by the hotel-spa complex over the last twelve months together with the corresponding number of average hotel guests using the health spa (internal); average number of external clients using the spa; average number having lunch; average number having dinner; and average number booked into the hotel. The revenues are according to the month but all the other data is average daily usage for that month. The purpose of analyzing this data for six-sigma management is to establish the relationship between outputs, revenues, and the various inputs; in this case facility usage.

Table HS-1

Month	Revenues (€) y	Using spa (internal) Daily (x_1)	Using spa (external) Daily (x_2)	Number for lunch Daily (x_3)	Number for dinner Daily (x_4)	Rooms sold/day (x_5)
January	1,607,038	87	34	23	33	75
February	2,023,158	107	46	25	37	92
March	2,852,314	141	51	31	42	104
April	2,785,400	132	85	33	47	92
May	2,410,028	136	108	37	42	90
June	3,262,133	159	112	41	49	110
July	4,160,460	191	125	46	51	120
August	4,656,401	186	141	51	57	120
September	3,022,607	103	104	35	39	87
October	2,636,192	84	56	27	35	99
November	1,530,458	88	42	24	32	87
December	1,684,160	92	29	30	37	77

Required

1   Using regression analysis between revenues and each of the x-variables separately, which of the variables shows the strongest correlation with revenues, if the coefficient of determination is the criterion for measurement?

2   Using regression analysis between revenues and each of the x-variables independently, which of the variables shows the weakest correlation with revenues, if the coefficient of determination is the criterion for measurement?

3   Using regression analysis between revenues and all of the five *x*-variables together, to the nearest four decimal places, what is the value of the coefficient of determination for this model?

4   Using the complete model from the immediate previous question, to the nearest €100,000, what is an estimate of monthly revenues if the average number of hotel guests (internal) using the spa is 100; average number of clients external using the spa is 80; average having lunch per day is 50; average having dinner per day is 55; and average number of rooms sold per day is 110?

5   Using the complete model, and to the nearest whole number, what would have to be the average number of dinner guests so that the estimated revenues for a given month are €5 million? The other variables remain the same as in the immediate previous question.

Answers

1   The variable "Number for lunch" is the strongest where $r^2 = 0.8478$.
2   The variable "Using spa (internal)" is the weakest where $r^2 = 0.7683$.
3   The coefficient of determination considering all of the variables is 0.9465.
4   The estimate of the monthly revenues is €4.8 million.
5   The average number of dinner guests would have to be 60.

8. Los Angeles restaurant

Objective

This case-exercise demonstrates the application of forecasting from correlated statistical data.

Situation

Felix and Camille own and operate a restaurant on Westwood Boulevard in Los Angeles, California. They are open for lunch and dinner and most of the time the restaurant is full. Their monthly revenues for the restaurant for 2015, in $US, are given in Table LA-1.

Table LA-1

Month	Revenues, $US
January	29,800
February	29,750
March	30,100
April	31,000
May	30,800
June	31,100
July	31,750
August	32,600
September	32,750
October	33,100
November	34,450
December	35,300

Required

1   Plot the scatter diagram for the data and show the regression line on the graph. Using linear regression, to the nearest whole number, what is an estimate of the restaurant annual revenue growth?
2   What are the values of the coefficient of determination and the coefficient of correlation? How would you interpret these results?
3   What is the linear regression equation that explains the progression of this information? Explain the terms in the equation.
4   Using linear regression, and to the nearest whole number, what is the forecast of revenues for September 2016? What are the 85% confidence limits for this forecast?
5   Using linear regression, and to the nearest whole number, what is the forecast of revenues for September 2018? What are the 85% confidence limits for this forecast? What are your comments concerning the values for this question and those for Question 4?
6   What are your general comments on this forecasting data analysis?

Answers

1   The scatter diagram with the regression line is given in Figure LA-1. The estimated annual revenue growth is $5,815.
2   The coefficient of determination is 0.9417; coefficient of correlation is 0.9704. Both of these values are close to 1.00 and greater than 0.80 so for this period the strength of revenues related to time is strong.
3   The linear regression equation is: $\hat{y} = 28,725 + 484.62x$. Here the value of 28,725 is the intercept on the y-axis and 484.62 is the slope of the regression line or the change in revenues with x, the month.

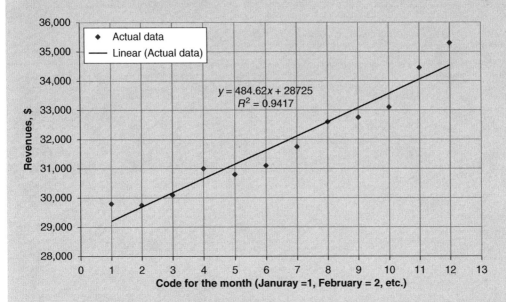

$$y = 484.62x + 28725$$
$$R^2 = 0.9417$$

Figure LA-1 Los Angeles restaurant.

4   The forecast for revenues in September 2016 is $38,902. The 85% confidence limits are $38,191 and $39,613.

5   The forecast for revenues in September 2018 is $44,717. The 85% confidence limits are $44,006 and $45,428. Since 2018 is further away from the base data than 2016 there is more uncertainty in the forecast. Markets, costs, and customer need may change.

6   This data for the period given indicates a strong linear growth of revenues with time. However, the restaurant business is notoriously precarious. Competition is strong; customers' tastes change; margins are not large so that for survival capacity utilization needs to always be close to 100%. Statistics show that the survival rate for new restaurants is two to three years.

9. Pub lunches

Objective

The objective of this exercise is to develop a forecast relationship when data exhibits a seasonal pattern.

Situation

Hugh Stevens owns a pub in Henley, on the River Thames in England. In the warmer months, Hugh has a large terrace where he can serve guests. The data in Table PL-1 gives historical data of number of lunches served in Hugh's pub on a monthly basis between 2011 and 2014.

Table PL-1

Month	2011	2012	2013	2014
January	91	98	102	104
February	227	273	285	318
March	773	1,045	1,273	1,409
April	1,045	2,273	2,491	2,864
May	2,439	2,452	3,227	3,991
June	3,758	3,333	3,636	3,827
July	4,015	4,749	4,955	5,015
August	3,997	3,930	4,017	4,464
September	3,773	3,636	3,736	4,045
October	455	500	591	636
November	91	106	121	136
December	45	52	50	59

Required

1   Plot the data using coded values for the months. That is, January 2011 has a code value of 1; February 2011 a value of 2; March 2011 a code value of 3, etc. What are your observations?

2   Group this data on a quarterly basis with winter being the months of January through March; spring April through June; summer July through September; and fall October

through December. Plot this data on a quarterly basis using again code values; winter quarter 2011 = 1; spring quarter 2011 = 2; summer quarter 2011 = 3; etc. = 1. What are your observations?

3 Using centered moving averages, what are the seasonal indices that describe this data after the seasonal effect has been removed? How would you interpret these indices?

4 Plot the data after the seasonal effect has been removed. What is the linear regression equation that describes the sales of lunches after the seasonal effect has been removed?

5 Is the regression line a good representation of sales from which the seasonal effect has been removed? Explain your reasoning.

6 Develop a forecast by quarter for 2015.

7 Develop a forecast by quarter for 2020.

8 What are your comments about the forecasts made for 2015 and 2020?

9 Make estimates for monthly sales for 2015.

Answers

1 There is a seasonal variation with highest sales in the spring and summer months and the lowest in the fall and winter months. In addition the sale of lunches is increasing over the years.

2 The data is smoother; more meals are served on the spring and summer quarters.

3 Winter 0.2694; spring 1.5206; summer 2.1700; fall 0.1145: winter sales are 73.06% less than yearly average; summer 52.06% more; summer 117.00% more; fall 88.55% less.

4 The linear regression equation is $\hat{y} = 167.53x + 4,421.22$.

5 Yes, the coefficient of determination, r^2, is 0.8662 or greater than 0.8.

6 In 2015: winter 1,958 meals; spring 11,308 meals; summer 16,502 meals; fall 890 meals for a total of 30,658 meals.

7 Winter 2,861 meals; spring 16,404 meals; summer 23,773 meals; fall 1,273 meals.

8 The year 2015 is closer than 2020, so the forecast for 2015 should be more reliable. In 2020 the pub may be reaching saturation as far as capacity is concerned—so possibly time for new capital investment?

9 In 2015: January 129 meals; February 357; March 1,443; April 2,760; May 3,889; June 4,759; July 6,089; August 5,352; September 4,961; October 706; November 147; December 67 meals for a total of 30,658 meals.

10. Public transport

Objective

The objective of this case-exercise is to illustrate causal forecasting

Situation

A community in Europe has for public transport: bus, metro, and tramway. Since 2002 it has been collecting data on the purchase of tickets for this transport system. Based on the annual number of tickets sold in that year it used this as an index of 100. In subsequent years it determined the number of tickets sold and related those to the base of 2002. In addition, it has a record of the population of the community. This information is given in Table PT-1.

Table PT-1

Year	Population	Index of public transport usage
2002	290,000	100
2003	298,750	102
2004	300,200	111
2005	302,200	108
2006	303,500	104
2007	304,450	115
2008	306,750	125
2009	308,900	119
2010	310,900	126
2011	311,500	131
2012	314,250	124
2013	316,000	127
2014	322,150	142
2015	329,400	145

An interest of the community management was to see if there was a correlation between the population growth and the use of public transport.

Required

1   Plot a scatter diagram of population against the usage index. What are your observations?
2   Develop the linear regression line that describes this situation. Define the variables used. How do you define the slope of the regression line?
3   Using the regression line, at a population of 310,000 (between 2009 and 2010) what is a forecast of the index of usage?
4   Using the regression line, at a population of 320,000 (between 2013 and 2014) what is a forecast of the index of usage?
5   Determine the geometric growth of the population and the geometric growth of public transport usage. What are your observations?
6   Using the geometric growth rate of the population, what is an estimate of the community population in 2020? Using this forecast estimate what would then be the index of public transport use?
7   What are your overall comments about an analysis of this nature?

Answers

1   The scatter diagram with the regression line is in Figure PT-1. There is a strong relationship between the population growth and the index of usage as the coefficient of determination is 0.8806. The coefficient of correlation is 0.9384.
2   The regression equation is $\hat{y} = 0.00013x - 285.6954$. The coefficient of x is the rate of change of the index with the population. For every 1,000 increase in the population, the usage index increases by 1.31.
3   The forecast of the usage index when the population is 310,000 is 121.91.
4   The forecast of the usage index when the population is 320,000 is 135.05.

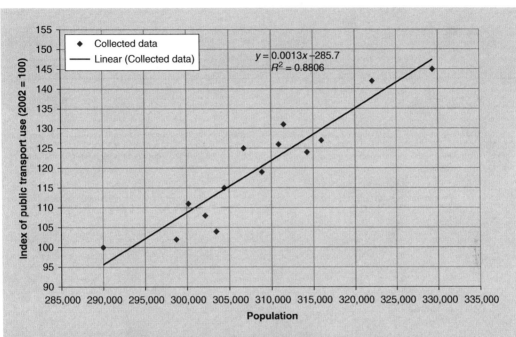

Figure PT-1 Public transport use and community population.

5   The geometric growth of the population is 0.98%; the geometric growth of public transport use is 2.90%. The use of public transport is increasing about three times faster than the population.
6   Using the geometric growth rate of the population in the given period, the population in 2020 is estimated to be 345,392. At this level the index of public transport use would be 169.16 or almost a 70% increase since the base period of 2002.
7   Many communities in Europe, and in other countries, are experiencing similar situations. The community population is increasing. This increases traffic congestion. With increase in traffic congestion more people use public transport to get to work and back. This saturates the public transport network during peak hours (London, Paris, Japan). This increases pressure on local communities to invest more capital to expand the transport network. London is an example where an East–West rail network is under construction (although London is a larger community than in this example).

11. Westwood Village restaurant

Objective

The objective of this exercise is to develop and apply linear and polynomial regression forecast models.

Situation

Christian has recently opened a restaurant in Westwood Village, Los Angles, offering French cuisine. Menu items include: escalope of salmon with tomatoes and sherry sauce; oven cooked Saint Jacques in a leek and saffron dressing; turkey rolls with mushrooms. His restaurant has 40 place settings and is open for lunch and dinner six days a week. At lunch he can only offer one service but at dinner since he offers two services: one at 19:00; the other at 21:30. Table WV-1 gives the number of meals sold for the first 12 weeks (three months) since opening. In the first week Christian had a promotion giving a free dessert for every meal ordered.

Table WV-1

Week	Meals served	Week	Meals served
1	84	7	125
2	61	8	131
3	85	9	127
4	92	10	201
5	105	11	199
6	117	12	237

Required

1 Plot this data on a time-series scatter diagram and insert the linear regression line. What is the equation to describe this line?
2 Plot this data on a time-series scatter diagrams and insert the first degree polynomial regression line. What is the equation to describe this line?
3 From the data from Questions 1 and 2, which of the regression lines seems to be the most appropriate for forecasting? Explain your reasoning.
4 What evidence is there from the data that you have plotted that Christian's restaurant is probably in the "growth" stage of its life cycle?
5 Using the linear regression equation what is the forecast for meals to be sold in Week 24. < that is, after six months of being opened?
6 Using the first degree polynomial regression equation what is the forecast for meals to be sold in Week 24, that is, after six months of being opened?
7 What are your comments concerning the forecast results from Questions 5 and 6?

Answers

1 Linear regression equation is $\hat{y} = 14.0839x + 39.2879$.
2 Polynomial regression equation is $\hat{y} = 1.3242x^2 - 3.1304x + 79.4545$.
3 The polynomial relationship looks best since it has a higher value for the coefficient of determination, $r^2 = 0.9472$, than for linear regression, $r^2 = 0.8750$.
4 The polynomial curve is the best fit for the data and has the characteristic profile of market growth for a new product.
5 The forecast in Week 24 for the linear regression is 377 meals.
6 The forecast using the polynomial regression is 767 meals.
7 The forecast using the polynomial function is beyond the weekly capacity of the restaurant of 720 (120 ∗ 6). This may be unreasonable. If not Christian has a problem to accommodate the demand of his clients. Is expansion a possibility?

12. Wine sales

Objective

The objective of this exercise is to examine a time series using regression analysis and then to re-look at the data using a causal relationship.

Situation

The Manzio family owns a small premium winery in the Florence area of Italy. The volume of wine in thousands of liters sold during 2013, 2014, and 2015 is given in Table WS-1. In addition in Table WS-2 is the tourist bureau published data on the number of tourists visiting the region for the same period.

Table WS-1

Month	Unit wine sales 2013	Unit wine sales 2014	Unit wine sales 2015
January	530	535	578
February	436	477	507
March	522	530	562
April	448	482	533
May	422	498	516
June	499	563	580
July	478	488	537
August	400	428	440
September	444	430	511
October	486	486	480
November	437	502	499
December	501	547	542

Table WS-2

	Tourists 2013	Tourist 2014	Tourist 2015
January	28,700	29,800	30,800
February	23,200	25,200	28,000
March	29,000	28,000	31,000
April	23,500	26,000	28,400
May	21,900	25,000	27,500
June	25,300	31,000	32,000
July	26,000	25,550	31,000
August	20,100	23,200	22,000
September	22,300	24,100	26,000
October	25,100	25,100	27,000
November	22,600	27,000	28,000
December	27,000	31,900	30,200

Required

1   Using a coded value for time, with January 2013 equal to 1, develop a time series scatter diagram of wine sales. What are your conclusions?
2   Develop a causal scatter diagram on wine sales against tourist bookings occupancy. What are your conclusions? What is the coefficient of determination to justify your answer.
3   Develop the linear regression equation based on the relationship from Question 2.
4   What is the relationship of wine sales with tourist bookings?
5   What is an estimate of wine sales by unit if tourist bookings were 30,000?
6   What is an estimate of wine sales by unit if tourist bookings were 60,000?
7   Why is the answer in Question 6 not reliable?

Answers

1   There is a very weak relationship. The coefficient of determination is 0.1607.
2   There appears to be a reasonable relationship. The coefficient of determination is 0.9111.
3   The linear regression equation is $\hat{y} = 14.0968x + 120,636$.

4   For every 1,000 increase in the number of hotel bookings, sales increase by €14,097.
5   The estimate of wine sales is 423,025 liters.
6   The estimate of wine sales is 845,930 liters.
7   There is no data covering tourist bookings of 60,000 and so this value is beyond the analytical range.

11 Business decisions and risk

Trying to avoid a disaster

No	Case-exercise	Statistical concept	Application
1	Adhesive product	Marginal analysis	Operations, inventory
2	Ardèche	Marginal productivity	Food, inventory
3	Art gallery	Expected values	Strategy and marketing
4	Bed and breakfast	Decision trees	Strategy, marketing
5	Computer manufacture	Maximax, maximin, regret	Project decision
6	Construk	Expected values, probability	Optimizing employee levels
7	Green coffee	Marginal analysis	Operations, food, inventory
8	Hanson	Decision trees	Strategy and marketing
9	Hotel complex, Kenya	Maximax, maximin, regret	Project, hospitality
10	Ipras	Decision trees	Refinery application
11	Polyethylene plant	Maximax + . . . decision trees	Strategy
12	Sierra Produce	Decision tables	Retail, food, inventory

1. Adhesive product

Objective

The objective of this case-exercise is to use maginal analysis in an inventory management situation.

Situation

Jameyson Brothers is a small company specializing in carpet laying. Jameyson purchases the organic-based adhesive it uses for carpet laying in drums from a large chemical company. This adhesive dries very quickly, allowing a newly carpeted room to be used within two hours after use. However, the adhesive product has a disadvantage in that if the drums, even if unopened, are not used within a week, the adhesive is unusable because it hardens in the drums. In this case, it is sold back to the supplier who recycles the material. Looking at past data, Jameyson developed Table AP-1 of the number of drums of adhesive demanded, against the probability of these levels being demanded. Also, in costing the carpet laying operation Jameyson calculated that the marginal profit on stocking an additional drum of adhesive as $1.70 and the marginal loss as $2.90.

Table AP-1

Drums of adhesive demanded	Probability of this level demanded	Drums of adhesive demanded	Probability of this level demanded
0	2.00%	6	16.00%
1	6.00%	7	13.00%
2	8.00%	8	10.00%
3	10.00%	9	7.00%
4	12.00%	10	2.00%
5	14.00%		

Required

1 What is the optimum stocking level for Jameyson?
2 What is the minimum probability at which Jameyson should stock drums of adhesive?
3 At the optimum stocking level what is the expected marginal profit, and the expected marginal loss?
4 Illustrate the expected marginal loss and expected marginal profit as a graph against stocking level.

Answers

1 Optimum stocking level is four drums.
2 Minimum probability at which Jameyson should stock drums of adhesive is 63.04%.
3 At optimum level expected marginal profit is $1.258, and expected marginal loss is $0.754.
4 The data from which the graph is drawn is in Table AP-2 and the graph is in Figure AP-1.

Table AP-2

Demand units, x	Probability of demand, P(x)	Probability of greater than demand, x	Expected MP ($)	Expected ML ($)
0	2.00%	100.00%	1.7000	0.0000
1	6.00%	98.00%	1.6660	0.0580
2	8.00%	92.00%	1.5640	0.2320
3	10.00%	84.00%	1.4280	0.4640
4	12.00%	74.00%	1.2580	0.7540
5	14.00%	62.00%	1.0540	1.1020
6	16.00%	48.00%	0.8160	1.5080
7	13.00%	32.00%	0.5440	1.9720
8	10.00%	19.00%	0.3230	2.3490
9	7.00%	9.00%	0.1530	2.6390
10	2.00%	2.00%	0.0340	2.8420
11	0.00%	0.00%	0.0000	2.9000
Total	100.00%			

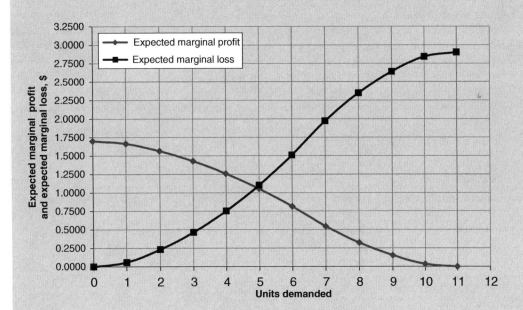

Figure AP-1 Adhesive product: expected marginal profit and loss.

2. Ardèche

Objective / Application

The objective of this case–exercise is to demonstrate the use of marginal analysis and its relationship to the normal distribution. It has application in inventory management and production levels.

Situation

The Cervelin family, in Ardèche, France, make cheese from the milk they obtain from their cows. The cheese making process is simple. To the milk is added rennet, which makes the milk curdle. Additives are added for flavor and then the product is air-dried and cut into portions. The cheese is sold at the local market. From past data the average daily sales of cheese is 25 kg with a standard deviation of 6.50 kg. The cheese is sold at €1.07 per kg and costs €0.47 per kg to make. The customers who buy the cheese want it fresh so that any taken to the market and not purchased is sold to the local farmer for pig feed for €0.19 per kg.

Required

1   Determine the marginal profit and the marginal loss for this production operation.
2   What is the probability at the optimum level?
3   What is the optimum cheese production level in kilograms?
4   Validate this production level by a plot of the expected profit against expected loss.
5   How would your answer change if the Cervelin family were unable to sell the cheese to the local farmer and the cheese had to be thrown out?

Answers

1   Marginal profit is €0.60; Marginal loss is €0.28.
2   The probability at the optimum level is 31.82%.
3   Optimum cheese production is 28.07 kg.
4   The plot is shown in Figure A-1.
5   The optimum cheese production level is 26 kg (25.99) and the minimum probability is 43.93%.

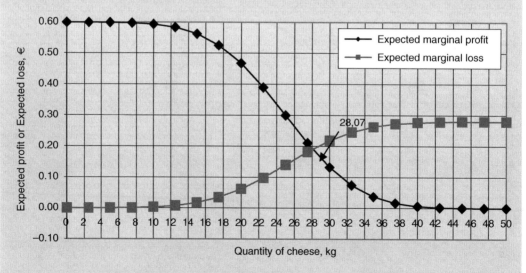

Figure A-1 Ardeche: cheese production.

3. Art gallery

Objective

The objective of this case-exercise demonstrates that when probabilities can be assigned to decision-making situations, using expected values is a way of selecting the best alternative.

Situation

Four colleagues from Los Angeles have had considerable success in selling art works of paintings and sculptures in California but their international exposure is limited. They are considering opening an art gallery in Europe where they believe they will have more international buyers. After doing considerable research and talking with specialists that came up with four possible locations of which one would be their choice: Lausanne, Switzerland; Florence, Italy; Madrid, Spain; and Vienna, Austria. In each of these cities they would purchase a disused warehouse and remodel the facility according to their needs. Table AG-1, the base case, gives an estimate of net income in €millions from these four locations over a five-year period from opening the gallery. These numbers are based on renovation costs, operating costs when the facility is open, and client revenues. Probability values are given that take into consideration the reliability of costs and also market demand and estimated prices of the art works.

Table AG-1

Location	Net income €millions according to market		
	Low market	Level market	High market
Lausanne, Switzerland	30	35	70
Florence, Italy	20	35	75
Madrid, Spain	15	40	80
Vienna, Austria	10	25	65
Probability	25.00%	40.00%	35.00%

Required

1   Using the concept of expected values determine the preferred choice. This would be the choice under risk (EVUR). Show the results on a decision tree.
2   Determine the expected value under certainty (EVUC).
3   Determine the expected value of certain information (EVPI). How is this information interpreted?
4   Two of the colleagues have family in Italy and have a preference for selecting Florence. In this case what would have to be the estimated net income from Florence in a high market such that the expected value under risk for Florence is 0.5% more than the value obtained from the determined choice from Question 1? In this case what would be the expected value of perfect information?
5   Using the base case, what would the decision be if the probability for a low market was 35%; level market 35%; and for a high market 30%.
6   What are your general comments on these methods of decision making?

Answers

1   Madrid with an expected value of €47.75 million.
2   Expected value under certainty is €51.50 million.
3   Expected value of perfect information is €3.75 million. The colleagues should not spend more than this amount on further research to be more certain of their estimated data.
4   The estimated net income from Florence, Italy in a high market such that this city is the preferred choice is €82.83 million.
5   Lausanne, Switzerland with an expected value under risk of €48.50 million; expected value under certainty of €43.75 million; expected value of perfect information of €4.75 million.
6   Extremely sensitive to given estimated financial data and probability values.

4. Bed and breakfast

Objective

The objective of this case-exercise is to illustrate the development of a decision tree for analysis.

Situation

Sally Stock is considering opening a bed and breakfast (B&B) operation in Sydney, Australia. This will be Sally's sole source of revenue. She is thinking of building either a small B&B, (12 rooms), a large B&B (50 rooms), or she may abandon the idea altogether. She is considering leasing, for five years, one of two facilities in the Sydney area for her project. A friend, Bill, will perform a market survey. Sally has estimated that if she establishes a large B&B, and the market is favorable, she can earn $Aus 60,000 over a five-year period, but will lose $Aus 40,000 if the market is unfavorable. The small B&B will return a $Aus 30,000 profit with a favorable market, and a $Aus 10,000 loss if the market is unfavorable.

At the present time, Sally believes there is a 50/50 chance that there will be a favorable market. Bill will charge Sally $Aus 5,000 for the market survey. Bill has estimated that there is a 60% probability that the market survey will be favorable. What is more he believes there will be a 90% probability that the market will be favorable given a favorable outcome of the market survey. However, Bill has warned Sally that there is a probability of only a 12% favorable market, if the results of the market survey are not favorable.

Required

1   What is the preferred strategic decision for Sally? Show how you arrive at your conclusion by developing a decision tree.

Answer

1   The best strategy is to use the survey since this produces a higher expected value than not using a survey. If the survey is favorable the subsequent strategy is to develop a large B&B. In this case:

- With a favorable market (90% probability) Sally will earn $Aus 55,000. With an unfavorable market she will lose $Aus 45,000.
- If the survey turns out to be unfavorable then do nothing. Sally will lose $Aus 5,000, the cost of the survey.

5. Computer manufacture

Objective

This case-exercise examines various approaches to decide on new production capacity.

Situation

A US based computer company is considering one of four possibilities in new facilities for increasing its production capacity: one in the USA that would be 100% owned: a joint venture in China; an expansion of an existing facility in India; or a relatively new small plant in Ireland. An estimation of the returns over five years from each facility, in $US, is given in Table CM-1 according to various market changes. The numbers take into account all construction costs, transportation, risks associated with currency changes, and financial assistance from government and regional authorities.

Required

1 Based on the data, what would be the preferred site if the decision maker was optimistic?
2 Based on the data, what would be the preferred site if the decision maker was pessimistic?
3 Based on the data, what would be the preferred site if the decision maker took a middle-of-the road approach?
4 Based on the data, what would be the preferred site if the concept of minimax regret was used?
5 Assume that the probability of the market changes of Table CM-1 were estimated according to Table CM-2, then using expected values, what decision would be made for site selection?

Table CM-1

Market change	50% increase	25% increase	No change: flat	5% decline
USA (100% owned)	34,500,000	22,250,000	2,000,000	−19,500,000
China (joint venture)	32,500,000	13,500,000	1,500,000	−5,000,000
India (expansion)	15,500,000	10,000,000	1,000,000	−4,500,000
Ireland (new)	10,000,000	6,500,000	0	−1,000,000

Table CM-2

Probability	40.00%	30.00%	25.00%	5.00%
Market change	50% increase	25% increase	No change: flat	5% decline

Answers

1 The preferred site if the decision maker was optimistic is the United States. This would potentially give the best return ($34,500,000); the financial risk would be higher (loss of $19,500,000).
2 The preferred site if the decision maker was pessimistic is a new facility in Ireland. This is the worst case with only a potential loss of $1,000,000.

3   The preferred site if the decision maker took a middle-of-the road approach is a joint venture with China as this indicates an average return of $10,625,000. Note this is only a quantitative approach to decision making. Only one market condition can occur.

4   The preferred site if the concept of minimax regret was used is a Chinese joint venture. It indicates the lowest "regret" of $8,750,000.

5   Based on the given probabilities, the site selection in the USA would be preferred.

6. Construk

Objective

The objective of this case-exercise is to demonstrate expected values in decision making.

Situation

Construk Inc. in Southern California performs engineering, design, and construction work for international and domestic clients. Next year it has a potential project from a French/Chinese consortium to perform design work on the reactor system of a nuclear power plant. The work is very specialized and Construk will need to hire the engineers trained in this area. The extent of the design work has not been completely defined but is expected to be between 19,600 hours and 27,440 hours. Table C-1 indicates the estimated probabilities of the work being released.

Engineers are paid an average $35/hour on a 2,080 hours per year basis. However, included in these 2,080 hours are 3 weeks paid vacation. (The workweek is 40 hours.) If engineers are hired, and the work level does not materialize, they will be retained by the company and assigned to developing equipment specifications (indirect work). This indirect work is not billed to the client. That is, these engineers are paid their hourly salary, and this represents a cost to Construk. It is budgeted that the work is finished within the year. The customer is billed at a rate of $42/hour but only for the time that the engineers work on the project.

Table C-1

Customer job hours	19,600	21,560	23,520	25,480	27,440
Probability of this level	20.00%	45.00%	30.00%	4.00%	1.00%

Required

1   Using the concept of expected values, without any financial considerations, estimate how many engineers should be hired.

2   Using an expected profit table (taking into account the financial information), how many engineers should be hired to maximize the expected profit? What is this expected profit?

3   What is the expected value of perfect information (EVPI)?

4   How might the concept of EVPI be interpreted in this particular situation?

Answers

1 Using the concept of expected values 12 engineers should be hired.
2 Ten engineers should be hired: that would give an expected profit of $95,200.
3 EVPI is $11,519.20.
4 You would not pay more than the EVPI in order to be more certain of your estimated information.

7. Green coffee

Objective

The objective of this case-exercise is to demonstrate the use of long and short costs for optimizing inventory management.

Situation

The raw material for producing ground coffee is green coffee. A producer, who purchases the green coffee from a local distributor, makes an estimate of the number of pallets of green coffee used per day over an 80-day period. This information is in Table GC-1. The long costs, the cost of stocking a unit that is not demanded, and the short cost, the cost of not stocking a unit that is demanded, are given in Table GC-2.

Table GC-1

Pallets demanded	Days this level occurred
20	15
40	20
60	12
80	10
100	23
Total days	80

Table GC-2

Long cost, per pallet, $US	30
Short cost per pallet, $US	70

Required

1 Using payoff tables, determine the optimum number of pallets of coffee to stock.

Answers

1 The optimum value is 80 pallets as this indicates the minimum cost. The payoff table is in Table GC-3.

Table GC-3

		Pallets demanded				Expected Cost, $
20	40	60	80	100		
Order point No. of pallets 20		0.00	1,400.00	2,800.00	4,200.00	5,600.00	2,905.00
40	600.00	0.00	1,400.00	2,800.00	4,200.00	1,880.00
60	1,200.00	600.00	0.00	1,400.00	2,800.00	1,355.00
80	1,800.00	1,200.00	600.00	0.00	1,400.00	1,130.00
100	2,400.00	1,800.00	1,200.00	600.00	0.00	1,155.00
Probability		18.75%	25.00%	15.00%	12.50%	28.75%	100.00%

8. Hanson

Objective

The objective of this case-exercise is to demonstrate the application of decision trees.

Situation

John Hanson is the Operations Manager of a manufacturing facility that makes automobile engine parts. The facility uses flexible manufacturing and John is able to change from one product range to another relatively easily. The capacity of John's facility is now saturated and he has to make a decision on adding new machines to handle the combined drilling, milling, and welding operations. This automated equipment is expensive. John knows he wants to purchase equipment manufactured by Brown–Boveri and he has narrowed his requirements down to one or two machines.

   If just one machine is purchased now, and demand for engine parts turns out to be more than expected, a second machine can be purchased at a later date. However, in this case, purchasing a second machine later would cost more than if both machines were purchased together, due to higher unit transportation and installation costs. Based on knowledge of the business, John estimates that the probability of low demand is about 40%, and high demand 60%. If John purchases two machines at the same time, and demand for engine parts is high, then the net present value (NPV) of profits associated with the machines would be $520,000. If demand turns out to be low, then the NPV of profits will be only $300,000, as one machine would be running below capacity. If John purchases only one machine now, and demand is low, then the NPV of profits is estimated at $360,000. However, if the demand turns out to be high, then John has three options:

1   Do nothing. This means losing potential business, and the NPV of profits would stay at $360,000.
2   Subcontract the additional business, which would yield a NPV of profits to John of $440,000.
3   Purchase another machine, which would result in a NPV of $400,000.

Required

1 Draw a decision tree showing the various options available to John Hanson. On the decision tree, show the value of all the expected outcomes. Based on the data given above, what decision should John make concerning the purchase of machines?
2 If the estimated NPV of profits for the subcontracting option were $480,000 instead of $440,000 what impact would this have on your decision?

Answers

1 Buy two machines now.
2 There is a breakeven situation regarding the expected values under risk.

9. Hotel complex, Kenya

Objective

The purpose of this case-exercise is to demonstrate the application of quantitative approaches to decision making, in particular to the type of facility to build.

Situation

A US based hotel company is considering establishing a hotel/spa/restaurant complex in Kenya. There are three alternative projects being considered by management: build a new grassroots facility on a virgin site; acquire an existing property and bring it up to standards; establish a joint venture with a Chinese firm. After some extensive studies, and initial designs, the estimated net present values of the payoff (net return) in $millions for the three alternatives, at three different scenario levels of future market demand, are in Table HK-1.

Table HK-1

Project decision	Low market demand	Stable market demand	High market demand
Grass roots on a virgin site	−50	30	100
Renovate an existing property	−25	50	75
Joint Venture with a Chinese company	10	30	45

Required

1 What would be the preferred decision if management was optimistic in its decision approach?
2 What would be the preferred decision if management was pessimistic in its decision approach?
3 What would be the preferred decision if management was middle-of-the-road in its decision approach?
4 What would be the preferred decision if management used the minimax regret approach to decision making?

5   After a further market study was performed, the company developed some probability estimates for the market outcome according to Table HK-2. Using this information what would be the preferred decision?

6   What is the expected value of perfect information (EVPI)? How is this interpreted?

Table HK-2

Future market	*Low market demand*	*Stable market demand*	*High market demand*
Probability (%)	20%	35%	45%

Answers

1   The preferred decision is the grass roots project that would give an estimated return of $100 million.

2   The preferred decision is the joint venture approach to give an expected return of $10 million.

3   The preferred decision is to renovate an existing property as the highest average is $33.33 million.

4   Using minimax regret the preferred decision is to renovate. Minimum regret is $35 million.

5   With probability values select renovation as this gives the highest expected value of $46.25 million.

6   EVPI is $18.25 million. You would not pay more than $18.25 million in research and analytical studies to obtain more precise information.

10. Ipras

Objective

The objective of this case-exercise is to use decision trees in order to judge alternatives in management choices.

Situation

Tugrul Osmal is the Chief Engineer at the Ipras Oil Refinery in Izmit, Turkey. He is considering adding a new Unit A to the present catalytic cracking unit to upgrade gasoline production. It is not certain that this Unit A will work because the crude used at Ipras is high in sulfur and excessive sulfur levels in the Unit A would poison the catalyst. However, if the Unit A works, Ipras could realize a return of $350,000 per year. If the unit does not work, the company stands to lose $150,000 per year. At the present time, Tugrul estimates there is a 40% chance that the Unit A will work.

An alternative option for Tugrul is to build a pilot plant to first test a smaller version of the Unit A. Based on the results of that, Tugrul could then decide whether to build a commercial version of Unit A. Constructing and operating the pilot plant would cost an annualized $45,000. There is a 50% chance that the pilot plant would work. If the pilot plants work, there is then a 90% probability that the commercial Unit A, if subsequently built, would perform correctly. If the pilot unit does not work, there is only a 20% chance that the commercial plant, if constructed, would work.

Required

1   Develop a decision tree based on the above information; show this quantitatively on the decision tree; and explain the best strategy for Tugrul.
2   Tugrul, after discussing the project with a US engineer, decides that the probability of the commercial facility working with no pilot plant could be increased to 55% from the original 40%. In this case, would there be a change in the decision for Tugrul? Justify your response.
3   What can you say about the sensitivity of this decision situation?

Answers

1   There are three options: 1. Add unit. This gives an expected value of $50,000 per year. Do nothing. This gives an EV of $0. Build pilot plant gives an EV of $105,000. Decision would be to build the pilot plant. If it works, commercialize. If it does not work, do not proceed any further.
2   In this case, the expected value of adding the unit directly, without going through the pilot plant phase, has the highest expected value ($125,000). The decision would be to add the unit without going through pilot plant phase. The refinery stands to make $350,000, but risks losing $150,000.
3   As is very often the case, the project is sensitive to probability changes. Thus, perhaps consider other steps before making a decision.

11. Polyethylene plant

Objective

The objective of this case–exercise is to demonstrate various approaches to decision making.

Situation

An international chemical company is considering increasing its worldwide polyethylene capacity. Polyethylene is a common plastic used in the manufacture of packaging materials, automobile components, kitchen utensils, etc. The following are the possible alternatives that management is reviewing for this project: build a new grassroots facility; expand an existing plant; or establish a joint venture with another overseas company. After some extensive studies, and initial designs, the estimated net present values of the payoff (net return) in $millions for the three alternatives, at three different scenario levels of future polyethylene demand, are as given in Table PE-1.

Table PE-1

Project decision	Low market demand	Stable market demand	High market demand
Grassroots	–50	30	100
Expansion	–25	50	75
Joint venture	10	30	45

Required

1   Based on the given information, what would be the preferred decision if management were pessimistic in its decision approach?

2   Based on the given information, what would be the preferred decision if management were optimistic in its decision approach?

3   Based on the information given in Table PE-1, what would be the preferred decision if management has a middle-of-the-road approach to decision making?

4   Based on the given information, what would be the preferred decision if management used the minimax regret approach to decision making?

5   After a further market study was performed, the company developed some probability estimates for the market outcome according to Table PE-2. Using this information what would be the preferred decision?

6   What is the expected value of perfect information (EVPI) for the situation presented in Question 5? How is this interpreted?

7   Develop a decision tree to illustrate the situation and solution, under a condition of risk as presented in Question 5. How would the decision be interpreted?

8   After further discussion, management came up with the following modified situation.

   • If the grassroots alternative was selected, and the market demand was high, the company had an option to expand the capacity. In this case, the estimated payoff would be $125 million instead of $100 million.
   • If the joint venture alternative was selected, and the market demand was high, the company had an option to increase its percentage share in the joint venture. In this case, the estimated payoff would be $65 million instead of $45 million.

   Use decision trees to illustrate the preferred decision for management.

9   What is the EVPI for this modified situation presented in Question 8?

Table PE-2

Future market	Low market demand	Stable market demand	High market demand
Probability (%)	20	35	45

Answers

1   Pessimistic is a maximin approach. Here, the decision would be to select the joint venture alternative since $10 million is "best" of the worst alternatives.

2   Optimistic is a maximax approach. Here the decision would be to select the grass roots facility, since $100 million is the maximum possible outcome.

3   Middle of the road is an equally likely approach. Here, expansion would be the preferred alternative as the average value of the outcome is the highest at $33.33 million.

4   Select expansion as this is the least painful regret.

5   The expansion project has the highest expected value at $46.25 million and thus this represents the best alternative.

6   EVPI is $18.25 million, which means that you would not expend more than this amount to refine your risk situation.
7   EVPI is $19.00 million.
    (Decision trees can be developed according to the criteria in the chapter.)

12. Sierra Produce

Objective

The objective of this case-exercise is to illustrate the development of conditional profit and loss tables. It has application in inventory management.

Situation

Jack Komiko owns Sierra Produce, a retail outlet for all types of vegetables, and fruit for the San Gabriel Valley residents in northeast Los Angeles. One of Sierra Produce's most exotic products is a mango imported from the Far East at a purchase cost price of $56.00 per case. Jack normally stocks 11, 12, 13, or 14 cases of mangoes each week. He sells the mangoes in his store for $91.00 per case. However, since Jack always wants to sell fruit of top quality, and as the mangoes soon get soft, Jack must dump the unsold fruit at the end of the week. Based on past experience, Jack knows that there is a probability of 35% of selling 11 cases, 35% also of selling 12 cases, 20% of selling 13 cases, and only 10% probability of selling 14 cases.

Required

1   Using a conditional profit table, what is the recommended inventory stocking level, if there is no value for the dumped fruit? Show the conditional profit table.
2   Using the results from Question 1 determine the expected value under risk (EVUR), expected value of under certainty (EVUC), and the expected value of perfect information (EVPI). How would the EVPI be interpreted?
3   If Jack could sell any unsold soft mangoes at the end of the week for $66.00 per case to a small company who makes fruit juices, instead of the $91.00 store level price, then using a conditional profit table what would be the recommended stocking level?
4   Using a conditional loss table taking into account both lost opportunities ("lost" profit) and real losses, what is the recommended stocking level, if there is no value for the dumped fruit?

Answers

1   Stocking level is 12 cases with an expected value of $388.15. The payoff is shown in Table SP-1.
2   EVUR is $388.15, the best value in Table SP-1. EVUC is $421.75. EVPI is $33.60. You would not pay more than this amount to be sure of obtaining perfect information.
3   Stocking level is 14 cases with an expected value of $441.25. The payoff is shown in Table SP-2.
4   Recommended stocking level is 12 cases to give a minimum expected loss of $33.60. The payoff is shown in Table SP-3.

Table SP-1

Stock level	Cases demanded—external event				Expected value
	11	12	13	14	
11	385.00	385.00	385.00	385.00	385.00
12	329.00	420.00	420.00	420.00	388.15
13	273.00	364.00	455.00	455.00	359.45
14	217.00	308.00	399.00	490.00	312.55
Probability	35.00%	35.00%	20.00%	10.00%	100.00%

Table SP-2

Stock level	Cases demanded—external event				Expected value
	11	12	13	14	
11	385.00	385.00	385.00	385.00	385.00
12	395.00	420.00	420.00	420.00	411.25
13	405.00	430.00	455.00	455.00	428.75
14	415.00	440.00	465.00	490.00	441.25
Probability	35.00%	35.00%	20.00%	10.00%	100.00%

Table SP-3

Stock level	Cases demanded—external event				Expected value
	11	12	13	14	
11	0.00	35.00	70.00	105.00	36.75
12	56.00	0.00	35.00	70.00	33.60
13	112.00	56.00	0.00	35.00	62.30
14	168.00	112.00	56.00	0.00	109.20
Probability	35.00%	35.00%	20.00%	10.00%	100.00%

12 Statistical process control (SPC)

Ensuring operations are according to specifications

No	Case-exercise	Statistical concept	Application
1	Automatic filling machine	x-bar and range chart	Operations, food
2	Carpet maker	c-chart	Manufacturing
3	Flow meters	x-bar and range chart	Manufacturing
4	Hotel housekeeping	p-chart	Operations, hotel, hospitality
5	Hotel occupancy	p-chart	Operations, hotel, hospitality
6	Internet response	x-bar and range chart	Information systems
7	Mustard filling machine	x-bar, R–chart	Operations, hospitality
8	Paper making	c-chart	Manufacturing
9	Steel bolts	x-bar and range chart	Manufacturing
10	Tablecloth material	c-chart	Suppliers, hospitality
11	Watch manufacture	p-chart	Manufacturing
12	Weight of food portions	x-bar, R–chart	Operations, food

1. Automatic filling machine

Objective

This case–exercise develops an *x*-bar and range statistical process control chart when there are variables in an operation.

Situation

A food factory has an automatic filling machine adjusted to drop 250 g of candies into plastic bags before they are sealed and labeled with the statement, "Net weight 250 grams." The company is concerned about the reliability of this machine and carries out a statistical process control on the equipment. First, the filling line is shut down, and adjusted according to specifications. The line is then put into operation and 20 random samples, each of size 25 units are taken during 8 hours of operation. For each of the 20 samples the maximum, minimum and sample average weight is recorded. This information is in Table AFM-1.

Table AFM-1

Sample No.	Sample average	Maximum weight	Minimum weight	Sample no.	Sample average	Maximum weight	Minimum weight
1	253.00	262.00	232.00	11	246.10	278.00	237.00
2	248.00	261.00	232.00	12	250.00	265.00	235.00
3	254.20	269.00	242.00	13	247.00	265.20	241.00
4	256.00	265.00	248.00	14	248.20	263.20	242.10
5	253.20	262.40	247.00	15	251.30	270.20	247.20
6	246.20	261.40	235.00	16	251.30	265.20	248.10
7	250.00	254.00	230.00	17	253.60	257.20	248.80
8	247.50	265.00	229.00	18	252.70	259.20	248.20
9	254.30	275.00	242.00	19	254.30	259.00	247.90
10	252.00	258.00	241.00	20	255.20	260.10	251.00

Required

1   Develop an *x*-bar control chart and a range chart to analyze this filling operation using the sample data as the control benchmark for control. Here the assumption is that this data is the best obtainable. How would you describe the results of your control charts?
2   Redo your *x*-bar and range chart using the net weight value as your benchmark *CL*. What are your observations?
3   What are the conclusions from these control charts?

Answers

1   The *x*-bar chart is shown in Figure AFM-1: *CL*, 251.21 g; *LCL*, 247.76 g; *UCL*, 254.65 g
    There are 6 data points outside the limits of the *x*-bar chart.
    The range chart is shown in Figure AFM-2: *CL*, 22.54 g; *LCL*, 10.37 g; *UCL*, 34.71 g
    There are 4 data points outside the limits of the range chart.

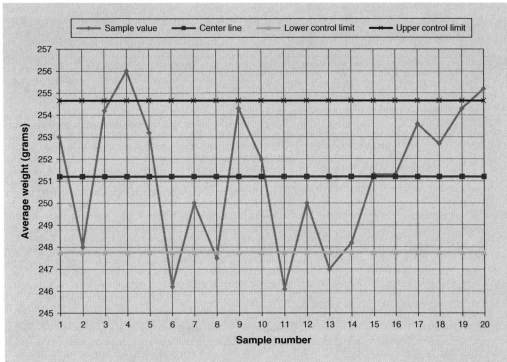

Figure AFM-1 Automatic filling machine: average control chart.

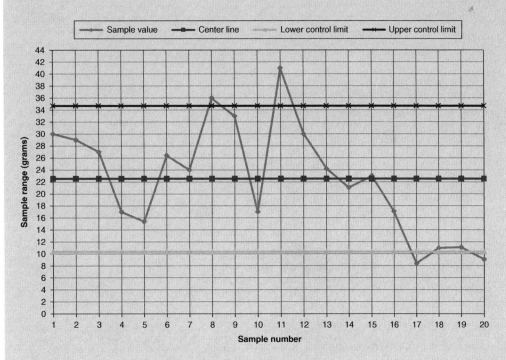

Figure AFM-2 Automatic filling machine: range control chart.

2   The *x*-bar chart is shown in Figure AFM–3: *CL*, 250.00 g; *LCL*, 246.55 g; *UCL*, 253.45 g There are 8 data points outside the limits of the *x*-bar chart.

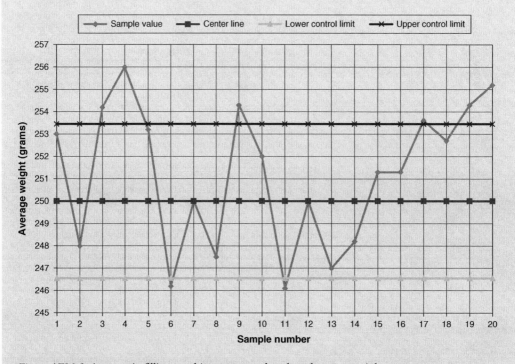

Figure AFM-3 Automatic filling machine: average chart based on net weight.

For the range chart: *CL*, 22.54 g; *LCL*, 10.37 g; *UCL*, 34.71 g There are 4 data points outside the limits of the range chart.

3   The filling process is out of control. When there is data outside the upper control limits and this is an excessive cost. When there is data below the lower control limit and this indicates that the customer is not receiving what is expected regarding quantity. The machine should undergo maintenance or be replaced.

2. Carpet maker

Objective

The objective of this case-exercise is to apply the development of a c-control chart for statistical process control.

Situation

A carpet maker in Iran will install a quality control program for the production of its carpets, which are made in two work centers. These carpets are sold directly to European retail stores and commercial outlets including company offices and hotels. For quality control purposes, the producer selects, at random, 250 carpets of similar size that have been recently produced by his most experienced workforce, and counts the total number of faults on these samples. The total number was 970 evenly distributed throughout the samples. A fault is considered to be slight color differences of threads or pattern misalignment. As these carpets were all handmade, and produced by his most experienced employees, these errors were considered an accepted condition of the operation. This base data was used as the benchmark for quality control.

After his benchmark was established, the producer compares the quality of the entire workforce in other areas of the carpet making operation. Thus at random for one week, Monday to Friday, the producer takes one carpet each from its two work centers and counts the number of blemishes on the carpet. These sample were taken at random, where the entire work force was involved. This new sampling information is given in Table CM-1. The number in parenthesis after each day indicates the work center from which the carpets were taken.

Table CM-1

Day	No. of faults
Monday (1)	2
Monday (2)	12
Tuesday (1)	3
Tuesday (2)	13
Wednesday (1)	3
Wednesday (2)	10
Thursday (1)	2
Thursday (2)	8
Friday (1)	2
Friday (2)	11

Required

1 Develop a c–control chart using 3–sigma control limits. Present the test data on a control chart. How many data points are outside the upper limit and what conclusion might you draw?
2 Develop a c–control chart using more rigorous 1–sigma control limits. Present the test data on a control chart. How many data points are outside the upper limit and what conclusion might you draw?

Answers

1 The c–control chart for 3–sigma control limits is in Figure CM-1. The *CL* is 3.88 faults; *LCL* is 0.00 faults; *UCL* is 9.79 faults. Four points are outside the upper limit in Work Center 2 and so here the process is out of control. Corrective action should be taken.
2 The c–control chart for 1–sigma control limits is in Figure CM-2. The *CL* is 3.88 faults; *LCL* is 1.97 faults; *UCL* is 5.85 faults. Five points, or all the data points are outside the upper limit in Work Center 2 and so here the process is out of control. Corrective action should be taken.

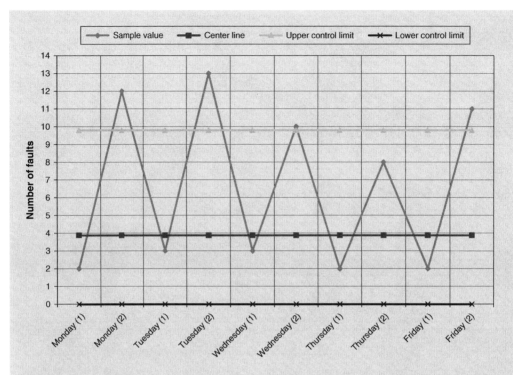

Figure CM-1 Carpet making: statistical process control chart (3–sigma limits).

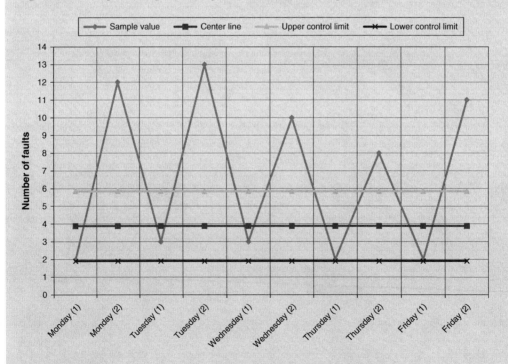

Figure CM-2 Carpet making: statistical process control chart (1-sigma limits).

3. Flow meters

Objective

This case-exercise illustrates the development of an *x*-bar and range chart for statistical process control.

Situation

The Gerber Company, in the Denver region of Colorado, manufactures flow meters for use in the gas and petroleum industry. The critical part of these meters is a rotating cylinder positioned inside a gas-tight casing. The cylinders are cut from aluminum tubing on a large automatic saw. It is critical that the lengths of these cylinders lie within the given specification, else the meters do not function correctly. Gerber operates on a 40-hour week, Monday to Friday. In the cutting operation the automatic machine is regulated twice each day, once in the morning at 08:00 and once in the afternoon at 13:00. The production manager has been unhappy with the quality of the meters being made and he authorized a quality control analysis to take place. This analysis was carried out over a one-month period when cylinder reference No.45983-M was being produced. This cylinder had a specified length of 3.8 cm. Samples of size 15 were taken twice a day at 09:30 and 15:30 and the sample average and the sample range were determined. This data is given in Table FM-1.

Table FM-1

Week 1	Time	x-bar (cm)	Range (cm)
Monday	09:30	3.8900	0.0855
Monday	15:30	3.8420	0.0950
Tuesday	09:30	3.8095	0.0950
Tuesday	15:30	3.8080	0.1045
Wednesday	09:30	3.8285	0.0855
Wednesday	15:30	3.8095	0.1045
Thursday	09:30	3.8285	0.1045
Thursday	15:30	3.8190	0.0950
Friday	09:30	3.8200	0.1045
Friday	15:30	3.8700	0.0950
Week 2	Time	x-bar(cm)	Range (cm)
Monday	09:30	3.7800	0.1140
Monday	15:30	3.7900	0.1045
Tuesday	09:30	3.8380	0.0855
Tuesday	15:30	3.8157	0.0950
Wednesday	09:30	3.8138	0.1045
Wednesday	15:30	3.8185	0.0855
Thursday	09:30	3.8157	0.0960

(Continued)

Table FM-1. (Continued)

Week 2	Time	x-bar(cm)	Range (cm)
Thursday	15:30	3.8275	0.0860
Friday	09:30	3.8340	0.0950
Friday	15:30	3.8700	0.0880

Week 3	Time	x-bar (cm)	Range (cm)
Monday	09:30	3.8640	0.0950
Monday	15:30	3.8043	0.1045
Tuesday	09:30	3.8245	0.0950
Tuesday	15:30	3.8190	0.0855
Wednesday	09:30	3.8123	0.0950
Wednesday	15:30	3.8190	0.1045
Thursday	09:30	3.8200	0.0855
Thursday	15:30	3.8100	0.0855
Friday	09:30	3.8280	0.0755
Friday	15:30	3.7810	0.0855

Week 4	Time	x-bar (cm)	Range (cm)
Monday	09:30	3.8960	0.0950
Monday	15:30	3.8640	0.1000
Tuesday	09:30	3.8040	0.0895
Tuesday	15:30	3.8190	0.0855
Wednesday	09:30	3.8320	0.0950
Wednesday	15:30	3.8190	0.1045
Thursday	09:30	3.8268	0.0855
Thursday	15:30	3.8170	0.0855
Friday	09:30	3.8050	0.0878
Friday	15:30	3.7700	0.0855

Required

1   Construct the *x*-bar control chart with the *CL*, and the *UCL* and *LCL*.
2   Construct the range chart indicating the *CL*, and the *UCL* and *LCL*.
3   Does either of the charts indicate any cause for concern? If so what do you suggest might be happening?

Answers

1   The *x*-bar control chart is in Figure FM-1. The benchmarks on the *x*-bar chart are: $CL = 3.8241$; $UCL = 3.8450$; $LCL = 3.8032$.
2   The range control chart is in Figure FM-2. The benchmarks on the range chart are: $CL = 0.0937$; $UCL = 0.1549$; $LCL = 0.0325$.
3   The *x*-bar chart indicates a problem on Monday and Friday as data points are outside the limits. Perhaps employees are anxious to leave work on Friday and so they are paying less attention to quality. Similarly on Monday, when they come back to work their minds are not initially on the job. In both cases there seems to be a problem of motivation.

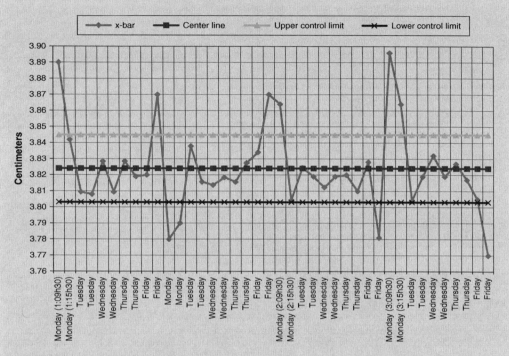

Figure FM-1 Flow meters: *x*-bar, statistical process control chart.

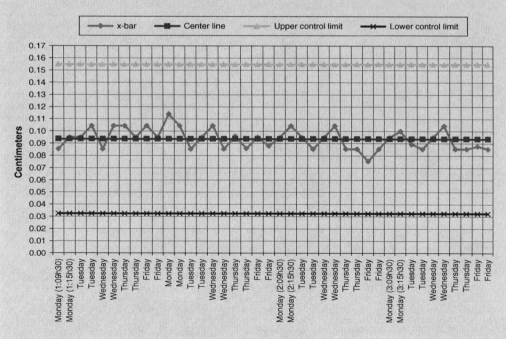

Figure FM-2 Flow meters: range, statistical process control chart.

4. Hotel housekeeping

Objective

The objective of this case-exercise is to develop a p–chart concerning the quality in a hotel for the conformity of rooms.

Situation

A large 250-room hotel in a major European city has just changed ownership. The new management wants to be sure that things are near perfect at the new hotel and for that reason a quality control check is carried out on all the rooms. Each time the room is prepared by the hotel cleaning staff, there is an inspection to see if everything conforms to management's expectation. A non-conforming room may mean that the telephone is not properly arranged, there are marks on the windows, there is no extra toilet paper, there is no shower soap, there is dust under the bed, the curtains are not arranged properly, there are no flowers in the room, etc. The quality control check was carried out over a 30-day consecutive period starting on Monday including weekends and for this first procedure all of the 250 rooms were inspected. The number of rooms non-conforming is given in Table HH-1 according to the day the inspection was performed. Day 1 is a Monday and Day 30 is a Tuesday.

Table HH-1

Day of inspection	No.rooms non-conforming	Day of inspection	No.rooms non-conforming	Day of inspection	No.rooms non-conforming
1	6	11	2	21	27
2	5	12	9	22	12
3	3	13	24	23	8
4	6	14	23	24	7
5	11	15	11	25	9
6	27	16	9	26	11
7	28	17	8	27	22
8	4	18	9	28	20
9	8	19	15	29	9
10	6	20	25	30	7

Required

1 Construct a p–chart for this operation, using a z value of 3. Use the average of the percentage non-conforming as the *CL*. What do you conclude from this control chart?
2 What might be an explanation for your conclusions from Question 1?
3 How would your response change if you were more rigorous in your analysis and used a z value of 1?
4 What are your comments on this quality control procedure as defined? Develop a table of random numbers for 10% of the 250 rooms. This is in part a response to this question.

Answers

1 The 3-sigma control chart is given in Figure HH-1. CL is 4.95% of rooms non-conforming; *LCL* is 0.83%; *UCL* is 9.06% not conforming. Six sample points are above the upper control point. Thus the conclusion is that the process is out of control based on requirements. A center, or target value, of 4.95% with an upper limit of 9% is high for a six-sigma operation.

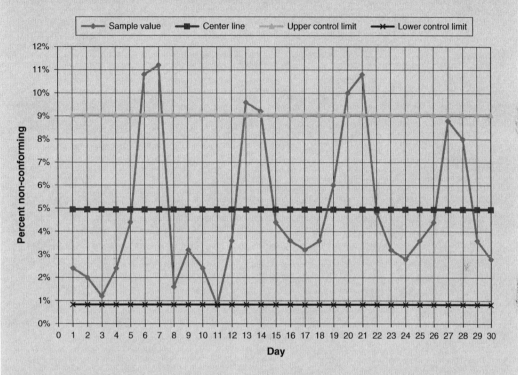

Figure HH-1 Hotel housekeeping: 3-sigma statistical process control chart.

2 The problem always occurs on a weekend. Could be insufficient help during this period; staff who are not properly trained; or no management personnel are available for proper supervision.

3 The 1-sigma limits are a more rigorous control. The 1-sigma control chart is given in Figure HH-2. *CL* is 4.95% of rooms non-conforming; *LCL* is 3.58%; *UCL* is 6.32% not conforming. Eight sample points are above the upper control point. Here again the conclusion is that the process is out of control based on requirements.

4 A quality control check of 250 rooms over a 30-day period (7,500 inspections) is time consuming and costly. It is a non-value added activity and to bring each hotel room up to conformity is part of the "hidden factor." When the situation has been stabilized a random check of the hotel rooms should be appropriate. For example, each day 10% of the hotel rooms are inspected (25 rooms) using random numbers generated for example with Microsoft Excel as given in Table HH-2.

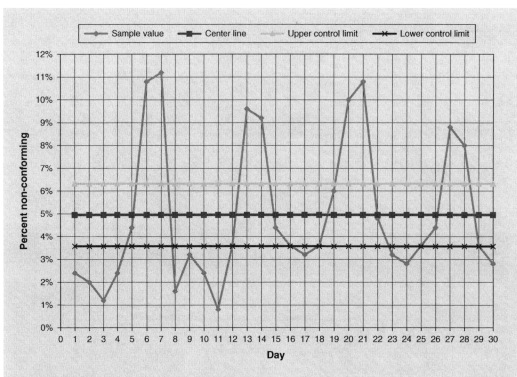

Figure HH-2 Hotel housekeeping: 1–sigma statistical process control chart.

Table HH-2

222	224	53	250	203
214	116	52	41	225
232	15	152	23	77
111	62	70	200	186
32	130	143	182	48

Note: these numbers are random so that they will change each time random data is generated.

5. Hotel occupancy

Objective

The objective of this case-exercise is to develop a p–chart using hotel room occupancy as the criteria.

Situation

A 380-room hotel in Shanghai, China is looking closely at revenue management for its facility. One of the variables considered is room occupancy. The new hotel manager, who has recently taken over operations, sets a target level of room occupancy of 80%, or 304 rooms filled each day. In order to establish the current situation, he asks a staff member to look at hotel occupancy for the last complete month. This data is given in Table HO-1.

Table HO-1

Day of month	No. of rooms filled	Day of month	No. of rooms filled	Day of month	No. of rooms filled
1	285	11	327	21	349
2	202	12	304	22	330
3	282	13	336	23	313
4	284	14	224	24	229
5	369	15	373	25	295
6	272	16	309	26	332
7	256	17	325	27	199
8	236	18	221	28	163
9	231	19	343	29	361
10	294	20	224	30	317

Required

1 Construct a p-chart for this operation, using a z value of 3. The p-value or *CL* for this chart is the proportion of rooms filled according to the sample data. The control limits are then determined from this value. What are your comments?

2 Construct a p-chart for this operation, using a z value of 1 or more rigorous limits. The p-value or *CL* for this chart is the proportion of rooms filled according to the sample data. The control limits are then determined from this value. What are your comments?

3 Construct a p-chart for this operation, using a z value of 3. The p-value or *CL* for this chart is now the target value of 80% of rooms filled. The control limits are then determined from this value of 80%. What are your comments?

4 Construct a p-chart for this operation, using a z value of 1 or more rigorous limits. The p-value or *CL* for this chart is now the target value of 80% of rooms filled. The control limits are then determined from this value of 80%. What are your comments?

5 How might the manager arrive at a process control chart according to the criteria in Question 4 such that the hotel occupancy is always above the target level of 80%?

6 What are your comments on the use of technology for closer control?

Answers

1 The control chart is in Figure HO-1. CL is 75.31% rooms filled: *UCL* is 98.93% rooms filled; *LCL* is 51.69% rooms filled. There is only one data point below the *LCL* specification level. There are none above the *UCL*. In any event this would not be a concern.

2 The control chart is in Figure HO-2. CL is 75.31% rooms filled: *UCL* is 83.18% rooms filled; *LCL* is 67.43% rooms filled. There are 10 data point below the *LCL* specification level and 11 data points above the *UCL*. In any event this would not be a concern.

3 The control chart is in Figure HO-3. CL is 80.00% rooms filled: *UCL* is 100.00%; *LCL* is 58.09%. There are 3 data point below the *LCL* specification level and no data points above the *UCL* (not possible).

4 The control chart is in Figure HO-4. CL is 80.00% rooms filled: *UCL* is 87.30% rooms filled; *LCL* is 72.70% rooms filled. There are 11 data point below the *LCL* specification level and 7 data points above the *UCL*. In any event this would not be a concern.

5 The sample data implies accepting the given situation. The target value is a management requirement and tighter control limits fit six-sigma management. Since in the sample

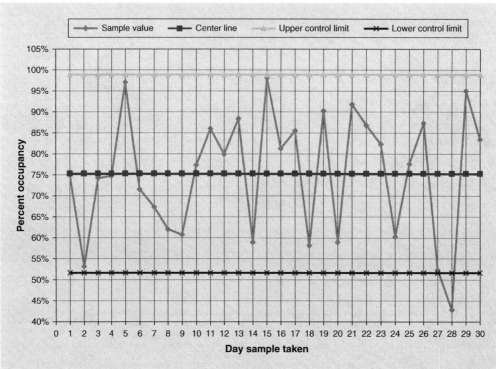

Figure HO-1 Hotel occupancy: 3-sigma p-control chart (Based on sample mean).

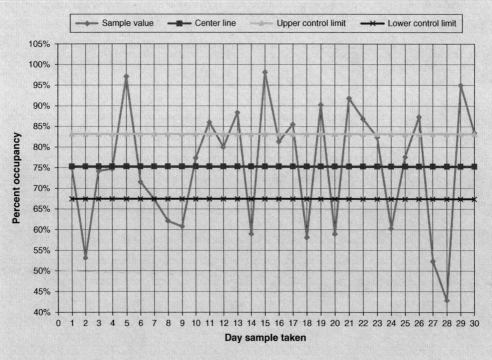

Figure HO-2 Hotel occupancy: 1-sigma p-control chart (Based on sample mean).

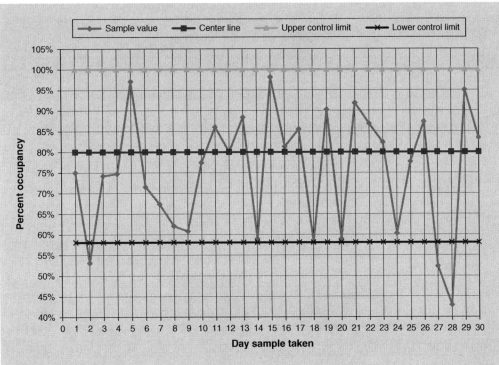

Figure HO-3 Hotel occupancy: 3-sigma p-control chart (Based on target mean).

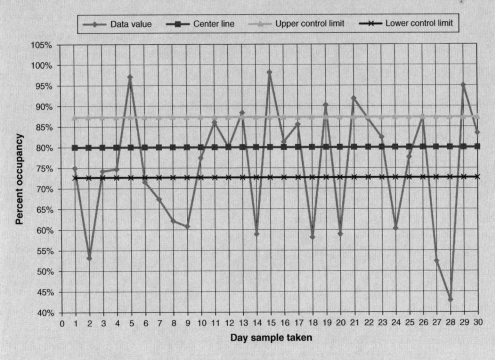

Figure HO-4 Hotel occupancy: 1-sigma p-control chart (Based on target mean).

there are 14 days when occupancy is greater than 80% the target is feasible. Management should look closer at revenue management variables such as room rates, promotions, advertising, etc.

6   The information system can be programmed to give a real-time read out of daily occupancy.

6. Internet response

Objective

This case-exercise demonstrates the development of an x-bar and range chart in statistical process control.

Situation

A consulting firm in France is looking into modifying its information systems network due to problems with the present arrangement. Before any action is taken, a control was made over a three-week period, Monday to Friday, for the time taken in seconds to access the first search screen of the Internet. During one-hour periods during the day consultants were asked to record the response time for connecting to the Internet. For each one-hour period data for 12 consultants picked at random was tabulated. This information is given in seconds, in Table IR-1, for the mean of the 12 response times, plus the range.

Table IR-1

Week 1			Week 2			Week 3		
Period	Mean	Range	Period	Mean	Range	Period	Mean	Range
08:00–09:00	8.25	5.20	08:00–09:00	2.25	14.50	08:00–09:00	5.20	23.60
09:00–10:00	10.65	12.20	09:00–10:00	5.86	26.20	09:00–10:00	4.60	45.20
10:00–11:00	9.45	22.90	10:00–11:00	13.25	15.20	10:00–11:00	3.50	12.60
11:00–12:00	9.86	18.00	11:00–12:00	10.56	32.00	11:00–12:00	3.60	23.60
12:00–13:00	12.24	14.60	12:00–13:00	13.46	14.60	12:00–13:00	4.50	25.40
13:00–14:00	20.25	23.70	13:00–14:00	20.45	41.20	13:00–14:00	13.60	25.60
14:00–15:00	32.40	21.00	14:00–15:00	30.25	12.60	14:00–15:00	9.50	12.60
15:00–16:00	40.20	25.00	15:00–16:00	35.21	23.50	15:00–16:00	12.60	18.60
16:00–17:00	45.70	26.30	16:00–17:00	40.56	18.50	16:00–17:00	13.50	21.60
17:00–18:00	55.23	30.20	17:00–18:00	47.50	19.60	17:00–18:00	20.60	23.60
18:00–19:00	56.20	23.60	18:00–19:00	48.60	21.30	18:00–19:00	25.60	5.60

Required

1   Construct the appropriate control charts to see if the response time of the process is in control according to the information provided.
2   What conclusions might you draw from your control charts?
3   What were your reasons for selecting the control chart you used for your interpretations?

Answers

1　The *x*-bar chart is in Figure IR-1: *CL* = 20.76 seconds; *UCL* = 26.40 seconds; *LCL* = 15.12 seconds.

　The range chart is in Figure IR-2: *UCL* = 21.21 seconds; *CL* = 36.42 seconds; *LCL* = 6.00 seconds.

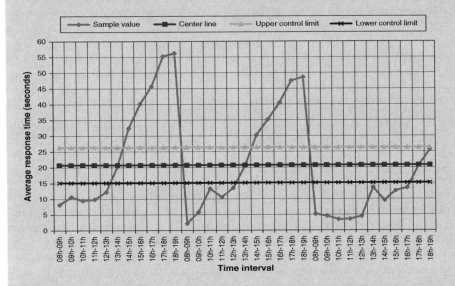

Figure IR-1　Internet response time: *x*-bar chart.

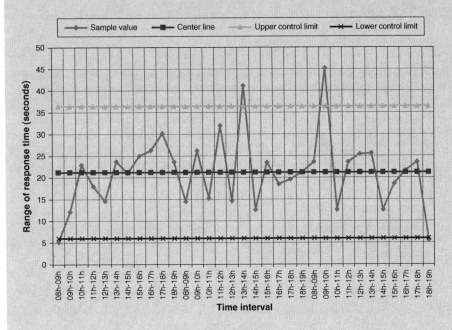

Figure IR-2　Internet response time: range chart.

2   From the *x*-bar chart between about 14:00 to 18:00 the situation is out of control. This may be explained by users in the USA starting to use the Web, starting slowly at first as the East Coast comes online and then more out of control as other States start waking up. It is also perhaps an indicator that the French users are plugging into more English language websites that at present heavily dominate the network.
3   The *x*-bar charts and range charts have to be used in conjunction since one chart alone may illustrate an in control situation, when in fact the process is out of control. This is the case here when the *R*-chart, except for a couple of points, is essentially in control.

7. Mustard filling machine

Objective

The objective of this case-exercise is to demonstrate the development of an *x*-bar and range chart and how conclusions may change depending on the established benchmark levels.

Situation

A subsidiary of Unilever has a production unit near Dijon, France that produces mustard that is distributed to restaurants and retail outlets in Europe. One of its production units fills 250 g bottles from an automatic filling machine. After the filling operation, a ticketing machine affixes a label, *Net weight 250 grams*. The company is concerned about the reliability of the filling machine and decides to carry out a statistical process control study. First, the filling line is shut down and adjusted according to specifications. The line is then put into operation and 20 random samples of size 25 units are taken during 8 hours of operation. For each of the samples taken, the average weight in grams is determined, and the maximum and minimum weight is recorded. This information is given in Table MFM-1.

Table MFM-1

Sample No.	Sample average	Maximum weight	Minimum weight	Sample No.	Sample average	Maximum weight	Minimum weight
1	253.00	262.00	232.00	11	246.10	278.00	237.00
2	248.00	261.00	232.00	12	250.00	265.00	235.00
3	254.20	269.00	242.00	13	247.00	265.20	241.00
4	256.00	265.00	248.00	14	248.20	263.20	242.10
5	253.20	262.40	247.00	15	251.30	270.20	247.20
6	246.20	261.40	235.00	16	251.30	265.20	248.10
7	250.00	254.00	230.00	17	253.60	257.20	248.80
8	247.50	265.00	229.00	18	252.70	259.20	248.20
9	254.30	275.00	242.00	19	254.30	259.00	247.90
10	252.00	258.00	241.00	20	255.20	260.10	251.00

Required

1   Develop an *x*-bar control chart and a range chart to analyze this filling operation using the sample data as your benchmark for control. Here the assumption is that your data is the best that you are able to obtain. How would you describe the results of your control charts?

2   Redo your *x*-bar and range chart using the net weight value as your benchmark *CL*.
3   What are the conclusions from these control charts?
4   How would the results for the *x*-bar chart change if you established a target value or *CL* of 250 g with specifications ± 1% of the target value rather than the sample data for your benchmark?
5   How would the results for the *x*-bar chart change if you established a target value or *CL* of 250 g with specifications ± 2% of the target value rather than the sample data for your benchmark?

Answers

1   The *x*-bar chart is in Figure MF-1. For this *x*-bar chart the *CL* is 251.21 g; *LCL* is 247.76 g; *UCL* is 254.65 g. A total of 6 data points are outside the limits of the *x*-bar chart.

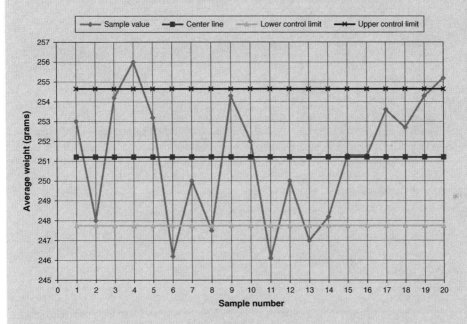

Figure MF-1   Mustard filling: *x*-bar chart using sample data.

The range chart is in Figure MF-2. For the range chart the *CL* is 22.54 g; *LCL* is 10.37 g; *UCL* is 34.71 g. A total of 4 data points are outside the limits of the range chart.

2   The *x*-bar chart is in Figure MF-3. For this *x*-bar chart the *CL* is 250.00 g; *LCL* is 246.55 g; *UCL* is 253.45 g. A total of 8 data points are outside the limits of the *x*-bar chart.

The range chart is the same as in Figure MF-2. There is no change. Again, a total of 4 data points are outside the limits of the range chart.

3   Filling process is out of control. Operation should be checked. Outside the upper limits represents an excessive cost to the firm. Those below lower limits may indicate that consumers are not obtaining what is advertised.

4   For the *x*-bar chart the *CL* is 250.00 g; *LCL* is 247.76 g; *UCL* is 254.65 g. There are 12 data points in total outside both of the control limits (9 above and 3 below).

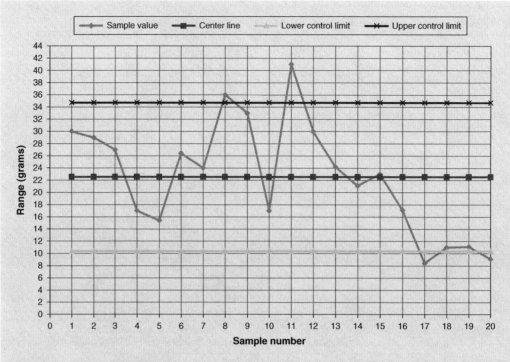

Figure MF-2 Mustard filling: range chart using sample data.

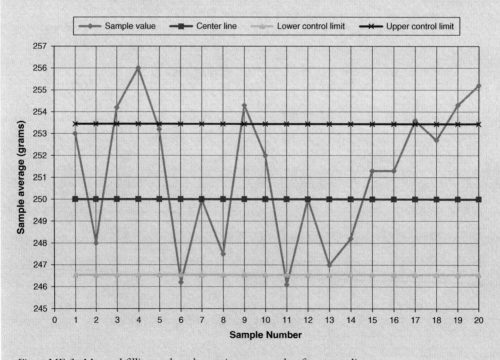

Figure MF-3 Mustard filling: *x*-bar chart using target value for center line.

5   For the *x*-bar chart the *CL* is 250.00 g; *LCL* is 245.00 g; *UCL* is 255.00 g. There are 2 data
     points in total outside both of the control limits 2 (2 above, 0 below).

8. Paper making

Objective

The objective of this case-exercise is to develop a c-control chart for statistical process control.

Situation

A company proposes to modify its paper-making process, using pulp from a foreign supplier,
rather than its normal domestic suppliers. It wants to be sure that it can use this raw material in its
paper-making operation and obtain the same quality paper as it does with domestic pulp. Prior
to running the foreign pulp, it makes paper from its normal suppliers for one week and analyaes
the blemishes on the paper. On a total of 32 m^2 of paper tested during one week it found
350 blemishes. The company then ran the paper-making machines with the foreign pulp and in
five bobbins of paper tested it obtained the average blemishes per m^2 on the paper according
to Table PM–1.

Table PM-1

Sample No.	Average blemishes per square m^2
1	30
2	19
3	6
4	25
5	3

Required

1   Develop a c-control chart for 3σ control limits using the domestic paper test. Show the
     information from the foreign paper on this control chart. What are your comments about
     the foreign paper?
2   Develop a c-control chart for more rigid 1σ control limits using the domestic paper test.
     Show the information from the foreign paper on this control chart. What are your
     comments about the foreign paper?
3   Do you have suggestions for the operation?

Answers

1   The statistical process control chart is shown in Figure PM-1. The $CL = 10.9375$ blemishes/
     m^2; $UCL = 20.8591$ blemishes/m^2; $LCL = 1.0159$ blemishes/m^2. There are two sample points
     above the *UCL*.

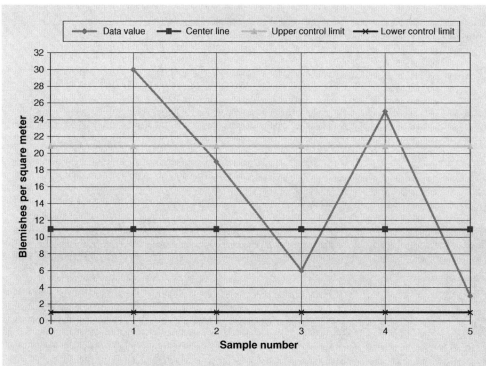

Figure PM-1 Paper making: statistical process control chart: 3–sigma limits.

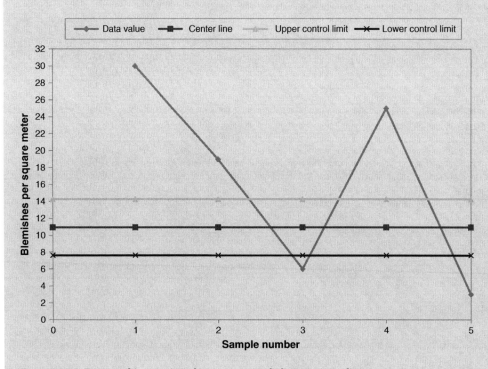

Figure PM-2 Paper making: statistical process control chart: 1–sigma limits.

2 The statistical process control chart is shown in Figure PM-2. The $CL = 10.9375$ blemishes/m^2; $UCL = 14.2447$ blemishes/m^2; $LCL = 7.6303$ blemishes/m^2. There are three sample points above the UCL.

3 The quality of paper seems erratic. Two averages are outside the limits, one is pretty close to the upper limit, and two are below the center line. The company should analysis the quality of the raw material, and perhaps re-evaluate the operating procedures such as drying time, tension etc. for paper making using this new raw material.

9. Steel bolts

Objective

The objective of this case-exercise is to illustrate the development of statistical process control charts for variable values.

Situation

Huntington and Sons is a small manufacturing firm which makes steel nuts, bolts, screws, and other metal fixation products for the automobile, aircraft, and household appliance industries. At the present time, business is good and the firm is working 24 hours per day in three eight-hour shifts (06h:00–14:00, 14:00–22:00, 22:00–06:00). In an effort to get a better control of the machining process for producing its threaded bolts, in accordance to specifications established by the client, Huntington carried out a statistical process control test over an entire 24-hour period covering the three shifts. The particular products in question were bolts of a nominal 10 cm diameter. Starting at 06:30, and then afterwards every hour, samples of bolts of size five were withdrawn at random from the production line and their diameter measured with a micrometer gauge. Thus, over the 24-hour shift, a total of 24 samples were taken (the first sample was at 06h:00 and the 24th sample at 05:00 the following day). The diameter of these bolts in cm is in the Table SB-1.

Table SB-1

Sample No.	Bolt No. 1	Bolt No. 2	Bolt No. 3	Bolt No. 4	Bolt No. 5	Time taken
1	10.0158	10.0244	10.0410	10.0024	10.0084	06h30
2	9.9661	10.0241	9.9667	10.0713	10.0087	07h30
3	9.9704	9.9707	9.9984	10.0370	9.9910	08h30
4	10.0436	10.0341	9.9967	9.9713	9.9533	09h30
5	10.0167	10.0321	10.0030	9.9350	9.9430	10h30
6	9.9933	9.9887	10.0141	10.0038	10.0170	11h30
7	9.9741	9.9750	9.9930	9.9516	10.0124	12h30
8	9.9973	9.9579	9.9513	10.0236	10.0267	13h30
9	10.0383	10.0563	10.0752	9.9980	10.0111	14h30
10	9.9757	10.0143	10.0369	10.0392	10.0518	15h30
11	9.9760	9.9777	10.0503	10.0601	9.9708	16h30
12	10.0223	9.9720	9.9946	10.0151	9.9482	17h30
13	10.0223	9.9439	9.9940	10.0140	9.9825	18h30
14	10.0111	10.0186	10.0335	10.0229	10.0769	19h30
15	9.9660	9.9923	10.0286	10.0355	10.0131	20h30

(Continued)

Table SB-1. (Continued)

Sample No.	Bolt No.1	Bolt No.2	Bolt No.3	Bolt No.4	Bolt No.5	Time taken
16	10.0515	9.9620	10.0638	9.9914	10.0069	21h30
17	10.0461	10.0571	10.0318	10.0507	10.0131	22h30
18	10.0332	9.9942	10.0347	10.0685	10.0080	23h30
19	10.0100	10.0060	10.0335	10.0651	10.0404	00h30
20	10.0361	10.0863	10.0602	10.0407	10.0120	01h30
21	10.0611	10.0866	10.0579	10.1122	10.0080	02h30
22	10.0100	10.0697	10.0863	10.0476	10.0536	03h30
23	9.9942	10.0347	10.0990	10.0263	10.0418	04h30
24	10.0008	10.0705	10.0482	10.0579	10.1122	05h30

Required

1 Develop the appropriate statistical process control charts for this operation. What conclusions would you draw from the charts you have developed?
2 Bearing in mind that these bolts are mated with nuts, what are your comments in accepting products with large specification ranges particularly if they are being used in the aircraft industry where safety is paramount?

Answers

1 The correct statistical process control chart is an x-bar chart, shown in Figure SB-1, and a range chart, shown in Figure SB-2. Here the benchmarks for the x-bar chart are; *CL* is

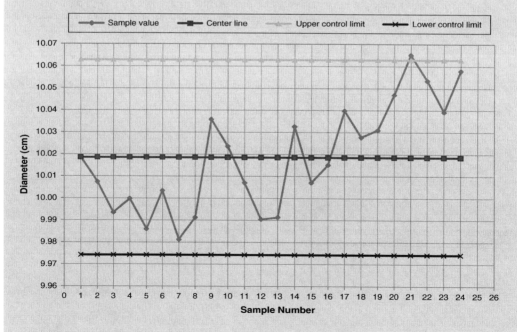

Figure SB-1 Steel bolts: x-bar chart.

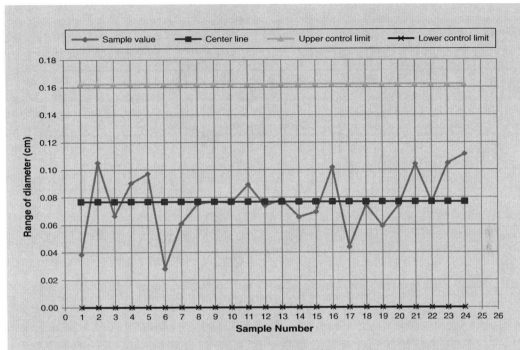

Figure SB-2 Steel bolts: range chart.

10.0185 cm; *LCL* is 9.9742 cm; *UCL* is 10.0628 cm. For the range chart the benchmarks are *CL* is 0.0768 cm; *LCL* is 0.00 cm; *UCL* is 0.1623 cm.

In the *x*-bar chart, except for Sample 21, all the data points are within the control limits. However, the night shift shows an upward trend from the mean value and so the night shift activity should be investigated. Perhaps there is insufficient supervision, operators are not motivated, or perhaps they are drinking or using drugs.

2   A bolt that is near the *LCL* and is subsequently mated with a nut near the *UCL* would have a sloppy fit.

10. Tablecloth material

Objective

The objective of this exercise is to demonstrate the development of a c–chart for statistical process control.

Situation

A large hotel under construction in Dubai is looking for a supplier of fabric for tablecloths. A prospective supplier from China provides samples of twelve different bobbins of fabric. A member of the purchasing department for the hotel inspects 10 m² of fabric from each of these bobbins and counts the number of blemishes. The criterion is that a blemish has a length of 1 mm. If a blemish is 2 mm in length then this would be considered two blemishes, a length of 3 mm, three blemishes, etc. The sample information is given in Table TCM-1, where the number of blemishes has been converted to the 1 mm criterion. Construct a c-control chart for this data where the x-axis is the number of the bobbin and the CL, benchmark, is the average number of 1 mm blemishes per square m^2 based on the analysis of the 12 bobbins. The standard deviation for a c-chart is the square root of the average blemishes per m^2.

Table TCM-1

Bobbin No.	Equivalent 1 mm blemishes on 10 m^2 fabric	Bobbin No.	Equivalent 1 mm blemishes on 10 m^2 fabric
1	51	7	42
2	33	8	14
3	30	9	62
4	40	10	11
5	69	11	15
6	18	12	15

Required

1   To the nearest two decimal places, what is the value of the CL in equivalent 1 mm length blemishes/m^2?

2   To the nearest four decimal places, with 3-sigma limits control limits, what are the benchmark control limits? How many of the bobbins indicate they have average blemishes above the UCL? Draw your 3-sigma control chart.

3   To the nearest four decimal places, with 1-sigma; limits control limits, what are the benchmark control limits? How many of the bobbins indicate they have average blemishes above the UCL? Draw your 1-sigma control chart. This is more rigorous control.

4   Assume now that you will only accept fabric from China with an average of 2 blemishes/m^2 with the appropriate UCL and LCL, then using 3-sigma limits, how many of your sample data points will be outside the UCL? Draw the 3-sigma limit control chart.

5   Assume now that you will only accept fabric from China with an average of 2 blemishes/m^2 with the appropriate UCL and LCL, then using 1-sigma control limits, how many of your sample data points will be outside the UCL? Draw the one-sig 1-sigma limit control chart.

6   How would you describe your results?

Answers

1   Value of CL is 3.3333 blemishes/m^2.

2   The LCL is 0 blemishes/m^2; UCL is 8.8106 blemishes/m^2. None of the bobbins indicate blemishes above the UCL. The control chart is given in Figure TCM-1.

3   The LCL is 0.5858 blemishes/m^2; UCL is 3.4142 blemishes/m^2. Five of the bobbins indicate blemishes above the UCL. The control chart is given in Figure TCM-2.

4   The CL is now 2.00 blemishes/m^2; LCL is 0 blemishes/m^2; UCL is 6.2426 blemishes/m^2. None of the bobbins indicate blemishes above the UCL. The control chart is given in Figure TCM-3.

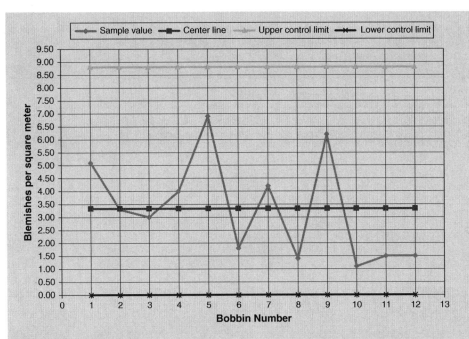

Figure TCM-1 Tablecloth material: 3–sigma statistical process control chart: center line = sample value.

5   The *CL* is 2.00 blemishes/m^2; *LCL* is 0.5858 blemishes/m^2; *UCL* is 3.4142 blemishes/m^2. Six of the bobbins indicate blemishes above the *UCL*. The control chart is given in

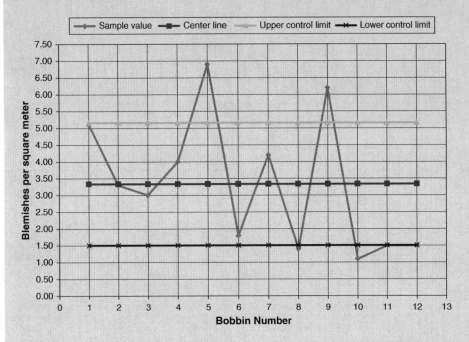

Figure TCM-2 Tablecloth material: 1–sigma statistical process control chart: center line = sample value.

6 At a Taguchi type limit of 1–sigma the process is not conforming in either situation, i.e. using the sample data or the defined criteria of 2 blemishes/m^2. In six–sigma management you would reject the supplier's offer. However, it will cost more with rigorous specifications.

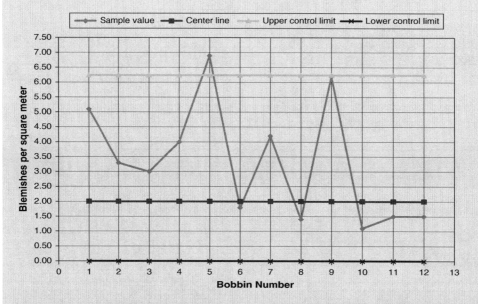

Figure TCM-3 Tablecloth material: 3-sigma control chart: center line = 2 blemishes/m^2.

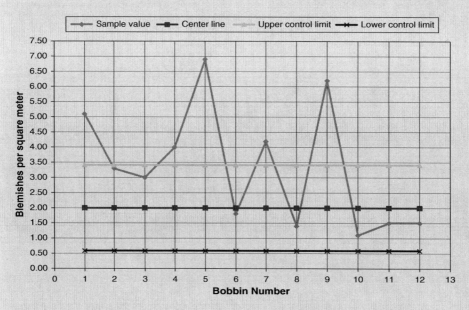

Figure TCM-4 Tablecloth material: 1-sigma control chart: center line = 2 blemishes/m^2.

11. Watch manufacture

Objective

This case-exercise illustrates the development of a p-chart for statistical process control.

Situation

The Picasso Company on the outside of Barcelona, Spain, assembles inexpensive watches for sale in the European market. Most of its watches are aimed at the market for children or young adolescents. The company imports the printed circuits from Singapore, and then assembles the watches with the frames made in its own factory. The watches are sold through distributors situated throughout Europe. Assembly is quite straightforward. It involves laying the printed circuit into the watch frame then making six solder connections. The battery is then inserted, and the back is snapped onto the frame. The company works seven days a week. The weekend employees are a mixture of temporary and permanent staff. Of late, Picasso has been receiving complaints from its distributors about defective watches being sold. As a result, Picasso decides to investigate its assembly operation. Picasso carries out sampling over a continuous 28-day period starting on a Monday. Each day, for the 28 days, it takes a random sample of 150 finished watches from that day's assembled lot, and tests them simply to see if they work. The criterion is that the watches operate, or they do not. The data for this sampling is given in Table WM-1.

Required

1 Develop the data and construct a 3-sigma p-chart for this operation.
2 Construct a new 3-sigma p-chart just using the Monday through Friday data (thus 20 pieces of data). What are your observations from new this p-chart?
3 Develop the data and construct a 1-sigma p-chart for this operation. What are your observations?

Table WM-1

Day	Quantity of defective watches	Day	Quantity of defective watches
1	3	15	1
2	2	16	4
3	5	17	3
4	4	18	2
5	1	19	9
6	12	20	13
7	16	21	19
8	3	22	5
9	0	23	3
10	2	24	1
11	0	25	0
12	7	26	5
13	11	27	13
14	14	28	18

4   Construct a new 1–sigma p–chart just using the Monday through Friday data (thus 20 pieces of data). What are you observations from new this p–chart?

5   Using this data what are your overall conclusions?

Answers

1   With the 3–sigma limits, from sampling for all days, the benchmark limits are $CL = 4.19\%$ defective; $LCL = 0.00\%$; $UCL = 9.10\%$ defective. Sundays are out of control and Saturday is higher than weekdays. The control chart is in Figure WM-1.

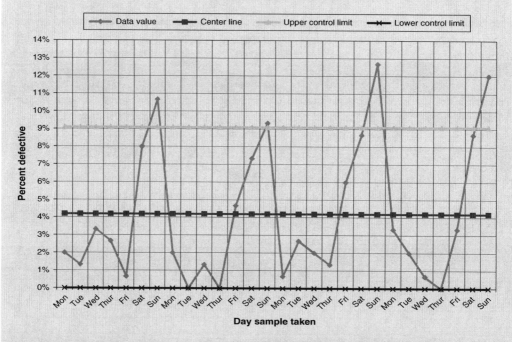

Figure WM-1 Watch manufacture: 3-sigma process control chart: all days.

2   With the 3-sigma limits just using the data for weekdays, the benchmark limits are $CL = 2.00\%$ defective; $LCL = 0.00\%$; $UCL = 5.43\%$ defective. There is only one piece of sample data that is above the UCL. The control chart is in Figure WM-2.

3   With the 1-sigma limits, from sampling for all days, the benchmark limits are $CL = 4.19\%$ defective; $LCL = 2.55\%$; $UCL = 5.83\%$ defective. With the 1-sigma limits, from sampling for all days, both Saturdays and Sundays and one Friday indicate sample information above the UCL. The control chart is in Figure WM-3.

4   With the 1-sigma limits, from sampling for just weekdays the benchmark limits are $CL = 2.00\%$ defective; $LCL = 0.86\%$; $UCL = 3.14\%$ defective. From sampling not including Saturday and Sunday using these 1-sigma limits the data shows five data points above the UCL. Thus on this criteria the process is out of control. The control chart is in Figure WM-4.

5   There seems to be a problem with the weekend, Saturday and Sunday, activity.

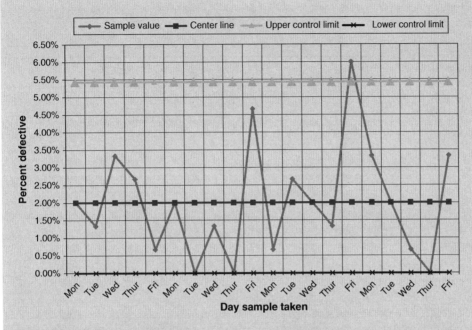

Figure WM-2 Watch manufacture: 3-sigma process control chart: weekdays.

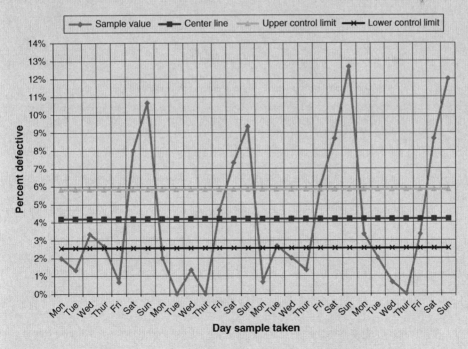

Figure WM-3 Watch manufacture: 1-sigma process control chart: all days.

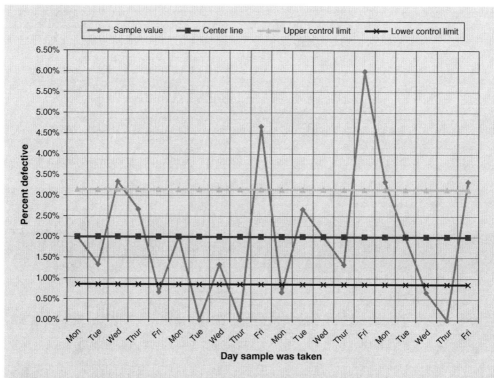

Figure WM-4 Watch manufacture: 1–sigma process control chart: weekdays.

12. Weight of food portions

Objective

The objective of this exercise is to demonstrate the development of an *x*-bar and range chart and how conclusions may change depending on the established benchmark levels.

Situation

A large restaurant in Spain is concerned about operating costs. One investigation it makes involves the weight of the portions of steak that it serves to customers. The benchmark value is 200 g. More than this increases the operating costs; less leads to customer dissatisfaction. It performs a statistical process control analysis over a period of three months by taking random samples of the raw steak portions that are cut before being cooked and served to clients. During this period 20 random samples are made and for each of these samples the weight of 25 client meat portions are noted. For each of the samples taken, the average weight is determined, and the maximum and minimum weight is recorded. This information is given in Table WFP-1.

Required

1   Develop an *x*-bar and a range chart to analyze this steak–cutting operation using the sample data as the benchmark for control. Here the assumption is that your data is the best that you are able to obtain. How would you describe the results of your developed control charts?

Table WFP-1

Sample No.	Sample average	Maximum weight	Minimum weight	Sample No.	Sample average	Maximum weight	Minimum weight
1	202	210	186	11	197	222	190
2	198	209	186	12	200	212	188
3	203	215	194	13	198	212	193
4	205	212	198	14	199	211	194
5	203	210	198	15	201	216	198
6	197	209	188	16	201	212	198
7	200	203	184	17	203	206	199
8	198	212	183	18	202	207	199
9	203	220	194	19	203	207	198
10	202	206	193	20	204	208	201

2 Redo your *x*-bar and range chart using the benchmark weight of 200 g as the *CL*.
3 What are the conclusions from these control charts?
4 How would the results for the *x*-bar chart change if you established a target value or *CL* of 200 g with specifications ± 0.5% of the target value rather than the sample data for your benchmark?
5 How would the results for the *x*-bar chart change if you established a target value or *CL* of 200 g with specifications ± 1.0% of the target value rather than the sample data for your benchmark?
6 How would the results for the *x*-bar chart change if you established a target value or *CL* of 200 g with specifications ± 2.0% of the target value rather than the sample data for your benchmark?

Answers

1 For the *x*-bar chart: *CL* is 200.95 g; *LCL* is 198.22 g; *UCL* is 203.68 g. In the *x*-bar chart there are two sample data points above the *UCL* and five below the *LCL*, for a total of 7. The *x*-bar control chart is in Figure WFP-1.

For the range chart: *CL* is 17.85 g; *LCL* is 8.21 g; *UCL* is 27.49 g. In the range cart there are two sample data points above the *UCL* and three below the *LCL*, for a total of 5. The *x*-bar control chart is in Figure WFP-2.

Here we have used the sample average as our *CL*. This is as experimental value and not a target value. Process is not-conforming given these criteria.

2 For the *x*-bar chart: *CL* is 200.00 g; *LCL* is 197.27 g; *UCL* is 202.73 g. In the *x*-bar chart there are seven sample data points above the *UCL* and two below the *LCL*, for a total of nine. There is no change in the range chart. Here we have used our benchmark value. The *x*-bar control chart is in Figure WFP-3

3 The benchmark value is your target and in six–sigma this is more meaningful as you have set the standard. With the empirical value of *x*-bar you are accepting the errors in your cutting operation. On average the food portion weight is 0.95 g above target. At 60 meals per day, 360 days per year this means an excess of 17,100 g or equivalent to 85.5 additional meals.

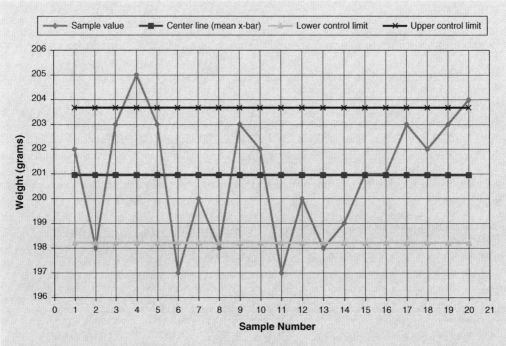

Figure WFP-1 x-bar chart: food portions (sample mean is center line).

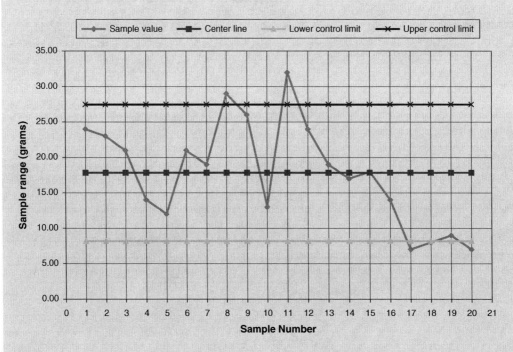

Figure WFP-2 Food portions: range chart – for mean of sample data.

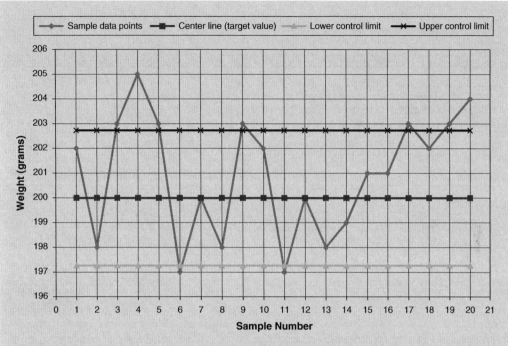

Figure WFP-3 x-bar chart: food portions (target value is center line).

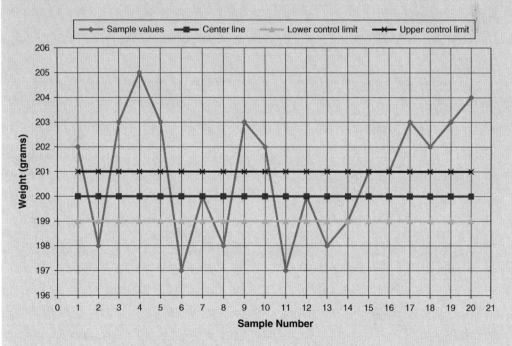

Figure WFP-4 x-bar chart: food portions: target value is center line: limits ± 0.5%.

4   For the *x*-bar chart: *CL* is 200.00 g; *LCL* is 199.00 g; *UCL* is 201.00 g. In the *x*-bar chart there are ten sample data points above the *UCL* and five below the *LCL*, for a total of 15. Only 5 sample data points conform. There is no change to the range chart. The *x*-bar control chart is in Figure WFP-4

5   For the *x*-bar chart: *CL* is 200.00 g; *LCL* is 198.00 g; *UCL* is 202.00 g. In the *x*-bar chart there are seven sample data points above the *UCL* and two below the *LCL*, for a total of nine. Only 11 sample data points conform. There is no change to the range chart.

6   For the *x*-bar chart: *CL* is 200.00 g; *LCL* is 196.00 g; *UCL* is 204.00 g. In the *x*-bar chart there is one sample data point above the *UCL* and none below the *LCL*, for a total of one. There are 19 sample data points that conform. We are not quite meeting six-sigma criteria in this case though cutting meat to a target weight is not easy. There is no change to the range chart.

13 Six-sigma management
Closing the loop to become world class

No	Case-exercise	Statistical concept	Application
1	Benchmarking in hospitality	Six-sigma measurement	Hospitality environment
2	Process capability in hospitality	Positioning for six-sigma	Operations, food

1. Benchmarking in hospitality

Objective

For an organization to operate according to a six-sigma philosophy it should make no more than 3.4 errors per million opportunities. This is the same as saying 3.4 errors as parts per million (ppm). The purpose of this case-exercise is to understand, in a hospitality environment, how to define a current process in terms of its sigma level and to benchmark what the characteristics of that process should be when operating at six-sigma. A process shift of 1.5σ is considered

Situation

A large hotel resort complex with restaurants, serving holiday groups, offers full pension of breakfast, lunch, and dinner. The owners of the complex wish to implement a more rigorous operation by applying six-sigma management concepts. As a first step it wants to establish its current base level operation and then to benchmark according to six-sigma criteria how it should be operating. Over a period of several months it carries out audits of six of the hotel activities, which are the restaurant service, bar service, housekeeping, room service, food and beverage management, and client invoicing. In these audits it records the number of errors incurred for a certain number of units analyzed. An error is anything that is not according to customer expectations, what has been promised to the customer, or an internal operating error. It could be a meal is late, room service is forgotten or is wrong, an incorrect or late bar order, a meal too cold, an invoice that is wrong or the clients name is spelled incorrectly, a room not clean, beauty products missing, inventory levels incorrect, late delivery of food, out of stock of certain wines, and other errors in stock management, etc. It is literally anything that does, or could, elicit a remark from the client of the type, "This is not what I expected, I am not happy," or a remark from management implying, "This is not how we should operate."

Required

1   In the restaurant there were 16 errors out of 240 meals prepared and served. To the nearest whole number, what is this in terms of errors in parts per million? What is the current sigma level expressed as z_{ST}? At a six-sigma level, giving your answer to the nearest thousand, the same number of 16 errors should correspond to how many meals being prepared and served? To the nearest whole number, for a six-sigma operation, how many errors would be tolerated for 4,000,000 meals prepared and served?

2   In the bar there were 5 errors out of 75 orders. To the nearest whole number, what is this in terms of errors in parts per million? What is the current sigma level expressed as z_{ST}? At a six-sigma level, giving your answer to the nearest thousand, the same number of 5 errors should correspond to how many bar orders? To the nearest whole number, for a six-sigma operation, how many errors would be tolerated for 300,000 bar orders?

3   In housekeeping there were 21 errors out of 802 rooms cleaned. To the nearest whole number, what is this in terms of errors in parts per million? What is the current sigma level expressed as z_{ST}? At a six-sigma level, giving your answer to the nearest thousand, the same number of 21 errors should correspond to how many rooms cleaned? To the nearest whole number, for a six-sigma operation, how many errors would be tolerated for 2,000,000 rooms cleaned?

4　In food and beverage management there were 11 errors out of 170 operations. To the nearest whole number, what is this in terms of errors in parts per million? What is the current sigma level expressed as z_{ST}? At a six-sigma level, giving your answer to the nearest thousand, the same number of 11 errors should correspond to how many operations in food and beverages? To the nearest whole number, for a six-sigma operation, how many errors would be tolerated for 2,500,000 operations in food and beverages?

5　In room service there were 15 errors out of 280 client orders. To the nearest whole number, what is this in terms of errors in parts per million? What is the current sigma level expressed as z_{ST}? At a six-sigma level, giving your answer to the nearest thousand, the same number of 15 errors should correspond to how many room service orders? To the nearest whole number, for a six-sigma operation, how many errors would be tolerated for 2,000,000 room service orders?

6　In invoicing there were 3 errors out of 45 client bills. To the nearest whole number, what is this in terms of errors in parts per million? What is the current sigma level expressed as z_{ST}? At a six-sigma level, giving your answer to the nearest thousand, the same number of 3 errors should correspond to how many invoices? To the nearest whole number, for a six-sigma operation, how many errors would be tolerated for 500,000 invoices?

Answers

1　No. of errors, 66,667 ppm; σ-level 3.00; units 4,706,000 (4,705,671); No. errors 14 (13.60).

2　No. of errors, 66,667 ppm; σ-level 3.00; units 1,471,000 (1,470,582); No. errors 1 (1.02).

3　No. of errors, 26,184 ppm; σ-level 3.44; units 6,175,000 (6,175,264); No. errors 7 (6.80).

4　No. of errors, 64,706 ppm; σ-level 3.02; units 3,235,000 (3,235,306); No. errors 9 (8.50).

5　No. of errors, 53,571 ppm; σ-level 3.11; units 4,412,000 (4,411,626); No. errors 7 (6.80).

6　No. of errors, 66,667 ppm; σ-level 3.00; units 882,000 (882,349) No. errors; 2 (1.70).

2. Process capability in hospitality

Objective

The objective of this case-exercise is to examine process capability in a restaurant environment.

Situation

A hotel and restaurant management school in Switzerland serves lunch to its students and staff in two shifts. The first shift is at 11:30 and the other is at 13:00. The students and staff have indicated they want to be served their lunch including first course, main meal, and dessert in 45 minutes but indicate that they will accept a tolerance level of ±10 min. These requirements are established as some of the staff have teaching commitments soon after lunch, some students wish to study, and some to play billiards as this reduces their stress level! The manager of the restaurant knows that with his current staff, which is principally students, the mean service time is 50 min with a standard deviation of 3 min. The restaurant manager is quite aware of the need to accommodate the staff and students, especially if the first shift is late, when the lost time spills over to the second shift.

Required

1 Determine the overall process capability ratio C_p and the process capability index C_{pk} under the current conditions. (Note that the mean value of the process *is not centered* on the target value of the client requirements. In this case we use a capability index C_{pk} when we consider one side of the distributions as the lower and upper values are different. There has not been a process shift in the usual sense but the process is already "shifted" based on the product target.)

2 What ar your comments on the results obtained in Question 1?

3 How would your answers to Question 1 change if the restaurant manager used an additional two students to reduce the mean service time down to 48 min with no change in the standard deviation of service time? What are your comments?

4 What needs to be the average service time such that the serving process is just capable of meeting the requirements of the staff and students, or the customers? Again the standard deviation of the process remains unchanged.

5 What needs to be the average service time to say that the restaurant service is able to meet six-sigma requirements? Again the standard deviation of the process remains unchanged.

Answers

1 Overall process capability ratio, C_p, is 1.11: Lower process capability index, C_{pk}, is 1.67: Upper process capability index, C_{pk}, is 0.56.

2 On the basis of just the process capability ratio it appears that the serving process is acceptable as it is greater than 1.00. However, when the upper specification of 0.56 is considered it is unacceptable. The process mean of 50 min is too close to the USL of 55 min, meaning there will be times when some staff and students are served late. (The lower specification is of no importance here as being served early is not a concern.)

3 Process capability ratio, C_p, is 1.11: Lower process capability index, C_{pk}, is 1.44: Upper process capability index, C_{pk}, is 0.78. There is an improvement in the lateness though 48 min is still too close to the USL of 55 min and some staff and students will be served late.

4 Mean total service time needs to be 46 min to give an upper process capability index, C_{pk}, of 1.00, meaning that the process is just able to avoid some staff and students being served late. This is a critical situation for six-sigma.

5 Mean total service time needs to be 41.5 min to be operating at six-sigma. Thus the upper process capability index, C_{pk}, is 1.50.

Appendix I   Glossary: key terminology and formulae in statistics

Expressions and formulae presented in **bold letters** in the textbook can be found in this section in alphabetical order. In this listing when there is another term in **bold letters** it means it is explained elsewhere in this *Appendix I*. At the end of this listing is an explanation of the symbols used in equations. Further, if you want to know the English equivalent of those that are Greek symbols, this is explained in *Appendix II, Mathematical relationships*.

A priori probability: making a probability estimate based on information already available.

Absolute frequency histogram: a vertical bar chart on an x-and y-axis. The x-axis is a numerical scale of the desired class width; y-axis gives the length of the bar proportional to the quantity of data in a class.

Absolute number: the numerical value of the value obtained; not a percentage.

Absolute value: the positive value of a number.

Acceptance range: in **statistical process control** that span of values that are acceptable.

Addition rule for mutually exclusive events: the sum of the individual probabilities.

Addition rule for non–mutually exclusive events: the sum of individual probabilities less the probability of the two events occurring together.

Alternative hypothesis: another value when the hypothesized value, or null hypothesis, is not valid at the given level of significance.

Arithmetic mean: sum of all the data values divided by the amount of data; same as the **average value.**

Assignable change: there is a reason that can be attributed to a change.

Asymmetrical data: numerical information that does not follow a **normal distribution**.

Average quantity weighted price index: given by $\dfrac{\sum P_n Q_a}{\sum P_0 Q_a} * 100$ where P_0 and P_n are prices in the base and current period respectively and Q_a is the average quantity consumed during the period under consideration. This index is also referred to as a **fixed weight aggregate price index.**

Average value: another term used for **arithmetic mean.**

Backup: an auxiliary unit in a system that can be used if the principal unit fails. In a **parallel arrangement** there are backup units.

Bar chart: type of histogram where the x- and y-axis have been reversed. It can be also called a Gantt chart after the American engineer Henry Gantt.

Bayes' theorem: gives the relationship for statistical probability under statistical dependence.

Bayesian decision making: this implies that if you have additional information, or based on the fact that *something has occurred*, certain probabilities may be revised to give *posterior* probabilities (*post* meaning afterwards).

Benchmark: the value of a piece of data to compare other data. It is the reference point.

Bernoulli process: where in each trial of an experiment there are only two possible outcomes, or **binomial**. The probability of any outcome remains fixed over time and the trials are statistically independent. The concept comes from Jacques (or Jacob) Bernoulli (1654–1705) a Swiss/French mathematician.

Bias: in sampling this is favoritism, purposely or unknowingly, present in sample data that gives lopsided, misleading, false, or unrepresentative results.

Bimodal: there are two values that occur most frequently in a dataset.

Binomial distribution: a table or graph showing all the possible outcomes of an experiment for a discrete distribution resulting from a **Bernoulli process**.

Binomial: there are only two possible outcomes of an event such as "yes or no"; "right or wrong"; "good or bad"; "works or does not work"; etc.

Bivariate data: involves two variables, x and y. Any data that is in graphical form is bivariate since a value on the x-axis has a corresponding value on the y-axis.

Body mass index: an indicator to describe a person's weight characteristics. It is calculated by weight in kilograms divided by the square of height in meters (kg/m^2).

Boundary limits of quartiles: these are denoted by Q_0, Q_1, Q_2, Q_3, and Q_4, where the indices indicate the quartile value going from the minimum value Q_0 to the maximum value Q_4.

Box and whisker plot: a visual display of quartiles. The box contains the middle 50% of the data. The first whisker on the left contains the first 25% of the data and the second whisker on the right contains the last 25%.

Box plot: an alternative name for the **box and whisker plot.**

Categorical data: information that includes a qualitative response according to a name, label, or category. For example: Asia, Europe, and the USA are geographical categories; men and women are gender categories; green, blue, yellow, and red are color categories. With categorical information there may be no quantitative data.

Categories: the **groups** into which data are organized.

Category: a distinct class into which information or entities belong.

Causal forecasting: when the change of the **dependent variable**, y, is caused or impacted by the amendment in value of the **independent variable**, x.

C-chart: a type of control chart in **statistical process control** where the number of errors from samples are considered in the analysis.

Central limit theorem: a rule in sampling that states as the size of the sample increases, there becomes a point when the **distribution of the sample means,** $\bar{x}$, can be approximated by the **normal distribution.** If the sample size taken is greater than 30, then the sample distribution of the means can be considered to follow a normal distribution even though the population itself is not normal.

Centered moving average: a term used in seasonal forecasting that is the linear average of four quarters around a given central time period. Moving forwards in time the average changes by eliminating the oldest quarter and adding the most recent.

Central tendency: how data clusters around a central measure such as the mean value.

Center line: in statistical process control is the middle, or target value for the operation.

Characteristic probability: that which is to be expected or that which is the most common in a statistical experiment.

Chebyshev's Theorem: states that no matter the shape of a data distribution at least 75% of the population values lie within ± 2 standard deviations of the mean and at least 89% lie within ± 3 standard deviations of the mean.

Chi-squared test: a method to determine if there is a dependency on some criteria between the proportions of more than two populations.

Chi-squared distribution: a continuous probability distribution to test a hypothesis associated with more than two populations. The chi-squared value is always positive.

Circular node: a symbol on a **decision tree** denoting that the outcome is subjected to the **states-of-nature** such that probabilities are involved.

Class range: the breadth or span of a given class.

Class width: an alternative description of the **class range**.

Class: a grouping into which data is arranged. The age groups, 20 to 29; 30 to 39; 40 to 49; 50 to 59 years are four classes that can be groupings used in market surveys.

Classical probability: the ratio of the number of favorable outcomes of an event divided by the total possible outcomes. Classical probability is also known as **marginal probability** or **simple probability**.

Closed-ended frequency distribution: where all data in the distribution is contained within the lower and upper limits.

Cluster sampling: where a population is divided into groups, or clusters, and each cluster is then sampled at random.

Coefficient of correlation, r: a measure of the strength of the relation between the **independent variable** x and the **dependent variable** y. The value of r can take any value between -1.00 and $+1.00$ and the sign is the same as the slope of the regression line for the bivariate values x and y.

Coefficient of determination, r^2: another measure of the strength of the relation between the variables x and y. The value of r^2 is always positive and less than the **coefficient of correlation**, r. If $r = 1.00$ then $r^2 = 1.00$.

Coefficient of variation: is the ratio of the **standard deviation** to the **mean** value, σ/μ for a given dataset.

Collectively exhaustive: gives all the possible outcomes of an experiment.

Combination: the arrangement of distinct items regardless of their order. The number of combinations is calculated by the expression, $^nC_x = \frac{n!}{x!(n-x)!}$.

Common cause variation: also known as **random variation** and occurs simply because of natural or unavoidable changes.

Conditional probability: the chance of an event occurring given that another event has already occurred.

Confidence interval: the range of the estimate at the prescribed confidence level.

Confidence level: the probability value for the estimate, such as 95%. Confidence level may also be referred to as the **level of confidence**.

Confidence limits of a forecast: given by, $\hat{y} \pm z.s_e$, when the sample size is greater than 30; and by $\hat{y} \pm t.s_e$, for sample sizes less than 30. The values of z and t are determined by the desired **level of confidence**.

Constant value: one that does not change with a variation in conditions. In an equation the beginning letters of the alphabet, a, b, c, d, e, f, etc., either lower or upper case, are typically used to represent a constant value.

Consumer price index: a measure of the change of prices from previous periods. It is used as a measure of inflation.

Consumer surveys: telephone, written, electronic, or verbal consumer responses concerning a given issue or product.

Contingency table: indicates data relationships when there are several categories present. It is also referred to as a **cross–classification table.**

Continuity correction factor: applied to a random variable in order to use the normal-binomial approximation.

Continuous data: has no distinct cut-off point and moves uninterrupted from one class to another. The volume of beer in a can may have a nominal value of 33 cl but the actual volume could be 32.3458 cl, 32.9584 or 33.5486 cl., etc. It is unlikely to be exactly 33.0000 cl.

Continuous probability distribution: a table or graph where the variable x can take any value within a defined range.

Continuous random variables: can take on any value within a defined range.

Control charts: used in statistical process control to ascertain if a system is operating according to specifications. The chart has an **upper control limit**, a **center line**, and a **lower control limit**.

Correlation: the measurement of the strength of the relationship between variables, often x and y.

Counting rules: mathematical relationships that describe the possible outcomes, or results, of various types of experiments, or trials.

Covariance: an application of the distribution of random variables often used to analyze the risk associated with financial investments.

Critical value: when used in hypothesis testing is that value outside of which the null hypothesis should be rejected. It is the **benchmark** value.

Cross–classification table: indicates data relationships when there are several categories present. It is also referred to as a **contingency table.**

Cumulative frequency distribution: a display of dataset values usually cumulated from the minimum to the maximum.

Curvilinear function: one that is not linear but curves according to the equation that describes its shape.

Cycle time: the elapsed time from start to finish of a given activity.

Data array: raw data that has been sorted in either ascending or descending order.

Data characteristics: the units of measurement that describe data such as the weight, length, volume, etc.

Data point: a single observation in a dataset.

Data: a collection of information.

Dataset: a collection of data either unsorted or sorted.

Deciles: fractiles that divide ordered data into ten equal parts.

Decision environment: the internal and external surroundings that impact judgment choices.

Decision making under certainty: making a choice in an environment when the parameters of the future outcome are known.

Decision making under risk: making a decision for an activity, or similar type, that has been done before such that probabilities can be applied.

Decision making under uncertainty: making a choice when there is no past data in the environment on which to base your judgment. "You have never done it, or you have never been there before."

Decision tree: a visual presentation of possible decisions to be made when probabilities can be considered.

Decision: a choice to be made from various possibilities.

Defects per opportunity: in **six-sigma management** refers to the number of imperfections that are deemed possible. Calculated by (No. of defects on a unit)/(No. of opportunities on a unit).

Defects per unit: in **six-sigma management** refers to the number of imperfections on a unit either product or service. Calculated by (No. of defects determined)/(total No. of units analyzed).

Degrees of freedom in a cross-classification table: (No. of Rows − 1) * (No. of Columns − 1).

Degrees of freedom in a Student's *t* distribution: $(n − 1)$ where n is the sample size.

Degrees of freedom: the choices available regarding taking certain actions.

Dependent variable: that value that is a function of or dependent on another variable. Graphically it is usually positioned on the y-axis.

Descriptive statistics: the analysis of sample data in order to describe the characteristics of that particular sample.

Deterministic: where outcomes or decisions made are based on data that are accepted and can be considered reliable or certain. For example, if sales for one month are $50,000 and costs are $40,000 then it is certain that net income is $10,000 ($50,000 − $40,000).

Deviation about the mean: in a dataset means that the relation $\sum(x − \bar{x})$ is zero where x is a value of any random variable and $\bar{x}$ is the mean of all the random variables in the dataset.

Discrete data: information that has a distinct cut-off point such as 10 students, 4 machines, or 144 computers. Discrete data comes from the counting process and comprises whole numbers or integer values.

Discrete random variables: those **integer values**, or **whole numbers**, that follow no particular pattern.

Dispersion: the spread or the variability in a dataset.

Distribution of the sample means: same as the **sampling distribution of the means.**

Empirical probability: the probability determined from an experiment or observation. **Relative frequency probability** is empirical.

Empirical rule for the normal distribution: states that no matter the value of the mean or the standard deviation, the area under the curve is always unity. As examples: 68.26% of all data falls within ± 1 standard deviations from the mean; 95.44% falls within ± 2 standard deviations from the mean; and 99.73% of all data falls within ± 3 standard deviations from the mean.

Estimate: in statistical analysis is that value judged to be equal to the population value.

Estimated standard deviation of the distribution of the difference between the sample means: given by $\hat{\sigma}_{\bar{x}_1 - \bar{x}_2} = \sqrt{\frac{\hat{\sigma}_1^2}{n_1} + \frac{\hat{\sigma}_2^2}{n_2}}$ where the indices refer to each sample.

Estimated standard error of the difference between two proportions: given by $\hat{\sigma}_{\bar{p}_1 - \bar{p}_2} = \sqrt{\frac{\bar{p}_1 \bar{q}_1}{n_1} + \frac{\bar{p}_2 \bar{q}_2}{n_2}}$ where the indices refer to each sample.

Estimated standard error of the proportion: given by $\hat{\sigma}_{\bar{p}} = \sqrt{\frac{\bar{p}(1 - \bar{p})}{n}}$ where $\bar{p}$ is the sample proportion and n is the sample size.

Estimating: forecasting or making a judgment about a future situation using entirely, or in part, quantitative information.

Estimator: that value of the statistic used to estimate the population number.

Event: the outcome of an activity or experiment that has been carried out.

Expected value of perfect information (EVPI): expected value under certainty (EVUC) less the expected value under risk (EVUR).

Expected value of the binomial distribution: given by $E(x)$ or the mean value, μ_x, is the product of the number of trials and the **characteristic probability,** or $\mu_x = E(x) = n.p.$

Expected value of the random variable: weighted average of the outcomes of an experiment according to the probabilities. It is the same as the **mean value** of the random variable and is given by the relationship, $\mu_x = \sum x.P(x) = E(x)$.

Expected value under certainty (EVUC): weighted outcome, based on the probabilities, of all the best possible outcomes.

Expected value under risk (EVUR): expected value calculated by taking into account the probabilities of the outcome.

Experiment: an activity, such as a sampling process, that produces an event. It will also give empirical results.

Exploratory data analysis (EDA): covers those techniques that give analysts a sense about data that is being examined. A **stem-and-leaf display** and a **box and whisker plot** are methods in EDA.

Exponential function: a special curvilinear relationship of the form $y = ae^{bx}$ where x and y are the **independent** and **dependent** variables respectively and a and b are constants.

Factorial rule: the calculation procedure for the arrangement of n different objects given by $n! = n(n-1)(n-2)(n-3)..(n-n)$ where $0! = 1$.

Finite population multiplier: a correction factor employed when data is considered to be finite. For a population of size N and a sample of size n, it is given by $\sqrt{\frac{N-n}{N-1}}$.

Finite population: a collection of data that has a stated, limited, or a small size. The number of playing cards (52) in a pack is considered finite.

First-time yield: the output before any **rework** has been considered.

Fixed weight aggregate price index: the same as the **average quantity weighted price index.**

Fractiles: divide data into specified fractions or portions. Quartiles and percentiles are specific fractiles.

Frequency distribution: groups data into defined classes. The distribution can be a table, polygon, or histogram. There is an **absolute** frequency distribution that gives the number of values or a **relative** frequency distribution that gives a proportion or percentage relative to the total amount of information in the dataset.

Frequency polygon: a line graph connecting the midpoints of the class ranges.

Functions in Excel: the built-in macros used for calculation. In this book it is principally the statistical functions that are employed but in Microsoft Excel there are financial, logic, database, and other functions.

Gaussian distribution: another name for the normal distribution after its German originator, Karl Friedrich Gauss (1777–1855).

Geometric mean: the average value when data is changing over time. It is calculated by the nth root of the growth rates for each year, where n is the number of years.

Graphs: visual displays of data such as line graphs, histograms, pie charts, or radar diagrams.

Greater than ogive: cumulative frequency distribution that illustrates data above certain values. It has a negative slope, where the y-values decrease from left to right.

Groups: the units or ranges into which data is organized.

Hidden activity: something performed that adds to cost but not to value.

Histogram: vertical bar chart showing data according to a named category or quantitative class range.

Historical data: information that has occurred, or has been collected in the past.

Horizontal bar chart: a **bar chart** in a horizontal form where the y-axis is the class and the x-axis is the proportion or amount of data in a given class.

Hypothesis testing: a way to test sample data and make on objective decision based on the results of the test using an appropriate significance level for the hypothesis test.

Hypothesis: a judgment about a situation, outcome, or population parameter based simply on an assumption or intuition with initially no concrete backup information or analysis.

Independent variable: that value upon which another value is dependent. Automobile accidents may be dependent on the amount of alcohol that is consumed. In a time series, sales revenues may be dependent on a function of time. Time is always independent. No matter what happens, if today is Monday, tomorrow will always be Tuesday. Graphically the independent variable is always plotted on the x-axis.

Index base value: the real value of a piece of data, which is then used as the reference point to determine the index number.

Index number: the ratio of a certain value to a base value usually multiplied by 100. When the base value equals 100 then the measured values are a percentage of the base. The index number may be called the **index value**.

Index value: an alternative name for the **index number**.

Inferential statistics: the analysis of sample data for the purpose of describing the characteristics of the population parameter from which that sample is taken.

Infinite population: a collection of data that has such a large size so that by removing or destroying some of the data elements it does not significantly impact the population that remains.

Insurance: A financial contract made for the purpose of attenuating a possible loss.

Integer values: whole numbers originating from the counting process.

Inter-quartile range: the difference between the values of the third and the first quartile in the dataset. It measures the spread of the middle half of an ordered dataset.

Interval estimate: a range for the estimate of the population parameter.

Interval measurement scale: where the difference between measurements is a meaningful quantity such as weight, height, length, volume, viscosity, etc.

Joint probability: the chance of two events occurring together or in succession.

Kurtosis: the numerical value describing the characteristic of the peak of a distribution curve.

Laspeyres weighted price index: this is given by $\dfrac{\sum P_n Q_0}{\sum P_0 Q_0} * 100$ where P_n is the price in the current period, P_0 is the price in the base period and Q_0 is the quantity consumed in the base period.

Law of averages: implies that the average value of an activity obtained in the long run will be close to the expected value, or the weighted outcome based on each probability of occurrence.

Least squares method: a calculation technique in **regression analysis** that determines the best straight line for a series of data that minimizes the error between the actual and forecast data from the regression calculation.

Leaves: the trailing or minority digits in a **stem-and-leaf display.**

Left tailed hypothesis test: used when asking the question, "Is there evidence that a value is less than?"

Left–skewed data: when the mean of a dataset is less than the median value, and the curve of the distribution tails off to the left side of the x-axis.

Leptokurtic: when the peak of a distribution is sharp. It is quantified by a small standard deviation and a small coefficient of variation.

Less than ogive: a cumulative frequency distribution that indicates the amount of data below certain limits. As a graph it has a positive slope such that the y-values increase from left to right.

Level of confidence: in estimating is $(1 - \alpha)$, where α is the proportion in the tails of the distribution, or that area outside of the confidence interval.

Line graph: shows bivariate data on an x- and y-axis. If time is included in the data this is always indicated on the x-axis.

Linear regression line: takes the generic form $\hat{y} = a + bx$. It is the equation of the best straight line for the data that minimizes the error between the data points on the regression line and the corresponding actual data from which the regression line is developed. The value $\hat{y}$ is the forecast for the dependent variable; a is a constant and the intercept on the y-axis; b is a constant and the slope of line; x is the value of the independent variable.

Lower control limit: the minimum boundary limit on a **control chart** below which a process would be considered **out of control**.

Margin of error: the $\pm$ range of the estimate from the true population value.

Marginal analysis: evaluates financial benefits when an activity increases by incremental amounts, usually the increase of one unit.

Marginal loss: that incremental loss realized when an additional unit is made or sold.

Marginal probability: the ratio of the number of favorable outcomes of an event divided by the total possible outcomes. Marginal probability is also known as **classical probability** or **simple probability**.

Marginal profit: that incremental profit realized when an additional unit is made or sold.

Maximax: choosing that outcome that indicates the highest of the best outcomes. Applies to an **optimistic** decision maker. A decision-approach for an **uncertain** environment.

Maximin: choosing that outcome that indicates the highest of the worst outcomes. Applies to a **pessimistic** decision maker. A decision-approach for an **uncertain** environment.

Mean proportion of successes: the average proportion of successes written $m_{\bar{p}}$ or simply p.

Mean value of random data: the weighted average of all the possible outcomes of the random variable.

Mean value: the same as the **arithmetic mean.**

Median: the middle value of an ordered set of data. It divides data into two halves. The second quartile and the fiftieth percentile are also the median value. The median is also the line that divides a road into two halves.

Mesokurtic: describes the curve of a distribution when it is intermediate between a sharp peak, **leptokurtic** and a relatively flat peak, or **playkurtic**.

Middle–of–the–road: choosing that outcome that considers the average outcomes of all the possible choices. Applies to a decision maker who has a **moderate** approach to outcomes. A decision-approach for an **uncertain** environment.

Mid–hinge: in quartiles is the average of the third and first quartile, $(Q_1 + Q_3)/2$.

Midpoint: in a class range is the maximum plus the minimum value divided by 2.

Midrange is the average of the smallest and the largest observations in a dataset.

Mid–spread: in quartiles, another term for the **inter-quartile range**.

Minimax regret: a decision method in an **uncertain** environment that considers the least painful, or the smallest regret of the outcome.

Mode: that value that occurs most frequently in a dataset.

Moderate: the characterization of an individual who is in between **optimistic** and **pessimistic** in decision making. A decision-approach for an **uncertain** environment.

Multiple regression: when the dependent variable y is a function of many independent variables. It can be represented by an equation of the general form, $y = a + b_1x_1 + b_2x_2 + b_3x_3 + \ldots + b_kx_k$ where a and b are constants and x are the values of the independent variables.

Mutually exclusive events: those that cannot occur together.

N in upper case is the size of the **population**. In lower case, **n**, is the size of the **sample**.

Node: a symbol used on a decision tree indicating a choice. There is a **square node** and a **circular node**.

Nominal measurement scale: when data is classified according to unique classifications or names.

Non-linear regression: when the dependent variable is represented by an equation where the power of some or all the independent variables is at least two. These powers of x are usually integer values.

Non-mutually exclusive events: those that can occur together.

Normal distribution density function: this is $f(x) = \frac{1}{\sqrt{2\pi}\sigma_x}e^{-(1/2)[(x-\mu_x)/\sigma_x]^2}$. It defines the shape of the normal distribution.

Normal distribution transformation relationship: this is $z = \frac{x-\mu_x}{\sigma_x}$ where z is the number of standard deviations; x is the value of the random variable; μ_x is the mean value of the dataset; and σ_x is the standard deviation of the dataset.

Normal distribution, or the **Gaussian distribution**: a continuous distribution of a random variable. It is symmetrical, has a single hump, and the mean, median and mode are equal. The tails of the distribution may not immediately cut the x-axis.

Normal-binomial approximation: can be is applied when $n.p \geq 5$ and $n.(1-p) \geq 5$. In this case substituting for the mean and the standard deviation of the binomial distribution in the normal distribution transformation relationship gives: $z = \frac{x-\mu}{\sigma} = \frac{x-n.p}{\sqrt{n.p.q}} = \frac{x-n.p}{\sqrt{n.p.(1-p)}}$. Here n is the sample size, p is the probability of "success", and q the probability of "failure."

Null hypothesis: The assumed or base value in an experiment or situation.

Numerical codes: those used to transpose qualitative or label data into numbers. This facilitates statistical analysis. For example if the time period is January, February, March, etc. These can be coded as 1, 2, 3, etc.

Odds: the chance of winning defined by the probability of losing to the chance of winning.

Ogive: a cumulative frequency distribution. A **less than ogive** gives data less than certain values; a **greater than ogive** shows data more than certain values. An ogive can illustrate **absolute** data, or **relative** data.

One-tailed hypothesis test: used when the interest is to know if something is less than or greater than a stipulated value. In the question, "Is there evidence that the value is greater than?" then this would be a **right-tailed hypothesis test.** Alternatively, if the question is: "Is there evidence that the value is less than?" then this would be a **left-tailed hypothesis test**.

One-arm-bandit: the slang or common term for slot machines found in gambling casinos. The game of chance where you put in a coin or chip, pull a lever and hope that you win a lucky combination! For this activity you only need "one-arm."

On-line statistical control: a computer-base system that measures automatically during an operation whether performance requirements are being met.

Optimistic: A decision maker who sees the world through rose-colored glasses. The person is always positive about the future. The **maximax** decision method applies to this individual.

Ordered dataset: values have been arranged in either increasing or decreasing order.

Ordinal measurement scale: a categorical scale in an ordered format. For example an ordinal scale may be, A • A– • B+ • B • B– • C+ • C • C– • D+ • D • D– • F, in the case of exam grades.

Outcome of a single type of event: given by k^n, where k is the number of possible events, and n is the number of trials.

Outcome of different types of events: given by $k_1 * k_2 * k_3 \ldots .k_n$, where $k_1, k_2 \ldots k_n$ are the number of possible events.

Outliers: numerical values either much higher or much lower than other values in a dataset. They can distort the value of the central tendency, such as the average, and the value of the dispersion such as the range or standard deviation.

Out-of-control: a situation where an activity, operation, or service is not functioning according to specifications as indicated in **statistical process control**.

P: in upper case or capitals is often the abbreviation used for probability written P(outcome). In lower case, p, is the probability of "success" or just probability.

Paired samples: those that are dependent or related, often in a before and after situation. Examples are the weight loss of individuals after a diet program or productivity improvement after a training program.

Parallel arrangement: a design system such that the components are connected allowing a choice to use one path or another. Whatever path is chosen the **system** continues to function. A parallel arrangement offers backup choices.

Parallel bar chart: similar to a parallel histogram but the x and y-axis have been reversed.

Parallel histogram: a vertical bar chart showing the data according to a category. Within a given category are sub-categories such as different periods. A parallel histogram is also referred to a **side-by-side histogram**.

Parameter: describes the characteristic of a population such as the weight, height, or length. It is usually considered a fixed value.

Pareto diagram: a combined histogram and line graph. The frequency of occurrence of data is indicated according to categories by a histogram and a line graph shows the cumulated data up to 100%. This diagram is a useful auditing tool.

Payoff: the financial outcome of an activity or project.

Percentage or p-chart: a type of control chart in **statistical process control** where the percentage or proportion of samples are considered in the analysis.

Percentiles: fractiles that divide ordered data into 100 equal parts. Sometimes referred to as centiles.

Permutation: combination of data arranged in a particular order. The number of ways, or permutations, of arranging x objects selected in order from a total of n objects is $^nP_x = \frac{n!}{(n-x)!}$.

Pessimistic: A characterization of a decision maker who is gloomy about a decision outcome. This is attributed to the individuals **utility**.

Pictogram: a diagram, picture, or icon that shows data in a relative form. It is not pragmatic for statistical analysis.

Pictograph: an alternative name for the **pictogram**.

Pie chart: a circular graphical presentation that shows the percentage of the data according to certain categories. The circle, or pie, contains 100% of the data.

Platykurtic: when the curve of a distribution has a flat peak. Numerically this is shown by a larger value of the coefficient of variation, σ/μ.

Point estimate: a single value used to estimate the population parameter.

Poisson distribution: describes events that occur during a given time interval and whose average value in that time period is known. The probability relationship is, $P(x) = \frac{\lambda^x e^{-\lambda}}{x!}$. Here λ is the average value, and x the value of the random variable.

Polynomial function: has the general form $y = a + bx + cx^2 + dx^3 + \ldots + kx^n$, where x is the independent variable, n is the order of the polynomial and a, b, c, d,...k are constants.

Population standard deviation: the square root of the **population variance**. It measures the spread or dispersion of population data.

Population variance: measures the dispersion of population data in squared terms. It is calculated by $\sigma^2 = \frac{\sum (x - \mu_x)^2}{N}$, where N is the amount of data, x is the particular data value, and μ_x is the mean value of the dataset.

Population: all of the elements under study and about which conclusions are required.

Portfolio risk: measures the exposure associated with financial investments.

Posterior probability: one that has been revised after additional information has been received.

Power of a hypothesis test: a measure of how well the test is performing.

Primary data: that which is collected directly from the source.

Probabilistic: where there is a degree of uncertainty, or probability of occurrence from the supplied data.

Probability: a quantitative measure, expressed as a decimal or percentage value, indicating the likelihood or chance of an event occurring. The value $[1 - P(x)]$ is the likelihood of the event not occurring.

Process capability index: a measurement indicating whether a process is able to meet required **specifications**.

Process capability: measuring if a process has to ability to meet product requirements.

Process variation: a fluctuation that occurs in an activity.

p-value: in hypothesis testing is the observed level of significance from the sample data or the minimum probably level to be tolerated in order to accept the null hypothesis of the mean or the proportion.

Q: in lower case, q, is the probability of failure or $(1 - p)$.

Quad-modal: there are four values in a dataset that occur most frequently.

Qualitative data: information that has no numerical response and cannot immediately be analyzed.

Quantitative data: information that has a numerical response.

Quartile deviation: one half of the inter-quartile range, or $(Q_3 - Q_1)/2$.

Quartiles: those three values, Q_1, Q_2, and Q_3 that divide ordered data into four equal parts.

Questionnaires: evaluations used to ascertain peoples' opinions of a subject, a product, service, etc.

Quota sampling: in market research is where the interviewer in the sampling experiment has a given quota or number of units to analyze.

Random sample: where each item of data in the sample has an equal chance of being selected.

Random variable: one that will have different values as a result of the outcome of a random experiment.

Random variation: also known as **common cause variation** that occurs simply because of natural or unavoidable changes.

Random: implies that any occurrence or value is possible.

Range chart: a type of control chart in **statistical process control** for variables that considers the sample ranges in the analysis.

Range: the numerical difference between the highest and lowest value in a dataset.

Ratio measurement scale: where the difference between measurements is based on starting from a base point to give a ratio. The **consumer price index** is usually presented on a ratio measurement scale.

Raw data: collected information that has not been organized.

Real value index: given by $RVI = \frac{\text{current value of commodity}}{\text{base value of commodity}} * \frac{\text{base indicator}}{\text{current indicator}} * 100$.

Regression analysis: a mathematical technique to develop an equation describing the relationship of variables. Useful in forecasting and estimating.

Relative frequency distribution: the percentage of data that appears in defined class ranges.

Relative frequency histogram: vertical bars that show the percentage of data that appears in defined class ranges.

Relative frequency probability: based on information or experiments that have previously occurred. Also known as **empirical probability**.

Relative price index: given by $I_P = \frac{P_n}{P_o} * 100$ where P_o is the price at the base period, and P_n is the price at another period.

Relative quantity index: given by $I_Q = \frac{Q_n}{Q_o} * 100$ where Q_o is the quantity at the base period and Q_n is the quantity at another period.

Relative regional index, RRI: compares the value of a parameter at one region to a selected base region. It is calculated by, $\frac{\text{Value at other region}}{\text{Value at base region}} * 100 = \frac{V_o}{V_b} * 100$.

Relative: in this textbook context is presenting data compared to the total amount collected. It can be expressed either as a percentage or fraction.

Reliability of a parallel system, R_S: given by one less the product of all the parallel components not working, or $R_S = 1 - (1 - R_1)(1 - R_2)(1 - R_3)(1 - R_4) \ldots (1 - R_n)$. The value of R_S is greater than the reliability of an individual component.

Reliability of a series system, R_S: the product of the reliability of all the components in the system, or $R_S = R_1 * R_2 * R_3 * R_4 * \ldots R_n$. The value of R_s is less than the reliability of a single component.

Reliability: the confidence in a product, process, service, work team, individual, etc. to operate under prescribed conditions without failure.

Replacement: taking an element from a population, noting its value, and then returning this element back into the population.

Representative sample: one that contains the relevant characteristics of the population and which occurs in the same proportion as in the population.

Research hypothesis: same as the alternative hypothesis and is a value that has been obtained from a sampling experiment.

Rework: repeating an activity. It adds to cost but not to value.

Right–tailed hypothesis test: used when asking the question, "Is there evidence that a value is greater than?"

Right–skewed data: when the mean of a dataset is greater than the median value, and the curve of the distribution tails off to the right side of the x-axis.

Risk: the possible loss, financial in business but can also be a personal or material loss, incurred when an activity or experiment is undertaken. Risk is associated with probability.

Rolled throughput yield: the yield taking into account successive throughputs. It is calculated using joint probability.

Rolling index number: the index value compared to a moving base value, often used to show the change of data each period.

Sample space: gives all the possible outcomes of an experiment. Used in a Venn diagram.

Sample standard deviation, s: the square root of the sample variation, $\sqrt{s^2}$. Measures the spread or the deviation of the sample values.

Sample variance, s^2: calculated by $S^2 = \frac{\Sigma(x-\bar{x})^2}{(n-1)}$, where n is the amount of data, x is the particular data value, and $\bar{x}$ is the mean value of the dataset. Measures dispersion in squared units.

Sample: the collection of a portion of population data elements.

Sampling distribution of the means: a distribution of all the means of samples withdrawn from a population.

Sampling distribution of the proportion: a probability distribution of all possible values of the sample proportion, $\bar{p}$.

Sampling error: the impression of a sample value in an experiment when used to estimate a population parameter.

Sampling from an infinite population: implies that even if the sample was not replaced, then the probability outcome for a subsequent sample would not significantly change.

Sampling with replacement: taking a sample from a population, and after analysis, the sample is returned to the population.

Sampling without replacement: taking a sample from a population, and after analysis not returning the sample to the population.

Sampling: the analytical procedure with the objective to estimate population parameters.

Scatter diagram: the presentation of time-series data in the form of dots on an x- and y-axis to illustrate the relationship between the x and y variables.

Score: a quantitative value given to a subjective response enabling statistical analysis.

Seasonal forecasting: making a forecast when in a time series the value of the dependent variable is a function of time but also varies often in a sinusoidal fashion according to the season.

Secondary data: published information collected by a third party.

Sensitivity: how data responds to changing circumstances. For example, "How is our market share sensitive to product price?"

Series arrangement: when in a system components are connected sequentially so it is necessary to pass through all the components in order that the system functions.

Shape of the sampling distribution of the means: is about normal if random samples of at least size 30 are taken from a non-normal population; if samples of at least 15 are withdrawn from a symmetrical distribution; or if samples of any size are taken from a **normal population**.

Side-by-side bar chart: where the data is shown as horizontal bars and within a given category there are sub-categories such as different time periods.

Side-by-side histogram: a vertical bar chart showing the data according to a category and within a given category there are sub-categories such as different time periods. A side-by-side histogram is also referred to as a **parallel histogram**.

Sigma limits: in **statistical process control,** are the values of z that set the benchmark to establish whether or not a process is in control.

Significance level: in **hypothesis testing** is how large, or important, is the difference before it can be concluded that a null hypothesis is invalid. It is denoted by α the area outside the distribution.

Significantly different: implies that in comparing data there is an important difference between values.

Significantly greater: a value is considerably greater than a hypothesized value.

Significantly less: a value is considerably smaller than a hypothesized value.

Simple probability: an alternative for **marginal** or **classical probability.**

Simple random sampling: each item in the population has an equal chance of being selected.

Six-sigma management: a rigorous form of management based on statistical measurements.

Skewed: data is not symmetrical.

Specifications: the benchmarks for an operation or activity in order to meet requirements.

Square node: an symbol in a **decision tree** to indicate that a decision maker can chose the outcome.

Stacked histogram: a presentation of data in vertical blocks according to categories. Within each category there are sub-categories. It is developed from a **cross–classification** or **contingency table**.

Standard deviation of a random variable: the square root of the variance or, $\sigma = \sqrt{\sum (x - \mu_x)^2 P(x)}$.

Standard deviation of the binomial distribution: the square root of the variance, or $\sigma = \sqrt{\sigma^2} = \sqrt{(n.p.q)}$.

Standard deviation of the distribution of the difference between sample means: given by $\sigma_{\bar{x}_1 - \bar{x}_2} = \sqrt{\frac{\sigma_1^2}{n_1} + \frac{\sigma_2^2}{n_2}}$.

Standard deviation of the Poisson distribution: the square root of the mean number of occurrences, or $\sigma = \sqrt{(\lambda)}$.

Standard deviation of the sampling distribution, $\sigma_{\bar{x}}$: related to the population standard deviation, σ_x, and sample size, n, from the **central limit theorem**, by the relationship, $\sigma_{\bar{x}} = \frac{\sigma_x}{\sqrt{n}}$.

Standard error of the difference between two means: given by $\sigma_{\bar{x}_1 - \bar{x}_2} = \sqrt{\frac{\sigma_1^2}{n_1} + \frac{\sigma_2^2}{n_2}}$.

Standard error of the difference between two proportions: given by $\sigma_{\bar{p}_1 - \bar{p}_2} = \sqrt{\frac{p_1 q_1}{n_1} + \frac{p_2 q_2}{n_2}}$.

Standard error of the estimate of the linear regression line: given by $S_e = \sqrt{\frac{\sum (y - \hat{y})^2}{n - 2}}$.

Standard error of the estimate: in forecasting is a measure of the variability of the actual data around the regression line.

Standard error of the proportion, $\sigma_{\bar{x}}$: given by $\sigma_{\bar{p}} = \sqrt{\frac{pq}{n}} = \sqrt{\frac{p(1-p)}{n}}$.

Standard error of the sample means: the error in a sampling experiment. It is the relationship $\sigma_{\bar{x}} = \frac{\sigma_x}{\sqrt{n}}$. Also known as the **standard error**.

Standard normal distribution: one which has a mean value of zero and a standard deviation of unity.

States–of–nature: implying that decisions are at the risk of the environment where probability outcomes are a consideration.

Statistic: describes the characteristic of a sample, taken from a population, such as the weight, volume, length, etc.

Statistical dependence: the condition when the outcome of one event impacts the outcome of another event.

Statistical independence: the condition when the outcome of one event has no bearing on the outcome of another event, such as in the tossing of a coin.

Statistical process control: using control charts to verify that a process is operating according to specifications.

Statistical quality control: using control charts to verify that a process is operating according to quality requirements.

Stem–and–leaf display: a frequency distribution where the data has a stem of principal values, and a leaf of minor values. In this display all data values are evident.

Stems: the principal data values in a **stem–and–leaf display.**

Strata: a homogeneous formation related to mineral extraction.

Stratified sampling: when the population is divided into homogeneous groups or strata and random sampling is made on the strata of interest.

Student's *t* distribution: a distribution of data from small sample sizes when the population standard deviation is unknown.

Subjective probability: based on the belief, emotion or "gut" feeling of the person making the judgment.

Symmetrical distribution: when one half of the distribution is a mirror image of the other half.

Symmetrical: in a **box and whisker plot** is when the distances from Q_0, to the median, Q_2, and the distance from Q_2 to Q_4, are the same; the distance from Q_0, to Q_1 equals the distance from Q_3 to Q_4 and the distance from Q_1 to Q_2 equals the distance from the Q_2 to Q_3, and the mean and the median value are equal.

System: the total of all components, pieces, or processes in an arrangement. Purchasing, transformation, and distribution are the processes of the supply chain system.

Systematic sampling: taking samples from a homogeneous population at a regular space, time or interval.

Target value: in **statistical process control** that value expected to be obtained.

Time-series deflation: a way to determine the real value in the change of a commodity using the **consumer price index**.

Time series: historical, past, or collected data that illustrates the progression of variables over time.

Transformation relationship: same as the **normal distribution transformation relationship**.

Tri-modal: when there are three values in a dataset that occur most frequently.

Two-tailed hypothesis test: used when asking the question, "Is there evidence of a difference?"

Type I error: occurs if the null hypothesis is rejected when in fact the null hypothesis is true.

Type II error: accepting a null hypothesis when in fact the null hypothesis is false.

Unbiased estimate: one that on average will be equal to the parameter that is being estimated.

Univariate data: composed of individual values that represent just one random variable, x.

Unreliability: when a system or component is unable to perform as specified.

Unweighted aggregate index: one that in the calculation each item in the index is given equal importance.

Upper control limit: the maximum boundary limit on a **control chart** above which a process would be considered **out of control**.

Utility: applies in statistics to the personality of a decision maker that influences their judgment.

Variable value: one that changes according to certain conditions. The ending letters of the alphabet, u, v, w, x, y, z, either upper or lower case, and in italics are typically used to denote variables.

Variance of a binomial distribution: the product of the number of trials n, the characteristic probability, p, of *success*, and the characteristic probability, q, of *failure*, or $\sigma^2 = n.p.q.$

Variance of a distribution of a discrete random variable: given by the expression $\sigma^2 = \sum (x - \mu_x)^2 P(x).$

Variance of a population: a measure of the dispersion of population data. In a distribution that is, or approximates, a normal distribution, it is the average of the squared distance between the mean and each item in the population.

Variance of a sample: a measure of the dispersion of sample data. In a distribution that is, or approximates, a normal distribution, it is the average of the squared distance between the mean and each item divided by the sample size, less 1.

Variation: similar to **variance** and indicates differences in performance.

Venn diagram: a visual representation of probability outcomes where the sample space gives all possible outcomes and a portion of the sample space represents an event or outcome.

Vertical histogram: a graphical presentation of vertical bars where the x-axis gives a defined class and the y-axis gives data according to the frequency of occurrence in a class.

Voice of the customer: the requirements demanded by clients.

Voice of the process: the requirements that are feasible by the process.

Weighted average: the mean value taking into account the importance, or weighting, of each value in the overall total. The total weightings must add up to 1 or 100%.

Weighted mean: an alternative for the **weighted average.**

Weighted price index: when different weights or importance are given to the items used to calculate the index.

What if?: the question asking, "What will be the outcome with different information?"

Whole numbers: those with no decimal or fractional components.

X: in lower case, x, a variable value and for bivariate data plotted as the independent variable on the horizontal axis in a graph.

X-bar chart: a type of control chart in **statistical process control** for variables that considers the average sample values in the analysis.

Y: in lower case, y, a variable value that is a function of other values, usually x. For bivariate data plotted on the vertical axis in a graph as a dependent value.

You have not been there before: implying that the action has not been experienced before, thus making it difficult to apply probabilities. This is the ground rule for **decision making under uncertainty**.

Z: in lower case, z, a unitless value indicating the number of standard deviations a quantity is from the mean. Can be positive or negative.

Symbols used in the equations

Symbol	Meaning
λ	Mean number of occurrences used in a Poisson distribution
μ	Mean value of population
n	Sample size in units
N	Population size in units
p	Probability of success, fraction or percentage
q	Probability of failure $= (1 - p)$, fraction or percentage
Q	Quartile value
r	Coefficient of correlation
r^2	Coefficient of determination
S	Standard deviation of sample
σ	Standard deviation of population
$\hat{\sigma}$	Estimate of the standard deviation of the population
S_e	Standard error of the regression line
t	Number of standard deviations in a Studen's t distribution
x	Value of the random variable. The independent variable in the regression line
$\bar{x}$	Average value of x
y	Value of the dependent variable
$\bar{y}$	Average value of y
$\hat{y}$	Value of the predicted value of the dependent variable
z	Number of standard deviations in a normal distribution

Subscripts or indices 0, 1, 2, 3 etc. indicate several data values in the same series.

Appendix II   Mathematical relationships

Subject matter

> Constants and variables • Equations • Integer and non-integer numbers • Arithmetic operating symbols and equation relationships • Sequence of arithmetic operations • Equivalence of algebraic expressions • Fractions • Decimals • The imperial, US, and metric measuring systems • Temperature • Conversion between fractions and decimals • Percentages • Rules for arithmetic calculations for non-linear relationships • Sigma, $\sum$ • Mean value • Addition of two variables • Difference of two variables • Constant multiplied by a variable • Constant summed n times • Summation of a random variable around the mean • Binary numbering system • Greek alphabet • Logarithms • Exchange rates.

Statistics involves numbers and the material in this textbook is based on many mathematical relationships, fundamental ideas, and conversion factors. Your memory of basic mathematical relationships may be rusty. The objective of this appendix is to give a detailed revision of arithmetic relationships, rules, and conversions.

Constants and variables

A constant is a value which does not change under any circumstances. The straight line distance from the centre of Trafalgar Square in London to the centre of the Eiffel Tower in Paris is constant. However, the driving time from these two points is a variable as it depends on road, traffic, and weather conditions. By convention, constants are represented algebraically by the beginning letters of the alphabet either in lower or upper case.

> Lower case **a, b, c, d, e**, ...
>
> Upper case **A, B, C, D, E** . . . . .

A variable is a number whose value can change according to various conditions. By convention variables are represented algebraically by the ending letters of the alphabet again either in lower or upper case, and in italics.

Lower case *u, v, w, x, y, z*

Upper case *U, V, W, X, Y, Z*

The variables denoted by the letters x and y are the most commonly encountered. Where two-dimensional graphs occur, x is the abscissa or horizontal axis, and y is the ordinate or vertical axis. This is bivariate data. In three-dimensional graphs, the letter z is used to denote the third axis. In textbooks, articles, and other documents you will see constants and variables written in either upper case or lower case. There seems to be no recognized rule; however, I prefer to use the lower case.

Equations

An equation is a relationship where the values on the left of the equals sign are equal to the values on the right of the equals sign. Values in any part of an equation can be variables or constants. The following is a linear equation, meaning that the power of the variables has the value of unity:

$$y = a + bx$$

This equation represents a straight line where the constant cutting the y-axis is equal to a and the slope of the curve is equal to b.

An equation might be non-linear, meaning that the power of any one of the variables has a value other than unity, as for example:

$$y = a + bx^3 + cx^2 + d$$

Integer and non-integer numbers

An integer is a whole number such as 1, 2, 5, 19, 25, etc. In statistics an integer is also known as a discrete number or discrete variable if the number can take any different value. Non-integer numbers are those that are not whole numbers such as the fractions ½, ¾ or 3½, 7¾ etc; or decimals such as 2.79, 0.56, and 0.75.

Arithmetic operating symbols and equation relationships

The following are arithmetic operating symbols and equation relationships.

+	addition
−	subtraction
±	plus or minus
=	equals
≠	not equal to
÷	divide
/	this means ratio but also divide. For example ¾ means the ratio of 3 to 4 but also 3 divided by 4
>	greater than
<	less than
≥	greater or equal to
≤	less than or equal to
≅	approximately equal to

For multiplication we have several possibilities to illustrate the operation. When we multiply two algebraic terms a and b together this can be shown as:

ab; a·b; a × b; or a * b

With numbers, and before we had computers, the multiplication or product of two values was written using the symbol × for multiplication:

$6 \times 4 = 24$

With Excel the symbol * is used as the multiplication sign and so the above relationship is written as,

$6 * 4 = 24$

It is for this reason that in this textbook, the symbol * is used for the multiplication sign rather than the historical × symbol.

Sequence of arithmetic operations

When we have expressions related by operating symbols the rule for calculation is to start first to calculate the terms in the **B**rackets, then **D**ivision and/or **M**ultiplication and finally **A**ddition and/or **S**ubtraction, (BDMAS) as shown in Table M-1.

If there are no brackets in the expression and only addition and subtraction operating symbols then you work from left to right. Table M-2 gives some illustrations.

Table M-1 Sequence of arithmetic operations

Symbol	Term	Evaluation sequence
B	Brackets	First
D	Division	Second
M	Multiplication	Second
A	Addition	last
S	Subtraction	last

Table M-2 Calculation procedures for addition and subtraction

Expression	Answer	Operation
$25 - 11 + 7$	21	Calculate from left to right
$9 * 6 - 4$	50	Multiplication before subtraction
$-22 * 4$	-88	A minus times a plus is a minus
$-12 * -6$	72	Minus times a minus equals a minus
$6 + 9 * 5 - 3$	48	Multiplication then addition and/subtraction
$7(9)$	63	A bracket is equivalent to a multiplication operation
$9(5 + 7)$	108	Addition in the bracket then the multiplication
$(7 - 4)(12 - 3) - 6$	12	Expression in brackets, multiplication, then subtraction
$20 * 3 \div 10 + 11$	17	Multiplication and divisions first then addition

Equivalence of algebraic expressions

Algebraic or numerical expressions can be written in various forms as Table M-3 illustrates.

Fractions

Fractions are units of measure expressed as one whole number divided by another whole number. The common fraction has the numerator on the top and the denominator on the bottom.

$$\text{Common fraction} = \frac{\text{Numerator}}{\text{Denominator}}$$

A common fraction is when the numerator is less than the denominator which means that the number is less than one as for example, $\frac{1}{7}$, $\frac{3}{4}$, and $\frac{5}{12}$. An improper fraction is when the numerator is greater than the denominator, which means that the number is greater than unity as for example $\frac{30}{7}$, $\frac{52}{9}$, and $\frac{19}{3}$. In this case these improper factions can be reduced to a whole number and proper fractions to give $4\frac{2}{7}$, $5\frac{7}{9}$ and $6\frac{1}{3}$. The rules for adding, subtracting, multiplying, and dividing fractions are given in Table M-4.

Decimals

A decimal number is a fraction whose denominator is any power of 10 so that it can be written using a decimal point, as for example:

$$7/10 = 0.70 \qquad 9/100 = 0.09 \qquad 7,051/1000 = 7.051$$

The imperial, US, and metric measuring system

The metric system, used in continental Europe, is based on the decimal system and changes in units of 10. Table M-5, Table M-6, Table M-7, and Table M-8, give the relationships for length, surface, volume, and weight.

Table M-3 Algebraic and numerical expressions

Arithmetic rule	Example
$a + b = b + a$	$6 + 7 = 7 + 6 = 13$
$a + (b + c) = a + b + c = (a + b) + c$	$9 + (7 + 3) = 9 + 7 + 3 = (9 + 7) + 3 = 19$
$a - b = -b + a$	$15 - 21 = -21 + 15 = -7$
$a * b = b * a$	$6 * 7 = 7 * 6 = 42$
$a * (b + c) = a * b + a * c$	$3 * (8 + 4) = 3 * 8 + 3 * 4 = 36$

Table M-4 Rules for treating fractions

$\dfrac{1}{a} + \dfrac{1}{b} = \dfrac{b + a}{ab}$	$\dfrac{1}{5} + \dfrac{1}{6} = \dfrac{6 + 5}{5 * 6} = \dfrac{11}{30}$
$\dfrac{a}{c} + \dfrac{b}{c} = \dfrac{a + b}{c}$	$\dfrac{4}{5} + \dfrac{16}{5} = \dfrac{4 + 16}{5} = \dfrac{20}{5} = 4$
$\dfrac{a}{c} - \dfrac{b}{c} = \dfrac{a - b}{c}$	$\dfrac{4}{7} - \dfrac{2}{7} = \dfrac{4 - 2}{7} = \dfrac{2}{7}$
$\dfrac{a}{c} * \dfrac{b}{d} = \dfrac{a * b}{c * d}$	$\dfrac{4}{5} * \dfrac{7}{6} = \dfrac{4 * 7}{5 * 6} = \dfrac{28}{30} = \dfrac{14}{15}$
$\dfrac{a}{c} \div \dfrac{b}{d} = \dfrac{a * d}{c * b}$	$\dfrac{4}{5} \div \dfrac{7}{6} = \dfrac{4 * 6}{5 * 7} = \dfrac{24}{35}$

Table M-5 Length or linear measure

Micrometre μm	millimetre mm	centimetre cm	decimetre dm	metre m	decametre dam	hectometre hm	kilometre km
10^9	1,000,000	100,000	10,000	1,000	100	10	1
10^8	100,000	10,000	1,000	100	10	1	0.1
10,000,000	10,000	1,000	100	10	1	0.1	0.01
1,000,000	1,000	100	10	1	0.1	0.01	0.001
100,000	100	10	1	0.100	0.01	0.001	0.0001
10,000	10	1	0.1	0.010	0.001	0.0001	0.00001
1,000	1	0.1	0.01	0.001	0.0001	0.00001	0.000001

Table M-6 Surface or area measure

Sq. micrometre μm²	Sq. millimetre mm²	Sq.Centimetre cm²	Sq. decimetre dm²	Sq. metre m²	Are a	Hectare Ha	Sq. km km²
10^{18}	10^{12}	10^{10}	10^8	1.000.000	10.000	100	1
10^{16}	10^{10}	10^8	1,000,000	10,000	100	1	0.01
10^{14}	10^8	1,000,000	10,000	100	1	0.01	0.0001
10^{12}	1,000,000	10,000	100	1	0.01	0.0001	0.000001
10^{10}	10,000	100	1	0.01	0.0001	0.000001	0.00000001
10^8	100	1	0.01	0.0001	0.000001	0.00000001	10^{-10}

Table M-7 Volume or capacity measure

micro-litre μl	milli-litre ml	centi-litre cl	deci-litre dl	litre l	deka-litre dal	hecto-litre hl	kilo-litre kl	cubic centi-metre cc or cm³	cubic deci-metre dm³	cubic metre m³
10^9	10^6	100,000	10,000	1,000	100	10	1	10^6	1,000	1
10^8	100,000	10,000	1000	100	10	1	0.1	100,000	100	0.1
10^7	10,000	1,000	100	10	1	0.1	0.01	10,000	10	0.01
10^6	1,000	100	10	1	0.1	0.01	0.001	1,000	1	0.001
100,000	100	10	1	0.1	0.01	0.001	0.0001	100	0.1	0.0001
10,000	10	1	0.1	0.01	0.001	0.0001	0.00001	10	0.01	0.00001
1,000	1	0.1	0.01	0.001	0.0001	0.00001	10^{-6}	1	0.001	10^{-6}
100	0.1	0.01	0.001	0.0001	0.00001	10^{-6}	10^{-7}	0.1	0.0001	10^{-7}
10	0.01	0.001	0.0001	0.00001	10^{-6}	10^{-7}	10^{-8}	0.01	0.00001	10^{-8}
1	0.001	0.0001	0.00001	10^{-6}	10^{-7}	10^{-8}	10^{-9}	0.001	10^{-6}	10^{-9}

The Imperial measuring system is used in the United States and partly in England, though there are efforts to change to the metric system. The Imperial numbering system is quirky with no apparent logic as compared to the metric system. Table M-9, Table M-10, Table M-11, Table M-12, and Table M-13 give approximate conversion tables for key measurements. The US system is not always the same as the Imperial measuring system.

Temperature

In Europe usually the Celsius system is used for recording temperature. Here the freezing point of water is measured at $0\,°C$ and the boiling point is $100\,°C$. In the USA, and sometimes in the UK the Fahrenheit system is used where the freezing point of water is given at $32\,°F$ and the boiling point as $212\,°F$.

Table M-8 Mass or weight measure

microgram μg	milligram mg	centigram cg	decigram dg	gram g	dekagram dag	hectogram hg	kilogram kg	metric ton t
10^{12}	10^9	10^8	10^7	1,000,000	100,000	10,000	1,000	1
10^9	1,000,000	100,000	10,000	1,000	100	10	1	0.001
10^8	100,000	10,000	1,000	100	10	1	0.1	0.0001
10^7	10,000	1,000	100	10	1	0.1	0.01	0.00001
1,000,000	1,000	100	10	1	0.1	0.01	0.001	0.000001
100,000	100	10	1	0.1	0.01	0.001	0.0001	10^{-7}
10,000	10	1	0.1	0.01	0.001	0.0001	0.00001	10^{-8}
1,000	1	0.1	0.01	0.001	0.0001	0.00001	0.000001	10^{-9}
100	0.1	0.01	0.001	0.0001	0.00001	0.000001	10^{-7}	10^{-10}
10	0.01	0.001	0.0001	0.00001	0.000001	10^{-7}	10^{-8}	10^{-11}
1	0.001	0.0001	0.00001	0.000001	10^{-7}	010^{-8}	10^{-9}	10^{-12}

Table M-9 Conversions for length or linear measurement

inches in	feet ft	yards yd	miles mi	millimetres mm	centimetres cm	metres m	kilometres km
1	0.0833	0.0278	1.5783×10^5	0.254	2.5400	0.0254	0.0000254
12	1	0.3333	0.0002	3.048	30.4800	0.3048	0.0003048
36	3	1	0.0006	9.144	91.4400	0.9144	0.0009144
63,360	5,280	1,760	1	16093.44	160,934.40	1,609.344	1.6093
0.3937	0.0328	0.0109	6.2137×10^6	0.1	1	0.01	0.00001
39.3701	3.2808	1.0936	0.0006	10	100	1	0.001
39,370.08	3,280.84	1,093.6133	0.6214	10,000	100,000	1,000	1

Table M-10 Conversions for surface or area measure

sq in in^2	sq ft ft^2	sq yd yd^2	sq mi mi^2	are a	sq cm cm^2	sq m m^2	are a	hectare ha	sq km km^2
1	0.0069	0.0008	0.3×10^{-9}	0.2×10^{-6}	6.4516	0.0006	0.7×10^{-5}	0.7×10^{-7}	0.7×10^{-9}
144	1	0.1111	0.4×10^{-7}	0.2×10^{-4}	929.03	0.0929	0.0009	0.9×10^{-5}	0.9×10^{-7}
1,296	9	1	0.3×10^{-6}	0.0002	8.4×10^3	0.8361	0.0084	0.8×10^{-4}	0.8×10^{-6}
4.01×10^9	0.28×10^8	3.1×10^6	1	640	0.3×10^{11}	0.26×10^7	0.26×10^5	259.00	2.59
0.63×10^7	43,560	4,840	0.0016	1	0.4×10^8	4.05×10^3	40.47	0.4047	0.004

Table M-11 Conversions for capacity or volume measure

USA gallon	USA quart	USA pint	Imperial gallon	Imperial quart	Imperial pint	cu inches in^3	Litres/1
1.0000	2.0000	4.0000	0.8326	1.6652	3.3304	231.0000	3.7850
0.5000	1.0000	2.0000	0.4163	0.8326	1.6652	115.5000	1.8925
0.2500	0.5000	1.0000	0.1041	0.2082	0.4163	28.8750	0.4731
1.2011	2.4021	4.8042	1.0000	2.0000	4.0000	277.4200	4.5460
0.6005	1.2011	2.4021	0.5000	1.0000	2.0000	138.7100	2.2730
0.3003	0.6005	1.2011	0.2500	0.5000	1.0000	69.3550	1.1365

Table M-12 Conversions for mass or weight measure

ounce oz	pound lb	short ton	long ton	grams g	kilograms kg	metric ton
1	0.0625	0.00003125	0.000027902	28.35	0.02835	0.00002835
16	1	0.0005	0.0004	453.60	0.4536	0.0005
32 000	2 000	1	0.8929	907 200	907.2	0.9072
35 840	2 240	1.1200	1	1 016 064	1016.064	1.0161

Table M-13 Conversion for pressure

psi	kg/cm^2	bar
1	0.07031	0.07031
14.2231	1.00	1.00

Table M-14 Conversion of fraction and decimals

Fraction	1/2	1/3	3/4	1/8	7/8	4 7/8	6 2/3	2 3/16
Decimal	0.50	0.33	0.75	0.125	0.875	4.875	6.6667	2.1875

The following is the relationship between the two scales.

$$C = \tfrac{5}{9}(F - 32) \qquad F = \tfrac{9}{5}C + 32$$

When $F = 212\,^\circ F$ then $C = \tfrac{5}{9}(212 - 32) = \tfrac{5}{9} * 180 = 100\,^\circ C$.

When $C = 100\,^\circ C$ then $F = \tfrac{9}{5}C + 32 = \tfrac{9}{5} * 100 + 32 = 212\,^\circ F$

When $C = 35\,^\circ C$ then $F = \tfrac{9}{5}C + 32 = \tfrac{9}{5} * 35 + 32 = 95\,^\circ F$

When $F = 104\,^\circ F$ then $C = \tfrac{5}{9}(104 - 32) = \tfrac{5}{9} * 72 = 40\,^\circ C$.

Conversion between fractions and decimals

To convert from a fraction to a decimal representation you divide the numerator, or upper value in the fraction, by the denominator or the lower value in the fraction. To convert from a two-decimal presentation to a fraction, you divide the numbers after the decimal point by 100 and simplify until both the numerator and denominator are the smallest possible integer values. That is, you find the lowest common denominator. To convert from a three-decimal presentation to a fraction you divide to numbers after the decimal point by 1,000, etc. Table M-14 gives some examples.

Percentages

The term percent means per 100 so that 50 percent, usually written, 50%, is 50 per 100. In order to change from one to another, see the next table.

A fraction to a percentage multiply by 100	¾ = 75% (¾ * 100 = 75%)
A decimal to a percentage multiply by 100	0.6712 = 0.6712 *100 = 67.12%
A percentage to a fraction divide by 100	25% = 25/100 = ¼
A percentage to a decimal move the decimal place two places to the left	25.86% = 0.2586

Rules for arithmetic calculations for non-linear relationships

Table M-15 gives algebraic operations when the power of the variable, or the constant, is non-linear.

Sigma, $\sum$

Very often in statistics we need to determine the sum of a set of data and to indicate this we use the Greek letter sigma, $\sum$. If we have a set of n values of data of a variable x then the sum of these is written as:

$$\sum x = x_1 + x_2 + x_3 + x_4 + \ldots + x_n \qquad \text{M(i)}$$

where x_1, x_2, x_3 etc. are the individual values in the dataset. If we have a dataset consisting of the values 7, 3, 2, 11, 21, and 9 then the sum of these values is:

$$\sum x = 7 + 3 + 2 + 11 + 21 + 9 = 53$$

Mean value

The mean or average of a dataset is equal to the sum of the individual values, $\sum x$ divided by the number of observations, n:

$$\bar{x} = \frac{\sum x}{n} \qquad \text{M(ii)}$$

If x has values of 1, −5, 9, −2, 6, then $n = 5$ and

$$\bar{x} = \frac{(1 - 5 + 9 - 2 + 6)}{5} = \frac{9}{5} = 1.80$$

Table M-15 Algebraic operations involving powers

Arithmetic rule	Example
$x^a \cdot x^b = x^{(a+b)}$	$5^3 * 5^2 = 5^{(3+2)} = 5^5 = 3{,}125$
$(x^a)^b = x^{ab}$	$(5^3)^2 = 5^6 = 15{,}625$
$1/x^a = x^{-a}$	$1/2^2 = 2^{-2} = 0.25$
$x^a/x^b = x^{(a-b)}$	$5^3/5^2 = 5^{(3-2)} = 5^1 = 5$
$x^a/x^a = x^{(a-a)} = x^0 = 1$	$6^2/6^2 = 6^{(2-2)} = 6^0 = 1$
$\sqrt{x.y} = \sqrt{x}.\sqrt{y}$	$\sqrt{16 * 25} = \sqrt{16}.\sqrt{25} = 4 * 5 = 20$
$\sqrt{\frac{x}{y}} = \frac{\sqrt{x}}{\sqrt{y}}$	$\sqrt{\frac{64}{144}} = \frac{\sqrt{64}}{\sqrt{144}} = \frac{8}{12} = \frac{2}{3} = 0.6667$

Addition of two variables

The total of the addition of two variables is equal to the total of the individual sum of each variable:

$$\sum (x + y) = \sum x + \sum y \qquad\qquad\qquad \text{M(iii)}$$

Assume the five values of the dataset (x, y) are: (1, 6); (–5, –2); (9, –8); (–2, 5); (6, 4). Then:

$$\sum (x + y) = (1 + 6) + (-5 - 2) + (9 - 8) + (-2 + 5) + (6 + 4)$$

$$\sum (x + y) = 7 - 7 + 1 + 3 + 10 = 14$$

$$\sum x + \sum y = (1 - 5 + 9 - 2 + 6) + (6 - 2 - 8 + 5 + 4)$$

$$\sum x + \sum y = 9 + 5 = 14$$

Difference of two variables

The sum of the difference of two variables is equal to the sum of the individual differences of each variable:

$$\sum (x - y) = \sum x - \sum y \qquad\qquad\qquad \text{M(iv)}$$

If we have the following five values:

$$(1, 6), (-5, -2), (9, -8), (-2, 5), (6, 4), \text{ then:}$$

$$\sum (x - y) = (1 - 6) + (-5 + 2) + (9 + 8) + (-2 - 5) + (6 - 4)$$

$$\sum (x - y) = -5 - 3 + 17 - 7 + 2 = 4$$

$$\sum x - \sum y = (1 - 5 + 9 - 2 + 6) - (6 - 2 - 8 + 5 + 4)$$

$$\sum x - \sum y = 9 - 5 = 4$$

Constant multiplied by a variable

The sum of a constant times a variable equals to the constant times the sum of the variables:

$$\sum (kx) = k \sum x \qquad\qquad\qquad \text{M(v)}$$

If k = 5, and x has values of 1, −5, 9, − 2, 6:

$$\sum(kx) = 5 \times 1 - 5 \times 5 + 5 \times 9 - 5 \times 2 + 5 \times 6 = 45$$

$$k\sum x = 5(1 - 5 + 9 - 2 + 6) = 45$$

Constant summed *n* times

A constant summed *n* times is equal to the *n* times the constant:

$$\sum k = n \times k \qquad\qquad\qquad \textbf{M(vi)}$$

If *n* = 6, and k = 5:

$$\sum k = 5 + 5 + 5 + 5 + 5 + 5 = 30 = 5 \times 6$$

Summation of a random variable around the mean

Summation rules can be used to demonstrate that the summation of a random variable around a mean is equal to zero. Or:

$$\sum(x - \bar{x}) = 0 \qquad\qquad\qquad \textbf{M(vii)}$$

From equation M(iv) equation M(vii) becomes:

$$\sum(x - \bar{x}) = \sum x - \sum \bar{x} = 0 \qquad\qquad\qquad \textbf{M(viii)}$$

For any fixed set of data, $\bar{x}$, is a constant and thus from equation M(vi):

$$\sum \bar{x} = n\bar{x} \qquad\qquad\qquad \textbf{M(ix)}$$

Thus from equation M(viii) we have:

$$\sum(x - \bar{x}) = \sum x - \sum \bar{x} = \sum x - n\bar{x} = 0 \qquad\qquad\qquad \textbf{M(x)}$$

From equation M(ii):

$$\bar{x} = \frac{\sum x}{n} \qquad\qquad\qquad \textbf{M(ii)}$$

or:

$$n\bar{x} = \sum x \qquad\qquad\qquad \textbf{M(xi)}$$

Thus substituting in equation M(x) we have:

$$\sum (x - \bar{x}) = \sum x - \sum \bar{x} = \sum x - n\bar{x} = \sum x - \sum x = 0 \qquad \text{M(xii)}$$

That is, the sum of the random variable about the mean is equal to zero.

Binary numbering system

In the textbook we introduced the binomial distribution. Related to this is the binary numbering system or binary code which is a system of arithmetic based on two digits, zero and one. The on/off system of most electrically appliances is based on the binary system where 0 = off and 1 = on.

The arithmetic of computers is based on the binary code. A digit, either the 0 or the 1 in the binary code, is called the bit, or binary digit. When a binary digit is moved one space to the left, and a zero is placed after it, the resulting number is **twice** the original number. In the binary code, the actual value of 1 depends on the position of the 1 in a binary number, reading from **right to left**. A digit doubles its value each time it moves one place further to the left as shown in Table M-16.

Table M-17 gives the equivalent between binary numbers and decimal numbers from 1 to 10. Note that we are reading from the right so that the first position is at the extreme right, the second position to the immediate left of the first position, etc.

Greek alphabet

In statistics, letters of the Greek alphabet are sometimes used as abbreviations to denote various terms. Table M-18 gives the Greek letters, both in upper and lower case and their English equivalent, and

Table M-16 Concept of binary numbering system

0001	value of 1 in first position is equal to	1
0010	value of 1 in second position is equal to	2
0100	value of 1 in third position is equal to	4
1000	value of 1 in fourth position is equal to	8

Table M-17 Relationship between decimal and binary numbers

Decimal	Binary	Explanation
0	0000	
1	0001	Value of 1 in the first position = 1
2	0010	Value of 1 in the second position = 2
3	0011	Value of 1 in the first position + value of 1 in the second position = 3
4	0100	Value of 1 in the third position = 4
5	0101	Value of 1 in the third position + value of 1 in the first position = 5
6	0110	Value of 1 in third position + value of 1 in second position = 6
7	0111	Value of 1 in the third position + value of 1 in second position + value of 1 in first position = 7
8	1000	Value of 1 in the fourth position = 8
9	1001	Value of 1 in fourth position + value of 1 in first position = 9
10	1010	Value of 1 in fourth position + value of 1 in second position = 10

Table M-18 Greek alphabet and English equivalent

Upper case	Lower case	Name	English Equivalent	Use
A	α	Alpha	a	hypothesis testing, exponential smoothing
B	β	Beta	b	hypothesis testing
Γ	γ	Gamma	g	
Δ	δ	Delta	d	Calculus as the derivative
E	ε	Epsilon	e	
Z	ζ	Zeta	z	
H	η	Eta	h	
Θ	υ	Theta	th (q)	angle of a triangle
I	ι	Iota	i	
K	κ	Kappa	k	
Λ	λ	Lambda	l	Poisson distribution, Queuing theory
M	μ	Mu	m	Mean value
N	ν	Nu	n	
Ξ	ξ	Xi	x	
O	o	Omicron	o	
Π	π	Pi	p	Circle constant
Γ	γ	Rho	r, rh	
Σ	σ	Sigma	s	total (upper case), standard deviation (lower case)
T	τ	Tau	t	
U	υ	Upsilon	u	
Φ	φ, φ	Phi	ph (f, j)	
X	χ	Chi	ch (c)	chi-squared test in hypothesis testing
Ψ	ψ	Psi	ps (y)	
Ω	ω	Omega	o (w)	Electrical resistance (upper case)

some of the areas where they are often used. To write these Greek letters in Microsoft Word you write the English letter and then change the font to *symbol*.

Logarithms

Up until about the 1970s logarithms were an important tool in mathematical calculations. They were conceived by the Scot, John Napier (1550–1617) in the early seventeenth century. Later in the mid–eighteenth century the Swiss Leonhard Euler (1707–83) developed the relationship between the natural logarithm and the exponential function. These mathematical ideas were adopted by engineers, architects, navigators, astronomers, and scientists to simplify tedious computations. They led to the development of log tables and the slide rule.[1] However, with the introduction of calculators in the 1970s, mainframe computers by IBM in the 1980s, and now the personal computer, the extensive daily use of logarithms has declined. However, logarithms have application in many areas of business, engineering, and other sciences: the Richter scale for earthquakes; the pH for acidity; decibel for sound measurement; and the learning curve concept are based on the logarithmic scale.

The logarithm of a number is the exponent, or power, to which another value, the base of the logarithm, must be raised in order to give that number. For example:

$$\text{Log}_{10}(10,000) = 4$$

This is because $10^4 = 10 * 10 * 10 * 10 = 10,000$.

or:

$$\text{Log}_2(8) = 3$$

This is because $2^3 = 2 * 2 * 2 = 8$

The general relationship for the logarithm is:

$$y = \log_b(x) \qquad\qquad\qquad \text{M(xiii)}$$

Here:

  b is the logarithm base

  x is a given variable

  y is the logarithmic value according to a given base

This can be written in the exponent form as:

$$x = b^y \qquad\qquad\qquad \text{M(xiv)}$$

A logarithm to the base 10, or $\log_{10}$ is known as the common logarithm and has application in science and engineering; the binary logarithm with a base = 2, or $\log_2$ is used in computer science and information technology. The natural logarithm has a base e ≈ 2.1783, or $\log_e$ often written ln and is used in pure mathematics and calculus. Logarithms of any base can be obtained using the function menu in Excel.

A logarithmic scale reduces large numbers to smaller values that are perhaps more manageable and often better conceptualized. Consider the Table M–19 that gives the log, to the base 10, of a value *x*.

Exchange rates

The Table M-20 gives exchange rates for various countries.[2] This is just used for illustration. Exchange rates change frequently.

Table M-19

Number	x	$y = \log_{10}(x)$
Ten	10	1
one hundred	100	2
One thousand	1,000	3
Ten thousand	10,000	4
One hundred thousand	100,000	5
One million	1,000,000	6
Ten million	10,000,000	7
One hundred million	100,000,000	8
One billion	1,000,000,000	9
Ten billion	10,000,000,000	10

Table M-20

Americas	Currency	Per $US
Argentina	Peso	9.0725
Brazil	Real	3.1129
Canada	Dollar	1.2359
Chile	Peso	631.3000
Colombia	Peso	2,555.2400
Ecuador	$US	1.0000
Mexico	Peso	15.5140
Peru	Sol	3.1679
Uruguay	Peso	26.9100
USA	Dollar	1.0000
Venezuela	Bolivar	6.3000

Europe	Currency	Per $US
Bulgaria	Lev	1.7470
Croatia	Kuna	6.7780
Czech Republic	Kuna	24.3070
Denmark	Krone	6.6634
Eurozone	Euro	0.8931
Hungary	Forint	279.2200
Iceland	Krona	131.9100
Norway	Krone	7.8177
Poland	Zloty	3.7263
Russia	Ruble	54.6530
Sweden	Krona	8.2465
Switzerland	Franc	0.9364
Turkey	Lira	2.6559
Ukraine	Hryvnia	21.1500
United Kingdom	Pound	0.6358

Asia-Pacific	Currency	Per $US
Australia	Dollar	1.2900
China	Yuan	6.2094
Hong Kong	Dollar	7.7522
India	Rupee	63.5137
Indonesia	Rupiah	13,305.0000
Japan	Yen	123.6400
Kazakhstan	Tenge	186.2100
Macau	Pataca	7.9842
Malaysia	Ringgit	3.7540
New Zealand	Dollar	1.4476
Pakistan	Rupee	101.7980
Philippines	Peso	45.0750
Singapore	Dollar	1.3433
South Korea	Won	1,111.9800
Sri Lanka	Rupee	133.8100
Taiwan	Dollar	30.9480
Thailand	Baht	33.7600

Middle East and Africa	Currency	Per $US
Bahrain	Dinar	0.3770
Egypt	Pound	7.6246
Israel	Shekel	3.7756
Kuwait	Dollar	0.3024
Oman	Sui rial	0.3850
Qatar	Rial	3.6410
Saudi Arabia	Riyal	3.7502
South Africa	Rand	12.1125
United Arab Emirates	Dirham	3.6728

Notes

1 The author started his math studies in the 1950s using log tables and then the slide rule as a chemical engineer through the 1970s.
2 Currency rates, London close June 25, 2015, *Wall Street Journal, European Edition*, June 26–28, p.23.

Appendix III   Microsoft Excel

Subject matter

Introduction • Table of statistical functions • Regression analysis

---

Introduction

Microsoft Excel is the working tool for this textbook either using a personal computer (PC), or a Macintosh computer (MAC). Note that in executing some functions and applications, keyboard commands are different between a PC and a MAC. The major use in this text is the development of graphs, particularly in **Chapter 2**, *Presenting and organizing data*, and the application of statistical functions. Some valuable functions are those related to the binomial distribution: **Chapter 4**, Discrete data and probability; normal distribution: **Chapter 5**, *Continuous distributions and probability*; those covering the Student's *t* distribution: **Chapter 7**, *Estimating population characteristics*; application of the the chi-squared test: **Chapter 9**, *Hypothesis testing from different populations*; and **Chapter 10**, *Forecasting from correlated data*. There is a useful on-line guide for all these applications.

Table of statistical functions

The main statistical functions used are in Table E-1. For good measure you have their French equivalent!

Regression analysis

The subject of **Chapter 10** is forecasting from correlated data. Here you need to use the **LINEST** to develop the appropriate statistics. If it is simple linear regression, then the procedure is to select a virgin block of cells of at least two columns and five lines and call up the function **LINEST**. The actual values for *x* and *y* are entered together with indicating "yes" or "1" for requiring the statistics. If you have a PC then pressing on control-shift-enter [[Ctrl – ↑ – ↵]] at the same time gives the framework indicated in Table E-2 that explains the meaning of the numerical value obtained. The framework is the same for a MAC.

English	French	Determines the…..
ABS	ABS	…absolute or positive value of a number.
AVERAGE	MOYENNE	…mean value of a dataset.
BINOMDIST (Excel 2007) or BINOM. DIST (Excel 2010 or later)	LOI.BINOMIALE	…binomial distribution given the random variable, x, and characteristic probability, p. If cumulative = 0, the individual value is determined. If cumulative = 1, the cumulative values are determined.
CEILING	ARRONDI.SUP	…rounded value of a number to the nearest integer value.
CHIDIST (Excel 2007) or CHISQ.DIST (Excel 2010 or later)	LOI.KHIDEUX	…area in the chi-squared distribution when the chi-squared value and degrees of freedom are entered.
CHIINV (Excel 2007) or CHISQ. INV (Excel 2010 or later)	KHIDEUX.INVERSE	…chi-squared value when the area under the distribution and the degrees of freedom are entered.
CHITEST (Excel 2007) or CHISQ.TEST (Excel 2010 or later)	TEST.KHIDEUX	…area in the chi-squared distribution when the matrix of the observed and expected frequency values are entered.
COMBIN	COMBIN	…number of combinations of x objects from a sample of n objects.
CONFIDENCE	INTERVALLE. CONFIANCE	…the confidence interval from the mean when the value α, standard deviation, and sample size are entered.
CORREL	COEFFICIENT; CORRELATION	…the coefficient of correlation for a bivariate dataset when the values of x and y are entered.
COUNT	NBVAL	…number of values in a given dataset.
COUNT.IF	NB.SI	…the number of times a selected value occurs in a given dataset.
EXPONDIST	LOI.EXPONENTIELLE	…exponential probability for a random variable, x, with a mean value λ given a cumulative value of 1.
FACT	FACT	…the factorial value $n!$ of a given number.
FLOOR	ARRONDI.INF	…rounded value of a given number down to the nearest integer value.
FORECAST (Excel 2007) or FORECAST.LINEAR (Excel 2016 or later)	PREVISION	…future value $\hat{y}$ of a dependent variable from known variables x and y in a linear relationship.
FREQUENCY	FREQUENCE	…how often values in occur in a given dataset according to defined class limits.
GEOMEAN	MOYENNE. GEOMETRIQUE	…the geometric mean growth rate from annual growth rate data. The percent geometric mean is the geometric mean growth rate less 1.
GOAL SEEK	VALEUR CIBLE	…required value according to given criteria. This function is in the tools menu.
IF	SI	…result according to specified logical conditions.
IRR	TRI	…the internal rate of return for a series of cash flows.
KURT	KURTOSIS	…the kurtosis value, of the hump of the distribution of a dataset.

(Continued)

English	French	Determines the.....
LINEST	DROITEREG	. . . the statistical parameters of a regression line.
MAX	MAX	. . . the highest value of a dataset.
MEDIAN	MEDIANE	. . . middle value of a dataset.
MIN	MIN	. . . the lowest value of a dataset.
MODE (Excel 2007) or MODE.MULT and MODE.SNGL (Excel 2010 or later)	MODE	. . . the value that occurs most frequently in a dataset. With the more recent version MODE.MULT returns an array of the most recently occurring values and MODE.SNGL returns the most common value.
NORMDIST (Excel 2007) or NORM.DIST (Excel 2010 or later)	LOI.NORMALE	. . . area under the normal distribution given the random variable, x, mean value, μ, standard deviation, σ, and set cumulative = 1.
NORMINV (Excel 2007) or NORM.INV (Excel 2010 or later)	LOI.NORMALE.INVERSE	. . . value of the random variable x given probability, p, mean value, μ, standard deviation, σ, and cumulative = 1.
NORMSDIST (Excel 2007) or NORM.S.DIST (Excel 2010 or later)	LOI.NORMALE. STANDARD	. . . probability, p, given the number of standard deviations z.
NORMSINV (Excel 2007) or NORM.S.DIST (Excel 2010 or later)	LOI.NORMALE.STAND-ARD.INVERSE	. . . number of standard deviations, z, given the value of the probability, p.
OFFSET	DECALER	. . . cell reference to another line or column according to the offset required.
PEARSON	PEARSON	. . . Pearson correlation, or the coefficient of correlation, r.
PERCENTILE (Excel 2007) or PERCENTILE.INC (Excel 2010 or later)	CENTILE	. . . percentile values of a given dataset.
PERMUT	PERMUTATION	. . . number of permutations of organizing x objects from a total of n objects
POISSON (Excel 2007) or POISSON.DIST (Excel 2010 or later)	LOI.POISSON	. . . Poisson distribution for a random variable, x, and mean value, λ. If cumulative = 0, individual values are determined; if cumulative = 1, cumulative values are determined.
POWER	PUISSANCE	. . . value of a number according to a specified power.
RAND	ALEA	. . . random number between 0 and 1.
RANDBETWEEN	ALEA.ENTRE.BORNES	. . . random number between specified values.
ROUND	ARRONDI	. . . specified number rounded to the nearest integer value.
RSQ	COEFFICIENT. DETERMINATION	. . . the coefficient of determination, r^2 or gives the square of the Pearson product moment correlation coefficient
SLOPE	PENTE	. . . the slope of the linear regression line.
SQRT	RACINE	. . . square root of a given value.
STDEV (Excel 2007) or STDEV.S (Excel 2010 or later)	ECARTYPE	. . . the standard deviation of a dataset on the basis of a sample
STDEVP (Excel 2007) or STDEV.P (Excel 2010 or later)	ECARTYPEP	. . . the standard deviation of a dataset on the basis of a population

Table E-1 (Continued)

English	French	Determines the.....
SUM	SOMME	... the total value of a defined dataset.
SUMPRODUCT	SOMMEPROD	... the sum of two columns or rows of data in a data matrix.
TDIST (Excel 2007) or T.DIST AND T.DIST.2T (Excel 2010 or later)	LOI.STUDENT	... probability of a random variable, x, given degrees of freedom, γ, and the number of tails. If the number of tails = 1 (use T.DIST in later versions), the area to the right is determined. If number of tails = 2 (use T.DIST.2T in later versions), the area in both tails is determined.
TINV (Excel 2007) or T.INV (Excel 2010 or later)	LOI.STUDENT.INVERSE	... the value of the Student's t given the probability or area outside the curve, p, and the degree of freedom, γ.
VAR (Excel 2007) or VAR.S (Excel 2010 or later)	VAR	... the variance of a dataset on the basis it is a sample.
VARP (Excel 2007) or VAR.P (Excel 2010 or later)	VAR.P	... the variance of a dataset on the basis it is a population.

Table E-2

Variable	Explanation	Variable	Explanation
b	Slope due to variable x	a	Intercept on the y-axis
s_{e_b}	Standard error for slope b	s_{e_a}	Standard error for intercept a
r^2	Coefficient of determination	s_e	Standard error of estimate
F	F-ratio for analysis of variance	df	Degrees of freedom (n − 2)
ss_{reg}	Sum of squares due to regression (explained variation)	ss	Sum of squares of residual (unexplained variation)

Table E-3

b_k Slope due to variable x_k	b_{k-1} Slope due to variable x_{k-1}	b_2 Slope due to variable x_2	b_1 Slope due to variable x_1	a intercept on y-axis
s_{e_k} standard error for slope, b_k	$s_{e_{k-1}}$ standard error for slope, b_{k-1}	s_{e_2} standard error for slope, b_2	s_{e_1} standard error for slope, b_1	s_{e_a} standard error for intercept a
r^2 coefficient of determination	s_e standard error of estimate			
F ratio	df degrees of freedom			
SS_{reg} sum of squares due to regression (explained variation)	SS_{resid} sum of squares of residual. (unexplained variation)			

If the interest is multiple regressions then for every additional x variable a new column is added. Thus if there are three variables for x then a virgin block of minimum size of four columns and five rows must be selected. The generic format for k variables with the meaning of the terms is in Table E-3.

In any event, to repeat, there is a useful Excel online guide for both the PC and MAC that explains the application of these functions and the appropriate syntax.

Index

Page numbers in *italics* refer to figures; page numbers in **bold** refer to boxes.